Generalized Matrix Inversion: A Machine Learning Approach

Predrag S. Stanimirović · Yimin Wei · Shuai Li ·
Dimitrios Gerontitis · Xinwei Cao

Generalized Matrix Inversion: A Machine Learning Approach

Springer

Predrag S. Stanimirović
University of Niš
Niš, Serbia

Shuai Li
University of Oulu
Oulu, Finland

Xinwei Cao
School of Business
Jiangnan University
Wuxi, China

Yimin Wei
School of Mathematical Sciences
Fudan University
Shanghai, China

Dimitrios Gerontitis
International Hellenic University
Thessaloniki, Greece

ISBN 978-3-032-01492-4   ISBN 978-3-032-01493-1 (eBook)
https://doi.org/10.1007/978-3-032-01493-1

This Springer imprint is published by the registered company Springer Nature Switzerland AG
The registered company address is: Gewerbestrasse 11, 6330 Cham, Switzerland

This monograph is dedicated to our families, whose unwavering support and encouragement have been the foundation of our motivation. Their belief in us has inspired us to pursue our goals with diligence.

We also thank our esteemed colleagues, whose invaluable cooperation, insights, comments, and guidance have shaped our research and encouraged our passion for exploring discoveries.

We are grateful to Springer Senior Editor Paul Drougas for the great support he gave us in the process of evaluation and in the process of publishing the manuscript.

Preface I

In the domain of linear algebra, the concept of the generalized inverse stands out as a pivotal tool for solving systems of equations, particularly when dealing with non-square matrices or ill-posed problems. Traditional inverse operations are unavailable when a matrix is singular or not invertible; however, generalized inverses, such as the Moore-Penrose pseudoinverse, provide a robust framework for addressing these challenges. The importance of generalized inverses extends beyond mere mathematical interest; they have practical applications in various fields, including statistics, signal processing, machine learning, and optimization.

This monograph delves into the theory and applications of generalized inverses, offering insights that bridge the gap between abstract mathematical concepts and tangible real-world implementations. We explore the fundamental properties of generalized inverses, elucidate their computational techniques, and highlight their relevance in both theoretical and applied contexts. The discussions are enriched by examples that illustrate the utility of these concepts in solving linear equations, data fitting, and regression analysis, among other areas.

As we embark on this exploration of generalized inverses, we pay tribute to the collaborative efforts of our families and colleagues, whose support and contributions have been instrumental in our pursuit of knowledge. Their encouragement motivates us to delve deeper into the intricacies of mathematics and its applications.

This monograph aims to describe the foundational concepts, methodologies, and applications of gradient neural networks (GNN) in solving linear algebra problems, mainly in solving matrix equations and computing generalized inverses. GNNs provide intelligent management of computing resources by creating adaptive models for task redistribution and optimization of computational flows.

The field of neural networks is marked by rapid advancements and innovations that continuously enhance understanding of machine learning. One significant development in this area is the Zhang neural network (ZNN), which represents a major evolution in developing neural network architecture and learning principles. ZNN dynamical evolution is used to solve real-time problems in data flow management, routing optimization and coordination in distributed computing systems. ZNNs are used for time-varying problems in a fixed-time or-predefined time. The ZNN design addresses some of the limitations found in traditional recurrent neural network (RNN) architectures when solving time-varying (TV) problems, offering a more robust frameworks for managing TV data patterns. By utilizing advanced algorithms and unique dynamical configurations, ZNNs have shown substantial performance improvements across various applications, from image and speech recognition, robot manipulation control, and financial forecasting and beyond. This monograph delves into foundations, principles, and applications of ZNNs, providing a comprehensive guide for researchers and practitioners.

We encourage readers, including students, researchers, and practitioners, to engage with the material presented in the monograph. We hope this monograph will inspire further inquiry and advancement in the understanding of generalized inverses and their vital role in modern mathematics and computer science.

Niš, Serbia — Predrag S. Stanimirović
Shanghai, China — Yimin Wei
Oulu, Finland — Shuai Li
Thessaloniki, Greece — Dimitrios Gerontitis
Wuxi, China — Xinwei Cao
May 2025

Preface II

Primary goal in this book is the application of dynamical system approach in numerical linear algebra, especially in computation of generalized inverses, solving systems of linear equations and solving linear matrix equations. Three main scientific areas are considered with the aim to answer properly to the stated goal: numerical linear algebra, recurrent neural networks (RNN) and nonlinear optimization.

RNN is one of the two broad types of artificial neural networks that allow the output from some nodes to affect subsequent input to the same nodes. A continuous-time recurrent neural network (CTRNN) uses a system of ordinary differential equations to define effects on incoming inputs on a neuron. We consider CTRNN dedicated to find zeros of equations or to minimize nonlinear functions. Two important classes of CTRNNs are known: Gradient Neural Networks (GNN) and Zhang (or Zeroing) Neural Networks (ZNN). GNN models are aimed to solving time-invariant problems, while ZNN models are able to solve time-varying problems. The aim of this book is to collect the latest developments in the theory and computation in numerical linear algebra by means of algorithm based on dynamical system approach. Main topics included in this book are: investigation of GNN and ZNN dynamical systems, computation of the usual matrix inverse and generalized inverses, solving systems of linear matrix and vector equations, solving specific matrix equations (Yang-Baxter, Lyapunov, Riccati, Sylvester), generating various matrix factorizations, generating the Takagi factorization of complex symmetric matrices. The dynamical system approach is a powerful tool for solving many kinds of matrix algebra problems because of:

(a) possibility to ensure a response within a predefined time-frame in real-time applications,
(b) its parallel distributed nature,
(c) convenience of hardware implementation,
(d) global convergence without conditions,
(e) in addition, dynamical system approach is applicable to online computation with time-varying matrices.

Considered algorithms are aimed to both time-invariant and time-varying, complex and real matrices. Continuous-time computation has attracted a great scientific research. In particular, there is an increasing interest in continuous-time computation, where the states of the dynamical system evolve continuously. Many biological systems, some control systems can be better described in the analog manner. The main motivation is the goal to derive efficient models for the variety of continuous dynamical systems we see in the world.

Two main dynamical systems are considered as appearances of CTRNNs: GNN and ZNN. GNNs are designed essentially for constant (or static, time-invariant) coefficient matrices and vectors. Objects involved in ZNN dynamics are time-varying. The ZNN method was proposed in 2001. ZNN dynamical evolution is based on an indefinite error function and an appropriate system of differential equations used to achieve exponential error decay. ZNNs are used to solve real-time problems in data flow management, routing optimization and coordination in distributed computing systems.

ZNN methods help in optimizing and controlling the behavior of robots and autonomous vehicles, localization of moving objects, and in solving various static or dynamic systems. On the other hand, generalized inverses are powerful tools and are applicable in many branches of mathematics, technics, and engineering. The most common and important application is in finding solutions to linear matrix equations and systems of linear equations. In addition to numerical linear algebra, there are many other mathematical and technical disciplines in which generalized inverses play an important role. Main application areas of generalized inverses are estimation theory (regression), computing polar decomposition, electrical circuits (networks) theory, automatic control theory, filtering, difference equations, pattern recognition, image restoration.

Many classes of generalized inverses have been proposed and investigated so far. The most popular are the Moore-Penrose inverse and the Drazin inverse. Computation models are defined in terms of continuous state space dynamical systems that appear as recurrent neural networks phenomena. In general, central topic of our research are algorithms focused on continuous-time computation for solving time-invariant and time-varying problems in real and complex domain. There is an increasing interest in continuous-time computation, where the states of a dynamical system evolve continuously. Many biological systems and some control systems can be better described in the analog manner. The main motivation is the goal to derive efficient models for the variety of continuous dynamical systems that are present in the world.

Therefore, time-varying matrix computations are a new, separate and still unexplored area of numerical linear algebra that is worth studying and learning.

All dynamical systems have discrete-time analogies, given in the form of appropriate difference equations. Main correlations between the continuous and discrete-time algorithms are considered in this book. Many computational problems (for example, structural analysis, electrical network analysis, weather forecasting) are based on matrix algorithms, such as solving linear systems, eigenvalue problem, singular value decomposition. These algorithms are often given in the discrete form of successive iterations $x(k+1) = F(x(k))$, where F is appropriate function.

Such iterations are *discrete-time* (DT) systems. On the other hand, the iterations $x(k+1) = F(x(k))$ can be considered as dynamical systems where the *state variable* x depends on the time k that takes discrete integer values. A sequence of points $\{x(k)\}_{k=-\infty}^{\infty}$ satisfying $x(k+1) = F(x(k))$ is called the orbit of F at $x(0)$. Nevertheless, most physical systems are dynamical continuous-time (CT) systems. CT systems are described in terms of ordinary differential equations (ODE) or partial differential equations (PDE) in the form $\dot{x}(t) = g(x(t))$, where $\dot{x}$ means the time derivative of $x(t)$ and g is appropriate time-varying function. In general, continuous-time algorithms and analog VLSI systems could be analyzed by means of time discretization that could destroy the main characteristics of these systems. There are important differences between CT and DT systems. The problem of the existence and uniqueness is an important issue for CT systems, but not for DT systems. Further, CT systems are easier for analysis that DT systems, because the orbits of many CT systems are continuous curves in the state space, while the orbits of DT systems are sequences of points, difficult for analysis. But, in spite of great difference, there is a great influence between the continuous-time and discrete-time computation. Results obtained in the research of dynamical systems can help in developing new iterations; and opposite, knowledge of derived iterations can help in better understanding and even in defining new dynamical evolutions.

On the other hand, time-varying problems appear frequently in scientific and engineering applications, such as circuit parameters in electronic circuits, aerodynamic coefficients in high-speed aircraft, and mechanical parameters in machinery [93, 381]. Our intention was the try to overcome sharp differences between engineering and mathematics discipline, in a way that would make the book interesting for engineers and adequately rigorous for mathematicians.

Nonlinear optimization is a method used to find the best possible solution for mathematical problems where the objective function or constraints are nonlinear, which makes it essential for complex real-world scenarios like machine learning and econometrics. Our research involves nonlinear optimization based on gradient-descent and quasi-Newton methods.

Our main interest is computation of major problems in numerical linear algebra using a dynamical system approach in both time-varying and time-invariant case. Algorithms for their computation are developed and implemented, and the performance comparison of such algorithms is given. Several applications are presented having in mind that generalized inverses are very powerful tools and are applicable in many branches of mathematics, technics and engineering. The most frequent and important is the application of generalized inverses in finding solution of many matrix equations and system of linear equations. Besides numerical linear algebra, there are a lot of other mathematical and technical disciplines in which generalized inverses play an important role. Some of them are: estimation theory (regression), computing polar decomposition, electrical circuits (networks) theory, automatic control theory, filtering, difference equations, pattern recognition, image restoration.

So far, many classes of generalized inverses have been proposed and investigated. The most popular are the Moore-Penrose inverse and the Drazin inverse.

Time-varying mathematical problems frequently appear in both scientific research and practical applications. For example, main appearances of time-varying problems are circuit parameters in electronic circuits, aerodynamic coefficients in high-speed aircraft, mechanical parameters in machinery, robot motion planning, chaotic noise rejection for sensors, numerical online problems, time-varying nonlinear optimization problems, etc. In 2001 Zhang et al. developed a special class of RNNs, known as Zhang or zeroing neural network (ZNN) dynamics for solving time-varying problems. ZNNs are able to perfectly track time-varying solutions by exploiting the time derivative of time-varying parameters. As a consequence, many researchers are making progresses in this direction by proposing various kinds of ZNN models to solve problems with different features. ZNN dynamical systems are new approach in numerical linear algebra, reliable and highly accurate for solving miscellaneous time-varying matrix, vector or scalar problems. ZNN dynamics differs from all other time-varying matrix methods in its use of an error equation and initiated model based on ordinary differential equation (ODE) that assures exponentially fast global convergence. The ODE used in the model can be transformed into discrete-time iterations by means of different finite difference schemes. Such an approach gives a new look to classical iterative methods, as well as to the interaction between continuous-time and discrete-time algorithms.

This book was primarily aimed to researchers in matrix theory, numerical linear algebra, particularly in the theory of generalized inverses and its applications. Researchers in recurrent neural networks and artificial intelligence may find useful material in this book. Also, the presented material should be of interest for readers interested in numerical analysis and symbolic computation. In addition, the book can be very useful for researchers interested in nonlinear programming. Generally speaking, dynamical systems are defined as continuous-time analogies of known nonlinear optimization algorithms, such as the class of gradient-descent algorithms or various modifications of the quasi-Newton method for solving nonlinear optimization problems. Both the time-varying and time-invariant cases are considered. This book is aimed to mathematics and engineering graduate students and researchers in the areas of optimization, dynamical systems, numerical linear algebra, control systems, signal processing. It can also be used as a text or reference for many graduate courses or as a reference for many courses in postgraduate levels in computer science, mathematics or in technical faculties. The reader should be familiar with basic linear algebra, matrix theory, ordinary differential equations, mathematical and functional analysis. Some knowledge of the algorithm theory, matrix theory and dynamical systems is recommended. Knowledge in *Matlab* programming and in *Simulink* modeling, which is a graphical extension to *Matlab* for modeling and simulation of systems. We believe that the book should be of use for many researchers, students in applied mathematics, statistics, engineering, and many other scientific disciplines.

According to MSC classification, the topics considered in this book are classified as follows:

- 15A09 Theory of matrix inversion and generalized inverses
- 15A10 Applications of generalized inverses
- 15A24 Matrix equations and identities
- 37N30 Dynamical systems in numerical analysis
- 65F45 Numerical methods for matrix equations
- 68T07 Artificial neural networks
- 90C30 Nonlinear programming.

This book consists of three chapters. In what follows, we present a short summary focusing on the key concepts of each chapter.

Chapter 1 is a short introduction into basic topics from the matrix theory, optimization theory, and gradient and zeroing dynamical systems. The first Sect. 1.1 is a review of basic notations, definitions and fundamental results from matrix theory. Definition of eigenvalues and eigenvector is given as well as the matrix trace function. Fundamental matrix decompositions are surveyed, like Jordan decomposition, singular value decomposition (SVD), QR decomposition, and LU decomposition [74]. Matrix operations such as Kronecker product, vectorization of matrices, and matrix derivatives, are also restated.

The second Sect. 1.2 provides definitions and properties of generalized inverses together with the most important applications to solving linear systems. Particularly, basic facts about the Moore-Penrose, weighted Moore-Penrose, Drazin and Group inverse, $\{i, j, k\}$ inverses, inner inverses, and outer inverses with prescribed range and null space, are presented. Related notions, such as idempotent matrices and projectors are stated. Least squares solutions and best approximate solution are investigated.

General information about unconstrained optimization is included in Sect. 1.3. Basic definitions of gradient and Hessian, matrix and vector norms are restated as well as basic principles and methods in nonlinear unconstrained optimization, overview of line search methods. Further, overview of gradient-descent methods, Newton method and quasi-Newton methods, conjugate gradient nonlinear optimization methods is presented.

Section 1.4 is dedicated to a brief introduction to gradient neural networks (GNNs) and zeroing neural networks (ZNNs). In addition we refer to stability and convergence properties of these networks according to Lyapunov theory.

Chapter 1 concludes with Sect. 1.5, which describes importance of generalized inverses and dynamical systems.

Chapter 2 presents a robust analysis of gradient neural networks. The focus of this chapter is in the computation of various kinds of generalized inverses and in solving matrix equations. The domains of constant real or complex matrices are considered.

The first Sect. 2.1 describes general GNN design and presents a short review of gradient neural networks for solving linear matrix equations, especially those that are focused on the computation of the matrix inverse and generalized inverses. The Moore-Penrose inverse is particularly investigated as well as the Drazin inverse and various outer inverses. Basic Simulink block and Simulink implementation are described. Detailed simulation examples are presented.

Section 2.2 is a short review of gradient neural networks for solving linear matrix equation (LME) $AXB = D$ for X. Several particular cases aimed to computing main generalized inverses, such as the Moore-Penrose inverse. Conditions for the existence and general solutions are investigated bith theoretically and through numerical examples. Influence of parameters involved in the GNN design on the convergence speed is considered.

GNN dynamics for computing {2, 3} and {2, 4} inverses are described in Sect. 2.3. Outer inverses of matrices are considered in particular cases.

Section 2.4 is aimed to GNN development of dynamics for solving the matrix equation $AXA = A$ and computing inner inverses. Main idea is to develop GNN matrix flow using specific error matrix equation. As a result, the set of solutions involves inner inverses of A.

Specific GNN model for solving the matrix equation $A^k XA = A^k$ is considered in Sect. 2.5. Particular case corresponding to appropriate values of k is the model for approximating the Drazin inverse.

Existence, representations and computation of matrix generalized inverses using corresponding GNN design are considered in Sect. 2.6. we consider algorithms based on the general framework consisting of two main steps: (a) solve a particular linear matrix equation; (b) If necessary, multiply the solution derived in the first step by some appropriate matrices. We consider several algorithms arising from this general framework whose results are certain inner or outer inverses with prescribed range and/or null space. This section represents an interesting combination of an algebraic approach and dynamical system approach in computation of various classes of generalized inverses with prescribed range and/or null space.

Section 2.7 investigates GNN for solving the matrix equation $BXCAB = B$ with the aim to compute outer inverses with prescribed range and null space. Conditions for existence, representations and computation of matrix outer generalized inverses with predefined range and null space are presented.

Section 2.8 studies recurrent neural networks arising from gradient neural networks and concludes with applications on the Drazin inverse and outer inverses based on the full-rank representation. The leading idea in defining new dynamical systems is to eliminate some constant terms in the standard GNN design. Obtained dynamical systems do not follow actual GNN design. They are simpler for numerical calculations compared to the standard GNN dynamics. Unfortunately, they lose global stability, which is the main characteristic of pure GNN models. A specific GNN model for solving linear systems is presented in Sect. 2.8.

Section 2.9 describes one specific GNN model for computing the Moore-Penrose inverse. The development of the model is based on the fact that the Moore-Penrose inverse is the unique inner inverse which additionally satisfies specific constraints on the range and null space.

A GNN dynamical system for solving online a set of simultaneous linear equations was considered in Sect. 2.10. The problem of solving systems of linear equations is considered as an appearance of the GNN model for solving linear matrix equations.

Section 2.11 investigates original and modified GNN dynamical systems for computing the composite generalized inverse $A^{d,\dagger} = A^{d}AA^{\dagger}$, known as the DMP inverse. The leading idea in this section is to show applicability of GNN design and corresponding modification in approximating composite generalized inverses. Such generalized inverses are defined as matrix expressions involving some known generalized inverses.

Section 2.12 presents a rigorous theoretical analysis of the main GNN models aimed at the computation of various generalized inverses. Also, the convergence behavior of a modification of the original GNN design is analyzed. Also, *Matlab Simulink* implementation of considered dynamics is presented and tested in terms of expected convergence results.

Application of gradient optimization methods in improving the GNN dynamics is investigated in subsequent Sect. 2.13. The aim of this section is to define and investigate some improvements of the GNN design based on the gradient descent methods of unconstrained optimization. Improvements of the GNN design are accordingly termed as GGNN design.

Chapter 3 presents a systematic treatment of new trends in zeroing neural networks (ZNN), with several recently published results from the authors. A unique feature of this chapter is the robust presentation of new ZNN models, complex ZNN models, integration-enhanced noise-tolerant ZNN models, Hybrid GNN-ZNN models, and varying parameter ZNN models. A number of simulation examples in time-dependent environment are presented.

The main topic of the first Sect. 3.1 is a review of fundamental ZNN models available in the literature. Complex-valued ZNN models and integration-enhanced models are presented.

Solving the time-varying inversion by ZNN neurodynamics is investigated in details in Sect. 3.2. Scalar, vector and matrix inversion are considered in details, including Simulink implementation and convergence analysis.

Section 3.3 investigates ZNN model for computing the time-varying (TV) matrix pseudoinverse. Left and right pseudoinverse are considered as well as the pseudoinverse of rank-defficient matrices.

ZNN models for solving TV linear systems $A(t)x(t) = b(t)$ are topic of Sect. 3.4.

Section 3.5 investigates several finite-time convergent ZNN models for computing the TV matrix inverse.

Section 3.6 surveys ZNN models for solving the general TV linear matrix equation $A(t)X(t)B(t) = C(t)$.

Some additional ZNN models for solving various TV problems are collected in Sect. 3.7.

Section 3.8 introduce complex ZNNs for computing time-varying outer inverse. ZNN models in this section are developed using the full-rank representation of {2}-inverses with prescribed range and null space.

Integration-enhanced and noise-tolerant ZNN (IENTZNN) models which are able to handle various types of noisy problems, are investigated in Sect. 3.9. Particular cases of IENTZNN are investigated theoretically and numerically. The leading idea is based on application of the integral term in the ZNN model. Developed models are defined on a general error function which is applicable in solving various types of time-varying problems.

Section 3.10 is aimed to development of dynamical systems based on hybridization of GNN and ZNN dynamics for real time matrix inversion and Hybrid GNN-ZNN models for solving linear matrix equations $AX = D$ and $XB = D$.

The last Sect. 3.11 is devoted to varying parameter ZNN models (complex or real) for computing time-varying outer inverses. However, since the design parameter represents the reciprocal of the capacitance or inductance coefficients in the circuit, the design parameter should be time-varying in order to enable as large as possible values under the permissible conditions of the circuit.

The authors are encouraging the readers for their comments, suggestions and corrections.

Niš, Serbia — Predrag S. Stanimirović
Shanghai, China — Yimin Wei
Oulu, Finland — Shuai Li
Thessaloniki, Greece — Dimitrios Gerontitis
Wuxi, China — Xinwei Cao
May 2025

Acknowledgements Predrag Stanimirović and Yimin Wei are supported by the Ministry of Science and Technology of China under grant H20240841.

Yimin Wei is supported by the National Natural Science Foundation of China under grants U24A2001 and 12271108, the Joint Research Project between China and Serbia under grant 2024-6-7, the Science and Technology Commission of Shanghai Municipality under grant 23JC1400501 with School of Mathematical Sciences and Key Laboratory of Mathematics for Nonlinear Sciences, Fudan University.

Predrag Stanimirović gratefully acknowledges support from the Research Project, grant no. 451-03-137/2025-03/ 200124 of the Serbian Ministry of Science and Education.

The research for this work has been carried out while visiting each other's institutions. Some of the work has been written in conjunction with the writing of research papers by the authors or in collaboration with many colleagues and Ph.D. students. During preparation of this book and, in general, during the many years long research in the area of generalized inverses, the authors have had frequent conversations and consultations with numerous colleagues.

Competing Interests The authors have no competing interests to declare that are relevant to the content of this manuscript.

Contents

About the Authors

Prof. Predrag S. Stanimirović has accomplished his Ph.D.in Computer Science at University of Niš, Serbia. He is a full Professor at University of Nis, Faculty of Sciences and Mathematics, Departments of Computer Science, Niš, Serbia. He acquired 36 years of experience in scientific research in diverse fields of mathematics and computer science, which span multiple branches of numerical linear algebra, recurrent neural networks, linear algebra, nonlinear optimization, symbolic computation and others. His main research topics include Numerical Linear Algebra, Operations Research, Recurrent Neural Networks and Symbolic Computation. He has published over 350 publications in scientific journals, including seven research monographs. He is the author in the World Rank List of 2% best authors in 2021–2023.

Prof. Yimin Wei received the B.S. degree in computational mathematics from Shanghai Normal University of Shanghai in China, and Ph.D. degree in computational mathematics from Fudan University of Shanghai in China, 1991 and 1997, respectively. He is currently a full Professor with the School of Mathematical Sciences, Fudan University. His current research interests include multilinear algebra and numerical linear algebra with their applications. He is author of more than 200 journal papers and five monographs published by Elsevier, Springer, World ScientifcEDP Science, and Science Press.

Prof. Shuai Li received the B.E. degree in precision mechanical engineering from the Hefei University of Technology, Hefei, China, in 2005, the M.E. degree in automatic control engineering from the University of Science and Technology of China, Hefei, in 2008, and the Ph.D. degree in electrical and computer engineering from the Stevens Institute of Technology, Hoboken, NJ, USA, in 2014. He is currently a Full Professor with the Faculty of Information Technology and Electrical Engineering, University of Oulu, Oulu, Finland, and also with the VTT-Technical National Research Center of Finland, Oulu. His current research interests include dynamic neural networks, robotics, machine learning, and autonomous systems.

Dimitrios Gerontitis received the B.S. degree in mathematics and the M.S. degree in theoretical informatics and systems and control theory from the Aristotle University of Thessaloniki, Thessaloniki, Greece, in 2013 and 2016, respectively. He is currently pursuing the Ph.D. degree in department of information and electronic engineering, in International Hellenic University, Greece. His main research interests include neural networks, optimization methods and analytical and numerical methods in linear algebra problems solving. (Detailed publication list is presented in the attached textual file)

Prof. Xinwei Cao received the bachelor's degree from Shandong University, China, in 2009, the master's degree from Tongji University, China, in 2012, and the Ph.D. degree from Fudan University, China, in 2017, all in management. She is currently a full Professor with School of Business, Jiangnan University, China. Her research interests include operations research, neural networks, operational management, computational and quantitative finance.

Acronyms

AF	Activation function
BFGS	Broyden-Fletcher-Goldfarb–Shanno
CFP-CDNN	Fixed-parameter convergent differential neural network
DFP	Davidon-Fletcher-Powell
ELS	Exact line search
GNN	Gradient neural network
NTZNN	Noise-tolerant zeroing neural network
PD	Positive definite
QN	Quasi-Newton
RNN	Recurrent neural network
SPD	Symmetric and positive definite
SR1	Symmetric Rank-One update
SVD	Singular Value Decomposition
ZNN	Zeroing neural network

List of Figures

List of Tables

List of Algorithms

Chapter 1
Background Information

This part of the book briefly introduces basic topics from matrix theory, optimization theory, and gradient and zeroing dynamical systems. Basic notations and notions presented in this part are necessary for understanding the results presented in subsequent chapters.

1.1 Review of Basic Notations, Definitions, and Fundamental Results from Matrix Theory

The set of natural numbers is denoted by $\mathbb{N}$, the set of real (resp. complex) numbers is denoted by $\mathbb{R}$ (resp. $\mathbb{C}$), while $\mathbb{R}^n$ (resp. $\mathbb{C}^n$) denotes the n-dimensional vector space of real (resp. complex) numbers, and $\mathbb{R}^{m\times n}$ denotes the space of real matrices with m types and n columns.

The vector (ordered n-tuple) $\mathbf{x} \in \mathbb{R}^n$ is represented as $\mathbf{x} = (x_1, \ldots, x_n)^{\mathrm{T}}$ and identified with the column matrix

$$\mathbf{x} = \begin{bmatrix} x_1 \\ \vdots \\ x_n \end{bmatrix}, \ x_1, \ldots, x_n \in \mathbb{R}.$$

If $\mathbf{x} \in \mathbb{R}^n$, $\mathbf{x}^{\mathrm{T}}$ denotes the transposed vector of the vector $\mathbf{x}$, while if $A = [a_{ij}] \in \mathbb{R}^{m\times n}$, $A^{\mathrm{T}} \in \mathbb{R}^{n\times b}$ denotes the transposed matrix of the matrix A, defined by $A^{\mathrm{T}} = [a_{ji}]$. Further, $\mathbf{I}_n$ denotes the $n \times n$ identity matrix.

P. Stanimirović et al., *Generalized Matrix Inversion: A Machine Learning Approach*,
https://doi.org/10.1007/978-3-032-01493-1_1

Definition 1.1.1 Let $\mathbb{V}$ be a vector space over $\mathbb{R}$. Mapping $\langle .,. \rangle : \mathbb{V} \times \mathbb{V} \to \mathbb{R}$ is a *scalar product* on the space $\mathbb{V}$ if it has the following properties:

- $(\forall \mathbf{x} \in \mathbb{V}) \quad \langle \mathbf{x}, \mathbf{x} \rangle \geq 0$ (*nonnegativity*);
- $\langle \mathbf{x}, \mathbf{x} \rangle = 0$ if and only if $\mathbf{x} = 0$ (*definiteness*);
- $(\forall \mathbf{x}, \mathbf{y} \in \mathbb{V}) \quad \langle \mathbf{x}, \mathbf{y} \rangle = \langle \mathbf{y}, \mathbf{x} \rangle$ (*symmetry*);
- $(\forall \mathbf{x}, \mathbf{y} \in \mathbb{V}) \quad (\forall \lambda \in \mathbb{R}) \quad \langle \lambda \mathbf{x}, \mathbf{y} \rangle = \lambda \langle \mathbf{x}, \mathbf{y} \rangle$ (*homogeneity*);
- $(\forall \mathbf{x}, \mathbf{y}, \mathbf{z} \in \mathbb{V}) \quad \langle \mathbf{x}, \mathbf{y} + \mathbf{z} \rangle = \langle \mathbf{x}, \mathbf{y} \rangle + \langle \mathbf{x}, \mathbf{z} \rangle$ (*additivity*).

Scalar product of two vectors $\mathbf{x}, \mathbf{y} \in \mathbb{R}^n$ is defined by

$$\langle \mathbf{x}, \mathbf{y} \rangle = \mathbf{x}^{\mathrm{T}} \mathbf{y} = \sum_{i=1}^{n} x_i y_i .$$

A natural way to generate the norm on $\mathbb{R}^n$ is to take the scalar product $\langle \cdot, \cdot \rangle$ on $\mathbb{R}^n$ and define the norm as follows:

$$\|\mathbf{x}\|_2 = \sqrt{\langle \mathbf{x}, \mathbf{x} \rangle} = \sqrt{\mathbf{x}^{\mathrm{T}} \mathbf{x}} = \sqrt{\sum_{i=1}^{n} x_i^2}, \quad \forall \mathbf{x} \in \mathbb{R}^n .$$

The norm defined in this way is called l_2-**norm** or *Euclidean norm*. The Euclidean norm implied in $\mathbb{R}^n$ is denoted by $\| \cdot \|_2$, so that the index 2 will be omitted.

Lemma 1.1.1 (Cauchy–Schwarz inequality) *In the Euclidean n-dimensional space for every two vectors* $\mathbf{x}, \mathbf{y} \in \mathbb{R}^n$ *the following inequality holds*

$$|\mathbf{x}^{\mathrm{T}} \mathbf{y}| \leq \|\mathbf{x}\| \cdot \|\mathbf{y}\|. \tag{1.1.1}$$

Lemma 1.1.2 *The following inequality holds for every two vectors* $\mathbf{x}, \mathbf{y} \in \mathbb{R}^n$ *in the Euclidean n-dimensional space*

$$\mathbf{x}^{\mathrm{T}} \mathbf{y} \leq \frac{1}{2} (\|\mathbf{x}\|^2 + \|\mathbf{y}\|^2). \tag{1.1.2}$$

Let $\mathbb{F}$ be a field. We call an ordered m-tuple of ordered n-tuples from $\mathbb{F}^n$ a *matrix* over a field $\mathbb{F}$ of type (or format) $m \times n$. Usually matrix A of type $m \times n$ is written in the form

$$A = \begin{bmatrix} a_{11} & a_{12} & \cdots & a_{1n} \\ a_{21} & a_{22} & \cdots & a_{2n} \\ \vdots & \vdots & & \vdots \\ a_{m1} & a_{m2} & \cdots & a_{mn} \end{bmatrix} . \tag{1.1.3}$$

The ordered n-tuple $(a_{i1}, a_{i2}, \ldots, a_{in})$ is called the i-th *row*, and the ordered m -tuple with entries $(a_{1j}, \ldots, a_{mj})$ is called the jth *column*. Instead of (1.1.3) we will write

$[a_{ij}]_{m\times n}$, i.e., $A=[a_{ij}]_{m\times n}$ or, if the matrix format is known, $A=[a_{ij}]$. The element in the intersection of ith row and j-th column is denoted by a_{ij}.

According to the standard notation, $\mathbb{C}(\mathbb{R})$ denotes the field of complex (real) numbers. Further, $\mathbb{C}^{m\times n}(\mathbb{R}^{m\times n})$ denote the vector space of $m \times n$ complex (real) matrices over $\mathbb{C}(\mathbb{R})$, $\mathbb{C}_r^{m\times n}(\mathbb{R}_r^{m\times n})$ the class of $m \times n$ complex (real) matrices of rank r and $\mathbb{C}^n(\mathbb{R}^n)$ the vector space of n-tuples of complex (real) numbers over $\mathbb{C}(\mathbb{R})$. The identity matrix of an appropriate order is denoted by I.

Let $A, B \in \mathcal{M}_{m\times n}$ be matrices over the field $\mathbb{F}$ and $\lambda \in \mathbb{F}$ be an arbitrary scalar.

The addition of two $m \times n$ matrices $A = [a_{ij}]$ and $B = [b_{ij}]$ is the$m \times n$ matrix $C = A + B = [c_{ij}]$ defined by

$$C = [c_{ij}] = A + B = [a_{ij}] + [b_{ij}] = [a_{ij} + b_{ij}]. \tag{1.1.4}$$

Multiplication of a scalar $\lambda \in \mathbb{F}$ and a matrix A is the matrix $D = \lambda A \in \mathcal{M}_{m\times n}$ defined by

$$D = \lambda A = \lambda[a_{ij}] = [\lambda a_{ij}]. \tag{1.1.5}$$

Consider matrices $A = [a_{ik}]_{m\times n} \in \mathbb{F}^{m\times n}$ and $B = [b_{kj}]_{n\times p} \in \mathbb{F}^{n\times p}$. *Product of matrices* A and B (in that order) is the matrix $C = AB \in \mathbb{F}^{m\times p}$ defined as

$$C = AB = \left[\sum_{k=1}^{n} a_{ik}b_{kj}\right]_{m\times p}. \tag{1.1.6}$$

The matrix obtained by omitting some type and/or some column from the matrix $A \in \mathcal{M}_{m\times n}$ is called a *submatrix* of the given matrix A.

The determinant of a square submatrix of order k of the matrix $A \in \mathbb{F}^{m\times n}$ is called the **minor** of order k of the given matrix A.

Let $\mathbb{N}^+ = \{0, 1, 2, \ldots\}$ denote the set of natural numbers. *Rank of a matrix*, denoted by rang, is the mapping

$$\text{rank} : \mathbb{F}^{m\times n} \to \mathbb{N}^+$$

defined by

(i) $\text{rank}(O) = 0$, where O is the zero matrix;
(ii) $\text{rank}(A) = r$, if there is a minor of order r of matrix A that is different from zero, and all minors of order k, such that $r + 1 \leq k \leq \min\{m, n\}$, if they exist, are equal to zero.

Each minor $M_r \neq 0$ of order r of a given matrix A for which $\text{rank}(A) = r$ is called a basic minor of that matrix. Rows and columns of the matrix A in whose intersections are elements of the basic minor M_r are called the basic rows and basic columns of the matrix A.

Proposition 1.1.1 *For a given matrix A we have:*
(i) *the basic rows (basis columns) of the matrix A are linearly independent;* (ii) *non-basic rows (columns) of matrix A, if any, are a linear combination of basic rows (basic columns).*

As a consequence, the rank of the matrix is equal to the maximum number of linearly independent species (columns).

The rank $\text{rank}(A)\mathbb{C}^{m\times n}$ satisfies the following basic properties, assuming that 0 is the zero matrix and I_n is the $n \times n$ identity matrix.

(a) $0 \leq rk(A) \leq min\{m, n\}$.
(b) $\text{rank}(A) = 0 \Longleftrightarrow A = O$.
(c) $\text{rank}(I_n) = n$.
(d) $\text{rank}(\lambda A) = \text{rank}(A)$ if $\lambda \neq 0$

The transpose, conjugate and conjugate-transpose of $A \in \mathbb{C}^{m\times n}$ are denoted by A^{T}, $\overline{A}$ and A^*, respectively. A matrix A is Hermitian if its conjugate transpose, A^*, satisfies $A^* = A$. The maximal number of linearly independent columns of a matrix A is called matrix *rank* of A and denoted by $\text{rank}(A)$.

For any $A \in \mathbb{C}^{m\times n}$, the range of A is defined by

$$\mathcal{R}(A) = \{y \in \mathbb{C}^m : y = Ax \text{ for some } x \in \mathbb{C}^n\},$$

and

$$\mathcal{N}(A) = \{x \in \mathbb{C}^n : Ax = 0\}$$

represents the null space of A.

The trace of a square matrix $A \in \mathbb{C}^{n\times n}$ is denoted by $\text{Tr}(A)$ and defined as the sum $\text{Tr}(A) = a_{11} + \cdots + a_{nn}$ of its diagonal entries. Further, the notion $\text{Re}\,(z)$ (resp. $\text{Im}\,(z)$) means the real (resp. imaginary) part of a complex number z. As usual, $\text{col}(A)$ and $\text{row}(A)$ denote the *column space* and the *row space* of A. The column space (resp. row space) of A represents the *span* (i.e., the set of linear combinations) of the columns (resp. rows) of A. The notation $\text{rank}(A)$ will denote the *rank* of A, i.e., the maximum number of linearly independent columns (rows): $\text{rank}(A) = \dim(\text{col}(A)) = \dim(\text{row}(A))$.

The reader is referred to the monographs [1, 6, 19, 266, 293] for further reading on main results and notions in matrix algebra.

Definition 1.1.2 defines the main classes of structured matrices.

Definition 1.1.2 It is said that a square matrix $A \in \mathbb{C}^{n\times n}$ (or $A \in \mathbb{R}^{n\times n}$) is:

(a) Hermitian (self-adjoint) if $A^* = A$ ($A^{\mathrm{T}} = A$),
(b) normal, if $A^*A = AA^*$ ($A^{\mathrm{T}}A = AA^{\mathrm{T}}$),
(c) lower-triangular , if $a_{ij} = 0$ for $i > j$,
(d) upper-triangular, if $a_{ij} = 0$ for $i < j$,
(e) positive semidefinite, if $\text{Re}\,(x^*Ax) \geq 0$ for all $x \in \mathbb{C}^n$,
(f) positive definite, if $\text{Re}\,(x^*Ax) > 0$ for all $x \in \mathbb{C}^n \setminus \{0\}$.

Definition 1.1.3 A mapping $\|\cdot\| : \mathbb{R}^{m\times n} \to \mathbb{R}$ is called a **matrix norm** if and only if it satisfies the following properties

- $\|A| \geq 0$ for any matrix $A \in \mathbb{R}^{m\times n}$ (*nonnegativity*);
- $\|A\| = 0$ if and only if $A = 0$;
- $\|\lambda A\| = |\lambda| \cdot \|A\|$ for any matrix $A \in \mathbb{R}^{m\times n}$ i $\lambda \in \mathbb{R}$ (*homogeneity*);
- $\|A + B\| \leq \|A\| + \|B\|$ for arbitrary matrices $A, B \in \mathbb{R}^{m\times n}$ (*triangle inequality*).

Definition 1.1.4 Let $A \in \mathbb{C}^{m\times n}$. A real or complex scalar λ which satisfies the equation

$$Ax = \lambda x \Longleftrightarrow (A - \lambda I)x = 0,$$

is the eigenvalue of A, and $x \in \mathbb{C}^n$ is the eigenvector of A corresponding to λ.

Since the equality $Ax - \lambda x$ is equivalent to

$$(A - \lambda I)x = 0, \tag{1.1.7}$$

it is clear that all eigenvectors corresponding to the scalar λ form the kernel of $\mathcal{N}(A - \lambda I)$. In addition, the eigenvectors represent solutions of the homogeneous linear system (1.1.7) with the system matrix

$$A - \lambda I = \begin{bmatrix} a_{11} - \lambda & a_{12} & \dots & a_{1n} \\ a_{21} & a_{22} - \lambda & \dots & a_{2n} \\ \vdots & \vdots & \dots & \vdots \\ a_{n1} & a_{n2} & \dots & a_{nn} - \lambda \end{bmatrix}, \tag{1.1.8}$$

which is called *eigen or characteristic matrix*. Homogeneous system (1.1.7) has nonzero solutions if and only if the matrix of that system $A - \lambda I$ be singular, i.e.,

$$P(\lambda) = \det(A - \lambda I) = 0. \tag{1.1.9}$$

The Eq. (1.1.9) is called the characteristic equation, and when its left side is developed, a polynomial of the nth degree in λ is obtained, which is called the characteristic polynomial of the matrix A:

$$\det(A - \lambda I) = P_A(\lambda) = \lambda^n - p_1\lambda^{n-1} + p_2\lambda^{n-2} + \cdots + (-1)^{n-1}\lambda p_{n-1} + (-1)^n p_n.$$

In the general case, (1.1.9) has p complex roots, $\lambda_1, \dots, \lambda_p$, $p \leq n$. Each root λ_i has multiplicity k_i, so $P(\lambda)$ can be represented in the form

$$P_A(\lambda) = \prod_{i=1}^{p} (\lambda - \lambda_i)^{k_i}.$$

In addition, the sum of multiplicities of all eigenvalues is equal to the power of $P(\lambda)$:

$$\sum_{i=1}^{p} k_i = n.$$

Koeficijent p_1 karakteristicnog polinoma jednak je tragu matrice A:

$$p_1 = \mathrm{Tr}(A) = \sum_{i=1}^{n} a_{ii}.$$

Definition 1.1.5 Let $A \in \mathbb{C}^{n\times n}$ and λ is an eigenvalue of A. A vector $x \in \mathbb{C}^n$ is called generalized eigenvector of A of a grade p corresponding to λ, or λ-vector of A of a grade p if it satisfies the equation

$$(A - \lambda I)^p x = 0.$$

Examples of matrix norms are as follows.

$$\begin{aligned}
\|A\|_1 &= \max_j \sum_i |A_{ij}| \\
\|A\|_2 &= \sqrt{\max \mathrm{eig}(A^*A)} \\
\|A\|_\infty &= \max_i \sum_j |A_{ij}| \\
\|A\|_{\max} &= \max_{i,j} |A_{ij}| \\
\|A\|_F &= \sqrt{\sum_{i,j} |A_{ij}|^2} = \sqrt{\mathrm{Tr}(AA^*)} \qquad \text{(Frobenius norm)}
\end{aligned} \tag{1.1.10}$$

Matrix index is another important characteristic of matrices. This notion has been frequently used in numerical linear algebra.

Proposition 1.1.2 *For every* $A \in \mathbb{C}^{n\times n}$ *there exists an integer* k *such that* $\mathrm{rank}(A^{k+1}) = \mathrm{rank}(A^k)$.

Definition 1.1.6 Let $A \in \mathbb{C}^{n\times n}$ be a given square matrix. The index of A is denoted by $\mathrm{ind}(A) = k$ and defined as the smallest integer k satisfying $\mathrm{rank}(A^{k+1}) = \mathrm{rank}(A^k)$.

Note that $\mathrm{ind}(A) = 0$ if A is regular and otherwise $\mathrm{ind}(A) \geq 1$.

1.1.1 Full Rank and Jordan Matrix Decompositions

The simplest matrix decompositions are LU and Cholesky factorization (decomposition) These notions are given in Proposition 1.1.3.

Proposition 1.1.3 (LU and Cholesky factorization) *For every regular square matrix $A \in \mathbb{C}^{n\times n}$ there exists lower-triangular matrix L and upper-triangular matrix U such that $A = LU$ and $l_{ii} = 1$ for every $i = 1, 2, \dots, n$. This factorization is known as LU factorization. Moreover, if A is Hermitian and positive definite, it holds that $U = L^*$, and the implied factorization $A = LL^*$ is called Cholesky factorization.*

The full rank factorization is one of the most important notions in matrix theory.

Definition 1.1.7 (*Full-rank factorization*) A full rank factorization of an arbitrary matrix $A \in \mathbb{C}^{m\times n}$ is defined as the decomposition $A = PQ$, where the matrix $P \in \mathbb{C}^{m\times r}$ is full column rank matrix and $Q \in \mathbb{C}^{r\times n}$ is full row rank matrix.

Proposition 1.1.4 *Let $A \in \mathbb{C}^{m\times n}$ such that A is of neither full column rank nor full row rank. Then, there exists at least one full rank factorization $A = PQ$ of the matrix A.*

The simplest matrix decompositions are LU and Cholesky factorizations. They are described in Proposition 1.1.3.

The LU decomposition of the square matrix A consists in its representation as a product of two matrices $A = LU$, where the matrix L is lower-triangular with units on the main diagonal, and the matrix U is upper-triangular. If such a decomposition is known, it can be used to solve the system of linear equations $Ax = b$. Namely, then it is

$$Ax = (LU)x = L(Ux) = b.$$

From here it can be seen that reš by folding the system

$$Ly = b$$

and then the system

$$Ux = y$$

it will be possible to get a solution to the starting system.

Relying on the LU decomposition, it will be possible to calculate it and the inverse of a matrix. Let $e_1, \dots, e_n$ be the columns of the identity $n \times n$ matrix I_n. Then the ith column a^i of the inverse A^{-1} is obtained by solving the system

$$Ly = e_i$$

and then the system

$$Ua^i = y.$$

If the square matrix A is symmetric and positive definite, there is a possibility of its more efficient triangular decomposition. The Cholesky decomposition in this case represents the matrix decomposition in the form

$$A = L\,L^*,$$

where L is a lower-triangular matrix with strictly positive diagonal elements. If the matrix A is real, then its Cholesky decomposition is defined by

$$A = L\,L^{\mathrm{T}},$$

The Cholesky decomposition of the $n \times n$ matrix is calculated by the following formulas:

$$l_{ii} = \left(a_{ii} - \sum_{k=1}^{n} l_{ik}\right)^{1/2}, \quad i = 1, \ldots, n;$$

$$l_{ji} = \frac{1}{l_{ii}}\left(a_{ij} - \sum_{k=1}^{n} l_{ik} l_{jk}\right), \quad i = 1, \ldots, n,\ j = 1, \ldots, n.$$

For each matrix there exists a basis of generalized eigenvectors with respect to which a matrix can be represented in the Jordan form. The Jordan decomposition is stated in the following proposition.

Proposition 1.1.5 (The Jordan decomposition) *Let $\{\lambda_1, \lambda_2, \ldots, \lambda_p\}$ be distinct eigenvalues of a square matrix $A \in \mathbb{C}^{n\times n}$. Then A is similar to a block diagonal matrix J with Jordan blocks on its diagonal, i.e., there exists a nonsingular matrix P such that*

$$A = PJP^{-1} = P \begin{bmatrix} J_{k_1}(\lambda_1) & 0 & \ldots & 0 \\ 0 & J_{k_2}(\lambda_2) & \ldots & 0 \\ \vdots & \vdots & \ddots & \vdots \\ 0 & 0 & \ldots & J_{k_p}(\lambda_p) \end{bmatrix} P^{-1},$$

where the Jordan blocks $J_{k_i}(\lambda_i)$ are defined by

$$J_{k_i}(\lambda_i) = \begin{bmatrix} \lambda_i & 1 & 0 & \ldots & 0 \\ 0 & \lambda_i & 1 & \ldots & 0 \\ \vdots & \vdots & \ddots & \ddots & \vdots \\ 0 & 0 & \ldots & \lambda_i & 1 \\ 0 & 0 & \ldots & 0 & \lambda_i \end{bmatrix}.$$

The Jordan matrix J is uniquely defined up to a rearrangement of its diagonal blocks.

Note that the characteristic polynomial of the Jordan block is equal to $(x - \lambda_i)^{k_i}$. This means that each Jordan block $J_{k_i}(\lambda_i)$ has its own eigenvalue λ_i, for each $i = 1, \ldots, p$.

Symmetric, positive definite, and positive semidefinite matrices play an important role in matrix theory.

A square matrix T of order n is symmetric and positive semidefinite (abbreviated SPSD and denoted by $T \succeq 0$) if

$$v^{\mathrm{T}} T v \geq 0 \text{ for all } v \in \mathbb{R}^n.$$

The matrix T is symmetric and positive definite (abbreviated SPD and denoted by $T \succ 0$) if

$$v^{\mathrm{T}} T v > 0 \text{ for all } v \in \mathbb{R}^n, v \neq 0.$$

Recall that a symmetric matrix T is positive definite (PD) if and only if all its eigenvalues are nonnegative.

Proposition 1.1.6 *If $T \in \mathbb{R}^{n\times n}$ is a symmetric PSD matrix, then the following statements are equivalent:*

(a) $T = MM^{\mathrm{T}}$, *for an appropriate matrix M satisfying $M \in \mathbb{R}^{n\times k}$, $k \geq 1$.*

(b) $v^{\mathrm{T}} T v \geq 0$ *for all $v \in \mathbb{R}^n$, $v \neq 0$.*

(c) *There exist vectors v_i, $i = 1, \ldots, n \in \mathbb{R}^k$ (for some $k \geq 1$) such that $T_{ij} = v_i^{\mathrm{T}} v_j$ for all $i, j = 1, \ldots, n$. The vectors v_i, $i = 1, \ldots, n$ are called a Gram representation of T.*

(d) *All principal minors of T are nonnegative.*

Proposition 1.1.7 *Let $T \in \mathbb{C}^{n\times n}$ is symmetric. Then $T \succeq 0$ and it is nonsingular if and only if $T \succ 0$.*

1.1.2 Singular Value Decomposition

Besides the Jordan decomposition, the Singular Value Decomposition (SVD) is another important decomposition which is based on the eigenvalues. More precisely, SVD is defined on the basis of *singular values* of a matrix, defined as in Definition 1.1.8.

Definition 1.1.8 Let $A \in \mathbb{C}^{m\times n}$ be a given matrix. Further, let the vectors $u \in \mathbb{C}^m$, $v \in \mathbb{C}^n$ and the real number $\sigma \geq 0$ be such that

$$Av = \sigma u, \quad A^* u = \sigma v. \tag{1.1.11}$$

Then σ is called the singular value of A. In addition, the vectors u and v are termed as the left and right singular vectors of A, respectively.

According to the identity (1.1.11), it follows that

$$A^*Av = \sigma^2 v, \quad AA^*u = \sigma^2 u,$$

which implies that the real number σ^2 is the eigenvalue of A^*A and AA^*.

Proposition 1.1.8 *Let $A \in \mathbb{C}^{m\times n}$ of rank r and $\{\lambda_1, \lambda_2, \ldots, \lambda_r\}$ be the nonzero eigenvalues of AA^* (i.e., A^*A). There are exactly p nonzero singular values of A, denoted by $\sigma_i(A)$, $i = 1, \ldots, p$, and equal to:*

$$\sigma_i(A) = \sqrt{\lambda_i}, \quad i = 1, \ldots, r.$$

Moreover, it can be shown that the singular values are positive stationary values of the (vector) function $f(x) = \|Ax\|/\|x\|$. Recall that a *stationary point* of the function $f : \mathbb{R}^n \to \mathbb{R}$ is every vector $x_0 \in \mathbb{R}^n$ which satisfies $\nabla f(x_0) = 0$, while $f(x_0)$ is the corresponding *stationary value*.

Theorem 1.1.1 (Singular Value Decomposition) *Let $A \in \mathbb{C}_r^{m\times n}$ be arbitrary $m \times n$ matrix of rank r. Then there exist unitary matrices $U \in \mathbb{C}^{m\times m}$ and $V \in \mathbb{C}^{n\times n}$ such that A can be partitioned as*

$$A = U \begin{bmatrix} \Sigma & O \\ O & O \end{bmatrix} V^*, \tag{1.1.12}$$

where Σ is a diagonal matrix

$$\Sigma = \begin{bmatrix} \sigma_1 & & \\ & \ddots & \\ & & \sigma_r \end{bmatrix} = \mathrm{diag}(\sigma_1, \sigma_2, \ldots, \sigma_r), \ \sigma_i = \sqrt{\lambda_i}$$

and $\lambda_1 \geq \lambda_2 \geq \cdots \geq \lambda_r > 0$ are the nonzero eigenvalues of A^*A.

Proof Since $A^*A \in \mathbb{C}_r^{n\times n}$ is a Hermitian positive semidefinite matrix, its eigenvalues are nonnegative. Let the eigenvalues of A^*A be the real quantities $\sigma_1^2, \ldots, \sigma_n^2$ satisfying inequalities $\sigma_1 \geq \cdots \geq \sigma_r > 0 = \sigma_{r+1} = \cdots = \sigma_n$. In addition, suppose that $v_1, \ldots, v_n$ is a set of orthonormal eigenvectors for $\sigma_1^2, \ldots, \sigma_n^2$, and let

$$V_1 = [v_1, \ldots, v_r], \ V_2 = [v_{r+1}, v_{r+2}, \ldots, v_n].$$

Consider the diagonal matrix

$$\Sigma = \mathrm{diag}(\sigma_1, \sigma_2, \ldots, \sigma_r).$$

Then

$$V_1^*(A^*A)V_1 = \Sigma^2, \quad V_2^* = (A^*A)V_2 = O$$

and, consequently,

$$\Sigma^{-1} V_1^* A^* A V_1 \Sigma^{-1} = I, \quad A V_2 = O.$$

Now let

$$U_1 = A V_1 \Sigma^{-1}.$$

Then $U_1^* U_1 = I$, that is the columns of U_1 are orthonormal. In accordance with that, let U_2 be chosen so that $U = [U_1 \ U_2]$ be a unitary matrix. Then A can be decomposed into the form

$$\begin{aligned} U^* A V &= \begin{bmatrix} U_1^* A V_1 & U_1^* A V_2 \\ U_2^* A V_1 & U_2^* A V_2 \end{bmatrix} \\ &= \begin{bmatrix} (A V_1 \Sigma^{-1})^* A V_1 & O \\ U_2^* (U_1 \Sigma) & O \end{bmatrix} \\ &= \begin{bmatrix} \Sigma & O \\ O & O \end{bmatrix}. \end{aligned}$$

Thus (1.1.12) holds. □

Remark 1.1.1 The quantities $\sigma_1 \geq \sigma_2 \geq \cdots \geq \sigma_r > 0$ are the nonzero singular values of A, and (1.1.12) is called the singular value decomposition of A.

Denote by U_r the submatrix consisting of the first r columns of the matrix U and by V_r the submatrix consisting of the first r types of the matrix V. Then (1.1.12) can be written in a more efficient form

$$A = U_r \Sigma V_r.$$

It will be referred to several times in this and further chapters (Fig. 1.1).

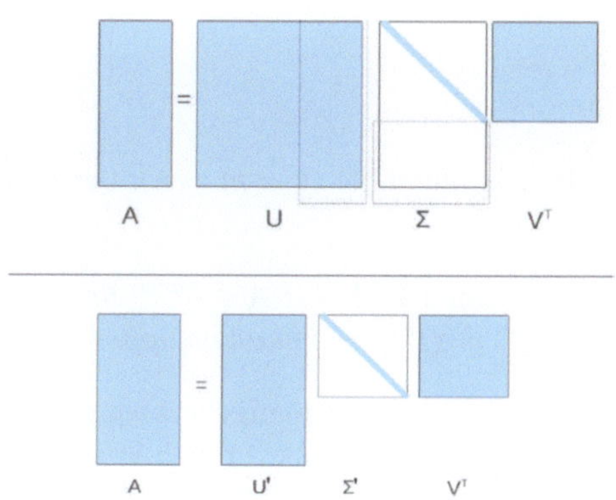

Fig. 1.1 Full SVD decomposition (up) and reduced (down) SVD decomposition

1.1.3 QR Decomposition

Proposition 1.1.9 *For any matrix $A \in \mathbb{C}^{m\times n}$ there exist unitary matrices $Q \in \mathbb{C}^{m\times m}$ and $P \in \mathbb{C}^{n\times n}$ such that holds $A = Q^* R P$ where matrix R has the form*

$$R = \begin{bmatrix} R_{11} & O \\ O & O \end{bmatrix} \in \mathbb{C}^{m\times n}, \quad R_{11} \in \mathbb{C}_k^{k\times k},$$

where O denotes a zero block of appropriate dimensions.

Proof Consider the SVD decomposition (1.1.12) and take $Q = U^*$, $P = V^*$ and $R_{11} = \Sigma = \operatorname{diag}(\sigma_1, \sigma_2, \ldots, \sigma_r)$. The matrix $R_1 \in \mathbb{C}^{r\times r}$ is regular since $\sigma_1, \sigma_2, \ldots, \sigma_r > 0$, while matrices Q and P are obviously unitary. Therefore, SVD decomposition is the special case of the decomposition stated in Proposition 1.1.9. □

In the case when A is a column of full rank ($r = n \leq m$), the following statement about QR factorization is valid.

Proposition 1.1.10 *Let the matrix $A \in \mathbb{C}_r^{m\times r}$ of rank r be given. Then there exist unitary matrices $Q \in \mathbb{C}^{m\times m}$ such that $A = Q^* R$ holds where R is a matrix of the form*

$$R = \begin{bmatrix} R_1 \\ O \end{bmatrix},$$

such that $R_1 \in \mathbb{C}^{n\times n}$ upper-triangular trapezoidal matrix of the for

$$R_1 = \begin{bmatrix} r_{11} & r_{12} & \cdots & r_{1n} \\ 0 & r_{12} & \cdots & r_{1n} \\ \vdots & \vdots & \ddots & \vdots \\ 0 & 0 & \cdots & r_{nn} \\ 0 & 0 & \cdots & 0 \\ \vdots & \vdots & \ddots & \vdots \\ 0 & 0 & \cdots & 0 \end{bmatrix}.$$

If A is invertible ($r = m = n$), the following statement on QR factorization is valid.

Proposition 1.1.11 *Let $A \in \mathbb{C}^{n\times n}$ be invertible. Then there exists a unitary matrix Q ($QQ^* = Q^*Q = I$) and an upper-triangular matrix R such that $A = QR$.*

In the case that the matrix A is square, the QR decomposition can be used to solve the system of equations $Ax = b$ in such a way that the upper-triangular system is solved

$$Rx = Qb.$$

1.1.4 The Trace Function

The *trace* of a square matrix $A \in \mathbb{C}^{n\times n}$ is the sum of its diagonal entries:

$$\mathrm{Tr}(A) = \sum_{i=1}^{n} a_{ii}.$$

Also, the trace is equal to the sum of the eigenvalues $\lambda_1, \ldots, \lambda_n$ of A:

$$\mathrm{Tr}(A) = \sum_{i=1}^{n} \lambda_i.$$

The trace of a matrix product $AB \in \mathbb{C}^{n\times n}$, where $A \in \mathbb{C}^{n\times k}$ and $AB \in \mathbb{C}^{k\times n}$, is equal to

$$\mathrm{Tr}(AB) = \sum_{i=1}^{n} (AB)_{ii} = \sum_{j=1}^{k} \sum_{i=1}^{n} a_{ij} b_{ji}.$$

If A and B are two square matrices, then the matrix trace satisfies the following properties:

$$\begin{aligned} \mathrm{Tr}(A) &= \mathrm{Tr}(A^{\mathrm{T}}); \\ \mathrm{Tr}(A + B) &= \mathrm{Tr}(A) + \mathrm{Tr}(B); \\ \mathrm{Tr}(\alpha A) &= \alpha \mathrm{Tr}(A), \ \alpha \in \mathbb{C}. \end{aligned}$$

If A and B are of the same order, then

$$\mathrm{Tr}(A^{\mathrm{T}} B) = \mathrm{Tr}(AB^{\mathrm{T}}) = \mathrm{Tr}(B^{\mathrm{T}} A) = \mathrm{Tr}(BA^{\mathrm{T}}).$$

The trace operation is also invariant under a similarity transformation

$$\mathrm{Tr}(BAB^{-1}) = \mathrm{Tr}(A).$$

The trace is not a commutative operation, but it is invariant under cyclic permutations:

$$\mathrm{Tr}(ABCD) = \mathrm{Tr}(BCDA) = \mathrm{Tr}(CDAB) = \mathrm{Tr}(DABC).$$

As a consequence

$$\mathrm{Tr}(AB) = \mathrm{Tr}(BA).$$

Arbitrary permutations are not allowed: in general, since

$$\mathrm{Tr}(ABC) \neq \mathrm{Tr}(ACB).$$

1.1.5 Kronecker Product and Vectorization

The Kronecker product of an $m \times n$ matrix A and an $r \times q$ matrix B, is an $mr \times nq$ matrix, denoted by $A \otimes B$ and defined as

$$A \otimes B = \begin{bmatrix} a_{11}B & a_{12}B & \dots & a_{1n}B \\ a_{21}B & a_{22}B & \dots & a_{2n}B \\ \vdots & & & \vdots \\ a_{m1}B & a_{m2}B & \dots & a_{mn}B \end{bmatrix}.$$

Main properties of the Kronecker product are as follows.

$$\begin{aligned} A \otimes (B + C) &= A \otimes B + A \otimes C; \\ A \otimes (B \otimes C) &= (A \otimes B) \otimes C; \\ \alpha A \otimes \beta B &= \alpha\beta(A \otimes B); \\ (A \otimes B)^{\mathrm{T}} &= A^{\mathrm{T}} \otimes B^{\mathrm{T}}; \\ (A \otimes B)^{-1} &= A^{-1} \otimes B^{-1}; \\ (A \otimes B)(C \otimes D) &= AC \otimes BD; \\ \mathrm{rank}(A \otimes B) &= \mathrm{rank}(A)\mathrm{rank}(B); \\ \mathrm{Tr}(A \otimes B) &= \mathrm{Tr}(A)\mathrm{Tr}(B); \end{aligned}$$

The *vectorization* (vec-operator) applied on a matrix A stacks the columns into a vector. If $a_1, \dots a_n$ are columns of a matrix A, then

$$\mathrm{vec}(A) = \begin{bmatrix} a_1 \\ a_2 \\ \vdots \\ a_n \end{bmatrix}.$$

The vectorization has the following properties:

$$\begin{aligned} \mathrm{vec}(A + B) &= \mathrm{vec}(A) + \mathrm{vec}(B) \\ \mathrm{vec}(ABC) &= \left(C^{\mathrm{T}} \otimes A\right) \mathrm{vec}(B) \\ \mathrm{vec}(AB) &= (I_p \otimes A)\mathrm{vec}(B) = (B^{\mathrm{T}} \otimes I_m)\mathrm{vec}(A), \end{aligned}$$

where $A \in \mathbb{C}^{m\times n}$, $B \in \mathbb{C}^{n\times p}$, $C \in \mathbb{C}^{p\times q}$.

1.1.6 Matrix Derivatives

The matrix differentiation is frequently used in the definition and analysis of dynamical systems. Assume that $A(t) = [a_{ij}(t)]\mathbb{R}^{m\times n}$ is a time-varying $m \times n$ matrix with elements $a_{ij}(t)$. Then $\frac{\mathrm{d}A}{\mathrm{d}t}$ is the matrix defined by element-wise differentiation of each element of A, i.e., $\frac{\mathrm{d}A(t)}{\mathrm{d}t} = [\dot{a}_{ij}]$. The following properties of the matrix time derivative are important:

(a) $\dfrac{\mathrm{d}(AX(t)B)}{\mathrm{d}t} = A\dfrac{\mathrm{d}X(t)}{\mathrm{d}t}B;$

(b) $\dfrac{\mathrm{d}\,\mathrm{vec}(X(t))}{\mathrm{d}t} = \mathrm{vec}\left(\dfrac{\mathrm{d}\,X(t)}{\mathrm{d}t}\right);$

(c) $\dfrac{\mathrm{d}(A(t)X(t)B(t))}{\mathrm{d}t} = \dfrac{\mathrm{d}A(t)}{\mathrm{d}t}X(t)B(t) + A(t)\dfrac{\mathrm{d}X(t)}{\mathrm{d}t}B(t) + A(t)X(t)\dfrac{\mathrm{d}B(t)}{\mathrm{d}t};$

(d) $\dfrac{\mathrm{d}(A(t)\otimes B(t))}{\mathrm{d}t} = \dfrac{\mathrm{d}A(t)}{\mathrm{d}t}\otimes B(t) + A(t)\otimes\dfrac{\mathrm{d}B(t)}{\mathrm{d}t}.$

Let $F(X)$ be a scalar function with respect to the elements x_{ij} of the matrix $X \in \mathbb{R}^{m\times n}$. Then the derivative of F with respect to X is denoted by $\frac{\partial F}{\partial X}$ and defined as the matrix with entries $\frac{\partial F}{\partial X} = \left[\frac{\partial F}{\partial x_{ij}}\right]_{m\times n}$.

Also, the following properties of the matrix derivatives of the matrix trace are frequently exploited.

(a) $\dfrac{\partial}{\partial X}\mathrm{Tr}(X) = I;$

(b) $\dfrac{\partial}{\partial X}\mathrm{Tr}(XA) = A^{\mathrm{T}};$

(c) $\dfrac{\partial}{\partial X}\mathrm{Tr}(AXB) = A^{\mathrm{T}}B^{\mathrm{T}};$

(d) $\dfrac{\partial}{\partial X}\mathrm{Tr}(AX^{\mathrm{T}}B) = BA;$

(e) $\dfrac{\partial}{\partial X}\mathrm{Tr}(A\otimes B) = \mathrm{Tr}(A)I;$

(f) $\dfrac{\partial}{\partial X}\mathrm{Tr}(X^2) = 2X^{\mathrm{T}};$

(g) $\dfrac{\partial}{\partial X}\mathrm{Tr}(X^{\mathrm{T}}AX) = AX + A^{\mathrm{T}}X;$

(h) $\dfrac{\partial}{\partial X}\mathrm{Tr}(X^{\mathrm{T}}X) = 2X;$

(i) $\dfrac{\partial}{\partial X}\mathrm{Tr}(X^{\mathrm{T}}AXB) = AXB + A^{\mathrm{T}}XB^{\mathrm{T}}.$

1.2 Definition and Properties of Main Generalized Inverses

A singular matrix Q is a square matrix that is not invertible. This occurs when the determinant of Q is zero, indicating that its rows (or columns) are linearly dependent. In such a case, at least one row (or column) can be expressed as a linear combination of the others. Singular matrices often arise in systems of linear equations where there are either no solutions or infinitely many solutions. If A is singular or rectangular, the notion of the usual matrix inverse is extended to various kinds of **generalized inverses**. A generalized inverse is a matrix which:
–exists even for rectangular and/or singular matrices;
–inherits some properties of the ordinary inverse;
–reduces to the ordinary inverse in the case of a square nonsingular matrix.

1.2.1 The Inverse of a Nonsingular Matrix

A nonsingular matrix, also known as an invertible (or regular) matrix, is a square matrix that has a nonzero determinant. This implies that its rows (or columns) are linearly independent, meaning no row (or column) can be written as a linear combination of the others. A nonsingular matrix has an inverse. As a result, nonsingular matrices play a crucial role in many areas of mathematics, including linear algebra, where they are used to solve equations, perform transformations, and analyze systems. It is well known that every nonsingular (regular) matrix A has a unique inverse, denoted by A^{-1}, such that

$$AA^{-1} = A^{-1}A = I.$$

Matrix inverses are used in various applications, including solving systems of linear equations, computer graphics transformations, and more.

Matrix inverses have numerous applications across various fields. Here are some key applications:

The inverse matrix finds many applications, primarily in finding solutions to matrix equations and systems of linear equations:

The inverse of a matrix is often used to solve linear equations in the form $Ax = b$, where A is the coefficient matrix, x is the vector of variables, and b is the constant vector. The solution can be found using $x = A^{-1}b$, provided that A is invertible. Similarly, the linear matrix equation $AX = B$ for X can be solved using the matrix inverse: $AX = B \Longrightarrow X = A^{-1}B$.

In computer graphics, transformations such as rotation, scaling, and translating of objects can be represented using matrices. The inverse of a transformation matrix can be used to reverse a transformation, which is useful in rendering and animation. In control theory, certain matrices are used to model dynamic systems. In this case, the inverse of such matrices is useful in designing controllers and observers that stabilize and control the system response. In multivariate statistics, the inverse of the covariance matrix is used in techniques such as multivariate regression and in calculating Mahalanobis distance, which serves as a measure of the distance between points and a distribution. Matrix inverses play an important role in solving constrained and unconstrained optimization problems, especially in constrained optimization, where Lagrange multipliers are used. The inverse of the Hessian matrix of second-order partial derivatives is useful in finding local maxima or minima in nonlinear optimization. In various machine learning algorithms, particularly those involving linear regression, the normal equation $A^{\mathrm{T}}Ax = A^{\mathrm{T}}b$ can be expressed using the inverse of a matrix to find the weights: $x = (A^{\mathrm{T}}A)^{-1}A^{\mathrm{T}}b$, where A is the design matrix in $Ax = b$. Certain encryption algorithms use matrices for encoding and decoding messages. The inverse of the matrix involved in the encryption process is used to decrypt the message.

We mention a few of the numerous properties of the inverse matrix:

$$\begin{aligned}(A^{-1})^{-1} &= A,\\ (A^{\mathrm{T}})^{-1} &= (A^{-1})^{\mathrm{T}},\\ (A^{*})^{-1} &= (A^{-1})^{*},\\ (AB)^{-1} &= B^{-1}A^{-1}.\end{aligned}$$

Another important property of the inverse A^{-1} is that its eigenvalues are the reciprocals of the corresponding eigenvalues of A.

A singular matrix Q is a square matrix that is not invertible. This occurs when the determinant of Q is zero, indicating that its rows (or columns) are linearly dependent. In such a case, at least one row (or column) can be expressed as a linear combination of the others. Singular matrices often arise in systems of linear equations where there are either no solutions or infinitely many solutions.

1.2.2 Solvability of Linear Systems

One of the most common applications of matrices is in finding the solution of a system of linear equations. A system of m equations and n unknowns over the field $\mathbb{R}$, denoted $(S_{m\times n})$, is a conjunction of equations

$$\begin{matrix} a_{11}x_1 & a_{12}x_2 & \cdots & a_{1n}x_n & b_1\\ a_{21}x_1 & a_{22}x_2 & \cdots & a_{2n}x_n & b_2\\ & & \vdots & & \\ a_{m1}x_1 & a_{m2}x_2 & \cdots & a_{mn}x_n & b_m \end{matrix}, \tag{1.2.1}$$

where $a_{ij}, b_i, i = 1, \ldots, m,\ j = 1, \ldots, n$ are given real numbers, and x_j are variables from the field $\mathbb{R}$. The numbers a_{ij} are called coefficients, while b_i are the free coefficients of the system.

Let

$$Ax = b \tag{1.2.2}$$

be the matrix form of the linear system (1.2.1), wherein b is a given vector and x is an unknown vector. If A is nonsingular, there is a unique solution with respect to the unknown vector x, given by

$$x = A^{-1}b.$$

In the general case (when A may be singular or rectangular), there may sometimes be no solutions or multiple solutions.

The consistent system of linear equations $Ax = b$, which satisfies the condition $b \in \mathcal{R}(A)$ and which is given in the form

$$Ax = b, \quad A \in \mathbb{C}^{m\times n}, \ m < n, \ b \in \mathcal{R}(A)$$

has many solutions. In the opposite case, $b \notin \mathcal{R}(A)$, the inconsistent system of linear equations

$$Ax = b, \quad A \in \mathbb{C}^{m\times n}, \ b \notin \mathcal{R}(A)$$

has no solution.

The following Kronecker–Capelli criterion statement gives one criterion for the existence of a solution to the system $(S_{m\times n})$. System $S_{m\times n}$ is solvable if and only if $\operatorname{rank}(A) = \operatorname{rank}([A \mid b])$.

Proposition 1.2.1 *Consistent system* $(S_{m\times n})$ *possesses:*

(a) *unique solution if* $\operatorname{rank}(A) = n$ *(n is the number of unknowns),*
(b) *infinitely many solutions if it* $\operatorname{rank}(A) < n$.

1.2.3 Definitions and Main Properties of Generalized Inverses

The idea to introduce a generalization of the usual inverse in the case of a singular matrix was first presented by Moore in 1920. But, no efficient investigation of the subject was made by the mid-1950s when the investigation of generalized inverses progressed rapidly. The study of generalized inverses has progressed significantly since its rebirth in the early 1950s. Namely, Roger Penrose, unfamiliar with previous work, reclassified the Moore "reciprocal inverse" in 1955. He concluded that Moore's inverse could be represented by four equations (known as Penrose equations). Another well-known kind of generalized inverse is the Drazin inverse, named after Michael P. Drazin (1958). A major extension of this field came in the fifties, when C. R. Rao and J. Chipman exploited the relation between generalized inverses, least squares, and statistics.

Various papers have arisen both in the theory of generalized inverses and its applications. The reason for developing a generalized inverse matrix is to get a matrix that can fill in as the inverse in some sense for a more extensive class of matrices than invertible matrices. Now, it is difficult to give even an approximate number of articles devoted to the theory, computation, and application of generalized inverses. It is justifiable to say that the theory of generalized inverses has extensively grown and has become an important part of pure and computational mathematics, as well as an important part of many applicable scientific areas, such as computer science, electrical engineering, etc.

Let $A \in \mathbb{C}^{m\times n}$ be arbitrary complex $m \times n$ matrix. Unlike the case of the nonsingular matrix, which has a single unique inverse for all purposes, there are various generalized inverses for different purposes. The notion of generalized inverses is closely related to the following four matrix equations with respect to the unknown matrix $X \in \mathbb{C}^{n\times m}$, called the Penrose equations:

$$(1)\quad AXA = A \qquad (2)\quad XAX = X \qquad (3)\quad (AX)^* = AX \qquad (4)\quad (XA)^* = XA. \tag{1.2.3}$$

For a subset $\mathcal{S} \subseteq \{1, 2, 3, 4\}$, the set of all matrices obeying the conditions contained in $\mathcal{S}$ is denoted by $A\{\mathcal{S}\}$. Any matrix from $A\{\mathcal{S}\}$ is called an $\mathcal{S}$-inverse of A and it is denoted by $A^{(\mathcal{S})}$. For example the set $A\{1\}$ contains all $\{1\}$-inverses (or inner inverses) of A. More precisely, $A\{1\} = \{X \in \mathbb{C}^{n\times m} |\ AXA = A\}$. An arbitrary $\{1\}$-inverse of A is denoted by $A^{(1)}$. Similarly, $A\{2\}$ denotes all $\{2\}$-inverses (or outer inverses) of A. An arbitrary $\{2\}$-inverse of A is denoted by $A^{(2)}$. It means certainly $A\{2\} = \{X \in \mathbb{C}^{n\times m} |\ XAX = X\}$. Moreover, $\mathcal{S}$-inverse of A, which satisfy some conditions on its rank as well as on its range and/or null space, are very important. By $A\{\mathcal{S}\}_s$ we denote the set of all $\mathcal{S}$-inverses of A whose rank is fixed and equal to s.

In 1955, Penrose showed that for every finite matrix A (square or rectangular) of real or complex elements, there is a unique matrix X satisfying all four equations. The matrix X satisfying all the matrix equations (1) – (4) is known as the Moore–Penrose inverse of A, denoted by $A^\dagger$. Theorem 1.2.1 verifies the uniqueness of the Moore–Penrose inverse.

Theorem 1.2.1 ([192]) *For any matrix $A \in \mathbb{C}^{m\times n}$ there exists a single element in the set $A\{1, 2, 3, 4\}$, called the Moore–Penrose inverse $A^\dagger$ of A.*

Proof The proof is based on the contradiction. In the case of the existence $X, Y \in A\{1, 2.3, 4\}$ it can be concluded

$$
\begin{aligned}
X &= XAX \\
&= X(AX)^* = XX^*A^* \\
&= X(AX)^*(AY)^* \\
&= XAY = (XA)^*(YA)^*Y \\
&= A^*Y^*Y \\
&= (YA)^*Y \\
&= YAY = Y,
\end{aligned}
$$

which completes the proof. □

If A is nonsingular, then the usual matrix inverse $X = A^{-1}$ satisfies the four Penrose equations. Since the Moore–Penrose inverse is unique, it follows that the Moore–Penrose inverse $A^{\dagger}$ of a nonsingular matrix is the same as its ordinary inverse A^{-1}.

In this way, we achieved the notion of $\{i, j, \ldots, k\}$-inverses, where $i, j, k \in \mathcal{S}$. For example, for a given matrix $A \in \mathbb{C}^{m\times n}$, if there exists a matrix such that it satisfies only the first Penrose equation, then this matrix is called an $\{1\}$-inverse of the matrix A and it is denoted by $A^{(1)}$. Similarly, if the generalized inverse satisfies the first and the third Penrose equations, it is an $\{1, 3\}$-inverse of A, denoted by $A^{(1,3)}$ while the corresponding set is denoted by $A\{1, 3\}$.

For a given subspaces T and S from $\mathbb{C}^n$, a projector from $\mathbb{C}^n$ on T along S is usually denoted by $P_{T,S}$. If $S = T^{\perp} = L^{\perp} := \{x \in \mathbb{C}^n :$ x is orthogonal to every vector in $L\}$ (i.e., if S is orthogonal complement of T), then P_T is the orthogonal projector from $\mathbb{C}^n$ on T. A matrix which corresponds to a is an idempotent matrix. An arbitrary matrix which corresponds to a linear map that is an orthogonal projector is a Hermitian idempotent matrix.

If A is a square matrix and is invertible, the solution of $AX = B$ can be computed directly using

$$X = A^{-1}B.$$

When A is rectangular or singular In cases where A is not square or is singular (non-invertible), we can use the Moore–Penrose pseudoinverse:

$$X = A^{\dagger}B$$

This approach provides the best approximate solution.

If A is a square matrix, we can solve $AX = B$ using LU decomposition, which decomposes A into a lower-triangular matrix L and an upper-triangular matrix U.
1. Decompose A in the form $A = LU$;
2. Solve $LY = B$ for Y (forward substitution);
3. Solve $UX = Y$ for X (back substitution).

Properties and Representations of {1} Inverses

Lemma 1.2.1 *Let $A \in \mathbb{C}_r^{m\times n}$, and let $E = E_k E_{k-1} \cdots E_1 \in \mathbb{C}_m^{m\times m}$ be the product of elementary row operations, and let P be a permutation matrix such and $P \in \mathbb{C}_n^{n\times n}$ be matrices which transform A into the form*

$$EAP = \begin{bmatrix} I_r & K \\ O & O \end{bmatrix}.$$

Since

$$A = E^{-1}\begin{bmatrix} I_r & K \\ O & O \end{bmatrix}P^{-1}$$

it can be verified that the $n \times m$ matrix

$$X = P\begin{bmatrix} I_r & O \\ O & L \end{bmatrix}E \tag{1.2.4}$$

is an $\{1\}$-inverse (or inner inverse) of A, for arbitrary block $L \in \mathbb{C}_m^{(n-r)\times(m-r)}$.

Lemma 1.2.2 *For a given matrix $A \in \mathbb{C}_r^{m\times n}$ the following statement are valid:*

(a) $(\lambda A)^{(1)} = \lambda^\dagger A^{(1)}$, *where* $\lambda \in \mathbb{C}$ *and* $\lambda^\dagger = \begin{cases} \frac{1}{\lambda}, & \lambda \neq 0 \\ 0, & \lambda = 0 \end{cases}$;
(b) $AA^{(1)}$ *is a projection from* $\mathbb{C}^m$ *on* $\mathcal{R}(A)$, *i.e.,* $AA^{(1)} = P_{\mathcal{R}(A),S}$ *where* $S \in \mathbb{C}^m$ *is such that* $\mathcal{R}(A) + S = \mathbb{C}^m$;
(c) $I - A^{(1)}A$ *is a projection from* $\mathbb{C}^n$ *on* $\mathcal{N}(A)$, *i.e.,* $I - A^{(1)}A = P_{\mathcal{N}(A),T}$ *where* $T \in \mathbb{C}^n$ *is a matrix satisfying* $T + \mathcal{N}(A) = \mathbb{C}^n$;
(d) $\operatorname{rank}(A^{(1)}) \geq \operatorname{rank}(A)$;
(e) $A^{(1)}A = I_n$ *if and only if* $r = n$;
(f) $AA^{(1)} = I_m$ *if and only if* $r = m$;
(g) *If* $X \in A\{1\}$, *then* $X \in A\{1, 2\}$ *if and only if* $\operatorname{rank}(A) = \operatorname{rank}(X)$;
(h) $(A^*A)^{(1)}A^* \in A\{1, 2, 3\}$;
(i) $A^*(AA^*)^{(1)} \in A\{1, 2, 4\}$;
(j) $A^{(1,4)}AA^{(1,3)} = A^\dagger$.

Lemma 1.2.3 establishes one very important relationship between $\{i, j, \ldots, k\}$-inverses and solutions of linear matrix equations. Further details can be found in monographs [6, 44, 266, 293].

Lemma 1.2.3 *Let $A \in \mathbb{C}^{m\times n}$, $B \in \mathbb{C}^{p\times q}$, $D \in \mathbb{C}^{m\times q}$. Then the matrix equation*

$$AXB = D$$

is consistent if and only if it holds

$$AA^{(1)}DB^{(1)}B = D,$$

for some $A^{(1)}$, $B^{(1)}$. *In this case, the general solution is*

$$X = A^{(1)}DB^{(1)} + Y - A^{(1)}AYBB^{(1)}$$

for arbitrary $Y \in \mathbb{C}^{n\times p}$.

Corollary 1.2.1 *Let* $A \in \mathbb{C}^{m\times n}$ *and* $A^{(1)} \in A\{1\}$ *be arbitrary inner inverse of A. Then the set of all inner inverses of A is included in the set defined by*

$$A\{1\} = \{A^{(1)} + Z - A^{(1)}AZAA^{(1)} \mid Z \in \mathbb{C}^{n\times m}\},$$

where Z is an arbitrary matrix of appropriate dimensions.

Lemma 1.2.4 *Two linear matrix equations*

$$AX = B, \quad XD = E \tag{1.2.5}$$

have a common solution if and only if each of these equations has a solution, i.e.,

$$AA^{(1)}B = B, \qquad ED^{(1)}D = E, \tag{1.2.6}$$

and in addition

$$AE = BD.$$

In this case,

$$X = A^{(1)}B + ED^{(1)} - A^{(1)}AED^{(1)}$$

is a common solution of both equations, for arbitrary $A^{(1)}$ *and* $D^{(1)}$.

Corollary 1.2.2 *Let* $A \in \mathbb{C}^{m\times n}$ *be a given matrix and consider the linear system*

$$Ax = b \tag{1.2.7}$$

with respect to unknown $x \in \mathbb{C}^n$, *where* $A \in \mathbb{C}^{m\times n}$ *and* $b \in \mathbb{C}^m$. *Then the system* (1.2.7) *is consistent if and only if it holds*

$$AA^{(1)}b = b$$

for some $A^{(1)}$. *In that case, the general solution of the system* (1.2.7) *is defined by*

$$x = A^{(1)}b + (I - A^{(1)}A)y,$$

wherein $y \in \mathbb{C}^n$ *is arbitrary.*

Lemma 1.2.5 (Consistency of $Ax = b$) *Let* $A \in \mathbb{C}^{m\times n}$ *and* $b \in \mathbb{C}^m$. *The following four statements are equivalent:*

(*a*) *the vector equation* $Ax = b$ *s solvable for* x,
(*b*) $b \in \mathcal{R}(A)$,
(*c*) $\operatorname{rank}([A|b]) = \operatorname{rank}(A)$,
(*d*) $AA^{\dagger}b = b$.

Lemma 1.2.6 *Let the equations given in* (1.2.5) *have a particular common solution* X_0. *Then the general solution of these equations is given by*

$$X = X_0 + (I - A^{(1)}A)Y(I - DD^{(1)}),$$

for arbitrary inner inverses $A^{(1)} \in A\{1\}$, $D^{(1)} \in D\{1\}$, *and arbitrary matrix* $Y \in \mathbb{C}^{m\times n}$.

The end of this subsection restates some useful properties of inner matrix inverses. These properties lead to extremely useful applications of these generalized inverses in solving matrix equations and systems of linear equations.

Proposition 1.2.2 *Let* $A \in \mathbb{C}^{m\times n}$, $X \in \mathbb{C}^{n\times m}$. *Then* $X \in A\{1\}$ *if and only if, for all* $b \in \mathcal{R}(A)$, $x = Xb$ *is a solution of the system* (1.2.7).

Proposition 1.2.3 *The identity* $AB(AB)^{(1)}A = A$ *holds if and only if* $\operatorname{rank}(AB) = \operatorname{rank}(A)$. *Similarly,* $B(AB)^{(1)}AB = B$ *is valid if and only if* $\operatorname{rank}(AB) = \operatorname{rank}(B)$.

Proposition 1.2.4 *Let* $A \in \mathbb{C}^{m\times n}$ *and let* $A^{(1)}$ *be an arbitrary element of* $A\{1\}$. *Further, denote by* $\mathcal{R}(A) = L$ *and* $\mathcal{N}(A) = M$. *The matrices* $AA^{(1)}$ *and* $A^{(1)}A$ *are idempotent and satisfy*

$$AA^{(1)} = P_{L,S}, \quad A^{(1)}A = P_{T,M},$$

where S *is a subspace of* $\mathbb{C}^m$ *complementary to* L, *and* T *is a subspace of* $\mathbb{C}^n$ *complementary to* M.

Properties and Representations of {1,2}-Inverses

It is known that the existence of a $\{1\}$-inverse of a matrix A implies the existence of its $\{1, 2\}$-inverse. An arbitrary $\{1, 2\}$-inverse is also termed a reflexive inner inverse. This fact is verified in Lemma 1.2.7, which shows that two arbitrary inner inverses can generate a $\{1, 2\}$-inverse.

Lemma 1.2.7 *Let $Y, Z \in A\{1\}$ be two arbitrary inner inverses of A. Then the matrix product YAZ satisfies $X = YAZ \in A\{1, 2\}$.*

According to Lemma 1.2.7, for any $L \in \mathbb{C}^{(n-r)\times(m-r)}$, the $n \times m$ matrix X defined in (1.2.4) belongs to $A\{1, 2\}$ if and only if X is given in the form (1.2.4).

Lemma 1.2.8 (Bjerhammar [11]) *For a given A and $X \in A\{1\}$, it follows that $X \in A\{1, 2\}$ if and only if* $\text{rank}(X) = \text{rank}(A)$.

Lemma 1.2.9 is a more detailed version of Lemma 1.2.8.

Lemma 1.2.9 *Any two of the following three statements imply the third:*

$$\begin{aligned} X &\in A\{1\}, \\ X &\in A\{2\}, \\ \text{rank}(X) &= \text{rank}(A). \end{aligned}$$

Proposition (1.2.5) shows that projectors can be generated using $\{1, 2\}$-inverses.

Proposition 1.2.5 *If A and X are $\{1, 2\}$-inverses of each other, then*

$$AX = P_{\mathcal{R}(A),\mathcal{N}(X)}, \quad XA = P_{\mathcal{R}(X),\mathcal{N}(A)}.$$

Basic Properties of the Moore–Penrose Inverse

The most important result related to the Penrose equations (1.2.3) is the statement that there always exists a unique matrix which satisfies the four Penrose equations. This result was shown by Penrose [191] in 1955. This matrix is called the Moore–Penrose inverse and denoted by $A^\dagger$.

The concept of generalized inverses of an arbitrary matrix $A \in \mathbb{C}^{m\times n}$ is originally due to Moore, in 1920, (called by him the *general reciprocal*). His definition was essentially as follows.

Definition 1.2.1 If $A \in \mathbb{C}^{m\times n}$, then the generalized inverse of A is the matrix $X \in \mathbb{C}^{n\times m}$ such that the following equations are satisfied:

$$1.\ \ AX = P_{\mathcal{R}(A)}; \qquad 2.\ \ XA = P_{\mathcal{R}(X)}. \tag{1.2.8}$$

Moore proved the existence and the uniqueness of the solution of such defined generalized inverse by proving the following result.

Proposition 1.2.6 *For every complex matrix $A \in \mathbb{C}^{m\times n}$ there exists a unique matrix $X \in \mathbb{C}^{n\times m}$ satisfying* (1.2.8).

Rado proved the equivalence of Moore's definition (1.2.8) and Penrose's definition of the generalized inverse (based on the matrix equations (1)–(4)). Today, the generalized inverse satisfying (1)–(4) is known as the Moore–Penrose pseudoinverse (shortly M-P inverse or pseudoinverse).

Although $\{1\}$-inverses and $\{1, 3\}$-inverses provide a solution of a given matrix equation, the Moore–Penrose inverse is the most popular and most similar to the ordinary inverse. This statement is justified by its uniqueness and the properties listed in the following two lemmas. Also, since the Moore–Penrose inverse is $\{1\}$-inverse, we should take into account that the properties from Lemma 1.2.2 are also valid for the Moore–Penrose inverse.

Proposition 1.2.7 (Penrose [191]) *Let $A \in \mathbb{C}^{m\times n}$ and $b \in \mathbb{C}^{m\times 1}$. The minimal-norm least-squares solution of the system $Ax = b$ is given by $x_* = A^\dagger b$. All other least-squares solutions are given by*

$$x = A^\dagger b + (I_n - A^\dagger A)z, \quad z \in \mathbb{C}^n.$$

Lemma 1.2.10 *Let $A \in \mathbb{C}^{m\times n}$ be an arbitrary matrix. The Moore–Penrose inverse $A^\dagger$ possesses the following properties:*

(a) $(A^\dagger)^\dagger = A$, $(A^\dagger)^* = (A^*)^\dagger$;
(b) $(AA^*)^\dagger = (A^*)^\dagger A^\dagger$, $(A^*A)^\dagger = A^\dagger (A^*)^\dagger$;
(c) $A^\dagger AA^* = A^* = A^*AA^\dagger$;
(d) $A^\dagger = (A^*A)^\dagger A^* = A^*(AA^*)^\dagger$;
(e) $\mathcal{N}(AA^\dagger) = \mathcal{N}(A^\dagger) = \mathcal{N}(A^*) = \mathcal{R}(A)$
(f) $\mathcal{R}(AA^*) = \mathcal{R}(AA^{(1)}) = \mathcal{R}(A)$;
(g) $\operatorname{rank}(AA^\dagger) = \operatorname{rank}(A^\dagger A) = \operatorname{rank}(A)$;
(h) $AA^\dagger = P_{\mathcal{R}(A^*),\mathcal{N}(A)}$ *and* $A^\dagger A = P_{\mathcal{R}(A),\mathcal{N}(A^*)}$.

Lemma 1.2.11 *Let $A \in \mathbb{C}^{m\times n}$ be an arbitrary matrix. Then the matrix A can be written in the form*

$$A \sim \begin{bmatrix} A_1 & O \\ O & O \end{bmatrix} : \begin{bmatrix} \mathcal{R}(A^*) \\ \mathcal{N}(A) \end{bmatrix} \to \begin{bmatrix} \mathcal{R}(A) \\ \mathcal{N}(A^*) \end{bmatrix}, \tag{1.2.9}$$

where A_1 is invertible. Hence,

$$A^\dagger \sim \begin{bmatrix} A_1^{-1} & O \\ O & O \end{bmatrix} : \begin{bmatrix} \mathcal{R}(A) \\ \mathcal{N}(A^*) \end{bmatrix} \to \begin{bmatrix} \mathcal{R}(A^*) \\ \mathcal{N}(A) \end{bmatrix}$$

Generalized inverses are used to find least-squares solutions to systems of linear equations that do not have a unique solution. For instance, if a system has more equations than unknowns (overdetermined), the pseudoinverse can provide the least-squares solution that minimizes the residual error.

The representation (1.2.9) of can be easily obtained from the SVD of A. In more precise terms, the SVD decomposition of A assumes that the block A_1 is a diagonal matrix containing the singular values of A on the main diagonal.

If the vector b in the system (1.2.7) satisfies $b \notin \mathcal{R}(A)$, then it is necessary to search for an approximate solution by trying to find a vector x which minimizes the norm of the vector $Ax - b$. More precisely, if the linear system $Ax = b$ is not consistent, then approximate solutions are called least-squares solutions. The notion of a least-squares solution is restated in Definition 1.2.2.

Definition 1.2.2 Let $A \in \mathbb{C}^{m\times n}$ and $b \in \mathbb{C}^m$. Arbitrary vector $\hat{x} \in \mathbb{C}^n$ which satisfies the minimization problem

$$\|A\hat{x} - b\|^2 = \min_{x\in\mathbb{C}^n} \|Ax - b\|^2. \tag{1.2.10}$$

is called a *least-squares solution* of the system (1.2.7).

The next lemma gives a characterization of all least-squares solutions of the system (1.2.7).

Lemma 1.2.12 *The vector x is a least-squares solution of the system of linear equations $Ax = b$ if and only if x is a solution of the linear system, called normal equation and defined by*

$$A^*Ax = A^*b. \tag{1.2.11}$$

The following Proposition 1.2.8, restated from [6], shows that $\|Ax - b\|$ is minimized by the vector $x = A^{(1,3)}b$. This statement establishes very important relation between the set of $\{1, 3\}$-inverses and the least-squares solutions to the system (1.2.7).

Proposition 1.2.8 *Let $A \in \mathbb{C}^{m\times n}$, $b \in \mathbb{C}^m$. Then the vector norm $\|Ax - b\|$ is smallest in the case $x = A^{(1,3)}b$, where $A^{(1,3)} \in A\{1, 3\}$ is arbitrary. On the other hand, if $X \in \mathbb{C}^{n\times m}$ possesses the characteristic that, for all b, the expression $\|Ax - b\|$ is minimized when $x = Xb$, then $X \in A\{1, 3\}$.*

Since $A^{(1,3)}$ inverse of a matrix is not unique, a consequence is that any system of linear equations has many least-squares solutions in general. However, among all least-squares solutions of a given system of linear equations, there exists only one such solution of minimum norm.

Definition 1.2.3 Let $A \in \mathbb{C}^{m\times n}$ and $b \in \mathbb{C}^m$. A vector $\hat{x}$, which satisfies the minimization problem

$$\|\hat{x}\|^2 = \min_{x\in\mathbb{C}^n} \|x\|^2 \tag{1.2.12}$$

is called a *minimal-norm solution* of the system (1.2.7).

Proposition 1.2.9 is restated from [6] and establishes a relation between $\{1, 4\}$-inverses and the minimum-norm solutions of the system (1.2.7).

Proposition 1.2.9 *Let $A \in \mathbb{C}^{m\times n}$, $b \in \mathbb{C}^m$. If $Ax = b$ has a solution for x, the unique solution x for which $\|x\|$ is smallest is given by $x = A^{(1,4)}b$, where $A^{(1,4)} \in A\{1, 4\}$. Conversely, if $X \in \mathbb{C}^{n\times m}$ is such that, whenever $Ax = b$ has a solution, $x = Xb$ is the solution of minimal-norm, then $X \in A\{1, 4\}$.*

The most important property of the Moore–Penrose inverse can be concluded by taking both the results from Propositions 1.2.8 and 1.2.9 in common. This property was originated by Penrose in 1955.

Corollary 1.2.3 (Penrose [191]) *For an arbitrary matrix $A \in \mathbb{C}^{m\times n}$ and an arbitrary vector $b \in \mathbb{C}^m$, the expression $A^\dagger b$ is the unique minimum-norm solution among the least-squares solutions of $Ax = b$. Conversely, if $X \in \mathbb{C}^{n\times m}$ has the property that, for all b, Xb is the* minimal-norm least-squares solution *of $Ax = b$, then $X = A^\dagger$.*

In essence, Lemma 1.2.3 shows that $A^\dagger b$ is the minimal-norm least-squares solution of the linear system $Ax = b$. This fact caused a dramatic increase in interest in the generalized inverses theory after 1955.

The result, restated in Proposition 1.2.10, establishes a useful correlation between the minimal-norm solution and least-squares solutions. More precisely, Proposition 1.2.10 characterizes the set of all least-squares solutions of a given system of linear equations.

Proposition 1.2.10 (Nashed 1970, 1976, [185]) *If $A \in \mathbb{C}^{m\times n}$ has a closed range $\mathcal{R}(A)$, then the set S of all least-squares solutions of the system $Ax = b$ is given by*

$$S = A^\dagger b \oplus \mathcal{N}(A) = \{A^\dagger b + (I - A^\dagger A)y \mid y \in \mathbb{C}^n\},$$

where $\mathcal{N}(A)$ denotes the null space of A.

Some additional properties of $A^\dagger$ and $A^{(1)}$ can be found in monographs [6, 266].

Proposition 1.2.11 shows the Moore–Penrose inverse can be computed using an arbitrary $\{1\}$-inverse.

Proposition 1.2.11 (Yanai et al. [326]) *The Moore–Penrose inverse $A^\dagger$ can be expressed by an arbitrary inner inverse, as*

$$\begin{aligned} A^\dagger &= A^{\mathrm{T}} A \left(A^{\mathrm{T}} A A^{\mathrm{T}} A\right)^{(1)} A^{\mathrm{T}} \\ &= A^{\mathrm{T}} \left(A A^{\mathrm{T}}\right)^{(1)} A \left(A^{\mathrm{T}} A\right)^{(1)} A^{\mathrm{T}}. \end{aligned}$$

The singular value decomposition is a very important notion in developing RNN models and even in numerical linear algebra.

Lemma 1.2.13 *Let* $A \in \mathbb{C}_r^{m\times n}$ *and let* $A = U\Sigma V^*$ *be the singular value decomposition of* A *where* $U \in \mathbb{C}^{m\times m}$ *and* $V \in \mathbb{C}^{n\times n}$ *are unitary matrices and* $\Sigma = \mathrm{diag}(\sigma_1, \sigma_2, \ldots, \sigma_r)$, $\sigma_i = \sqrt{\lambda_i}$ *and* $\lambda_1 \geq \lambda_2 \geq \cdots \geq \lambda_r > 0$ *are the nonzero eigenvalues of* A^*A. *If*

$$A = U \begin{bmatrix} \mathrm{diag}(\sigma_1, \ldots, \sigma_r) & O \\ O & O \end{bmatrix} V^* = U \begin{bmatrix} \Sigma & O \\ O & O \end{bmatrix} V^* \in \mathbb{C}^{m\times n},$$

then

$$A^\dagger = V \begin{bmatrix} \mathrm{diag}(1/\sigma_1, \ldots, 1/\sigma_r) & O \\ O & O \end{bmatrix} U^* = V \begin{bmatrix} \Sigma^{-1} & O \\ O & O \end{bmatrix} U^* \in \mathbb{C}^{n\times m}.$$

Moreover, let S_1, S_2, S_3 *be arbitrary* $r \times m - r, n - r \times r,$ *and* $m - r \times n - r$ *matrices, respectively. Then an arbitrary inner inverse of* A *is given by*

$$A^{(1)} = V \begin{bmatrix} \mathrm{diag}(1/\sigma_1, \ldots, 1/\sigma_r) & S_1 \\ S_2 & S_3 \end{bmatrix} U^* \in \mathbb{C}^{n\times m}.$$

Theorem 1.2.2 *Let* $A \in \mathbb{C}_r^{m\times n}$ *be* $m\times$ *matrix of rank* r *and*

$$A = U \begin{bmatrix} \Sigma & O \\ O & O \end{bmatrix} V^* \tag{1.2.13}$$

be the SVD decomposition of A, *where* $U \in \mathbb{C}^{m\times m}$ *and* $V \in \mathbb{C}^{n\times n}$ *are unitary matrices and* $\Sigma = \mathrm{diag}(\sigma_1, \sigma_2, \ldots, \sigma_r)$, *such that* $\sigma_i = \sqrt{\lambda_i}$ *and* $\lambda_1 \geq \lambda_2 \geq \cdots \geq \lambda_r > 0$ *are the nonzero eigenvalues of* A^*A. *Then* $\sigma_1 \geq \sigma_2 \geq \cdots \geq \sigma_r > 0$ *denote ordered nonzero singular values of* A *and*

$$\|A\|_2 = \sigma_1, \quad \|A^\dagger\|_2 = \frac{1}{\sigma_r}. \tag{1.2.14}$$

Proof From (1.2.13), we conclude

$$A^*A = V \begin{bmatrix} \Sigma^2 & O \\ O & O \end{bmatrix} V^*.$$

Thus, the eigenvalues of A^*A are $\sigma_i^2 = \lambda_i(A^*A)$, $i = 1, 2, \ldots, n$ and

$$\|A\|_2^2 = \|A^*A\|_2 = |\lambda_1(A^*A)| = \sigma_1^2.$$

Clear conclusion is that $\|A\|_2 = \sigma_1$ holds. It is easy to verify the following very important representation of the matrix pseudoinverse:

$$A^\dagger = V \begin{bmatrix} \Sigma^{-1} & O \\ O & O \end{bmatrix} U^*. \tag{1.2.15}$$

Hence the nonzero singular values of $A^\dagger$ are

$$\frac{1}{\sigma_r} \geq \frac{1}{\sigma_{r-1}} \geq \cdots \geq \frac{1}{\sigma_1} > 0.$$

Thus $\|A^\dagger\|_2 = \dfrac{1}{\sigma_r}$ holds. □

A nonzero matrix that is of neither full column rank nor full row rank can be expressed as an appropriate product of a matrix of full column rank and a matrix of full row rank. Such kind of matrix factorizations is known as the full rank factorization.

Theorem 1.2.3 ([266]) *Let $A \in \mathbb{C}_r^{m\times n}$ be of rank $r > 0$. Then there exist matrices $P \in \mathbb{C}_r^{m\times r}$ and $Q \in \mathbb{C}_r^{r\times n}$ such that*

$$A = PQ. \tag{1.2.16}$$

Lemma 1.2.14 shows that the full rank factorization of a matrix A leads to an explicit formula for its Moore–Penrose inverse $A^\dagger$. This formula is known as the full rank representation of the Moore–Penrose inverse. As usual, by A_R^{-1} and A_L^{-1} we denote a right and a left inverse of A, respectively.

Lemma 1.2.14 (MacDuffe [174]) *Let $A \in \mathbb{C}_r^{m\times n}$ and $A = PQ$, $P \in \mathbb{C}_r^{m\times r}$, $Q \in \mathbb{C}_r^{r\times n}$ be an arbitrary full rank factorization of A. Then the Moore–Penrose inverse of A can be computed using its full rank representation defined by*

$$A^\dagger = Q^*(P^*AQ^*)^{-1}P^* = Q^*(QQ^*)^{-1}(P^*P)^{-1}P^*.$$

In addition, Q is right invertible and P is left invertible:

$$Q_R^{-1} = Q^*(QQ^*)^{-1}, \quad P_L^{-1} = (P^*P)^{-1}P^*.$$

The right inverse of $A \in \mathbb{C}_n^{m\times n}$ will be denoted by $A_R^{-1} = (A^*A)^{-1}A^*$, while the left inverse of $A \in \mathbb{C}_m^{m\times n}$ will be denoted by $A_L^{-1} = A * (AA^*)^{-1}$.

The Weighted Moore–Penrose Inverse and the Drazin Inverse

Definition 1.2.4 For a given matrix $A \in \mathbb{C}^{m\times n}$, its weighted Moore–Penrose inverse $A^{\dagger}_{M,N} \in \mathbb{C}^{n\times m}$ satisfies the following matrix equations:

$$A^{\dagger}_{M,N}A = A, \quad A^{\dagger}_{M,N}AA^{\dagger}_{M,N} = A^{\dagger}_{M,N},$$

$$(AA^{\dagger}_{M,N})^{\sharp} = AA^{\dagger}_{M,N}, \quad (A^{\dagger}_{M,N}A)^{\#} = A^{\dagger}_{M,N}A$$

where M and N are the symmetric positive definite matrices of order m and n, respectively, and $A^{\#} = N^{-1}A^{T}M$. In particular, when M and N are both equal to the identity matrix I, then $A^{\dagger}_{M,N}$ reduces to the Moore–Penrose inverse $A^{\dagger}$.

Definition 1.2.5 Let $A \in \mathbb{C}^{n\times n}$ be a given matrix. The smallest nonnegative integer k for which the condition

$$\operatorname{rank}(A^{k+1}) = \operatorname{rank}(A^{k}) \tag{1.2.17}$$

holds is called the index of A, and it is denoted by $\operatorname{ind}(A) = k$.

Further the four Penrose equations (1.2.3) are supplemented by the following equations applicable only to square matrices:

$$(1^{k})\ \ A^{l+1}X = A^{l},\ l \geq \operatorname{ind}(A), \qquad (5)\ \ AX = XA. \tag{1.2.18}$$

Definition 1.2.6 The Drazin inverse of a square matrix $A \in \mathbb{C}^{n\times n}$ is the unique matrix $X \in \mathbb{C}^{n\times n}$ which fulfills the matrix equations (2), (1^{k}) and (5), and it is denoted by $X = A^{\mathrm{D}}$.

More details about the Drazin inverse can be found in [266].

Definition 1.2.7 The Drazin inverse in the case $\operatorname{ind}(A) = 1$ is called the group inverse and is denoted by $X = A^{\#}$.

Definition 1.2.8 The outer inverse of $A \in \mathbb{C}^{m\times n}_{r}$ with prescribed range T and null space S is the matrix X which satisfies the Penrose equation (2) and two additional properties: $\mathcal{R}(X) = T$ and $\mathcal{N}(X) = S$, and is denoted by $A^{(2)}_{T,S}$,

Outer Generalized Inverses

It is well known that the Moore–Penrose inverse $A^{\dagger}$ and the weighted Moore–Penrose inverse $A^{\dagger}_{M,N}$, the Drazin inverse A^{D} and the group inverse $A^{\#}$ can be presented as particular generalized inverses $A^{(2)}_{T,S}$ for appropriate choice of the matrices $T = \mathcal{R}(G)$ and $S = \mathcal{N}(G)$. For example, the next statements are valid for a rectangular matrix A:

$$A^{\dagger} = A^{(2)}_{\mathcal{R}(A^{*}),\mathcal{N}(A^{*})}, \quad A^{\dagger}_{M,N} = A^{(2)}_{\mathcal{R}(A^{\sharp}),\mathcal{N}(A^{\sharp})}, \tag{1.2.19}$$

where M, N are positive definite matrices of appropriate orders and $A^{\sharp} = N^{-1}A^*M$. More details can be found in [6]. For a given square matrix A the next identities (see [6, 266]) are satisfied:

$$A^{\mathrm{D}} = A^{(2)}_{\mathcal{R}(A^k),\mathcal{N}(A^k)}, \quad A^{\#} = A^{(2)}_{\mathcal{R}(A),\mathcal{N}(A)}, \tag{1.2.20}$$

where $k = \mathrm{ind}(A)$.

Generalized inverses are included in an extensive variety of mathematical fields, for example, matrix theory and operator theory. Many types of generalized inverses have various applications, such as linear estimation, differential equations, and difference equations. Markov chains, graphics, cryptography, coding theory, robotics, incomplete data recovery, sociology, demography, and many other fields. A special case of the Drazin inverse, called the Group inverse, has found application in characterizing the sensitivity of the stationary probabilities to perturbations in the underlying transition probabilities. Finally, the group inverse has recently proven to be fundamental in the analysis of Google's PageRank search engine. Generalized inverses play an important role in finding solutions of many stochastic models, in particular Markov chains in discrete or continuous time and Markov renewal processes [96]. The Drazin inverse has been successfully and extensively applied in different fields of science; for example, in finding closed-form solutions of singular differential equations with matrix coefficients, in solving difference equations, in Markov chains, multibody system dynamics, as well as in finding solutions of various iterative methods [18, 19]. The Moore–Penrose inverse has found a wide range of applications in many areas of science and has become a useful tool in finding least-squares solutions of linear systems, in optimization problems, in data analysis, in finding the solution of linear integral equations, etc. A global overview of various applications of generalized inverses can be found in [7]. The intrinsic estimator (IE) has become a widely used tool for the analysis of age-period-cohort (APC) data in sociology, demography, and other fields. It was recently observed in [59] that the IE is a subtype of a larger class of estimators based on the Moore–Penrose generalized inverse (MP estimators). Also, different estimators can lead to radically divergent estimates of the true, unknown APC effects [59]. In statistics, the pseudoinverse is commonly used in linear regression analysis. It allows for the computation of the regression coefficients when fitting a model to data, especially when there are more data points than parameters to estimate. In signal processing, generalized inverses are employed in problems such as filtering and system identification, where the system equations may not be square or may be ill-conditioned. Generalized inverses are used in control theory to design controllers for systems that may be underactuated or have fewer inputs than outputs. They can be crucial for finding appropriate control inputs in such systems. Generalized inverses are used in image reconstruction, particularly in scenarios involving incomplete or corrupted data, allowing for an estimation of the original image. Generalized inverses are used in multivariate statistical analysis, including methods such as canonical correlation analysis and factor analysis, where the data may not meet the assumptions required for standard matrix inversions. In game theory, the pseudoinverse is used to analyze strategies and equilibria in games involving mixed strategies, helping to find optimal decisions for players in nonlinear contexts. In neural network training, particularly in single-layer networks, the Moore–Penrose pseudoinverse can be employed to compute weights for linear models. For more information of he history of generalized inverses, the reader is referred to two survey papers [7, 8].

Outer generalized inverses with prescribed range and null space are very important in matrix theory. The {2}-inverses have application in the iterative methods for solving the nonlinear equations [6] as well as in statistics [67, 92]. In particular, outer inverses play an important role in stable approximations of ill-posed problems and in linear and nonlinear problems involving rank-deficient generalized inverses [185, 395].

Outer inverses with prescribed range and null space are useful in solving the restricted system of linear equations. This application is based on the following essential result from [32]:

Proposition 1.2.12 ([32]) *Let $A \in \mathbb{C}^{m\times n}$ be of rank r, let T be a subspace of $\mathbb{C}^n$, and let the condition*

$$b \in AT, \quad \dim(AT) = \dim(T)$$

be satisfied. Then the unique solution of

$$Ax = b, \quad x \in T$$

is given by

$$x = A^{(2)}_{T,S} b,$$

for any subspace S of $\mathbb{C}^m$ satisfying $AT \oplus S = \mathbb{C}^m$.

For other important properties of generalized inverses see [6, 266].

Core-EP Inverse

Let $\mathbb{C}_n^{CM}$ be the set of $n \times n$ matrices of index one, that is,

$$\mathbb{C}_n^{CM} = \left\{A \in \mathbb{C}^{n\times n} |\ \mathrm{rank}(A^2) = \mathrm{rank}(A)\right\}.$$

The matrices of the unit index are also called "core matrices" or "group matrices" in the literature. Baksalary and Trenkler [4] introduced the notion of core inverse on the set $\mathbb{C}_n^{CM}$.

Definition 1.2.9 Let $A \in \mathbb{C}_n^{CM}$. A matrix $A^{\#} \in \mathbb{C}^{n\times n}$ satisfying

$$AA^{\#} = P_A, \qquad \mathcal{R}(A^{\#}) \subseteq \mathcal{R}(A) \tag{1.2.21}$$

is called the core inverse. Here, P_A $(= AA^{\dagger})$ denotes an orthogonal projection onto $\mathcal{R}(A)$.

As long as A belongs to $\mathbb{C}_n^{CM}$, its core inverse always exists and is unique and reflexive. The paper [4] has attracted a lot of attention, and the theory of core inverse has been further fortified by several authors. Perturbation bounds for the core inverse are presented in [172].

Recently, Manjunatha Prasad and Mohana presented the core-EP inverse as follows.

Definition 1.2.10 ([199]) A matrix X, denoted by $A^{\textcircled{\dagger}}$ is called the core-EP inverse of A if it satisfies

$$XAX = X, \quad \mathcal{R}(X) = \mathcal{R}(X^*) = \mathcal{R}(A^k).$$

Wang in [267] introduced a new decomposition for the core-EP inverse by the Schur decomposition.

Lemma 1.2.15 ([61, 267]) *Let* $A \in \mathbb{C}^{n\times n}$ *with* $\operatorname{ind}(A) = k$ *and* $\operatorname{rank}(A^k) = r$. *There is a Schur form of A,*

$$A = U\begin{bmatrix} T_1 & T_2 \\ O & T_3 \end{bmatrix}U^*, \tag{1.2.22}$$

where $U \in \mathbb{C}^{n\times n}$ *is unitary,* T_1 *is a* $r \times r$ *nonsingular upper-triangular matrix and* T_3 *is a nilpotent matrix with index k. The core-EP decomposition of A is defined as the sum* $A = A_1 + A_2$, *where* [267]

$$A_1 = U\begin{bmatrix} T_1 & T_2 \\ O & O \end{bmatrix}U^* = A^k(A^k)^\dagger = U\begin{bmatrix} I_{\operatorname{rank}(A^k)} & O \\ O & O \end{bmatrix},$$

$$A_2 = U\begin{bmatrix} O & O \\ O & T_3 \end{bmatrix}U^* = A - A^k(A^k)^\dagger A.$$

Then the core-EP inverse can be expressed by [61]

$$A^{\textcircled{\dagger}} = U\begin{bmatrix} T_1^{-1} & O \\ O & O \end{bmatrix}U^* = A^dA^k(A^k)^\dagger = A^k(A^{k+1})^\dagger.$$

Lemma 1.2.16 ([202]) *If* PAQ *is a non-null matrix, an outer inverse* X *with* $\mathcal{R}(X) = \mathcal{R}(Q)$ *and* $\mathcal{N}(X) = \mathcal{N}(P)$ *is uniquely determined if and only if* $\operatorname{rank}(P) = \operatorname{rank}(Q) = \operatorname{rank}(PAQ)$. *In this case, the outer inverse X is given by*

$$X = Q(PAQ)^{(1,2)}P = Q(PAQ)^{(1)}P,$$

for arbitrary choices of reflexive g-inverse $(PAQ)^{(1,2)}$ *of* PAQ *and g-inverse* $(PAQ)^{(1)}$ *of* PAQ.

Also, the following representation from [199] will be useful:

$$A^{\textcircled{\dagger}} = A^{(2)}_{\operatorname{col}(A^k),\operatorname{col}(A^k)} = A^{(2)}_{\mathcal{R}(A^k),\mathcal{N}((A^k)^*)} = A^k\left((A^k)^*A^{k+1}\right)^- A^{*k}. \tag{1.2.23}$$

Main publications on the core-EP inverse are [57, 58, 183, 199, 200, 204, 267]. A number of numerical methods and theoretical investigations devoted to the computation and representation of the core-EP inverse have been presented recently. Prasad and Raj in [200] introduced a bordering method for computing the core-EP inverse by relating this generalized inverse with a suitable bordered matrix. Prasad, Raj, and Vinay in [201] introduced an iterative method to approximate the core-EP inverse. Three limit representations of the core-EP inverse were proposed in [397].

1.2.4 Idempotent Matrices and Projectors

Idempotent matrices and projectors are very important notions and appear in numerous problems concerning various generalized inverses.

Lemma 1.2.17 *Let $E \in \mathbb{C}^{n\times n}$ be idempotent. Then E possesses the following properties:*

(a) *E^* and $I - E$ are idempotent.*
(b) *The eigenvalues of E are 0 and 1 and the multiplicity of the eigenvalue 1 is equal to* $\mathrm{rank}(E)$.
(c) $\mathrm{rank}(E) = \mathrm{Tr}(E)$.
(d)*] $E(I - E) = (I - E)E = O$.*
(e)*] $Ex = x$ holds if and only if $x \in \mathcal{R}(E)$.*
(f) $E \in E\{1, 2\}$.
(g) $\mathcal{N}(E) = \mathcal{R}(I - E)$.

The transformation denoted by $P_{L,M}$ maps any n-dimensional $x \in \mathbb{C}^n$ into its projection on L along M. The transformation $P_{L,M}$ is called the projector on L along M, or the oblique projector. It is a well-known result that the projector $P_{L,M}$r is a linear transformation.

In addition, well-known Proposition 1.2.13 points out that an arbitrary idempotent matrix is a certain projector.

Proposition 1.2.13 *The subspaces $\mathcal{R}(E)$ and $\mathcal{N}(E)$ corresponding to arbitrary idempotent matrix $E \in \mathbb{C}^{n\times n}$ are complementary and satisfy the identity*

$$E = P_{\mathcal{R}(E),\mathcal{N}(E)}.$$

The generalized *core–EP (or *GCEP) inverse of A is represented as

$$A_{\textcircled{2}}^{T,S} = \left(A_{T,S}^{(2)}A\right)^{\dagger} A_{T,S}^{(2)} \in A\{2, 4\}.$$

Table 1.1 Particular cases of Φ-GCEP inverse

Constraints	Φ	$A^{\textcircled{2},\Phi}$	$A_{\textcircled{2},\Phi}$
$A \in \mathbb{C}^{m\times n}_{T,S}$	$A^{(2)}_{T,S}$	$A^{\textcircled{2},\Phi} = A^{\textcircled{2}}_{T,S}$	$A_{\textcircled{2},\Phi} = A_{\textcircled{2}}^{T,S}$
	$A^{(2)}_{T,S} = A^{\dagger}$	$A^{\textcircled{2},\Phi} = A^{\textcircled{2}}_{T,S} = A^{\dagger}$	$A_{\textcircled{2},\Phi} = A_{\textcircled{2}}^{T,S} = A^{\dagger}$
	A^{*}	$A^{\textcircled{2},\Phi} = A^{\textcircled{2}}_{T,S} = A^{\dagger}$	$A_{\textcircled{2},\Phi} = A_{\textcircled{2}}^{T,S} = A^{\dagger}$
$m = n$	$A^{(2)}_{T,S} = A^{\mathrm{d}}$	$A^{\textcircled{2},\Phi} = A^{\textcircled{2}}_{T,S} = A^{\textcircled{\dagger}}$	$A_{\textcircled{2},\Phi} = A_{\textcircled{2}}^{T,S} = A_{\textcircled{\dagger}}$
$m = n$	A^{k}	$A^{\textcircled{2},\Phi} = A^{\textcircled{2}}_{T,S} = A^{\textcircled{\dagger}}$	$A_{\textcircled{2},\Phi} = A_{\textcircled{2}}^{T,S} = A_{\textcircled{\dagger}}$
$m = n,\ 1 = \mathrm{ind}(A)$	$A^{(2)}_{T,S} = A^{\#}$	$A^{\textcircled{2},\Phi} = A^{\textcircled{2}}_{T,S} = A^{\textcircled{\#}}$	$A_{\textcircled{2},\Phi} = A_{\textcircled{2}}^{T,S} = A_{\textcircled{\#}}$
$m = n$	$A^{(2)}_{T,S} = A_{\textcircled{\dagger}}$	$A^{\textcircled{2},\Phi} = A^{\textcircled{2}}_{T,S} = A_{\diamond}$	–
$m = n$	$A^{(2)}_{T,S} = A^{\textcircled{\dagger}}$	–	$A_{\textcircled{2},\Phi} = A_{\textcircled{2}}^{T,S} = A^{\diamond}$

Composite Generalized Inverses

Using an appropriate matrix $\Phi \in \mathbb{C}^{n\times s}$ instead of $A^{(2)}_{T,S}$ in the definition of the GCEP inverse, the more general Φ-GCEP inverse of A was proposed. In particular, for $A \in \mathbb{C}^{m\times n}$ and $\Phi \in \mathbb{C}^{n\times s}$, the Φ-GCEP inverse of A is represented as

$$A^{\textcircled{2},\Phi} = \Phi(A\Phi)^{\dagger}.$$

In dual case, for $A \in \mathbb{C}^{m\times n}$ and $\Phi \in \mathbb{C}^{s\times m}$, the Φ-*GCEP inverse of A is introduced as

$$A_{\textcircled{2},\Phi} = (\Phi A)^{\dagger}\Phi.$$

Particular cases of $A^{\textcircled{2},\Phi}$ and $A_{\textcircled{2},\Phi}$ are surveyed in Table 1.1. The assumption $k = \mathrm{ind}(A)$ holds.

1.3 Basic Facts About Unconstrained Nonlinear Optimization

Unconstrained optimization is a very popular area in applied and computational mathematics as well as in operations research. I have wide applications in science, engineering, business management, military, and many other disciplines. The main mathematical model of the unconstrained minimization problem is the following:

$$\min f(\mathbf{x}),\quad \mathbf{x} = (x_1, \ldots, x_n) \in \mathbb{R}^n, \tag{1.3.1}$$

wherein $\mathbb{R}^n$ denotes the n-dimensional Euclidean space and $f : \mathbb{R}^n \to \mathbb{R}$ is a given objective function that is the object of the minimization.

For convenience in the notation, the following notations are used:

$$\mathbf{g}(\mathbf{x}) = \nabla f(\mathbf{x}),\ G(\mathbf{x}) = \nabla^2 f(\mathbf{x}),\ \mathbf{g}_k = \nabla f(\mathbf{x_k}),\ G_k = \nabla^2 f(\mathbf{x}_k),$$

where $\nabla f(\mathbf{x})$ denotes the gradient vector $\left[\frac{\partial f}{\partial x_1}(\mathbf{x}), \ldots, \frac{\partial f}{\partial x_n}(\mathbf{x})\right]$ of f and $\nabla^2 f(\mathbf{x})$ denotes the Hessian of f in $\mathbf{x}$, defined as

$$\left(\nabla^2 f(\mathbf{x})\right)_{ij} = \left(\frac{\partial f}{\partial x_i \partial x_j}(\mathbf{x})\right)_{ij},\quad 1 \le i, j \le n.$$

The gradient is a vector whose elements are all the partial derivatives of the function with respect to the variables. The gradient vector $\nabla f(\boldsymbol{x})$ at every point $\boldsymbol{x} \in \mathbb{R}^n$ is normal to the surface with constant value $f(\boldsymbol{x})$ and passes through the given point. In this way, the negative gradient points in the direction of the steepest descent.

For given vectors $\mathbf{x}, \mathbf{y} \in \mathbb{R}^n$, by $\mathbf{x}^{\mathrm{T}}\mathbf{y}$ we denote the Frobenius inner product, and $\|\mathbf{x}\| = \sqrt{\mathbf{x}^{\mathrm{T}}\mathbf{x}}$. The set of all nonnegative integers is denoted by $\mathbb{N}_0$.

The most desirable solution of the problem is its global minimizer, i.e., a point $\mathbf{x}^*$ for which $f(\mathbf{x}^*) \le f(\mathbf{x})$ for all $\mathbf{x} \in \mathbb{R}^n$. However, optimization algorithms seek only local minima.

Definition 1.3.1 Let $f : \mathbb{P} \to \mathbb{R}$, $\mathbb{P} \subset \mathbb{R}^n$ be a continuously differentiable function on the open set $\mathbb{P}$ and let the vector $\mathbf{d} \in \mathbb{R}^n$ such that it holds

$$\nabla f(\mathbf{x})^{\mathrm{T}}\mathbf{d} < 0 \tag{1.3.2}$$

Then the vector $\mathbf{d} \in \mathbb{R}^n$ is called the **descent direction** of the function f at the point $\mathbf{x} \in \mathbb{P}$.

Based on the Taylor approximation of the function f in the point $\mathbf{x}_{k+1} = \mathbf{x}_k + \alpha\mathbf{d}$ it follows that

$$f(\mathbf{x}_{k+1}) = f(\mathbf{x}_k + \alpha\mathbf{d}) \approx f(\mathbf{x}_k) + \alpha\mathbf{g}_k^{\mathrm{T}}\mathbf{d} + o(\alpha). \tag{1.3.3}$$

Now evidently follows

$$\exists\delta > 0,\ \text{such that}\ f(\mathbf{x}_k + \alpha\mathbf{d}) < f(\mathbf{x}_k),\ \forall\alpha \in (0, \delta),$$

if and only if $\mathbf{d}$ is the decreasing direction of the function f at the point $\mathbf{x}_k$, i.e., if the condition (1.3.2) is fulfilled.

Definition 1.3.2 Point $\mathbf{x}^*$ is said to be a **local minimum** of the function f if there exists $\delta > 0$ tako da je

$$f(\mathbf{x}^*) \le f(\mathbf{x})\ \text{for all}\ \mathbf{x} \in \mathbb{R}^n\ \text{for whiche}\ \|\mathbf{x} - \mathbf{x}^*\| < \delta.$$

Definition 1.3.3 Point $\mathbf{x}^*$ is said to be a **strong local minimum** of the function f if there exists $\delta > 0$ such that

$$f(\mathbf{x}^*) < f(\mathbf{x}) \text{ for all } \mathbf{x} \in \mathbb{R}^n \text{ for which } \|\mathbf{x} - \mathbf{x}^*\| < \delta \text{ and } \mathbf{x} \neq \mathbf{x}^*.$$

Definition 1.3.4 Point $\mathbf{x}^*$ is said to be a **global minimum** of the function f if

$$f(\mathbf{x}^*) \leq f(\mathbf{x}) \text{ for all } \mathbf{x} \in \mathbb{R}^n.$$

Definition 1.3.5 Point $\mathbf{x}^*$ is said to be a **strong global minimum** of the function f if

$$f(\mathbf{x}^*) < f(\mathbf{x}) \text{ for all } \mathbf{x} \in \mathbb{R}^n \text{ and } \mathbf{x} \neq \mathbf{x}^*.$$

In the sequel, we give some basic theorems for the necessity and sufficiency, a point $\mathbf{x}^*$ to be a local minimizer of the function f [161].

Theorem 1.3.1 (First-order necessary conditions) *If $\mathbf{x}^*$ is a local minimizer of the function f, which is continuously differentiable in an open neighborhood of $\mathbf{x}^*$, then $g(\mathbf{x}^*) = 0$.*

Theorem 1.3.2 (Second-order necessary conditions) *If x^* is a local minimizer of the function f and G is continuous in an open neighborhood of $\mathbf{x}^*$, then $g(\mathbf{x}^*) = 0$ and $G(\mathbf{x}^*)$ is positive semidefinite.*

Theorem 1.3.3 (Second-order sufficient conditions) *Suppose that G is continuous in an open neighborhood of $\mathbf{x}^*$ and that $g(\mathbf{x}^*) = 0$ and $G(x^*)$ is positive definite. Then $\mathbf{x}^*$ is a strict local minimizer of the function f, i.e., $f(\mathbf{x}^*) < f(\mathbf{x})$ for all $\mathbf{x}$ in some neighborhood of $\mathbf{x}^*$.*

Definition 1.3.6 Function $f : \mathbb{P} \to \mathbb{R}$, $\mathbb{P} \subset \mathbb{R}^n$ is **convex function** if its domain $\mathbb{P}$ is convex set and for all $\mathbf{x}, \mathbf{y} \in \mathbb{P}$ it follows

$$f(\alpha\mathbf{x} + (1-\alpha)\mathbf{y}) \leq \alpha f(\mathbf{x}) + (1-\alpha)f(\mathbf{y}), \text{ za svako } \alpha \in (0, 1).$$

Function f je **strongly (strictly) convex function** if for any $\mathbf{x}, \mathbf{y} \in \mathbb{P}$, $\mathbf{x} \neq \mathbf{y}$ it follows

$$f(\alpha\mathbf{x} + (1-\alpha)\mathbf{y}) < \alpha f(\mathbf{x}) + (1-\alpha)f(\mathbf{y}), \text{ za svako } \alpha \in (0, 1).$$

Function f je **uniformly convex function** if there exists a constant $c > 0$ such that for any $\mathbf{x}, \mathbf{y} \in \mathbb{P}$ it follows

$$f(\alpha\mathbf{x} + (1-\alpha)\mathbf{y}) \leq \alpha f(\mathbf{x}) + (1-\alpha)f(\mathbf{y}) - \frac{1}{2}c\alpha(1-\alpha)\|\mathbf{x} - \mathbf{y}\|^2, \text{ for all } \alpha \in (0, 1).$$

Definition 1.3.7 Function $F : \mathbb{R}^n \to \mathbb{R}^n$ je **monotonic function**, if for all $\mathbf{x}, \mathbf{y} \in \mathbb{R}^n$ it follows

$$(F(\mathbf{x}) - F(\mathbf{y}))^{\mathrm{T}}(\mathbf{x} - \mathbf{y}) \geq 0.$$

Definition 1.3.8 Function $f : \mathbb{R}^n \to \mathbb{R}$ is Lipschitz continuous function, if there exists a constant $L > 0$ such that

$$\|f(\mathbf{x}) - f(\mathbf{y})\| \leq L\|\mathbf{x} - \mathbf{y}\|, \text{ for all } \mathbf{x}, \mathbf{y} \in \mathbb{R}^n.$$

If a function is Lipschitz continuous, then it is uniformly continuous on $\mathbb{R}^n$.

Theorem 1.3.4 *Let $f : \mathbb{P} \to \mathbb{R}$, $\mathbb{P} \subset \mathbb{R}^n$ such that $\mathbb{P}$ is a nonempty convex set and let $\mathbf{x}^* \in \mathbb{P}$ be a local minimum of the function f:*

- *If f is convex function, then $\mathbf{x}^*$ is global minimum of f.*
- *If f is strongly (strictly) convex function, then $\mathbf{x}^*$ is the unique global minimum of f.*

Definition 1.3.9 Sequence of vectors $\{\mathbf{x}_k\} \subset \mathbb{R}^n$ converges to the vector $\mathbf{x}^*$, if it holds that

$$\lim_{k\to\infty} \|\mathbf{x}_k - \mathbf{x}^*\| = 0.$$

Definition 1.3.10 Sequence of vectors $\{\mathbf{x}_k\} \subset \mathbb{R}^n$ is said to be **Cauchy sequence** if it holds that

$$\lim_{i,j\to\infty} \|\mathbf{x}_i - \mathbf{x}_j\| = 0,$$

i.e., for a given $\delta > 0$, there exists an integer M such that $\|\mathbf{x}_i - \mathbf{x}_j\| < \delta$, for all $i, j > M$.

Definition 1.3.11 Let the iterative sequence $\{\mathbf{x}_k\} \subset \mathbb{R}^n$ converges to a vector $\mathbf{x}^*$. If there exist positive constants α i β, such that $\alpha \geq 1$, and β is independent on k, and fulfills

$$\lim_{k\to\infty} \frac{\|\mathbf{x}_{k+1} - \mathbf{x}^*\|}{\|\mathbf{x}_k - \mathbf{x}^*\|^\alpha} = \beta,$$

then the sequence $\{\mathbf{x}_k\}$ possesses the convergence order α.

Particularly:

- If $\alpha = 1$ i $\beta \in (0, 1)$, then the sequence $\{\mathbf{x}_k\}$ converges **linerarly**.
- If $\alpha = 1$ i $\beta = 0$ ili $1 < \alpha < 2$ i $\beta > 0$, then the sequence $\{\mathbf{x}_k\}$ converges **superlinerarly**.
- If $\alpha = 2$, then the sequence $\{\mathbf{x}_k\}$ converges **quadratically**.

Definition 1.3.12 Iterative method for find the solution $\mathbf{x}^* \in \mathbb{R}^n$ of the nonlinear system

$$F(\mathbf{x}) = 0, \quad \mathbf{x} \in \mathbb{R}^n, \tag{1.3.4}$$

such that $F : \mathbb{R}^n \longrightarrow \mathbb{R}^n$ is continuous is locally (linearly, superlinearly, quadratically) convergent if the sequence of iterates $\{\mathbf{x}_k\}$ converges to $\mathbf{x}^*$ (linearly, superlinearly, quadratically), where the initial iteration is close enough to the solution $\mathbf{x}^*$.

The most frequently used general iterative scheme aimed to solve the multivariable unconstrained minimization problem (1.3.1) is **the line search method**:

$$\mathbf{x}_{k+1} = \mathbf{x}_k + \alpha_k \mathbf{d}_k, \quad k \geq 0, \tag{1.3.5}$$

where $\mathbf{x}_{k+1}$ is a new iterative point, $\mathbf{x}_k$ is the previous iterative point, $\alpha_k > 0$ is a *step size* (or learning rate), and $\mathbf{d}_k$ is a search direction. The line search approach finds a descent direction $\mathbf{d}_k$ along which the objective function f will be reduced and then computes a step size α_k that determines the step length along $\mathbf{d}_k$ from $\mathbf{x}_k$. A small learning rate means the algorithm is careful and precise, but the convergence can be slow. On the other hand, a large learning rate can speed up convergence but can increase the risk of overshooting.

A general rule for defining appropriate search directions arises from the Taylor expansion. Taylor approximation of the function f at the point $\mathbf{x}_{k+1} = \mathbf{x}_k + \alpha_k \mathbf{d}_k$ is defined by

$$f(\mathbf{x}_{k+1}) \approx f(\mathbf{x}_k) + \alpha_k \mathbf{g}_k^{\mathrm{T}} \mathbf{d}_k. \tag{1.3.6}$$

Therefore, an appropriate descent search direction $\mathbf{d}_k$ must be determined on the basis of the *descent condition*

$$\mathbf{g}_k^{\mathrm{T}} \mathbf{d}_k < 0, \quad \text{for all } k. \tag{1.3.7}$$

The gradient-based methods rely on the gradient vector to determine the search direction in finding an optimal solution. Primary choice for descent direction is the negative gradient $\mathbf{d}_k = -\mathbf{g}_k$, which reduces the general line search iterations (1.3.5) into the *gradient descent* (GD) iterative scheme

$$\mathbf{x}_{k+1} = \mathbf{x}_k - \alpha_k \mathbf{g}_k. \tag{1.3.8}$$

(see, for example, [21, 161, 257]). The search direction d_k is usually required to satisfy the descent condition

$$\mathbf{g}_k^{\mathrm{T}} \mathbf{d}_k < 0, \quad \text{for all } k, \tag{1.3.9}$$

In order to ensure the global convergence, it is sometimes required that $\mathbf{d}_k$ satisfies the sufficient descent condition

$$\mathbf{g}_k^{\mathrm{T}} \mathbf{d}_k \leq -c_1 \|\mathbf{g}_k\|^2, \tag{1.3.10}$$

where c_1 is a positive constant [77].

As for the choice of search direction, one of the possible choices for the search direction in unconditional optimization is to move from the current point along the negative gradient in each iteration, which corresponds to $\boldsymbol{d}_k = -\boldsymbol{g}_k$. This choice of search direction leads us to a class of methods known as *gradient descent methods*. One negative feature of gradient methods is the relatively frequent occurrence of so-called zig-zagging phenomenon, which initiates very slow convergence of GD algorithms to the optimal point, or even divergence [187]. To remedy this property, some accelerated gradient descent methods have been proposed [97].

The advantages of GD methods can be summarized as follows.

1. Negative gradient is a very accurate direction.
2. GD methods generate stable updates. As a result, these methods are globally convergent, i.e., converge to a local minimizer regardless of the starting point.
3. Many optimization methods switch to the GD rule when they do not make sufficient progress toward convergence.

Disadvantages of GD methods can be summarized as follows.

1. The convergence is linear and usually very slow.
2. Numerically, GD methods are often not convergent.

Another important direction of the search is the *Newton's direction* $\boldsymbol{d}_k = -G_k^{-1}\boldsymbol{g}_k$, obtained from the second-order Taylor-development, assuming that the Hessian G_k is positive definite. The pure Newton method (without line search) for minimization of a function $f : \mathbb{R}^n \to \mathbb{R}$ is defined using a quadratic approximation of $f(\mathbf{x}_{k+1})$:

$$\Phi(\mathbf{d}) := f(\mathbf{x}_k + \mathbf{d}) \approx f(\mathbf{x}_k) + \mathbf{g}_k^{\mathrm{T}}\mathbf{d} + \frac{1}{2}\mathbf{d}^{\mathrm{T}}G_k\mathbf{d}. \tag{1.3.11}$$

The solution $\mathbf{d}_k = \min_{\mathbf{d}}(\Phi(\mathbf{d}))$ is obtained using the matrix calculus

$$\frac{\partial \mathbf{d}^{\mathrm{T}}G_k\mathbf{d}}{\partial \mathbf{d}} = (G_k + G_k^{\mathrm{T}})\mathbf{d} = 2G_k\mathbf{d}$$
$$\frac{\partial \mathbf{g}_k^{\mathrm{T}}\mathbf{d}}{\partial \mathbf{d}} = \mathbf{g}_k,$$

which implies

$$\frac{\partial \Phi(\mathbf{d})}{\partial \mathbf{d}} = 0 \Longleftrightarrow \mathbf{d} = -G_k^{-1}\mathbf{g}_k.$$

So, the pure Newton method is defined by

$$\mathbf{x}_{k+1} = \mathbf{x}_k - G_k^{-1}\mathbf{g}_k. \tag{1.3.12}$$

The Newton method with line search uses an appropriate step-size α_k in (1.3.12) with the aim to ensure global stability. The resulting iterations are of the form

$$\mathbf{x}_{k+1} = \mathbf{x}_k - \alpha_k G_k^{-1}\mathbf{g}_k, \tag{1.3.13}$$

wherein the step-size α_k is computed by performing a line search.

The Newton method exhibits three major drawbacks in practical applications.

1. The descent (and convergence) may not be achieved if the iterations (1.3.12) are started far away from the local minimizer.
2. Another drawback is the numerically expensive and tedious necessity to compute the second derivative matrix (Hessian) and its inverse in every iteration. Moreover, the second derivatives may sometimes be unavailable.
3. The main disadvantages of the Newton method are the possibility that the Hessian G_k is not positive definite.

Due to that, numerous modifications of it were created, which can be globally divided into two large groups: modified Newton's methods and *quasi-Newton* (QN) methods. The QN methods are aimed at addressing all the above difficulties of the Newton method. The first drawback is overcome by taking an appropriately defined positive definite matrix B_k that approximates the Hessian G_k or an appropriately defined positive definite matrix H_k that approximates the true Hessian inverse G_k^{-1} and then performing a line search at each iteration. For a given initial point $\boldsymbol{x}_0 \in \mathbb{R}^n$ and a symmetric positive definite matrix H_0, the search direction in the kth iteration of the quasi-Newton method is defined by $\boldsymbol{d}_k = -H_k \boldsymbol{g}_k$, where H_k is a symmetric and positive definite matrix.

QN methods and modified Newton methods belong to most powerful methods for solving unconditional optimization and are applicable in many nonlinear optimization problems. A survey of QN methods for solving nonlinear least-squares problems was considered in [182]. See [187] for more details.

After the search direction $\mathbf{d}_k$ is established, we need to determine a steplength α_k which will ensure a sufficient decrease of f. There are two main line search strategies: exact line search methods and inexact line search methods.

In the exact line search, the step-size α_k is chosen according to the one-dimensional minimization problem

$$f(\mathbf{x}_k + \alpha_k \mathbf{d}_k) = \min_{t>0} f(\mathbf{x}_k + \alpha \mathbf{d}_k). \tag{1.3.14}$$

Since the exact minimization in (1.3.14) is expensive and of no practical value, the inexact line search algorithms are preferable. This class of algorithms can be divided according to the changes of the objective function f. In fact, if we force monotone decreasing of the function f in each iteration, then we have the monotone line search methods; otherwise, we consider the non-monotone line search techniques. The first one generates a limited number of trial steplengths until it finds one that provides a sufficient decrease of the objective function f. Many inexact line search algorithms have been proposed: Armijo, Goldstein, Wolfe, Powell, Fletcher, and others (see [3, 42, 209, 298]). Among them, the most popular is the algorithm known as backtracking line search (Armijo line search).

The search direction d_k in the nonlinear class of *conjugate gradient methods* is defined by the rule

$$\mathbf{d}_k := \mathfrak{d}_k := \mathfrak{d}(\beta_k, \mathbf{g}_k, \mathbf{d}_{k-1}) = \begin{cases} -\mathbf{g}_0, & k=0, \\ -\mathbf{g}_k + \beta_k \mathbf{d}_{k-1}, & k \geq 1, \end{cases} \tag{1.3.15}$$

where β_k is the real value which is known as the conjugate gradient update parameter (CGUP). More precisely, the search direction $\mathbf{d}_k$ of the CG method is defined as a proper linear combination of the gradient descent direction and a positive multiple of the direction used in the previously finished iteration. From (1.3.15) and (1.3.5), it clearly follows that the CG methods are defined simply only by the gradient direction $\mathbf{g}_k$ and by the CGUP β_k. Different CG methods arise from proper choices of the scalar β_k. According to the common agreement, $\beta_k^{\mathcal{M}}$ denotes the parameter β_k of the CG method $\mathcal{M}$. It is important to mention that some researchers propose the usage of $\beta_k^{\mathcal{M}+} = \max\{\beta_k^{\mathcal{M}}, 0\}$. So, it is possible to use $\beta_k^{\mathcal{M}+}$ instead of $\beta_k^{\mathcal{M}}$ and generate corresponding dual method.

1.3.1 Global Algorithms and Line Search Variants

First, we present an algorithm that describes the general scheme of line search methods

Algorithm 1.3.1 Global line search algorithm

Input: Objective $f(\boldsymbol{x})$, initial point $\boldsymbol{x}_0 \in \mathbb{R}^n$ and the tolerance $0 < \varepsilon \ll 1$.

1: $k := 0$.
2: **while** $\|\boldsymbol{g}_k\| > \varepsilon$ **do**
3: Determine the vector $\boldsymbol{d}_k$ which represents the search direction.
4: Compute the step length α_k
5: Compute the new approximation of the minimum $\boldsymbol{x}_{k+1} := \boldsymbol{x}_k + \alpha_k \boldsymbol{d}_k$
6: $k := k + 1$
7: **end while**

Output: $\boldsymbol{x}_{k+1}$, $f(\boldsymbol{x}_{k+1})$

Line Search Procedures

To achieve the global convergence of iterative methods, an appropriate step-size α_k is required. The most promising at first glance is the exact line search (ELS), which assumes the unidimensional function

$$\Phi(\alpha) := f(\boldsymbol{x}_k + \alpha \boldsymbol{d}_k) \tag{1.3.16}$$

and the step-size is defined after the unidimensional optimization of the form

$$f(\mathbf{x}_k + \alpha_k \mathbf{d}_k) = \min_{\alpha > 0} \Phi(\alpha). \tag{1.3.17}$$

The ELS rule may give the most precise minimum. However, ELS is too expensive in practice or even impossible to implement, especially in situations where the iteration is far from the exact solution.

Applying the iterative procedure (1.3.5), it is most logical to choose a new point so that the step length α_k reduces the value of the goal function:

$$\Phi(\alpha_k) = f(\boldsymbol{x}_{k+1}) < \Phi(0) = f(\boldsymbol{x}_k). \tag{1.3.18}$$

Methods that in each iterative step require the fulfillment of conditions (1.3.18), that is, the reduction of the value of the objective function, define iterations that, in each step, approach the minimum of the given function. The methods conceived in this way belong to the class of methods of *monotone line search*. Many variants of *inexact line search* (ILS) rules are proposed and dominant in nonlinear optimization. The most commonly used ILS techniques are Armijo, Goldstein, Wolfe, Powell, Fletcher, and other [2, 3, 42, 69, 88, 89, 209, 298]. In most conjugate gradient methods, one of the next ILS procedures is used to calculate the step length α_k: Wolfe line search developed in [88, 89], strong Wolfe line search, or backtracking line search from [2].

In contrast to the monotonic line search, the *non-monotonic line search* is also known in the literature, where it is not necessary to reduce the value of the objective function in each iteration [77–79]. Although non-monotone techniques do not provide a minimum approach to the function in each iteration, they are very common in practical applications and have very good convergence properties. A number of non-monotonic linear search methods have been proposed recently (see, for example, [256, 262]).

Backtracking Line Search

A backtracking line search scheme based on the Armijo condition is aimed at determining the maximal value while moving along a given search vector. It starts with a relatively large step-size estimate and iteratively reduces the step-size value until a decrease in the value of the objective function is observed, according to the local gradient of the goal function. Let $\beta \in (0, 1)$, $\varphi \in (0, 1)$ and $\alpha > 0$ be given. Then there exists a smallest nonnegative integer m_k satisfying

$$f(\boldsymbol{x}_k + \beta^{m_k} t \boldsymbol{d}_k) \le f(\boldsymbol{x}_k) + \varphi \beta^{m_k} t \boldsymbol{g}_k^{\mathrm{T}} \boldsymbol{d}_k, \; t > 0. \tag{1.3.19}$$

The procedure for backtracking line search proposed in [2] starts from the initial value $\alpha = 1$ and its output values are defined such that it decreases the goal function.

Consequently, Algorithm 1.3.2 from [235] is used in numerical experiments as the implementation of the ILS, which defines the principal step-size α_k.

Algorithm 1.3.2 The backtracking line search.

Input: Nonlinear multivariate function $f(\mathbf{x})$, the vector $\mathbf{d}_k$, previous approximation $\mathbf{x}_k$, and the real numbers $0 < \omega < 0.5$ and $\varphi \in (0, 1)$.

1: $\alpha = 1$.

2: While $f(\mathbf{x}_k + \alpha \mathbf{d}_k) > f(\mathbf{x}_k) + \omega\alpha \mathbf{g}_k^{\mathrm{T}} \mathbf{d}_k$, take $\alpha := \alpha\varphi$.

3: Return $\alpha_k = \alpha$.

Goldstein Line Search

In order to ensure a sufficient decrease of the objective function, the Goldstein rule for ILS requires the following conditions:

$$f(\boldsymbol{x}_k + \alpha \boldsymbol{d}_k) \le f(\boldsymbol{x}_k) + \rho t \boldsymbol{g}_k^{\mathrm{T}} \boldsymbol{d}_k \tag{1.3.20}$$

and

$$f(\boldsymbol{x}_k + \alpha \boldsymbol{d}_k) \ge f(\boldsymbol{x}_k) + (1 - \rho) t \boldsymbol{g}_k^{\mathrm{T}} \boldsymbol{d}_k, \tag{1.3.21}$$

where $0 < \rho < \frac{1}{2}$ and $t > 0$. Conditions (1.3.20) and (1.3.21) define the Goldstein rule for inexact line search.

Wolfe Line Search

Wolfe line search conditions are well known, and these are given by

$$f(\mathbf{x}_k + \alpha_k \mathbf{d}_k) \le f(\mathbf{x}_k) + \eta \alpha_k \mathbf{g}_k^{\mathrm{T}} \mathbf{d}_k, \tag{1.3.22}$$

$$\mathbf{g}(\mathbf{x}_k + \alpha_k \mathbf{d}_k)^{\mathrm{T}} \mathbf{d}_k \ge \sigma_1 \mathbf{g}_k^{\mathrm{T}} \mathbf{d}_k, \tag{1.3.23}$$

where $0 < \eta < \sigma_1 < 1$. In addition, for conjugate gradient methods, the generalized strong Wolfe conditions, which are a conjunction of (1.3.22) and

$$-\sigma_2 \mathbf{g}_k^{\mathrm{T}} \mathbf{d}_k \ge g(\mathbf{x}_k + \alpha_k \mathbf{d}_k)^{\mathrm{T}} \mathbf{d}_k \ge \sigma_1 \mathbf{g}_k^{\mathrm{T}} \mathbf{d}_k \tag{1.3.24}$$

are often used, where $\sigma_1 > 0$. In the case $\sigma_1 = \sigma_2$, the generalized strong Wolfe conditions reduce to the strong Wolfe conditions, which are a conjunction of (1.3.22) and

$$|\mathbf{g}(\mathbf{x}_k + \alpha_k \mathbf{d}_k)^{\mathrm{T}} \mathbf{d}_k| \le -\sigma_1 \mathbf{g}_k^{\mathrm{T}} \mathbf{d}_k. \tag{1.3.25}$$

The condition (1.3.22) of the Wolfe conditions is called the Armijo condition, which is often used apart or in the form of its variants.

1.3.2 Quasi-Newton Methods and their Classification

The importance and success of the Newton method is reflected in the use of the Hessian, which gives us useful information about the curvature of the objective function. However, the use of the Hessian in Newton's method also represents its weakness. Since the calculation of the Hessian matrix and its inverse in each iteration can be very complicated, or we may have a case where the Hessian is not analytically available. The aforementioned weakness of the Newton method has led to a class of Newton methods that use only values of the objective function and its gradient. Quasi-Newton methods use function values and gradients instead of the Hessian matrix and its inverse. Also, it should be noted that the class of quasi-Newton (QN) method belongs to the class of gradient methods of the first order, because it uses only the first derivative of the objective function.

In general, most of the problems that occur with Newton methods are overcome with QN methods by constructing a series of approximations of the Hessian or inverse of the Hessian, whereby it is required that each of the approximations of the Hessian or the inverse of the Hessian satisfy a quasi-Newton equation that guarantees good approximations.

The most general iterative rule of QN type with line search is of the form

$$\mathbf{x}_{k+1} = \mathbf{x}_k - \alpha_k H_k \mathbf{g}_k, \tag{1.3.26}$$

such that H_k is an approximation of the inverse Hesiian G_k^{-1}. Further, it is assumed that B_k is an appropriately generated symmetric positive definite approximation of G_k [257]. The following notations are used in defining an appropriate updating formula

$$\mathbf{s}_k = \mathbf{x}_{k+1} - \mathbf{x}_k, \quad \mathbf{y}_k = \mathbf{g}_{k+1} - \mathbf{g}_k \tag{1.3.27}$$

are typical. The update B_{k+1} of B_k is defined using the rule

$$B_{k+1} = B_k + E_k, \tag{1.3.28}$$

where E_k is defined on the basis of the quasi-Newton property (*secant condition*)

$$B_{k+1}\mathbf{s}_k = \mathbf{y}_k. \tag{1.3.29}$$

The quasi-Newton condition for the matrix H_k is given by

$$H_{k+1}\mathbf{y}_k = \mathbf{s}_k. \tag{1.3.30}$$

Methods that require the calculation or approximation of the Hessian matrix and its inverse belong to the class of QN methods, as well as their numerous modifications. The pure Newton method requires the calculation of the second derivative matrix, which is avoided in QN methods. As a consequence, the Newton method is computationally expensive and exhibits slow computation, while QN methods are computationally cheap and faster in computation. On the other hand, the Newton method requires fewer iterative steps and generates more precise convergence than QN methods.

Symmetric Rank-One Update

The Symmetric Rank-One update (SR1) assumes the matrix H_{k+1}in the form

$$H_{k+1} = H_k + E_k,$$

where E_k is assumed to be a symmetric rank-one matrix. Therefore,

$$H_{k+1} = H_k + \boldsymbol{u}_k \boldsymbol{v}_k^{\mathrm{T}}, \tag{1.3.31}$$

where $\boldsymbol{u}_k, \boldsymbol{v}_k \in \mathbb{R}^n$. The quasi-Newton condition (1.3.30) initiates

$$H_{k+1}\boldsymbol{y}_k = (H_k + \boldsymbol{u}_k \boldsymbol{v}_k^{\mathrm{T}})\boldsymbol{y}_k = \boldsymbol{s}_k,$$

that is

$$(\boldsymbol{v}_k^{\mathrm{T}} \boldsymbol{y}_k)\boldsymbol{u}_k = \boldsymbol{s}_k - H_k \boldsymbol{y}_k. \tag{1.3.32}$$

The conclusion is that $\boldsymbol{u}_k$ must be in the direction $\boldsymbol{s}_k - H_k \boldsymbol{y}_k$. Suppose that

$$\boldsymbol{s}_k - H_k \boldsymbol{y}_k \neq 0,$$

(otherwise H_k would satisfy the quasi-Newton equation) and the vector $\boldsymbol{v}_k$ satisfies $\boldsymbol{v}_k^{\mathrm{T}} \boldsymbol{y}_k \neq 0$. Then, on the basis of (1.3.31) and (1.3.32), it follows that

$$H_{k+1} = H_k + \frac{1}{\boldsymbol{v}_k^{\mathrm{T}} \boldsymbol{y}_k}(\boldsymbol{s}_k - H_k \boldsymbol{y}_k)\boldsymbol{v}_k^{\mathrm{T}}. \tag{1.3.33}$$

The condition that the approximation H_{k+1} of the Hessian inverse is symmetric requires us to take $\boldsymbol{v}_k = \boldsymbol{s}_k - H_k \boldsymbol{y}_k$, so it was obtained

$$H_{k+1} = H_k + \frac{(\boldsymbol{s}_k - H_k \boldsymbol{y}_k)(\boldsymbol{s}_k - H_k \boldsymbol{y}_k)^{\mathrm{T}}}{(\boldsymbol{s}_k - H_k \boldsymbol{y}_k)^{\mathrm{T}} \boldsymbol{y}_k}, \tag{1.3.34}$$

which is the Symmetric Rank-One (SR1) update.

For general functions, Conn, Gould, and Toint in [43] proved that the sequence of SR1 Hessian approximations converges to the true Hessian provided that the steps are uniformly linearly independent.

DFP Update

The Hessian update B_{k+1} is defined as the solution to the problem

$$\min_{B} \|B - B_k\|, \quad \text{s.t. } B = B^{\mathrm{T}}, \; B\mathbf{s}_k = \mathbf{y}_k. \tag{1.3.35}$$

The solution to (1.3.35) is equal to

$$B_{k+1}^{\mathrm{DFP}} = (I - \rho_k \mathbf{y}_k \mathbf{s}_k^{\mathrm{T}}) B_k^{\mathrm{DFP}} (I - \rho_k \mathbf{s}_k \mathbf{y}_k^{\mathrm{T}}) + \rho_k \mathbf{y}_k \mathbf{y}_k^{\mathrm{T}}, \quad \varrho_k = \frac{1}{y_k^{\mathrm{T}} s_k}.$$

The inverse Hessian update can be generated using the Sherman–Morrison–Woodbury formula.

Moreover, the DFP update is known as a method of updating, of rank 2, that is, H_{k+1} is formed by adding the matrix H_k with two symmetric matrices, each of which is rank 1:

$$H_{k+1} = H_k + a\boldsymbol{u}_k\boldsymbol{u}_k^{\mathrm{T}} + b\boldsymbol{v}_k\boldsymbol{v}_k^{\mathrm{T}},$$

where $\boldsymbol{u}_k, \boldsymbol{v}_k \in \mathbb{R}^n$, and a, b are scalars. From the quasi-Newton condition (1.3.30) it follows that

$$H_k\, \boldsymbol{y}_k + a\, \boldsymbol{u}_k\boldsymbol{u}_k^{\mathrm{T}}\boldsymbol{y}_k + b\, \boldsymbol{v}_k\boldsymbol{v}_k^{\mathrm{T}}\boldsymbol{y}_k = \boldsymbol{s}_k. \tag{1.3.36}$$

The vectors $\boldsymbol{u}_k$ and $\boldsymbol{v}_k$ are not unique, but they can obviously be determined in the following way

$$\boldsymbol{u}_k = \boldsymbol{s}_k, \quad \boldsymbol{v}_k = H_k\boldsymbol{y}_k.$$

Now, (1.3.36) implies

$$a = \frac{1}{\boldsymbol{u}_k^{\mathrm{T}}\boldsymbol{y}_k} = \frac{1}{\boldsymbol{s}_k^{\mathrm{T}}\boldsymbol{y}_k}, \quad b = -\frac{1}{\boldsymbol{v}_k^{\mathrm{T}}\boldsymbol{y}_k} = -\frac{1}{\boldsymbol{y}_k^{\mathrm{T}}H_k\boldsymbol{y}_k}.$$

So we get

$$H_{k+1}^{\mathrm{DFP}} = H_k + \frac{\boldsymbol{s}_k\boldsymbol{s}_k^{\mathrm{T}}}{\boldsymbol{s}_k^{\mathrm{T}}\boldsymbol{y}_k} - \frac{H_k\boldsymbol{y}_k\boldsymbol{y}_k^{\mathrm{T}}H_k}{\boldsymbol{y}_k^{\mathrm{T}}H_k\boldsymbol{y}_k}. \tag{1.3.37}$$

Formula (1.3.37) was proposed by Davidon and later was developed by Fletcher and Powell, so that it is called DFP update.

BFGS Update

One famous broadly used updating formula is the Broyden–Fletcher–Goldfarb–Shanno (BFGS) rule. The inverse Hessian update H_{k+1} is defined as the solution to

$$\min_{H} \|H - H_k\|, \quad \text{s.t. } H = H^{\mathrm{T}}, \; H\mathbf{y}_k = \mathbf{s}_k. \tag{1.3.38}$$

Certainly, the BFGS update is overtly known as

$$\begin{aligned} H_{k+1}^{\mathrm{BFGS}} &= (I - \rho_k \mathbf{s}_k \mathbf{y}_k^{\mathrm{T}}) H_k^{\mathrm{BFGS}} (I - \rho_k \mathbf{y}_k \mathbf{s}_k^{\mathrm{T}}) + \rho_k \mathbf{s}_k \mathbf{s}_k^{\mathrm{T}}, \quad \rho_k = \frac{1}{y_k^{\mathrm{T}} s_k} \\ &= H_k^{\mathrm{BFGS}} - \frac{H_k \mathbf{y}_k \mathbf{s}_k^{\mathrm{T}} + \mathbf{s}_k \mathbf{y}_k^{\mathrm{T}} H_k}{\mathbf{s}_k^{\mathrm{T}} \mathbf{y}_k} + \left(1 + \frac{\mathbf{y}_k^{\mathrm{T}} H_k \mathbf{y}_k}{\mathbf{s}_k^{\mathrm{T}} \mathbf{y}_k}\right) \frac{\mathbf{s}_k \mathbf{s}_k^{\mathrm{T}}}{\mathbf{s}_k^{\mathrm{T}} \mathbf{y}_k}, \end{aligned} \tag{1.3.39}$$

where $\mathbf{s}_k$ and $\mathbf{y}_k$ are defined in (1.3.27).

The updated BFGS formula for the Hessian matrix can be generated using the Sherman–Morrison–Woodbury formula. A rank-one modification (or perturbation) $M = A + \mathbf{b}\mathbf{c}^*$ of a matrix $A \in \mathbb{C}^{m \times n}$ uses two vectors $\mathbf{b} \in \mathbb{C}^{m \times 1}$ and $\mathbf{c} \in \mathbb{C}^{n \times 1}$. The Sherman-Morrison formula establishes a relationship between M^{-1} and A^{-1} as follows [55]:

$$M^{-1} = A^{-1} - \left(1 + \mathbf{c}^* A^{-1} \mathbf{b}\right)^{-1} A^{-1} \mathbf{b}\mathbf{c}^* A^{-1}. \tag{1.3.40}$$

As a result, the following update for B_k is obtained:

$$B_{k+1}^{\mathrm{BFGS}} = B_k^{\mathrm{BFGS}} - \frac{B_k \mathbf{s}_k \mathbf{s}_k^{\mathrm{T}} B_k}{\mathbf{s}_k^{\mathrm{T}} B_k \mathbf{s}_k} + \frac{\mathbf{y}_k \mathbf{y}_k^{\mathrm{T}}}{\mathbf{s}_k^{\mathrm{T}} \mathbf{y}_k}. \tag{1.3.41}$$

Broyden Family of Methods

The weighted combinations of DFP and BFGS updates give the whole update class, which is known as the Broyden class. This class of update is defined by

$$H_{k+1}^{\phi} = (1 - \phi) H_{k+1}^{\mathrm{DFP}} + \phi H_{k+1}^{\mathrm{BFGS}}, \tag{1.3.42}$$

where ϕ is a real parameter. If $\phi \in [0, 1]$, then (1.3.42) is called the Broaden convex update class. It is obvious that Broyden's class (1.3.42) satisfies the quasi-Newton equation (1.3.29). Also, the expression (1.3.42) can be rewritten in the following form

$$\begin{aligned} H_{k+1}^{\phi} &= H_{k+1}^{\mathrm{DFP}} + \phi \boldsymbol{v}_k \boldsymbol{v}_k^{\mathrm{T}} \\ &= H_{k+1}^{\mathrm{BFGS}} + (\phi - 1) \boldsymbol{v}_k \boldsymbol{v}_k^{\mathrm{T}} \\ &= H_k + \frac{\boldsymbol{s}_k \boldsymbol{s}_k^{\mathrm{T}}}{\boldsymbol{s}_k^{\mathrm{T}} \boldsymbol{y}_k} - \frac{H_k \boldsymbol{y}_k \boldsymbol{y}_k^{\mathrm{T}} H_k}{\boldsymbol{y}_k^{\mathrm{T}} H_k \boldsymbol{y}_k} + \phi \boldsymbol{v}_k \boldsymbol{v}_k^{\mathrm{T}}, \end{aligned} \tag{1.3.43}$$

where

$$\boldsymbol{v}_k = (\boldsymbol{y}_k^{\mathrm{T}} H_k \boldsymbol{y}_k)^{1/2} \left[\frac{\boldsymbol{s}_k}{\boldsymbol{s}_k^{\mathrm{T}} \boldsymbol{y}_k} - \frac{H_k \boldsymbol{y}_k}{\boldsymbol{y}_k^{\mathrm{T}} H_k \boldsymbol{y}_k} \right]. \tag{1.3.44}$$

If we put in (1.3.43):

- $\phi = 0$, then we will obtain DFP update (1.3.37)
- $\phi = 1$, then we will obtain BFGS update (1.3.39)
- $\phi = \frac{\boldsymbol{s}_k^{\mathrm{T}} \boldsymbol{y}_k}{(\boldsymbol{s}_k - H_k \boldsymbol{y}_k)^{\mathrm{T}} \boldsymbol{y}_k}$, then we will obtain SR1 update (1.3.34).

The Broyden class of methods can be derived directly from the quasi-Newton equation. Consider the general formula for updating rank 2, which contains $\boldsymbol{s}_k$ and $H_k \boldsymbol{y}_k$:

$$H_{k+1} = H_k + a \boldsymbol{s}_k \boldsymbol{s}_k^{\mathrm{T}} + b(H_k \boldsymbol{y}_k \boldsymbol{s}_k^{\mathrm{T}} + \boldsymbol{s}_k \boldsymbol{y}_k^{\mathrm{T}} H_k) + c H_k \boldsymbol{y}_k \text{`} \mathrm{y}_k^{\mathrm{T}} H_k, \tag{1.3.45}$$

where a, b, c are unknown scalars. We obtain

$$\begin{aligned} 1 &= a \boldsymbol{s}_k^{\mathrm{T}} \boldsymbol{y}_k + b \boldsymbol{y}_k^{\mathrm{T}} H_k \boldsymbol{y}_k, \\ 0 &= 1 + b \boldsymbol{s}_k^{\mathrm{T}} \boldsymbol{y}_k + c \boldsymbol{y}_k^{\mathrm{T}} H_k \boldsymbol{y}_k. \end{aligned} \tag{1.3.46}$$

Here we have two equations with three unknowns, so we can introduce the replacement

$$b = -\phi / \boldsymbol{s}_k^{\mathrm{T}} \boldsymbol{y}_k,$$

where ϕ is a parameter. Solving system (1.3.46) and substituting the obtained result in (1.3.45), we obtain

$$H_{k+1}^{\phi} = H_k + \frac{\boldsymbol{s}_k \boldsymbol{s}_k^{\mathrm{T}}}{\boldsymbol{s}_k^{\mathrm{T}} \boldsymbol{y}_k} - \frac{H_k \boldsymbol{y}_k \boldsymbol{y}_k^{\mathrm{T}} H_k}{\boldsymbol{y}_k^{\mathrm{T}} H_k \boldsymbol{y}_k} + \phi \boldsymbol{v}_k \boldsymbol{v}_k^{\mathrm{T}} = H_{k+1}^{\mathrm{DFP}} + \phi \boldsymbol{v}_k \boldsymbol{v}_k^{\mathrm{T}},$$

where $\boldsymbol{v}_k$ is defined by means of (1.3.44). Previous expression is identical to (1.3.43).

Some Modifications of Quasi-newton Methods

A great effort has been invested to discover QN methods that do not merely possess convergence, but it is also better than the BFGS update in the numerical performance. Table 1.2 shows the main modifications of the quasi-Newton equations.

Also, it is important to state the spectral gradient method. Therein the updating of the formula for B_{k+1} is done in the following way [211]

$$B_{k+1} = \mathrm{diag}(r_k^{(i)}) \tag{1.3.47}$$

Table 1.2 Some modifications of quasi-Newton equations

Quasi-Newton Equations	$\tilde{\mathbf{y}}_{k-1}$	References
$B_k\mathbf{s}_{k-1}=\tilde{\mathbf{y}}_{k-1}$	$\tilde{\mathbf{y}}_{k-1}=\varphi_{k-1}\mathbf{y}_{k-1}+(1-\varphi_{k-1})B_{k-1}\mathbf{s}_{k-1}$	[198]
$B_k\mathbf{s}_{k-1}=\tilde{\mathbf{y}}_{k-1}$	$\tilde{\mathbf{y}}_{k-1}=\mathbf{y}_{k-1}+t_{k-1}\mathbf{s}_{k-1},\ t_{k-1}\le 10^{-6}$	[127]
$B_k\mathbf{s}_{k-1}=\tilde{\mathbf{y}}_{k-1}$	$\tilde{\mathbf{y}}_{k-1}=\mathbf{y}_{k-1}+\frac{2(f_{k-1}-f_k)+(\mathbf{g}_k+\mathbf{g}_{k-1})^{\mathrm{T}}\mathbf{s}_{k-1}}{\|\mathbf{s}_{k-1}\|^2}\mathbf{s}_{k-1}$	[297]
$B_k\mathbf{s}_{k-1}=\tilde{\mathbf{y}}_{k-1}$	$\tilde{\mathbf{y}}_{k-1}=$ $\mathbf{y}_{k-1}+\frac{\max(0,2(f_{k-1}-f_k)+(\mathbf{g}_k+\mathbf{g}_{k-1})^{\mathrm{T}}\mathbf{s}_{k-1})}{\|\mathbf{s}_{k-1}\|^2}\mathbf{s}_{k-1}$	[337]
$B_k\mathbf{s}_{k-1}=\tilde{\mathbf{y}}_{k-1}$	$\tilde{\mathbf{y}}_{k-1}=$ $\mathbf{y}_{k-1}+\frac{\max(0,6(f_{k-1}-f_k)+3(\mathbf{g}_k+\mathbf{g}_{k-1})^{\mathrm{T}}\mathbf{s}_{k-1})}{\|\mathbf{s}_{k-1}\|^2}\mathbf{s}_{k-1}$	[338]
$B_k\mathbf{s}_{k-1}=\tilde{\mathbf{y}}_{k-1}$	$\tilde{\mathbf{y}}_{k-1}=\frac{1}{2}\mathbf{y}_{k-1}+\frac{(f_{k-1}-f_k)-\frac{1}{2}\mathbf{g}_k^{\mathrm{T}}\mathbf{s}_{k-1}}{\mathbf{s}_{k-1}^{\mathrm{T}}\mathbf{y}_{k-1}}\mathbf{y}_{k-1}$	[90]

with

$$r_k^{(i)}=\frac{1}{1+\frac{\sum_{i=1}^n(\hat{s}_k^{(i)})^2-\sum_{i=1}^n(\hat{\mathbf{s}}_k^{(i)})(\hat{\mathbf{y}}_k^{(i)})}{\sum_{i=1}^n(\hat{\mathbf{s}}_k^{(i)})^4}(\hat{\mathbf{s}}_k^{(i)})^2},\quad \hat{\mathbf{s}}_k=\mathbf{s}_k-\frac{7}{11}\mathbf{s}_{k-1}+\frac{2}{11}\mathbf{s}_{k-2},$$

$$\hat{\mathbf{y}}_k=\mathbf{y}_k-\frac{7}{11}\mathbf{y}_{k-1}+\frac{2}{11}\mathbf{y}_{k-2}.$$

The general framework of the QN algorithm is given in Algorithm 1.3.3.

Algorithm 1.3.3 General framework of the quasi-Newton algorithm

Input: Objective $f(\boldsymbol{x})$, initial point $\boldsymbol{x}_0\in\mathbb{R}^n$, initial inverse Hessian approximation $H_0\in\mathbb{R}^{n\times n}$, and the tolerance $0<\varepsilon\ll 1$.

1: $k:=0$.
2: **while** $\|\boldsymbol{g}_k\|>\varepsilon$ **do**
3: Compute

$$\boldsymbol{d}_k=-H_k\boldsymbol{g}_k. \qquad (1.3.48)$$

4: Compute $\alpha_k>0$ using exact or inexact line search.
5: Compute $\boldsymbol{x}_{k+1}:=\boldsymbol{x}_k+\alpha_k\boldsymbol{d}_k$.
6: Update H_k into H_{k+1}, such that (1.3.30) holds.
7: $k:=k+1$.
8: **end while**

Output: $\boldsymbol{x}_{k+1}$, $f(\boldsymbol{x}_{k+1})$.

1.4 Introduction to GNN and ZNN Dynamics

There are two general approaches in computing generalized inverses:
1. classical numerical algorithms (discrete time) and
2. continuous-time neural network algorithms.

Numerical algorithms can be divided into two main categories: direct and iterative methods.

–Direct methods typically compute the solution in a finite sequence of operations, which are able to generate exact solutions in the absence of rounding errors.

–Iterative methods start with an appropriate initial solution and recursively improve previously generated approximate solutions until an approximation to the theoretical solution is obtained with a desired accuracy.

Direct methods for the computation of generalized inverses can be divided as follows.
The singular value decomposition (SVD) algorithm.
Algorithms based on other types of matrix factorizations, such as the QR decomposition, LU factorization.
Methods based on the application of the Gauss-Jordan elimination process to an appropriate augmented matrix.
Greville's Partitioning method, the Leverrier–Faddeev algorithm.

Detailed description of direct and iterative methods for computing generalized inverses is available in [293]. The SVD algorithm is the most commonly used method, but it requires a large amount of computational resources. The presence of very small singular values is therefore a potential drawback of this method. Usually, traditional numerical algorithms are of a serial-processing nature and may not be efficient enough for online or real-time simulations, which must guarantee a response within specified time constraints. A demanding benchmark test for the pseudoinverse computation was introduced in [121]. Also, five known algorithms from the literature were tested on these test matrices in [121].

An important approach in numerical linear algebra includes continuous-time parallel processing methods based on *dynamical systems* and initiated *recurrent neural networks* (RNN). In mathematics, a dynamical system is a system in which is a function that describes the time dependence of a point in an ambient space. Recurrent Neural Networks (RNN) are one of the two broad types of artificial neural networks that allow the output from some nodes to affect subsequent input to the same nodes. A continuous-time recurrent neural network (CTRNN) uses a system of ordinary differential equations to define effects on incoming inputs on a neuron.

The continuous-time neural learning algorithms have emerged in recent years as parallel distributed computational models for a variety of real-time applications. In many real-time systems, the requirement for real-time solutions of the matrix pseudoinverse is typically imperative. An example of such applications in robotics is the solution to the manipulator inverse kinematics problem in real-time motion control of kinematically redundant robots [49]. Reported theoretical and numerical results show that the dynamical system approach and initiated neural network models are more suitable for real-time applications than the conventional numerical algorithms. We consider RNN models dedicated to find zeros of equations or to minimize nonlinear functions. These models represent optimization networks.

We consider CTRNN dedicated to find zeros of equations or to minimize nonlinear functions. Two important classes of CTRNNs are known:
Gradient Neural Networks (GNN) and
Zhang Neural Networks (ZNN).

GNN models are aimed to solving *time-invariant* problems. ZNN models are able to solve *time-varying* problems.

Main applications of the ZNN time-varying dynamical approach can be listed as follows:

–applications in motion planning and control of robot manipulators;
–tracking control of chaotic systems;
–usage in populations control in mathematical biosciences;
–solving time-varying problems in economics.

1.4.1 Global Overview of GNN Dynamics

GNN neural dynamics can be defined in two global steps. Essentially, GNN models are based on the gradient descent nonlinear optimization method. The first step is to formulate a cost function which will be minimized using the gradient descent optimization method in order to give the result of the considered problem.

Step1GNN. In the first step, it is necessary to define an appropriate error matrix $E(t)$ on the basis of the matrix equation which is currently being solved. The error matrix $E(t)$ is defined by replacing the unknown matrix from the considered problem by the time-varying matrix $V(t)$ which will be approximated during the time $t \geq 0$.

For example, the matrix equation $AX = I$ is appropriate for computing the inverse. In this case, the error matrix is defined by $E(t) = AV(t) - I$. The goal function of a Gradient Neural Network (GNN) model is the scalar function which is defined by the Frobenius norm of $E(t)$:

$$\varepsilon(t) = \frac{\|E(t)\|_F^2}{2}, \quad \|E\|_F = \sqrt{\mathrm{Tr}(E^{\mathrm{T}}E)}.$$

Step2GNN. In the second step, it is necessary to apply the dynamic evolution, which is based on direct proportionality between the time derivative $\dot{V}(t)$ and the negative gradient of the goal function $\varepsilon(t)$:

$$\dot{V}(t) = \frac{\mathrm{d}V(t)}{\mathrm{d}t} = -\gamma \frac{\partial \varepsilon(t)}{\partial V}, \quad V(0) = V_0. \tag{1.4.49}$$

Here, $t \in [0, +\infty)$ is the time, $V(t)$ is the matrix of activation state variables, and $\dot{V}(t)$ is the time derivative of $V(t)$. It is known that larger values of the scaling parameter γ enable faster convergence. This fact will be explained later during the convergence analysis of GNN design.

For example, in the case $E(t) = AV(t) - I$, the corresponding GNN dynamics is defined by the system of differential equations

$$\begin{aligned}\dot{V}(t) = \frac{\mathrm{d}V(t)}{\mathrm{d}t} &= -\gamma \frac{\partial \varepsilon(t)}{\partial V} \\ &= -\gamma A^{\mathrm{T}}(AV(t) - I), \\ & V(0) = V_0.\end{aligned}$$

Consequently, the gradient direction for the least-squares minimization problem $\min \|AX - I\|_F^2$ is given by $\frac{\partial \varepsilon(t)}{\partial V} = \nabla \varepsilon(t) = A^{\mathrm{T}}(AV(t) - I)$ [91, 261]. As a result, the GNN design is an adaptation of the gradient descent method for nonlinear optimization in the continuous-time case

$$\dot{V}(t) = -\gamma \nabla \varepsilon(t). \tag{1.4.50}$$

The equilibrium points satisfy

$$\nabla \varepsilon(t) = 0$$

and the convergence rate is exponential with the speed defined by $\lambda_{\min}(A^{\mathrm{T}}A)$ [91].

Now, consider the analogy of (1.4.50) in the discrete-time case. Assume that the discretization of the continuous-time GNN model is performed by using the Euler forward-difference rule

$$\dot{V}(t) \approx (V_{k+1} - V_k)/\tau,$$

where τ denotes the sampling time and $V_k = V(t = k\tau)$, $k = 1, 2, \ldots$.

Later, taking into account $\frac{\partial \varepsilon(V_k)}{\partial V} = \nabla \varepsilon(V_k)$, the Euler-type discrete-time analogy of the GNN model (1.4.50) is just a gradient descent method for nonlinear optimization:

$$V_{k+1} = V_k - \beta \nabla \varepsilon(V_k), \tag{1.4.51}$$

where $\beta = \tau\gamma > 0$ is the step size. Clearly, a smaller time τ can be compensated using larger values γ.

The nonlinear gradient-based neural dynamics are defined by

$$\dot{V}(t) = -\gamma \mathcal{F}(\nabla \varepsilon(t)). \tag{1.4.52}$$

and exploits acceleration by the nonlinear activation function array $\mathcal{F}(\cdot)$. The real activation function $\mathcal{F}$ is defined on a real matrix $C = (c_{ij}) \in \mathbb{R}^{m \times n}$ by the elementwise application $\mathcal{F}(C) = (f(c_{ij}))$ of an odd and monotonically increasing function f. The complex activation function array $\mathcal{H}$ is defined by $\mathcal{H}(A + \imath B) = \mathcal{F}(A) +$

$\iota\mathcal{F}(B)$, where $\mathcal{F}(\cdot) : \mathbb{R}^{m\times n} \to \mathbb{R}^{m\times n}$ denotes a real activation function array. If $A = (a_{ij}) \in \mathbb{R}^{m\times n}$, then $\mathcal{F}\left(a_{ij}\right) = f(a_{ij})$, where $f(\cdot)$ is an odd and monotonically increasing function.

1.4.2 Brief Review of Basic Activation Functions

An efficient activation function is an essential element in structuring a neural network, and its choice largely determines the performance and the complexity of the neural network. By definition, an activation function is a function used to transform the activation level of a unit (neuron) into an output signal. One can say, an activation function serves as a threshold, alternatively called classification or a partition. More simply, the purpose of the activation function is to ensure that the representation in the input space is mapped to a different space in the output. In its most general sense, a neural network layer performs a projection that is followed by a selection. Without selection and only projection, a network will thus remain in the same space and be unable to create higher levels of abstraction between the layers.

A neural network is composed of a set of inputs, a set of outputs, and a set of simple processing elements called neurons. A signal can be sent from any neuron in the neural network to others, and all these signals are rescaled by the connection weights. These neurons usually sum a number of the weighted input variables and pass the result through a mapping procedure, which in fact is the input–output relationship of the transformation. While each neuron performs a simple transformation concurrently in a parallel distributed manner, the strength of interaction among neurons is reflected by the connection weights, which determine the functional behavior of the neural network.

Various types of activation functions are proposed in the literature. The most common ones are the linear function, the bipolar-sigmoid function, the power-sigmoid function, and the smooth power-sigmoid function. Three common properties of all these functions are that they are monotonically increasing, odd, and differentiable functions. The following widely used real-valued linear and nonlinear functions satisfy these requirements.

(1) *Linear activation function*:

$$f(x) = x. \tag{1.4.53}$$

(2) *Bipolar-sigmoid activation function*:

$$f(x) = \frac{1 - \exp(-\varrho x)}{1 + \exp(-\varrho x)}, \quad \varrho > 2. \tag{1.4.54}$$

(3) *Power-sigmoid function*:

$$f(x)=\begin{cases} x^{\rho}, & if\ |x|\geq 1\\ \dfrac{1+exp(-\varrho)}{1-\exp(-\varrho)}\cdot\dfrac{1-\exp(-\varrho x)}{1+\exp(-\varrho x)}, & otherwise,\end{cases} \tag{1.4.55}$$

where $\varrho>2,\ \rho\geq 3$.
(4) *Smooth power-sigmoid activation function*:

$$f(x)=\frac{1}{2}x^{\rho}+\frac{1+\exp(-\varrho)}{1-\exp(-\varrho)}\cdot\frac{1-\exp(-\varrho x)}{1+\exp(-\varrho x)},\quad \rho\geq 3,\ \ \varrho>2. \tag{1.4.56}$$

(5) *Sign-bi-power function* (Li function):

$$f(x)=\frac{1}{2}|x|^{r}\operatorname{sign}(x)+\frac{1}{2}|x|^{\frac{1}{r}}\operatorname{sign}(x),$$

where $0<r<1$ and $\operatorname{sign}(x)$ is the signum function

$$\operatorname{sign}(x)=\begin{cases}1, & x>0,\\ 0, & x=0,\\ -1, & x<0.\end{cases}$$

(6) *Tunable sign-bi-power function* (Tunable function):

$$f(x)=\frac{1}{2}k_1|x|^{r}\operatorname{sign}(x)+\frac{1}{2}k_2x+\frac{1}{2}k_3|x|^{\frac{1}{r}}\operatorname{sign}(x), \tag{1.4.57}$$

where $k_1>0,\ k_2>0,\ k_3>0,\ 0<r<1$.
(7) *Power-sum activation function*: $\phi(e_i)=\sum_0^N e_i^{2N-1}$, where $N>1$.
(8) *Hyperbolic sine activation function*: $\phi(e_i)=\frac{\exp(e_i m)}{2}-\frac{\exp(-e_i m)}{2}$, where m is an odd integer.
(9) *Li array activation function* is defined as $\Psi(E(t)))=\psi(e_{ij})$, where e_{ij} denotes the (ij)th element of the error matrix $E(t)$ and ψ is defined by

$$\psi(a_{ij})=\operatorname{Lip}^{\sigma}(a_{ij})+\operatorname{Lip}^{\frac{1}{\sigma}}(a_{ij})$$

on each element of a given matrix $A=(a_{ij})\in\mathbb{R}^{m\times n}$ The design parameter satisfies $\sigma\in(0,1)$ and the function $\operatorname{Lip}^{\sigma}(a_{ij})$ is defined as

$$\operatorname{Lip}^{\sigma}(a_{ij})=\begin{cases}|a_{ij}|^{\sigma}, & a_{ij}>0,\\ 0, & a_{ij}=0,\\ -|a_{ij}|^{\sigma}, & a_{ij}<0.\end{cases}$$

For the illustration and comparison, the four types of activation functions f are illustrated in Fig. 1.2.

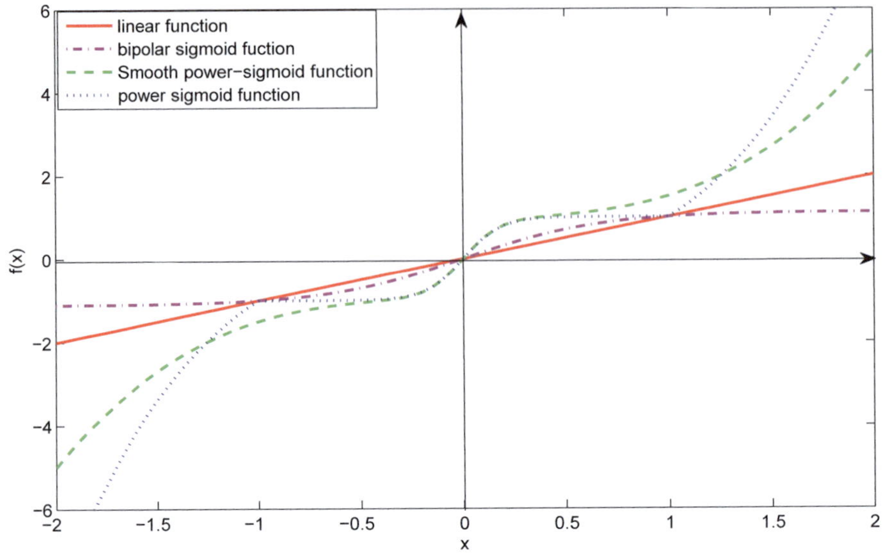

Fig. 1.2 Behavior of the four basic types of activation functions

Note that other new activation functions can be generated based on these basic types of activation functions. Also, all of them must be monotonically increasing and odd functions.

Remark 1.4.1 The activation function array on complex matrices can be defined in two ways: one approach is to handle separately the real parts and imaginary parts of the entries of the input complex matrix. The other one is capable of operating separately on the modulus and the arguments of complex values included in the input matrix.

The following definition introduces two types of activation functions in the complex domain.

Definition 1.4.1 ([129]) Let $A \in \mathbb{R}^{n\times n}$, $B \in \mathbb{R}^{n\times n}$ be given. Type I activation function array is defined by

$$\mathcal{H}_1(A + \iota B) = \mathcal{F}(A) + \iota\mathcal{F}(B), \tag{1.4.58}$$

where $\iota = \sqrt{-1}$ denotes the imaginary unit.

Type II activation function array is defined as

$$\mathcal{H}_2(A + \iota B) = \mathcal{F}(\Gamma) \circ \exp(\iota\Theta), \tag{1.4.59}$$

where $\Gamma = |A + \iota B| \in \mathbb{R}^{n\times n}$ and $\Theta = \Theta(A + \iota B) \in (-\pi, \pi]^{n\times n}$ represents element-wise modulus and the element-wise arguments, respectively, of the complex matrix $A + \iota B$.

Additionally, $\mathcal{F}(A)$ is defined to be an element-wise odd and monotonically increasing function for a real matrix A, i.e., $\mathcal{F}(A) = (f(a_{kj}))$ for $A = (a_{kj}) \in \mathbb{R}^{n\times n}$, with an odd and monotonically increasing function $f(\cdot)$.

1.4.3 Global Overview of ZNN Dynamics

Nowadays, it is considered normal in academia and industry to handle intractability problems and solve complex computational problems by using neural networks. These problems can be formulated as a set of equations, and the goal is to find the zeros of these equations. Zeroing neural networks (ZNNs) are a class of neural networks which are particularly dedicated to find zeros of equations. In the past years, ZNNs have played an essential role in finding online solutions to time-varying problems. As a result, many useful research results have been reported in the literatures.

Zhang Neural Network or Zeroing Neural Network (ZNN) dynamics has been proposed in 2001 by Zhang et al. for solving **time-varying** problems, such as for matrix inversion and pseudoinversion, solving systems of matrix and linear equations, solving convex quadratic optimization problems, finding different matrix functions.

A ZNN model is often designed in an **implicit dynamics** instead of an explicit dynamics which is assumed in GNNs. An arbitrary ZNN dynamics can be defined in two global steps.

Step 1ZNN. The first step assumes construction of an appropriate error function, also termed *Zhang function (ZF)*, denoted as $E(t)$. The zero point of the error function $E(t)$ should be equal to the exact solution of the problem. The Zhang error function can be scalar valued, vector valued, or matrix valued. Zhang and Guo in the monograph [351] presented a comprehensive overview of different Zhang functions, classifying them in four categories: scalar-valued ZF in real domain, vector-valued ZF in real domain, matrix-valued ZF in real domain, and ZFs in complex domain.

Step 2ZNN. The second step is defining ZNN dynamics, which assumes the application of a defined ZF in the ZNN dynamics defined by the evolution law

$$\dot{E}(t) = \frac{\mathrm{d}E(t)}{\mathrm{d}t} = -\gamma\mathcal{F}\left(E(t)\right), \tag{1.4.60}$$

where $\dot{E}(t)$ is the time derivative of $E(t)$, $\gamma \in \mathbb{R}^+$ is a positive scalar used to scale the convergence rate and $\mathcal{F}(\cdot) : \mathbb{R}^{n\times n} \to \mathbb{R}^{n\times n}$ denotes an appropriate *activation function*.

The first-order linear differential equation $\dot{E}(t) = -\gamma E(t)$ which defines the linear ZNN evolution law has an analytic solution $E(t) = E(0)\mathrm{e}^{-\gamma t}$. So, $E(t)$ is exponentially convergent with convergence rate $\gamma > 0$ to the zero point $E(t) = 0$ as $t \to +\infty$.

1.4.4 GNN Versus ZNN Dynamics

According to the number of results related to the computation of inverses, generalized inverses, and other matrix algebra problems, we could split neural network models into two main types: more traditional gradient-based neural networks (or termed gradient neural networks, GNN), usually aimed at constant (i.e., time-invariant) matrices and recently proposed new types of neural networks called Zhang neural networks (ZNNs), introduced to generalize to the solution of online time-varying problems. Time-varying systems often exhibit dynamic behavior which is subject to changes in time. For example, the response of an electrical circuit or mechanical system may vary based on external influences. Time-varying systems are typically modeled using differential equations or difference equations that include variables that explicitly depend on time. The ZNN differs from conventional gradient-based neural networks designed to solve static problems in several important things.

- The design of GNN models is based on the elimination of the norm-based scalar-valued error function, which could only be positive or at least lower-bounded. In contrast, the design of ZNN models is based on zeroing of every entry of the matrix-valued, vector-valued or scalar-valued error function, which could be positive, negative, bounded or even unbounded. The matrix-valued error function could make the resultant ZNN models monitor and force every entry of the error to zero. Thus, more information are used for network learning and better performance can be achieved for the ZNN model, as compared to the GNN model.
- The ZNN models exploit the time-derivative information of the error function during the real-time solving process. This is the reason why ZNN models could globally exponentially converge to the theoretical solution of the time-varying problem. In contrast, GNN models have not exploited such important information, and thus may not be effective in solving such a time-varying problem. In essence, the ZNN method is based on a prediction thought, while the GNN method belongs to a conventional tracking approach.
- The GNN models are depicted in explicit dynamics, i.e., in the general form $\dot{X}(t) = \ldots$, which are usually associated with classic Hopfield-type recurrent neural networks. In contrast, the ZNN models are depicted in implicit dynamics (e.g., $\dot{X}(t)A(t)A^{\mathrm{T}}(t) = \ldots$), which frequently arise in analog electronic circuits and systems due to Kirchhoff's rules [353]. In addition, the implicit dynamical equations could preserve physical parameters in the coefficient matrices. They could describe the usual and unusual parts of a dynamic system in the same form. Thus, implicit systems have higher abilities in representing dynamic systems, as compared to explicit systems. If needed, the implicit dynamic systems could be transformed into explicit dynamic systems.

1.4.5 Lyapunov Stability and the Convergence of Dynamical Systems

Analyzing time-varying systems can be more complex than time-invariant ones due to the changing nature of the system parameters. This requires specialized methods for stability analysis and control design. In this part, we review the main results about Lyapunov stability theory. In order to prove the convergence of these nonlinear function-activated RNN models, a common approach is to construct a Lyapunov function candidate.

Consider a dynamical system

$$\dot{x} = f(x, t), \quad x(t_0) = x_0, \quad x \in \mathbb{R}^n. \tag{1.4.61}$$

It will be assumed that $f(x, t)$ satisfies the standard conditions for the existence and uniqueness of solutions. Such conditions are that $f(x, t)$ is Lipschitz continuous with respect to x, uniformly in t, and piecewise continuous in t. A point $x_e \in \mathbb{R}^n$ is an *equilibrium point* of (1.4.61) if $f(x_e, t) \equiv 0$.

Intuitively speaking, an equilibrium point is *locally stable* if every trajectory which starts near the equilibrium point x_e (meaning that the initial $x(t_0)$ is in a neighborhood of x_e) remain near x_e for all time.

The equilibrium point x_e is said to be *locally asymptotically stable* if x_e is locally stable and, furthermore, all solutions starting near x_e converge toward the isolated equilibrium point x_e as $t \to \infty$.
In formal way: a system is *locally asymptotically stable* (L.A.S.) near or at x_e if there is an $\delta > 0$ such that
$\|x(t_0) - x_e\| \le \delta \Longrightarrow x(t) \to x_e$ as $t \to \infty$.

An equilibrium point x_e of (1.4.61) is *asymptotically stable at* $t = t_0$ if
1. x_e is locally stable, and
2. x_e is *locally attractive*; i.e., there exists $\delta(t_0)$ such that

$$\|x(t_0) - x_e\| < \delta \Longrightarrow \lim_{t\to\infty} \|x(t) - x_e\| = 0.$$

The equilibrium point $x_e = 0$ of (1.4.61) is *stable in the sense of Lyapunov* at $t = t_0$ if for any $\varepsilon > 0$ there exists a real number $\delta(t_0, \varepsilon) > 0$ such that

$$\|x(t_0) - x_e\| < \delta \Longrightarrow \|x(t) - x_e\| < \varepsilon, \quad \forall t \ge t_0.$$

A system is *globally asymptotically stable* if every trajectory $x(t)$ satisfies $x(t) \to x_e$ as $t \to \infty$ (i.e., x_e is the unique equilibrium point).

The equilibrium point $x_e = 0$ is an *exponentially stable equilibrium point* of (1.4.61) if there exist constants $m, \alpha > 0$, such that

$$\|x(t) - x_e\| \le m\mathrm{e}^{-\alpha(t-t_0)}\|x(t_0) - x_e\| \tag{1.4.62}$$

for all $\|x(t_0)\| \le \varepsilon$ and $t \ge t_0$. The largest constant α which may be used in (1.4.62) is called the *rate of convergence*.

A dynamical system is said to be *globally convergent* if every state trajectory of the dynamical system converges to a non-isolated equilibrium point, starting from arbitrary initial point $x(t_0)$.

In the literature, more frequent practice is to say that an equilibrium point being stable or not instead that a dynamic system is stable or not.

1.5 Importance of Generalized Inverses and Dynamical Systems

There are many applications of outer generalized inverses with known range and null space in many fields of engineering. For example, {2}-inverses were used in the investigation of some iterative methods for solving nonlinear equations [6]. Also, outer inverses are important in some statistical investigations [67, 92] as well as in solving certain systems of linear equations [32], restated in Proposition 1.2.12.

Particularly, the Drazin inverse is applicable in the theory of finite Markov chains, in finding solutions of differential equations or singular linear difference equations [17] as well as in cryptography [140]. According to available literature, various outer inverses are useful in solving linear and nonlinear problems which involve the usage of various generalized inverses [185, 289, 335, 395]. One additional confirmation of the importance of outer inverses is the fact that the Moore–Penrose inverse $A^{\dagger}$, the weighted Moore–Penrose inverse $A^{\dagger}_{M,N}$, the Drazin inverse A^{D} and the group inverse $A^{\#}$ are exact particular cases of inverses $A^{(2)}_{T,S}$. Moreover, the Bott-Duffin inverse $A^{(-1)}_{(L)}$ [33] and the generalized Bott-Duffin inverse $A^{(\dagger)}_{(L)}$ are also outer inverses. For more details about these classical results, the reader is referred to the well-known monographs [6, 266]. More precisely, the next identities are satisfied:

$$A^{\dagger} = A^{(2)}_{\mathcal{R}(A^*),\mathcal{N}(A^*)}, \quad A^{\dagger}_{M,N} = A^{(2)}_{\mathcal{R}(A^{\sharp}),\mathcal{N}(A^{\sharp})}. \tag{1.5.1}$$

For a given square matrix A, the Drazin and the group inverse are defined by [6, 266]:

$$A^{\mathrm{D}} = A^{(2)}_{\mathcal{R}(A^k),\mathcal{N}(A^k)},\ k \ge \operatorname{ind}(A), \quad A^{\#} = A^{(2)}_{\mathcal{R}(A),\mathcal{N}(A)}. \tag{1.5.2}$$

Finally, the core-EP inverse $A^{\textcircled{\dagger}}$ and the core inverse $A^{\textcircled{\#}}$ of a square matrix $A \in \mathbb{C}^{n\times n}$ satisfy

$$A^{\textcircled{\dagger}} = A^{(2)}_{\mathcal{R}(A^k),\mathcal{N}((A^k)^*)}, \quad k = \operatorname{ind}(A), \quad A^{\textcircled{\#}} = A^{(2)}_{\mathcal{R}(A),\mathcal{N}(A^*)}.$$

Computation of {2}-inverses with predefined image and kernel has attracted the attention of many researchers. Importance of generalized inverses in solving various matrix approximation and optimization problems was surveyed in [237].

The main efforts in the pseudoinverse computation can be divided into two main types: numerical algorithms and continuous-time neural network algorithms. The numerical algorithms can be divided into two categories: direct and iterative methods. The singular value decomposition (SVD) algorithm is the most well known among the direct methods [6]. Also, other types of matrix factorizations have been exploited in the computation of generalized inverses, such as the QR decomposition [117, 225], LU factorization [246]. Methods based on the application of the Gauss-Jordan elimination process to an appropriate augmented matrix were investigated [87, 230]. The SVD algorithm is more accurate and is thus the most commonly used method, but it requires a large amount of computational resources. The iterative methods, such as the orthogonal projection algorithms, the Newton iterative algorithm, and the higher-order convergent iterative methods, are more suitable for implementation. The Newton iterative algorithm has a fast convergence rate, but it requires an initial condition for its convergence. All iterative methods, in general, require initial conditions which are ultimate, rigorous, and sometimes cannot be fulfilled easily. A number of direct and iterative methods were proposed in [34, 51, 144, 194, 218, 219, 230, 243, 285, 294, 295] and many other references. The Newton iterative method for block matrices was developed in [184].

Iterative methods for computing generalized inverses assume certain conditions for their convergence, and the convergence is theoretically established. An iterative algorithm applicable in the complex matrix case was proposed in [300]. Compared with existing iterative algorithms for computing generalized inverses, the neural network models GNN and ZNN are globally convergent, without any condition.

Time-varying problems can be described as a class of problems in which unknown values change as functions of time. In recent years, time-varying problems have attracted great attention in scientific research. As a consequence, the continuous-time neural learning algorithms have emerged as parallel distributed computational models for real-time applications. In many real-time systems, real-time solutions of the matrix inverse and pseudoinverse are usually imperative. An example of such applications in robotics is the solution to the manipulator inverse kinematics problem in real-time motion control of kinematically redundant robots. The reported results indicate that they are more suitable for real-time applications and dynamic systems than for conventional numerical algorithms.

Unknown parameters in many engineering problems are complex-valued vectors and matrices and, often, the task of the system designer is to find the pseudoinverse and outer inverses of complex-valued matrices. Such examples might appear in telecommunications, where digital filters can contain complex-valued coefficients, adaptive filters and data compression, etc. [14, 94, 169, 214, 301, 339]. Traditionally, it is required to split complex-valued problems into their real and imaginary parts separately [13]. However, it is usually difficult to convert the complex-valued problem into a real-valued one in the large-sized problem.

Neural networks have shown a huge potential as a parallel distributed computational models for solving many computationally challenging problems, such as various types of constrained optimization problems [93, 154], or the Winner-Take-All (WTA) competition phenomena [132]. A number of results related to the application of neural networks in solving a variety of matrix algebra problems have been published recently. Different types of neural networks have been introduced to solve systems of linear algebraic equations. The authors of the papers [39, 274–277] designed recurrent neural networks for solving simultaneous linear algebraic equations. In [277], it was verified that the proposed recurrent neural networks are asymptotically stable in the large and capable of computing inverse matrices and solving Lyapunov matrix equations. Two three-dimensional structured networks for solving linear equations and the Lyapunov equation were developed in [279]. Neuron-like network architectures for computing eigenvalues and eigenvectors of real matrices were investigated in [38, 206]. Two recurrent neural networks for computing LU decomposition and Cholesky factorization were presented in [278]. A variety of other matrix algebra problems have been solved by using neural networks, see for example [15, 36, 280].

The dynamical system approach is a powerful tool for solving many kinds of matrix algebra problems because of its parallel distributed nature, possibility to provide output within a predefined time frame in real-time applications, convenience of hardware implementation, global convergence without any condition [302], applicability to online computation with time-varying matrices. Various nonlinearly activated GNNs for approximating the Lyapunov matrix equation with constant coefficients were proposed in [333].

Being motivated by these advantages, the usage of a recurrent neural network (RNN) is becoming a common practice in academia and industry. Two main research streams in developing RNN neural dynamics are known as gradient neural networks (GNNs) and zeroing neural networks (ZNNs). Both of these streams belong to the class of optimization neural networks. The GNN neural evolution solves the matrix equation $E(t) = 0$ by minimizing the scalar error function, which is defined by the Frobenius norm of $E(t)$ using the gradient descent optimization technique. On the other hand, the ZNN design tends to solve $E(t) = 0$. According to the leading design pattern, the ZNN evolution is implicit, while the GNN evolution is always explicit from [375].

In order to illustrate the main results derived using the dynamical system approach, our intention is to provide a brief overview of the most important models. GNN neural design models for computing the inverse or the Moore–Penrose inverse were proposed in [162, 272, 277]. Further, various dynamical systems aimed to approximating the pseudoinverse of rank-deficient matrices originated in [273]. Wei in [290] proposed three RNN models for approximation of the weighted Moore–Penrose inverse. Cichocki in [37] proposed a feed-forward neural network for approximating the Drazin inverse. Online matrix inversion in the complex matrix case was considered in [321].

Also, a number of dynamical systems which are based on a modification of the GNN evolution were proposed. The modification is based on the exception of the constant term of the gradient. The first of these models is the dynamical system for computing the Drazin inverse, which was established in [251]. The influence of nonlinear activation on the convergence rate of GNNs for computing the Drazin inverse was considered in [240]. In [285], the authors defined an extension of the modified GNN design from [251] and introduced analogous and more general design for computing the W-weighted Drazin inverse. Further, as a generalization of previous modified GNN models, it is possible to emphasize two gradient-based models for the approximation of outer inverses with given image and kernel, which were introduced in [399]. The GNN evolution, which is based on a modified gradient descent direction, does not exhibit global convergence. A globally convergent dynamical system defined as the modification of the dynamical system from [251] was proposed in [252].

The leading idea in [254] was the usage of the full rank representation $G = QP$ and the full rank representation of outer inverses of the form $P(QAP)^{-1}Q$. In [240], the authors used the Frobenius norm of $E(t) = A - AV(t)A$ and $E_D(t) = A^k - A^kV(t)A$, and defined globally convergent GNN models for computing $\{1\}$-inverses, the Drazin inverse and the Moore–Penrose inverse.

It is known that the evolution along the gradient descent direction of the Frobenius norm of the error matrix is not appropriate in time-varying problems. More precisely, the GNN evolution in the time-varying case cannot converge even after infinite time [149]. In order to overcome this drawback of GNN dynamics, Zhang neural network (or zeroing neural network) (ZNN) dynamics was proposed for finding solutions to online time-varying problems. The ability of ZNN dynamics to overcome the deficiency of GNN dynamics in the time-varying case arises from a different principle, which is based on an indefinite error-monitoring function, called the Zhang function (ZF) [375].

Zeroing neural networks, which are able to eliminate various kinds of noise and resolve the redundancy in kinematic control, have attracted great attention in recent years. Last year, an integration-enhanced noise-tolerant ZNN design (shortly IENTZNN) was investigated and applied. The aim of IENTZNN models is the ability to avoid the unavoidable presence of noise. An IENTZNN model for computing the time-varying matrix inverse was defined in [110]. Zhang et al. in [111] applied a noise-tolerant zeroing neural network (NTZNN) in solving some problems of control. A kind of IENTZNN model for computing outer inverses was proposed in [223]. The model uses an input matrix A and two appropriate matrices F, G. For this purpose, it was termed as IENTZNN(A, F, G). The evolution design IENTZNN(A, F, G) without the integral part and without noise was termed as ZNN(A, F, G).

In addition, different kinds of discretization of the continuous-time ZNN design have been investigated. The discretization transforms the continuous-time RNN design into some variants of iterative methods. A Discretization of continuous-time ZNN models was investigated in [111]. Various ZNN models able to resolve the redundancy of robotic manipulators for kinematic control in the presence of noise were introduced in [134–136]. A noise-tolerant Zeroing neural network (NTZNN) applicable in approximating the time-varying matrix pseudoinverse under noisy environments was proposed in [303]. A summarization of noise-tolerant ZNN models was given in [106, Table 2]. An enhanced discrete-time Zhang neural network (EDTZNN) model was for time-varying matrix inversion was presented in [176].

So far, the listed ZNN models possess a so-called complex fixed-parameter convergent-differential neural network (CFP-CDNN) design. The definition of a

CFP-CDNN design assumes constant values of all included parameters. Moreover, it is desirable to choose values of the constant scaling parameter as hardware permits in order to achieve the fastest convergence. But, according to numerical experience, too large values of the scaling parameter could increase the CPU time for the simulation and even cause numerical instability. As an appropriate solution to this drawback, the parameters included in complex varying-parameter convergent-differential neural networks (CVP-CDNN) vary in time and are not too large. The idea is that time-varying parameters will be able to trace the behavior of dynamical systems during the time. CVP-CDNN neural design becomes dominant in ZNN research. Varying-parameter ZNNs for finding solutions to the time-varying Sylvester equation were proposed in [388, 389]. In [385], the authors considered an application of a constrained quadratic programming dual neural network design in redundant robot manipulators. An application of CVP-CDNN design in controlling the movement of an aerial vehicle was considered in [390]. Zhang et al. in [384] proposed a CVP-CDNN model in order to solve the joint-angular-drift problems of robot manipulators. A variant of CVP-CDNN efficient in solving online continuous time-varying convex quadratic programming with a time-varying linear equality constraint was originated in [383].

In many real-time systems, real-time solutions of generalized inverses are usually imperative. An example of such applications in robotics is the solution to the manipulator inverse kinematics problem in real-time motion control of kinematically redundant robots. Reported results show that neural network architectures are more suitable for real-time applications than conventional numerical algorithms. Thereinto, the neural approach is now regarded as a powerful alternative for scientific computing because of its parallel distributed nature and convenience of hardware implementation. A number of nonlinear and linear recurrent neural network models have been developed for the inversion and generalized inversion of square and full rank rectangular matrices (for more details, see, e.g., [98, 162, 272, 277]). Various recurrent neural networks for computing generalized inverses of rank-deficient matrices were designed in [273, 290]. Three recurrent neural networks for computing the weighted Moore–Penrose inverse of rank-deficient matrices are presented in [290]. A neural network approach to compute the Drazin inverse A^{d} was developed in [37]. The approach from [37] is based on a feed-forward multi-layer neural network, the gradient optimization technique, and the standard back propagation learning algorithm. A new type of complex-valued ZNN, based on a complex Zhang function, was proposed and investigated in [126]. ZNN models for online time-varying full rank matrix Moore–Penrose inversion are generalized, investigated, and analyzed in [370].

In [221] the authors proposed conditions for the existence and representations of $\{2\}$, $\{1, 2\}$, and $\{1\}$-inverses, along with a new computational framework for computing these generalized inverses. Proposed representations are applicable to complex constant matrices. Four nonlinear gradient-based recurrent neural networks for computation of the Drazin inverse for real constant matrices, based on the limiting representations of the Drazin inverse, are investigated in [282]. The global convergence performance of defined neural networks is ensured by any monotonically increasing odd activation function. Two gradient-based recurrent neural networks for computing the W-weighted Drazin inverse of a real constant matrix were presented in [285]. Complex neural network models for computing time-varying Drazin Inverse computation are considered in [283]. The proposed network models are based on limit representations of the Drazin inverse and error-monitoring functions, which exploit the Tikhnov regularization method. Two gradient-based recurrent neural networks for generating various inner inverses, including the Moore–Penrose and the Drazin inverse, are investigated in [240]. Extension of the ZNN algorithmic conceptual framework, which was used for the computation of the regular matrix inverse, pseudoinverse, and the Drazin inverse, to the class of time-varying complex outer inverses with prescribed range and null space can be found in [284]. Moreover, the same paper investigates a hybrid combination of ZNN and GNN models for computing outer inverses of real constant matrices.

Chapter 2
Gradient Neural Network (GNN) and Modifications

This chapter investigates gradient neural networks (GNNs) used for solving linear matrix equations (LME) and to compute different generalized inverses of constant real or complex matrices with respect to continuous-time models. In addition, GNN models for solving systems of linear equations are considered. Several modified recurrent neural networks (RNN) are considered. These models are related to a certain number of independent subnetworks, which can operate in parallel. The introduced GNNs are convenient for an implementation in an electronic circuit. An extensive collection of gradient-based continuous-time neural networks is presented in [238].

2.1 General GNN Dynamics

In order to complete our motivation, we survey necessary known results regarding solving the matrix equations as well as regular and generalized inverses computation using both GNN-based recurrent neural networks.

Construction of GNN dynamics is based on two general steps.

Step 1GNN
The GNN neural dynamics is based on the usage of the scalar-valued norm-based error (objective) function $\varepsilon(t) = \varepsilon(V(t)) = \frac{1}{2}\|E(t)\|_F^2$, where $E(t)$ is appropriately defined error matrix, $\|A\|_F := \sqrt{\mathrm{Tr}(A^{\mathrm{T}}A)}$ denotes the Frobenius norm of the matrix A and $\mathrm{Tr}(\cdot)$ denotes the trace of a matrix. The error function $E(t)$ is defined replacing unknown matrix in the considered problem by the time-varying state-variables matrix $V(t)$. Since minimization of $\varepsilon(t)$ assumes $\varepsilon(t)$ and $E(t) = 0$, it is clear that solution of the the equation $E(t) = 0$ represents solution of the starting problem.

P. Stanimirović et al., *Generalized Matrix Inversion: A Machine Learning Approach*,
https://doi.org/10.1007/978-3-032-01493-1_2

Step 2GNN

The general GNN design is defined in the second step, as a movement during the time along the negative gradient $-\partial\varepsilon(V(t))/\partial V$ of $\varepsilon(V(t))$, until the minimum is reached. It gives the following dynamics in the linear case:

$$\dot{V}(t) = \frac{\mathrm{d}V(t)}{\mathrm{d}t} = -\gamma\left(\frac{\partial\varepsilon(V(t))}{\partial V}\right). \tag{2.1.1}$$

Here, $V(t)$ is the activation state variables matrix, $t \in [0, \infty)$ is a time parameter, and γ is a positive scaling constant whose values must be harmonized with the chosen time interval.

Further, nonlinear GNN dynamics is defined using the nonlinear activation function array $\mathcal{F}$, applicable on matrices, as follows:

$$\dot{V}(t) = \frac{\mathrm{d}V(t)}{\mathrm{d}t} = -\gamma\mathcal{F}\left(\frac{\partial\varepsilon(t)}{\partial V}\right). \tag{2.1.2}$$

In general $\mathcal{F}$ is an odd and monotonically increasing function array, element-wise applicable to elements of a real matrix $R = (r_{ij}) \in \mathbb{R}^{m\times n}$, i.e., $\mathcal{F}(R) = (f(r_{ij}))$, $i = 1, \ldots, m$, $j = 1, \ldots, n$, wherein $f(\cdot)$ is an odd and monotonically increasing function.

In order to approve gradient essence of GNN models, we look inside the discretization of GNN dynamics. Assume that the discretization of the continuous-time linear GNN model is performed by using the Euler forward-difference rule

$$\dot{V}(t) \approx (V_{k+1} - V_k)/\tau,$$

where τ denotes the sampling time and $V_k = V(t = k\tau)$, $k = 1, 2, \ldots$. Using the gradient $\frac{\partial\varepsilon(V(t))}{\partial V} = \nabla\varepsilon(t)$, the discrete-time analogy of the GNN model becomes just a gradient descent method for nonlinear optimization of the form

$$V_{k+1} = V_k - \beta\,\nabla\varepsilon(V_k), \tag{2.1.3}$$

where $\beta = \tau\,\gamma > 0$ is the step size defined as a trade-off between a small and a large value. Clearly, smaller sampling time τ is enabled using larger values γ.

2.1.1 GNN for Computing Matrix Inverse

Wang in [272] proposed the dynamics of the linear recurrent neural network for computing the inverse of a nonsingular matrix A. The introduced GNN dynamics is initiated by the error matrix $E(t) = AV(t) - I$. The gradient of the goal function $\varepsilon(V(t)) = \frac{1}{2}\|AV(t) - I\|_F^2$ is defined by

$$\frac{\partial \varepsilon(V(t))}{\partial V} = \frac{1}{2}\frac{\partial \|AV(t) - I\|_F^2}{\partial V} = A^{\mathrm{T}}\left(AV(t) - I\right).$$

Here, $\|A\|_F := \sqrt{\mathrm{Tr}(A^{\mathrm{T}}A)}$ denotes the Frobenius norm of the matrix A and $\mathrm{Tr}(\cdot)$ denotes the trace of a matrix. Now, using the general design rule (2.1.1) with linear activation function $\mathcal{F} \equiv I$, the dynamics can be defined as

$$\begin{aligned}\frac{\mathrm{d}V(t)}{\mathrm{d}t} &= -\gamma A^{\mathrm{T}}AV(t) + \gamma A^{\mathrm{T}} \\ &= -\gamma A^{\mathrm{T}}\left(AV(t) - I\right), \; V(0) = V_0,\end{aligned} \tag{2.1.4}$$

where $V(t)$ is a matrix of activation state variables aimed to approximate the inverse matrix of A during time t in predefined time interval $[0, T]$, and γ is a positive gain parameter. In addition, $V(0) = V_0$ denotes an arbitrary initial state matrix.

The following three types of activation functions $f(\cdot)$ are commonly used in numerical experiments: Linear (1.4.53), power-sigmoid (1.4.55), and smooth power-sigmoid (1.4.56). The parameters γ, initial state $V(0)$ as well as the parameters ρ and ϱ of the nonlinear activation functions (1.4.55) and (1.4.56), are entered directly in the model. On the other hand, matrices A, B, and D are defined from the workspace. We will select mainly $\rho = \varrho = 3$ in subsequent examples.

Proposition 2.1.1 *Consider a nonsingular matrix $A \in \mathbb{R}^{n\times n}$. Starting from any initial state $X(0) \in \mathbb{R}^{n\times n}$ of gradient-based neural network (2.1.4) will exponentially converge to theoretical inverse A^{-1}. In addition, the exponential convergence rate is equal to $\gamma\, \lambda_{\min}(A^{\mathrm{T}}A)$, where $\lambda_{\min}(A^{\mathrm{T}}A)$ denotes the minimum eigenvalue a of $A^{\mathrm{T}}A$.*

Basic Simulink Blocks and Simulink Implementation

Simulink is a MATLAB-based graphical programming environment appropriate for modeling, simulating, and analyzing dynamic systems. It allows users to create models using a block diagram approach. MATLAB Simulink environment contains an inclusive block library of sources, sinks, mathematical functions, connectors, and a lot of toolboxes which are related with the control systems domain.

Some key features of Simulink are as follows.

Graphical modeling: Users can create models by dragging and dropping blocks from libraries, representing functions, components, and systems.

Why Simulink is used?
Simulink is used to model, simulate, and analyze dynamic systems.

Design; simulate; deploy.
Simulink is a block diagram environment used to design systems, simulate before moving to hardware, and develop without writing code.

Why use Simulink instead of MATLAB?

Simulink blocks provide a visual representation of your system, which can help you to verify its logic and behavior.

On the other hand, MATLAB code requires you to write and edit text commands, which can be more complex and error-prone. Figure 2.1 presents basic Simulink blocks.

- The **Constant** block creates a complex or real constant value signal. The use of this block is to give a constant signal input and can generate scalar, vector, or matrix output.
- The **Product** block generates the result of multiplying two inputs and can be of two scalars, product of a scalar and a nonscalar, or product of two nonscalars of appropriate dimensions. The **Product1** block can be derived from **Product** block and it is used for the multiplication of two (or more) matrices.
- The **Abs** block returns the absolute value of the input.
- The **Sum** block returns the sum of two inputs and can be used for more than two inputs.
- The **Gain** block multiplies the input by a constant value.
- The **Display** block represents the value of the input.
- The **Scope** block is used for the presentation of time domain signals.
- The **To Workspace** block is used to save the input to specified time series, array, or structure in a workspace in MATLAB.
- The **Trigonometric Function** block generates the trigonometric and hyperbolic functions.
- The **Integrator** block outputs the value of the integral of its input signal with respect to time.

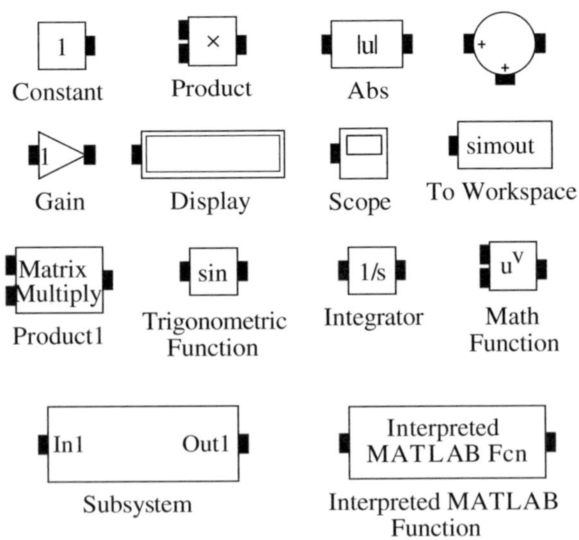

Fig. 2.1 Basic Simulink blocks

- The **Math Function** block includes different mathematical functions such as logarithmic, exponential, power, modulus functions, and, etc.
- The **Subsystem** block includes a subset of blocks within a model or system and can be used for the presentation of a virtual subsystem or a nonvirtual subsystem.
- The **Interpreted MATLAB Functions** block applies specified MATLAB function or expression to the input data.

Creating a MATLAB Simulink model involves several steps, including defining the system you want to simulate, building the model using blocks, and configuring the parameters.

The Simulink implementation of (2.1.4) is presented in Fig. 2.2.

It was proven in [272] that the GNN model (2.1.4) is asymptotically stable in the large and the steady-state matrix of the recurrent neural network is equal to the inverse matrix of A, i.e., $\lim_{t\to\infty} V(t) = A^{-1}$, for arbitrary $V(0)$, which is formally stated in Proposition 2.1.2.

Proposition 2.1.2 ([272]) *If A is nonsingular matrix, then the equilibrium state of the recurrent neural network (2.1.4) is asymptotically stable in the large. Further, the equilibrium state matrix of (2.1.4) coincides with A^{-1}.*

The global exponential convergence of gradient neural network (2.1.5) in the case when A is nonsingular as well as its global stability when A is singular was verified in [363].

Example 2.1.1 Consider the matrix

$$A = \begin{bmatrix} 1 & 10 & 10 \\ 3 & 10 & 10 \\ 6 & 2 & 5 \end{bmatrix}$$

generated by the command `A = randi([1 10],3,3)`. Using $\gamma = 10^4$ and the zero initial state Simulink generates elementwise trajectories as in Fig. 2.3 in the time interval [0.0.1] (left) and in the time interval [0.0.01] (right).

Graphs in Fig. 2.3 indicate fast convergence of trajectories in $V(t)$. The convergence is achieved in the time interval $[0, 10^{-2}]$ and continues in the rest of the interval $[0, 10^{-1}]$.

The error norm $\|I - AV(t)\|_F$ in the time interval $[0.10^{-5}]$ is presented in Fig. 2.4, left, for $\gamma = 10^3$,$\gamma = 10^4$ and $\gamma = 10^6$. The error norm in the time interval $[0.10^{-1}]$ is presented in Fig. 2.4, right, for $\gamma = 10^2$,$\gamma = 10^3$ and $\gamma = 10^4$.

Graphs in Fig. 2.4 confirm that larger values of the gain parameter γ initiate faster convergence $\|I - AV(t)\| \to 0$. Left figure indicates that the time frame $[0, 10^{-5}]$ is not enough in the cases $\gamma = 10^3$ and $\gamma = 10^4$, and only $\gamma = 10^6$ is enough. Right figure indicates that the time frame $[0, 10^{-1}]$ is appropriate for the convergence in the cases $\gamma = 10^3$ and $\gamma = 10^4$, while $\gamma = 10^2$ requires a longer time frame.

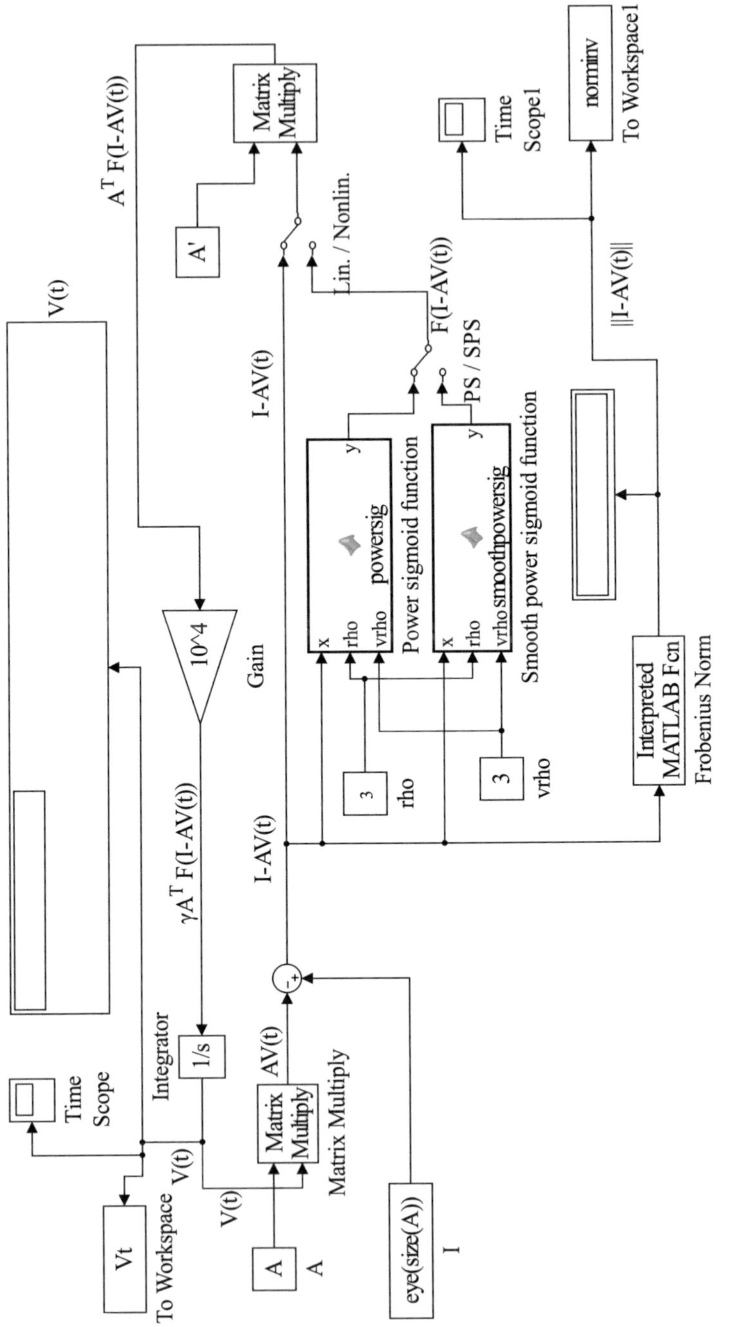

Fig. 2.2 Simulink implementation of GNN (2.1.4) for computing A^{-1}

2.1.2 GNN for Computing the Moore–Penrose Inverse of Full-Rank Matrix

Recurrent neural network defined in (2.1.4) can be used for computing the Moore–Penrose inverse of a full-column rank rectangular matrix $A \in \mathbb{R}_n^{m\times n}$, by simply allowing the activation state matrix to be rectangular. In the full-row rank rectangular matrix case $A \in \mathbb{R}_m^{m\times n}$, the recurrent neural network

$$\begin{aligned}\frac{\mathrm{d}V(t)}{\mathrm{d}t} &= -\gamma V(t)AA^{\mathrm{T}} + \gamma A^{\mathrm{T}} \\ &= -\gamma\,(V(t)A - I)\,A^{\mathrm{T}},\ \ V(0) = V_0,\end{aligned} \tag{2.1.5}$$

which is dual with respect to (2.1.4), can be also used to compute the Moore–Penrose inverse of A. The closed-form solution of the state matrix $V(t)$ in (2.1.4) and (2.1.5) is expressed as follows (see [273]):

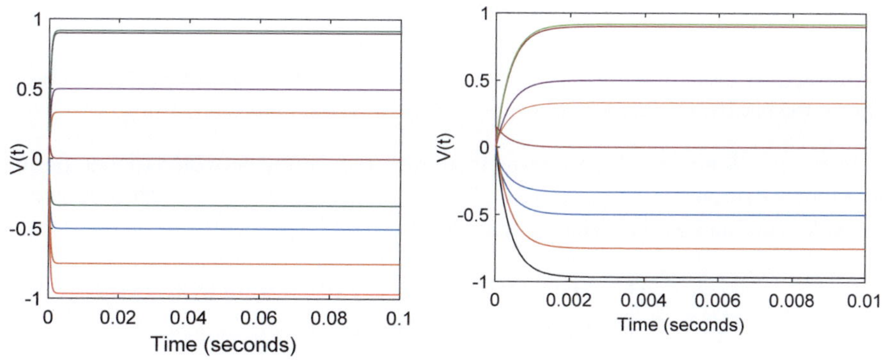

Fig. 2.3 Elementwise trajectories in [0, 0.1]

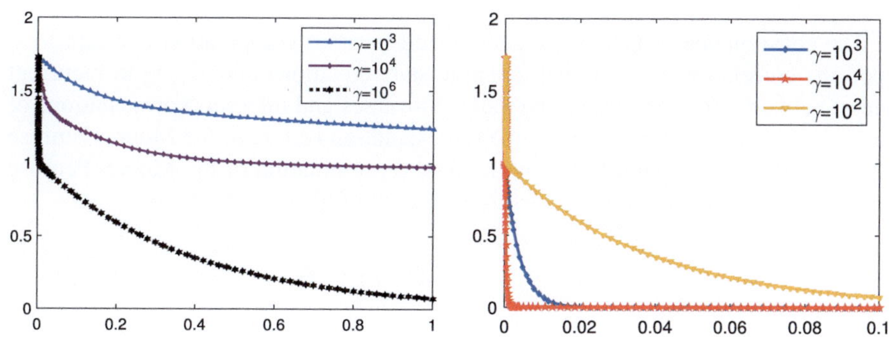

Fig. 2.4 The error norm $\|I - AV(t)\|_F$ for different γ: in $[0, 10^{-5}]$ (left) and $[0, 10^{-1}]$ (right)

$$V(t)=\begin{cases} e^{-\gamma A^{\mathrm{T}}At}V(0)+\gamma e^{-\gamma A^{\mathrm{T}}At}\int_0^t e^{\gamma A^{\mathrm{T}}A\tau}A^{\mathrm{T}}\,\mathrm{d}\tau, & m\geq n,\\ V(0)e^{-\gamma AA^{\mathrm{T}}t}+\gamma A^{\mathrm{T}}e^{-\gamma AA^{\mathrm{T}}t}\int_0^t e^{\gamma AA^{\mathrm{T}}\tau}\,\mathrm{d}\tau, & m<n. \end{cases} \tag{2.1.6}$$

In the case of a full-column rank rectangular A, Wang in [273] derived the following representation of $A^{\dagger}$, which is independent of $V(0)$:

$$A^{\dagger}=\begin{cases} \lim\limits_{t\to\infty}\gamma e^{-\gamma A^{\mathrm{T}}At}\int_0^t e^{\gamma A^{\mathrm{T}}A\tau}A^{\mathrm{T}}\,\mathrm{d}\tau, & m\geq n,\\ \lim\limits_{t\to\infty}\gamma A^{\mathrm{T}}e^{-\gamma AA^{\mathrm{T}}t}\int_0^t e^{\gamma AA^{\mathrm{T}}\tau}\,\mathrm{d}\tau, & m<n. \end{cases} \tag{2.1.7}$$

According to (2.1.6) and (2.1.7), the models (2.1.4) and (2.1.5) can be used in computing the Moore–Penrose inverse of an arbitrary matrix under the zero initial state $V(0)=O$, where O denotes appropriate zero matrix.

Remark 2.1.1 The parameter $\gamma>0$ might be chosen as large as possible. Larger values of γ accelerate the convergence speed, given that the expression

$$\lim_{t\to\infty} e^{-\gamma A^{\mathrm{T}}At}$$

will disappear more quickly for larger γ. Conventionally, values of γ ought to be as large as the hardware allows, or chosen suitably for simulations and testing objectives.

Moreover, Wang in [273] proposed three recurrent neural networks for computing the Moore–Penrose inverse of rank-deficient matrices. The original recurrent neural network is defined by the dynamic equation

$$\frac{\mathrm{d}V(t)}{\mathrm{d}t}=\begin{cases} -MA^{\mathrm{T}}AV(t)+MA^{\mathrm{T}}, & V(0)=0,\ m\geq n,\\ -V(t)AA^{\mathrm{T}}M+A^{\mathrm{T}}M, & V(0)=0,\ m<n, \end{cases} \tag{2.1.8}$$

where M is a positive diagonal matrix satisfying $M\in\mathbb{R}^{n\times n}$ if $m\geq n$ and $M\in\mathbb{R}^{m\times m}$ if $m<n$.

The convergence of GNNs for generalized inverses computation is established upon the correlation between the dynamic state equations underlying in recurrent neural networks for computing generalized inverses and integral representations of generalized inverses. For example, the representation (2.1.7) of the Moore–Penrose inverse corresponds to the following integral representation of the Moore–Penrose inverse for bounded linear operators, introduced in [80]:

$$A^{\dagger}=\int_0^{\infty}\exp(-A^{\mathrm{T}}A\tau)A^{\mathrm{T}}\,\mathrm{d}\tau. \tag{2.1.9}$$

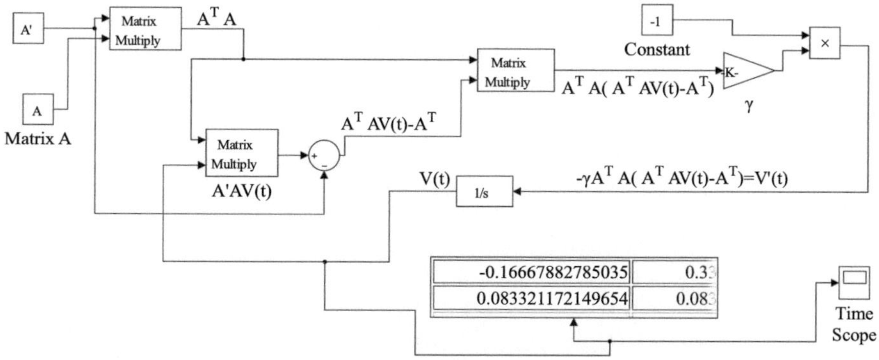

Fig. 2.5 Simulink implementation of GNN (2.1.10) for computing $A^{\dagger}$

2.1.3 *GNN for Computing the Moore–Penrose Inverse of Full-Rank Matrix*

Two improved GNN (IGNN) nonlinearly activated models were developed and investigated for computing the Moore–Penrose inverse of a full-rank matrix in [170]. The principal idea is that in the case $A \in \mathbb{C}_n^{m\times n}$ the matrix $A^{\mathrm{T}}A$ is nonsingular, and A has the unique Moore–Penrose inverse (left pseudoinverse) given by $A^{\dagger} = (A^{\mathrm{T}}A)^{-1}A^{\mathrm{T}}$. Then $A^{\mathrm{T}}AA^{\dagger} - A^{\mathrm{T}} = O$, so that the scalar goal function is equal to

$$\varepsilon(t) = \frac{1}{2}\|(A^{\mathrm{T}}A)V(t) - A^{\mathrm{T}}\|_F^2.$$

Therefore, the linear GNN model (called LGNN-L) for approximating the Moore–Penrose inverse was defined as

$$\dot{V}(t) = -\gamma\frac{\partial\varepsilon(V(t))}{\partial V} = -\gamma A^{\mathrm{T}}A\left(A^{\mathrm{T}}AV(t) - A^{\mathrm{T}}\right). \tag{2.1.10}$$

Simulink implementation of GNN (2.1.10) is presented in Fig. 2.5.

In the dual case $A \in \mathbb{C}_m^{m\times n}$ the matrix AA^{T} is nonsingular, and A has the unique Moore–Penrose inverse (right pseudoinverse) given by $A^{\dagger} = A^{\mathrm{T}}(AA^{\mathrm{T}})^{-1}$. Then $A^{\dagger}AA^{\mathrm{T}} - A^{\mathrm{T}} = O$, so that the error function is equal to

$$\varepsilon(t) = \frac{1}{2}\|V(t)\,AA^{\mathrm{T}} - A^{\mathrm{T}}\|_F^2. \tag{2.1.11}$$

Therefore, the linear GNN model (called LGNN-R) for approximating the Moore–Penrose inverse was defined as

$$\dot{V}(t) = -\gamma\left(V(t)AA^{\mathrm{T}} - A^{\mathrm{T}}\right)AA^{\mathrm{T}}. \tag{2.1.12}$$

After some proper nonlinear activations with the activation $\mathcal{F}$, the authors proposed the following improved versions, termed as IGNN-L and IGNN-R, respectively. The nonlinear IGNN-L dynamics for approximating the Moore–Penrose inverse in the case $n \leq m$ was defined as

$$\dot{V}(t) = -\gamma A^{\mathrm{T}} A \mathcal{F}\left(A^{\mathrm{T}} A V(t) - A^{\mathrm{T}}\right). \tag{2.1.13}$$

The nonlinear IGNN-R design for approximating the Moore–Penrose inverse in the dual case $m < n$ is defined as

$$\dot{V}(t) = -\gamma \mathcal{F}\left(V(t) A A^{\mathrm{T}} - A^{\mathrm{T}}\right) A A^{\mathrm{T}}. \tag{2.1.14}$$

Theorem 2.1.1 ([170]) *Assume that $A \in \mathbb{R}_n^{m\times n}$. If a monotonically-increasing odd activation function array $\mathcal{F}$ is used, the state matrix $V(t)$ of IGNN-L is globally convergent to the theoretical Moore–Penrose inverse $A^{\dagger} = (A^{\mathrm{T}}A)^{-1}A^{\mathrm{T}}$, no matter what initial value $V(0) \in \mathbb{R}_n^{n\times m}$ is.*

Theorem 2.1.2 *In addition to Theorem 2.1.1, the convergence of the LGNN-L model converges exponentially with rate being $\gamma\alpha^2$, where α is the minimal eigenvalue of $I_m \otimes (A^{\mathrm{T}}A)$. In addition, if IGNN-L model is activated by the power sigmoid function superior convergence is obtained in comparison to LGNN-L.*

2.1.4 GNN for Computing the Weighted Moore–Penrose Inverse

Wei in [290] introduced the following dynamic state equation of the first recurrent neural network (called $NN1$) for computing the weighted Moore–Penrose inverse of a rank-deficient matrix:

$$\frac{\mathrm{d}V(t)}{\mathrm{d}t} = \begin{cases} -DA^{\sharp}AV(t) + DA^{\sharp}, & V(0)=0, \ m \geq n, \\ -V(t)AA^{\sharp}D + A^{\sharp}D, & V(0)=0, \ m < n, \end{cases} \tag{2.1.15}$$

where D is a positive diagonal matrix of proper dimensions and $A^{\sharp} = N^{-1}A^{\mathrm{T}}M$ for appropriately chosen positive definite matrices M and N. The simplest choice for D is the constant diagonal matrix $D = \gamma I$, where $\gamma > 0$ is a real large scalar value [290].

Corresponding integral representation of the weighted Moore–Penrose inverse of a linear operator between Hilbert spaces was introduced in [288]:

$$A^{\dagger}_{M,N} = \int_0^{\infty} \mathrm{e}^{-A^{\sharp}A\tau} A^{\sharp}\, \mathrm{d}\tau. \tag{2.1.16}$$

2.2 GNN Dynamics for Solving Linear Matrix Equations

We will describe application of continuous-time RNNs in solving linear matrix equations and computing generalized inverses.

The most general linear matrix equation is given in the form $AXB = D$, wherein A, B, D are proper given matrices and X is unknown. The solution of this matrix equation and its particular appearances were investigated in [231]. It is important to mention that so far considered GNN models can be defined as particular appearances of the GNN for solving $AXB = D$. This model will be investigated in next section.

2.2.1 *GNN for Solving the Linear Matrix Equation $AXB = D$*

A gradient-based neural dynamical design corresponding to the matrix equation $AXB = D$ was investigated in [231]. The model is based on the matrix-valued error function $E(t) = D - AV(t)B$ and the scalar-valued residual error norm $\varepsilon(t) = \frac{1}{2}\|D - AV(t)B\|_F^2$. The derivative of the goal function $\varepsilon(t)$ is equal to

$$\frac{\partial \varepsilon(V(t))}{\partial V} = -A^{\mathrm{T}}\left(D - AV(t)B\right)B^{\mathrm{T}}.$$

Using the general evolution design (2.1.1), the standard linear GNN (LGNN) dynamic, termed as LGNN(A, B, D), is defined as follows:

$$\dot{V}(t) = \gamma A^{\mathrm{T}}(D - AV(t)B)B^{\mathrm{T}}. \tag{2.2.1}$$

Stanimirović and Petković in [231] defined and investigated the following dynamical evolution design with nonlinear activation $\mathcal{F}$:

$$\dot{V}(t) = \gamma A^{\mathrm{T}}\mathcal{F}(D - AV(t)B)B^{\mathrm{T}} = \gamma A^{\mathrm{T}}\mathcal{F}\left(E(t)\right)B^{\mathrm{T}}. \tag{2.2.2}$$

The model (2.2.2) was termed as GNN(A, B, D). The investigation of activation functions in gradient neural network for online solving linear matrix equation was investigated in [260].

Remark 2.2.1 1. It is important to mention the difference between the nonlinear dynamics (2.2.2) and the original GNN dynamics

$$\dot{V}(t) = \gamma\mathcal{F}\left(A^{\mathrm{T}}(D - AV(t)B)B^{\mathrm{T}}\right). \tag{2.2.3}$$

2. Main motivation to use the nonlinear design (2.2.2) instead of the dynamics (2.2.3) can be explained by the fact that the right-hand side of (2.2.2) can be related with the ZNN design $\dot{E}(t) = \gamma\mathcal{F}(E(t))$ as follows:

$$\dot{V}(t) = \gamma A^{\mathrm{T}}\mathcal{F}(E(t))B^{\mathrm{T}} = \gamma A^{\mathrm{T}}\dot{E}(t)B^{\mathrm{T}}.$$

The generalized nonlinearly activated GNN model (G-GNN model) is applicable in both time-varying and time-invariant case and it can operate with time-varying coefficient matrices $A(t)$, $B(t)$, $D(t)$ in solving $A(t)X(t)B(t) = D(t)$:

$$\dot{V}(t) = -\gamma A(t)^{\mathrm{T}}\mathcal{F}\left(D(t) - A(t)V(t)B(t)\right) B(t)^{\mathrm{T}}. \tag{2.2.4}$$

But, gradient-based numerical algorithms and neural-dynamic schemes are substantially designed for solving problems with time-invariant coefficient matrices [84].

Convergence properties of GNN(A, B, D) design are considered in [231].

Theorem 2.2.1 ([231]) *Assume that real matrices $A \in \mathbb{R}^{m\times n}$, $B \in \mathbb{R}^{p\times q}$ and $D \in \mathbb{R}^{m\times q}$ satisfy the constraint*

$$AA^{(1)}DB^{(1)}B = D, \tag{2.2.5}$$

for some inner inverses $A^{(1)}$ and $B^{(1)}$. If an array activation function $\mathcal{F}(\cdot)$ based on the elementwise application of appropriate odd, monotonically increasing and differentiable function $f(\cdot)$ is used, then the neural state matrix $V(t) \in \mathbb{R}^{n\times p}$ of the GNN(A, B, D) model (2.2.2) asymptotically and globally converges to the theoretical solution of the matrix equation $AXB = D$, i.e., $AV(t)B \to D$ as $t \to +\infty$, for an arbitrary initial state matrix $V(0)$.

Proof Firstly, Fulfillment of (2.2.5) for some $A^{(1)} \in A\{1\}$ and $B^{(1)} \in B\{1\}$ ensures solvability of the matrix equation $AXB = D$. The substitution $V(t) = \overline{V}(t) + A^{(1)}DB^{(1)}$ transforms the dynamics (2.2.2) into the equivalent form

$$\begin{aligned}\frac{\mathrm{d}\overline{V}(t)}{\mathrm{d}t} = \frac{\mathrm{d}V(t)}{\mathrm{d}t} &= \gamma A^{\mathrm{T}}\mathcal{F}\left(D - AV(t)B\right) B^{\mathrm{T}} \\ &= \gamma A^{\mathrm{T}}\mathcal{F}\left(D - A\left(\overline{V}(t) + A^{(1)}DB^{(1)}\right) B\right) B^{\mathrm{T}}.\end{aligned}$$

According to (2.2.5), it follows

$$\frac{\mathrm{d}\overline{V}(t)}{\mathrm{d}t} = \gamma A^{\mathrm{T}}\mathcal{F}\left(D - AA^{(1)}DB^{(1)}B - A\overline{V}(t)B\right) B^{\mathrm{T}} = -\gamma A^{\mathrm{T}}\mathcal{F}\left(A\overline{V}(t)B\right) B^{\mathrm{T}}. \tag{2.2.6}$$

The Lyapunov function candidate is

$$L(\overline{V}(t), t) := \frac{1}{2}\left\|\overline{V}(t)\right\|_F^2 = \frac{1}{2}\mathrm{Tr}\left(\overline{V}(t)^{\mathrm{T}}\,\overline{V}(t)\right). \tag{2.2.7}$$

Evidently, the inequality $L(\overline{V}(t), t) \geq 0$ holds for $\overline{V}(t) \neq 0$. According to (2.2.7), assuming (2.2.6) and using $\mathrm{dTr}\left(X^{\mathrm{T}}X\right) = 2\mathrm{Tr}\left(X^{\mathrm{T}}\mathrm{d}X\right)$ in conjunction with basic properties of the matrix trace function, one can express the time derivative of $L(\overline{V}(t), t)$ as in the following:

$$\begin{aligned}\frac{\mathrm{d}L(\overline{V}(t),t)}{\mathrm{d}t} &= \mathrm{Tr}\left(\overline{V}(t)^{\mathrm{T}}\frac{\mathrm{d}\overline{V}(t)}{\mathrm{d}t}\right)\\ &= -\gamma\mathrm{Tr}\left[\overline{V}(t)^{\mathrm{T}}A^{\mathrm{T}}\mathcal{F}\left(A\overline{V}(t)B\right)B^{\mathrm{T}}\right]\\ &= -\gamma\mathrm{Tr}\left[B^{\mathrm{T}}\left(A\overline{V}(t)\right)^{\mathrm{T}}\mathcal{F}\left(A\overline{V}(t)B\right)\right]\\ &= -\gamma\mathrm{Tr}\left[\left(A\overline{V}(t)B\right)^{\mathrm{T}}\mathcal{F}\left(A\overline{V}(t)B\right)\right].\end{aligned} \tag{2.2.8}$$

Since the scalar-valued function $f(\cdot)$ is an odd and monotonically increasing, immediately follows $f(-x) = -f(x)$ and

$$f(x)\begin{cases} > 0 \text{ if } x > 0,\\ = 0 \text{ if } x = 0,\\ < 0 \text{ if } x < 0,\end{cases}$$

which in conjunction with $\gamma > 0$ implies for $W(t) = A\overline{V}(t)B = A\left(V(t) - A^{(1)}DB^{(1)}\right)B = AV(t)B - D$

$$\begin{aligned}\frac{\mathrm{d}L(\overline{V}(t),t)}{\mathrm{d}t} &= -\gamma\mathrm{Tr}\left[\left(W^{\mathrm{T}}\mathcal{F}(W)\right)\right]\\ &= -\gamma\sum_{i=1}^{m}\sum_{j=1}^{n} w_{ij}\, f(w_{ij})\begin{cases} < 0 \text{ if } A\overline{V}(t)B \neq 0,\\ = 0 \text{ if } A\overline{V}(t)B = 0.\end{cases}\end{aligned} \tag{2.2.9}$$

Now, the conclusion is

$$\frac{\mathrm{d}L(\overline{V}(t),t)}{\mathrm{d}t}\begin{cases} < 0 \text{ if } W(t) \neq 0,\\ = 0 \text{ if } W(t) = 0.\end{cases} \tag{2.2.10}$$

This further implies
$-\frac{\mathrm{d}L(\overline{V}(t),t)}{\mathrm{d}t} < 0$ at any non-equilibrium state $V(t)$ satisfying $W(t) = AV(t)B - D \neq 0$;
$-\frac{\mathrm{d}L(\overline{V}(t),t)}{\mathrm{d}t} = 0$ at the equilibrium state $V(t)$ satisfying $W(t) = AV(t)B - D = 0$.
According to the Lyapunov stability theory, $W(t) = AV(t)B - D$ globally converges to the zero matrix, from arbitrarily chosen $V(0)$. □

Theorem 2.2.2 ([231]) *Assume that the real matrices $A \in \mathbb{R}^{m\times n}$, $B \in \mathbb{R}^{p\times q}$, and $D \in \mathbb{R}^{m\times q}$ satisfy the condition*

$$AA^{\dagger}DB^{\dagger}B = D. \tag{2.2.11}$$

Then the activation state variables matrix $V(t)$ of the model GNN(A, B, D)*, defined by (2.2.2), is convergent as $t \to +\infty$ and approaches the equilibrium state*

$$V(t) \to \tilde{V} = A^{\dagger}DB^{\dagger} + V(0) - A^{\dagger}AV(0)BB^{\dagger} \tag{2.2.12}$$

for every initial state matrix $V(0) \in \mathbb{R}^{n\times p}$.

Proof In view of (2.2.2), the matrix $V_1(t) = A^{\dagger}AV(t)BB^{\dagger}$ satisfies

$$\frac{\mathrm{d}V_1(t)}{\mathrm{d}t} = \gamma A^{\dagger}A\frac{\mathrm{d}V(t)}{\mathrm{d}t}BB^{\dagger} = \gamma A^{\dagger}AA^{\mathrm{T}}\mathcal{F}\left(E(t)\right)B^{\mathrm{T}}BB^{\dagger}.$$

According to the basic properties of the Moore–Penrose inverse, it follows that $B^{\mathrm{T}}BB^{\dagger} = B^{\mathrm{T}}$, $A^{\dagger}AA^{\mathrm{T}} = A^{\mathrm{T}}$ which further implies

$$\begin{aligned}\frac{\mathrm{d}V_1(t)}{\mathrm{d}t} &= \gamma A^{\mathrm{T}}\mathcal{F}\left(E(t)\right)B^{\mathrm{T}}\\ &= \frac{\mathrm{d}V(t)}{\mathrm{d}t}.\end{aligned}$$

Consequently, $V_2(t) = V(t) - V_1(t)$ satisfies $\frac{\mathrm{d}V_2(t)}{\mathrm{d}t} = 0$, which implies

$$V_2(t) = V_2(0) = V(0) - V_1(0) = V(0) - A^{\dagger}AV(0)BB^{\dagger},\quad t \geq 0. \tag{2.2.13}$$

Furthermore, according to Theorem 2.2.1, $AV(t)B \to D$ and $V_1(t)$ converge to

$$V_1(t) = A^{\dagger}(AV(t)B)B^{\dagger} \to A^{\dagger}DB^{\dagger},\quad t \to +\infty.$$

Accordingly, having in mind (2.2.13), $V(t)$ is convergent and its equilibrium value is

$$\begin{aligned}V(t) = V_1(t) + V_2(t) \to \tilde{V} &= A^{\dagger}DB^{\dagger} + V_2(0)\\ &= A^{\dagger}DB^{\dagger} + V(0) - A^{\dagger}AV(0)BB^{\dagger},\end{aligned}$$

which is just a confirmation of (2.2.12). □

Remark 2.2.2 It is worth to mention that the constraints (2.2.11) are not caused by the GNN evolution design. This condition is just the general condition for solvability of the general linear matrix equation $AXB = D$.

Inner inverses $A^{(1)}$ and $B^{(1)}$ are arbitrary, but the most convenient choice is $A^{\dagger}$ and $B^{\dagger}$. This choice is used in Theorem 2.2.2.

According to Theorem 2.2.2, the limiting value $\tilde{V}$ of $V(t)$ is determined by the choice of $V(0)$. For this purpose, the limiting value initiated by the initial stage $V(0)$ is denoted by $\tilde{V}_{V(0)}$.

The statement of Corollary 2.2.1 is directly implied by Theorem 2.2.1.

Corollary 2.2.1 ([231]) *Assume that the real matrices* $A \in \mathbb{R}^{m\times n}$, $B \in \mathbb{R}^{p\times q}$, *and* $D \in \mathbb{R}^{m\times q}$ *satisfy (2.2.11). Further, let an odd and monotonically increasing function* $f(\cdot)$ *be used to define the array activation function* $\mathcal{F}(\cdot)$ *and the design parameter* γ *satisfy* $\gamma > 0$. *Denote by* $\tilde{V}_{V(0)}$ *the limit* $\tilde{V} = \lim_{t\to\infty} V(t)$ *corresponding to the initial state* $V(0)$. *Then the equilibrium state matrix* $\tilde{V}_{V(0)}$ *of* $GNN(A, B, D)$ *satisfies*

$$\left\{A\,\tilde{V}_{V(0)}\,B|\;V(0)\in\mathbb{R}^{n\times m}\right\}=\{D\}. \tag{2.2.14}$$

for each $V(0)\in\mathbb{R}^{n\times m}$.

Remark 2.2.3 According to [175, 191], it is known that

$$\|AXB-D\|_2\geq\|AA^{\dagger}DB^{\dagger}B-D\|_2,$$

where the equality is valid if and only if

$$X=A^{\dagger}DB^{\dagger}+Y-A^{\dagger}AYBB^{\dagger}, \tag{2.2.15}$$

wherein Y is arbitrary matrix of appropriate dimensions. Accordingly, (2.2.15) defines the set of least squares solutions to $AXB=D$.

In addition, $A^{\dagger}DB^{\dagger}$ is the unique minimizer of minimal norm between least squares minimizers:

$$\|A^{\dagger}DB^{\dagger}\|_2\leq\|A^{\dagger}DB^{\dagger}+Y-A^{\dagger}AYB^{\dagger}B\|_2.$$

Solution (2.2.12) to the GNN model (2.2.2) coincides with the general solution

$$A^{(1)}DB^{(1)}+Y-A^{(1)}AYBB^{(1)}$$

to the matrix equation $AXB=D$, wherein Y is an arbitrary matrix. Also, two important details should be mentioned:
–The arbitrary matrix Y from (2.2.15) is replaced by the initial state $V(0)$ in (2.2.12).
–In general, the solutions (2.2.12) are the least squares solutions of $AXB=D$. The approach which assumes the zero initial condition $V(0)=0$ produces the Moore–Penrose solution $A^{\dagger}DB^{\dagger}$ of the matrix equation $AXB=D$, i.e., the least squares solution of minimal norm, which possesses the minimal norm attribute in $\|\cdot\|_2$ matrix norm with respect to all least squares solutions.

Corollary 2.2.2 *The general solution of the* GNN(A,I,I) *model, given in (2.1.4), is equal to*

$$\tilde{V}_{V(0)}=A^{\dagger}+V(0)-A^{\dagger}AV(0).$$

The GNN(A,I,I) model can be used in:
–finding the inverse of nonsingular A, starting from arbitrary $V(0)$;
–approximating the left inverse of a full-column rectangular matrix A, starting from arbitrary $V(0)$;
–computing the pseudoinverse of rank-deficient matrices under the zero initial condition $V(0)=0$.

Simulink Implementation of GNN (A, B, D) and Numerical Examples

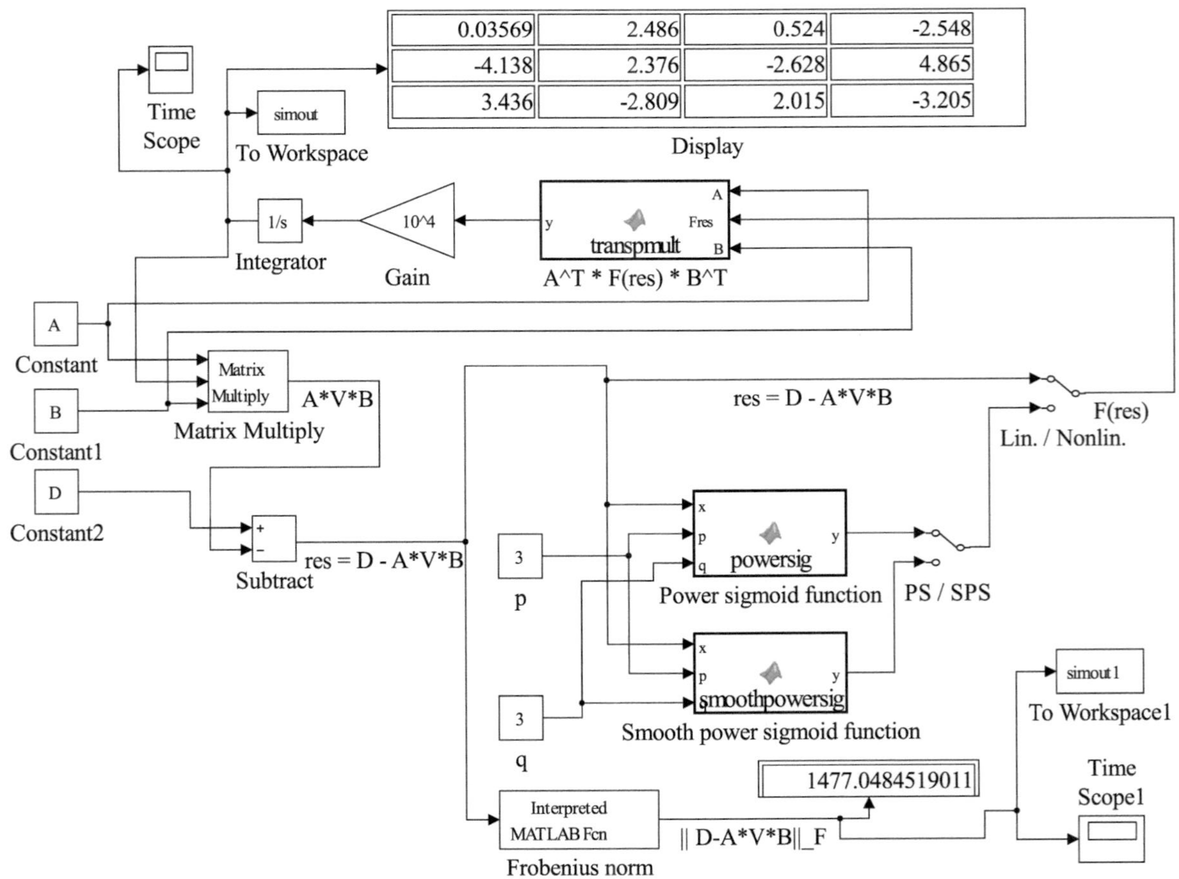

Fig. 2.6 Simulink implementation of GNN(A, B, D)

The Simulink implementation of the GNN(A, B, D) model is presented in Fig. 2.6. The following three types of activation functions $f(\cdot)$ are used in numerical experiments: Linear (1.4.53), power-sigmoid (1.4.55), and smooth power-sigmoid (1.4.56). The parameters γ, initial state $V(0)$ as well as the parameters ρ and ϱ of the nonlinear activation functions (1.4.55) and (1.4.56), are entered directly in the model. On the other hand, matrices A, B, and D are defined from the workspace. We assume that $\rho = \varrho = 3$ in all examples.

The blocks *powersig*, *smoothpowersig*, and *transpmult* include the following codes:

```
function y = powersig(x,p,q)
    [m,n] =size(x);
    y=x;
    for i=1:m
        for j=1:n
            y(i,j)=sign(x(i,j))*ps(abs(x(i,j)),p,q);
        end
    end
end
function y = ps(x,p,q)
    if (abs(x)>1)
        y = x^p;
    else
        y=(1+exp(-q))/(1-exp(-q))*(1-exp(-q*x))/(1+exp(-q*x));
    end
end

function y = smoothpowersig(x,p,q)
    [m,n] =size(x);
    y=x;
    for i=1:m
        for j=1:n
            y(i,j)=sign(x(i,j))*sps(abs(x(i,j)),p,q);
        end
    end
end
function y = sps(x,p,q)
    y = 0.5*x^p + (1+exp(-q))/(1-exp(-q))*(1-exp(-q*x))/(1+exp(-q*x));
end

function y = transpmult(A,Fres,B)
    y = A'*Fres*B';
```

Example 2.2.1 Consider the following matrices:

$$A = \begin{bmatrix} -8 & 8 & -4 \\ 11 & 4 & -7 \\ 1 & -4 & 3 \\ 0 & 12 & -10 \\ 6 & 12 & -12 \end{bmatrix}, \quad B = \begin{bmatrix} 12 & -4 & -16 \\ -35 & -10 & 45 \\ 2 & -5 & -3 \\ 14 & 17 & -17 \end{bmatrix}, \quad D = \begin{bmatrix} -84 & 2524 & 304 \\ -2252 & -623 & 2897 \\ 484 & -885 & -701 \\ -1894 & 2278 & 2652 \\ -2778 & 1524 & 3750 \end{bmatrix}.$$

All three matrices are of rank $r = 2$ and satisfy the condition $AA^{\dagger}DB^{\dagger}B = D$. We use the GNN$(A, B, D)$ model (2.2.2) to compute the minimum norm least squares solution

$$X = A^{\dagger}DB^{\dagger} = \begin{bmatrix} \frac{197968}{75725} & \frac{2043}{1165} & \frac{35007}{15145} & -\frac{479239}{75725} \\ -\frac{17404}{3029} & \frac{660}{233} & -\frac{11337}{3029} & \frac{21877}{3029} \\ \frac{296594}{75725} & -\frac{3431}{1165} & \frac{71137}{30290} & -\frac{592049}{151450} \end{bmatrix}$$
$$\approx \begin{bmatrix} 2.6143 & 1.75365 & 2.31146 & -6.32868 \\ -5.74579 & 2.83262 & -3.74282 & 7.22252 \\ 3.91672 & -2.94506 & 2.34853 & -3.9092 \end{bmatrix}.$$

of the matrix equation $AXB = D$. The gain parameter of the model is $\gamma = 10$ and the initial state is the zero matrix $V(0) = 0$.

The command line should be as follows:

```
A =[-8 8 -4; 11 4 -7;  1 -4 3;  0 12 -10; 6 12 -12];
B =[12 -4 -16; -35 -10 45; 2 -5 -3; 14 17 -17];
D =[-84 2524 304; -2252 -623 2897; 484 -885 -701; -1894 2278 2652; -2778 1524 3750];
```

(a) Consider the time span $[0, 10^{-9}]$, select the MATLAB `ode` solver `ode15s` and the zero initial state $V0 = V(0) = 0$. Then apply the following procedure:

```
k=10^3; print(simout1); hold on;
k=10^4; print(simout1); hold on;
k=10^6; print(simout1);
```

The resulting figure contains 3 graphs in Fig. 2.7.

Simulink implementation of GNN(A, B, D) with modules for computing activation and computing the residual norm exact solution $V^* = A^{\dagger}DB^{\dagger} + V(0) - A^{\dagger}AV(0)B^{\dagger}B$ are presented in Fig. 2.8.

Subsystem for computing Powersigmoid and SmmothPowersigmoid activation functions activations is presented in Fig. 2.9.

Subsystem for computing the exact solution $V^* = A^{\dagger}DB^{\dagger} + V(0) - A^{\dagger}AV(0)B^{\dagger}B$ is presented in Fig. 2.10.

Tasks–Perform numerical experiments with different initial states $V(0)$. For example, investigate $V(0) = B^{\mathrm{T}}$. Final results for $V(0) = B^{\mathrm{T}}$ are observable in Fig. 2.11.

–Perform experiments for different initial states $V0$, γ and time intervals. Elementwise trajectories of $V(t)$ for $\gamma = 10^3$, $V(0) = B^{\mathrm{T}}$ and $[0, 10^{-6}]$ are shown in Fig. 2.12 (left). Values $\|V(t) - V^*\|_F$ are shown in Fig. 2.12 (right).

(b) Using the linear activation function $f(x)=x$, we obtain the solution up to the accuracy 10^{-12} in time $t_{tot} = 2 \cdot 10^{-3}$. Elementwise trajectories of $V(t)$ are shown in Fig. 2.13 (left). It is evident that the trajectories indicate a usual convergence behavior, so the system is globally asymptotically stable. Figure 2.13 (right) shows

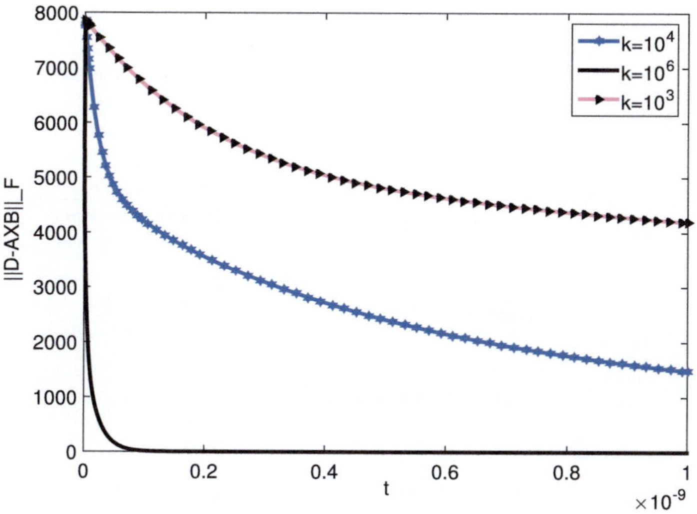

Fig. 2.7 Trajectories of the error norm $\|D - ABD\|_F$ for different gain parameters in Example 2.2.1

the error norm $\|D - AV(t)B\|_F$ when both linear and nonlinear activation functions are used. According to the presented graphs in the left figure, a fast convergence to the equilibrium point can be observed. According to the presented graphs in the right figure, it can be concluded that nonlinear activation functions accelerate the convergence.

Final residual errors after the total time $t_{tot} = 2 \times 10^{-3}$ are 3.34×10^{-12}, 1.85×10^{-12} and 8.03×10^{-13}, respectively for linear, power-sigmoid, and smooth power-sigmoid activation function.

(c) Residual errors for different choice of parameters γ, t_{tot}, given $V(0)$ and ODE solver `ode15s` are arranged in Table 2.1. The best results are marked in bold. These results again confirm that nonlinear activations accelerate the convergence.

2.3 GNN Dynamics for Computing {2, 3} and {2, 4} Inverses

There exist many full-rank representations for different generalized inverses of prescribed rank. For convenience some of them are comprised in the following proposition [41, 52, 205, 208, 233, 234, 239].

Proposition 2.3.1 *Let $A \in \mathbb{C}_r^{m\times n}$ be an arbitrary matrix, $0 < s \leq r$ and $A = PQ$ is a full-rank factorization of A. The following general representations for some classes of generalized inverses are valid:*

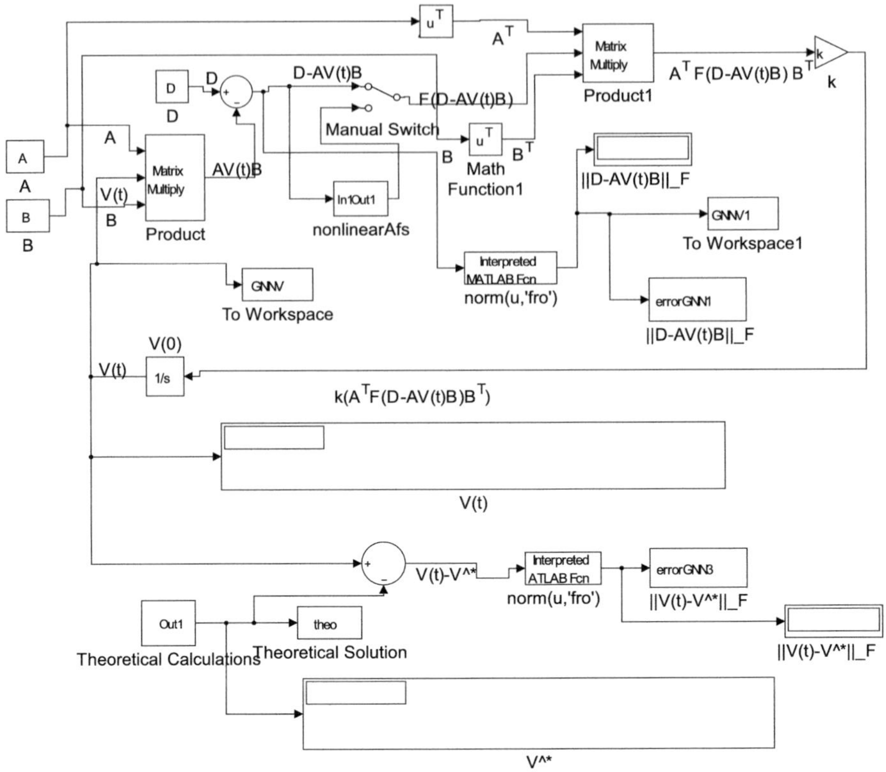

Fig. 2.8 Simulink implementation of GNN(A, B, D) with modules

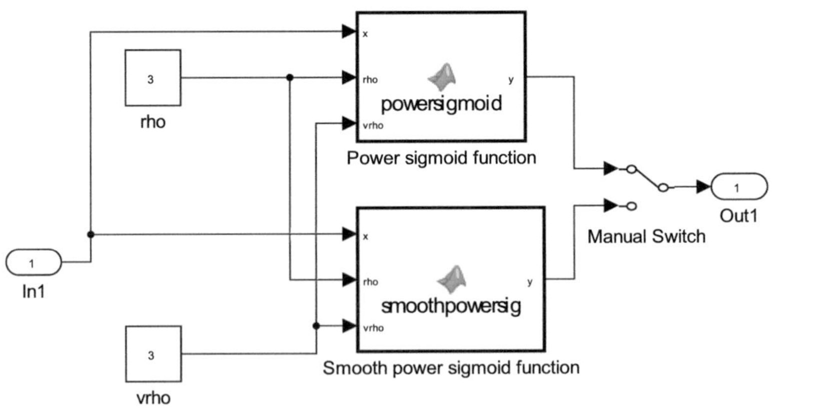

Fig. 2.9 Subsystem for computing Powersigmoid and SmmothPowersigmoid activation functions

$$
\begin{aligned}
A\{2\}_s &= \{F(GAF)^{-1}G \mid F\in\mathbb{C}^{n\times s},\ G\in\mathbb{C}^{s\times m},\ \operatorname{rank}(GAF)=s\};\\
A\{2\} &= \cup_{s=0}^{r} A\{2\}_s;\\
A\{2,4\}_s &= \left\{(VA)^{\dagger}V \mid V\in\mathbb{C}^{s\times m},\ VA\in\mathbb{C}_s^{s\times n}\right\};\\
A\{2,3\}_s &= \left\{U(AU)^{\dagger} \mid U\in\mathbb{C}^{n\times s},\ AU\in\mathbb{C}_s^{m\times s}\right\};
\end{aligned}
$$

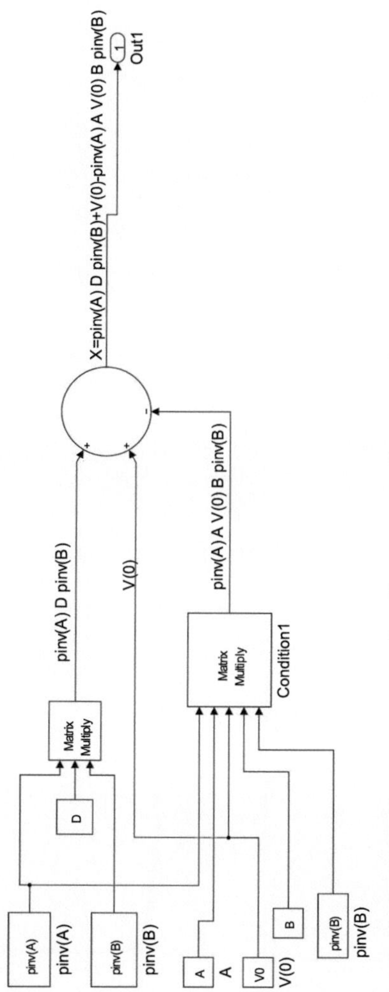

Fig. 2.10 Subsystem for computing the exact solution $V^* = A^\dagger D B^\dagger + V(0) - A^\dagger A V(0) B B^\dagger$

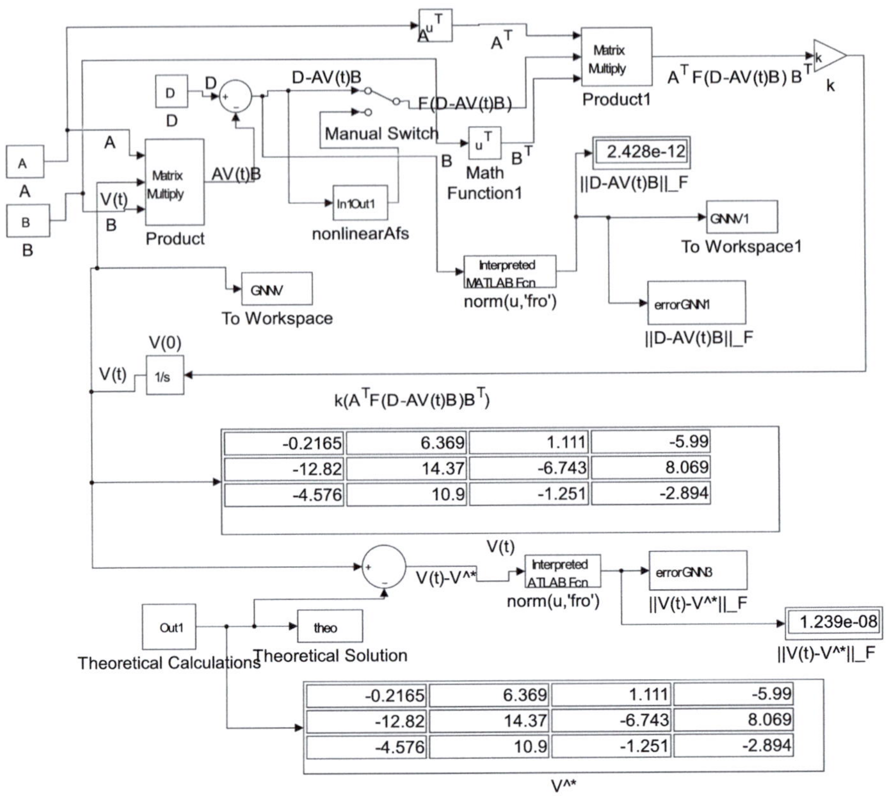

Fig. 2.11 Results in the case $V(0) = B^{\mathrm{T}}$

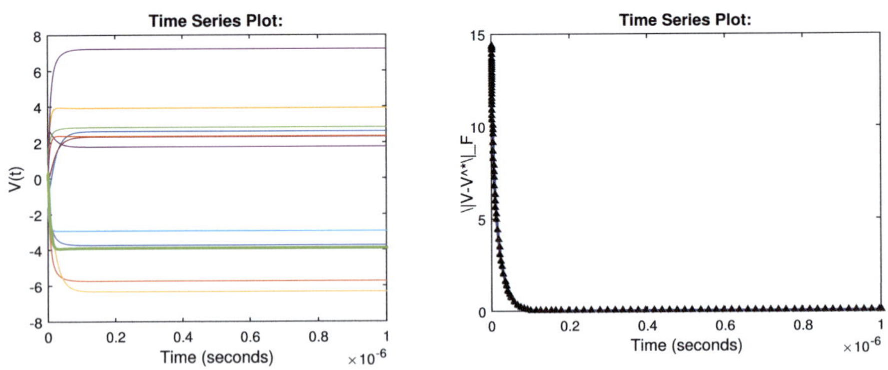

Fig. 2.12 Elementwise convergence trajectories of GNN(A, B, D) (left) and the trajectory of the residual norm $\|V(t) - V^*\|_F$ for linear activation (right) in Example 2.2.1

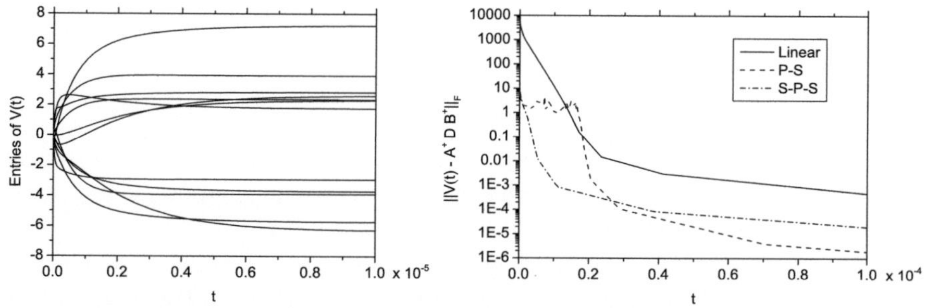

Fig. 2.13 Elementwise convergence trajectories of GNN(A, B, D) network with linear activation function (left) and the trajectory of the error norm for different activation functions (right) in Example 2.2.1

Table 2.1 Performance of GNN(A, B, D) in Example 2.2.1

Activation	γ	$\|AV(t)B - D\|_F$	Activation	γ	$\|AV(t)B - D\|_F$
$(2 \cdot 10^{-3}, ode15s)$, $V(0) = 0$			$(2 \cdot 10^{-3}, ode15s)$, $V(0) = B^{\mathrm{T}}$		
lin	1	3.5951e-13	lin	1	2.0714e-12
ps	1	2.8422e-13	ps	1	**2.2737e–13**
sps	1	**1.6078e–13**	sps	1	1.2085e-12
$(2 \cdot 10^{-4}, ode15s)$, $V(0) = 0$			$(2 \cdot 10^{-4}, ode15s)$, $V(0) = B^{\mathrm{T}}$		
lin	1	0.083172	lin	1	0.020686
ps	1	**1.054869e–06**	ps	1	**1.14258e–06**
sps	1	0.000366	sps	1	1.88123e-06
$(10^{-4}, ode15s)$, $V(0) = 0$			$(10^{-4}, ode15s)$, $V(0) = B^{\mathrm{T}}$		
lin	1	12.23908	lin	1	3.774600
ps	1	0.017396	ps	1	1.366278
sps	1	**0.000146**	sps	1	**0.031365**
$(10^{-5}, ode15s)$, $V(0) = eye(3, 2) \cdot eye(2, 4)$			$(10^{-5}, ode15s)$, $V(0) = magic(3) \cdot eye(3, 4)$		
lin	10^3	7.668596e-13	lin	10^3	**1.59972e–12**
ps	10^3	**5.11591e–13**	ps	10^3	6.996003e-07
sps	10^3	0.000369	sps	10^3	1.64158e-12

$$
\begin{aligned}
A\{1,2\} &= \{F(GAF)^{-1}G \mid F \in \mathbb{C}^{n\times r},\ G \in \mathbb{C}^{r\times m},\ \mathrm{rank}(GAF) = r\} = A\{2\}_r;\\
A\{1,2,4\} &= \{Q^*(VAQ^*)^{-1}V \mid V \in \mathbb{C}^{r\times m},\ \mathrm{rank}(VAQ^*) = r\} = \left\{(VA)^\dagger V \mid VA \in \mathbb{C}_r^{r\times n}\right\};\\
A\{1,2,3\} &= \{U(P^*AU)^{-1}P^* \mid U \in \mathbb{C}^{n\times r},\ \mathrm{rank}(P^*AU) = r\} = \left\{U(AU)^\dagger \mid AU \in \mathbb{C}_r^{m\times r}\right\};\\
A^\dagger &= Q^*(P^*AQ^*)^{-1}P^*;
\end{aligned}
$$

$$
if\, m = n,\quad A^D = F_{A^l}(G_{A^l}\ A\ F_{A^l})^{-1}G_{A^l},\ \ A^l = F_{A^l}G_{A^l},\ \ l \geq \mathrm{ind}(A),\ \ A^l = F_{A^l}G_{A^l}.
$$

A square matrix A has a group inverse if and only if QP is nonsingular, in which case [41]

$$A^{\#} = P(QP)^{-2}Q.$$

It is known that the sets $A\{2\}_0$, $A\{2,3\}_0$, $A\{2,4\}_0$, and $A\{2,3,4\}_0$ are identical and contain a single element, the $n \times m$ zero matrix. For this purpose, it suffices to consider only positive ranks s.

If $A \in \mathbb{C}_r^{m\times n}$, T is a subspace of $\mathbb{C}^n$ of dimension $t \le r$ and S is a subspace of $\mathbb{C}^m$ of dimension $m-t$, then A has a $\{2\}$-inverse X such that $\mathcal{R}(X) = T$ and $\mathcal{N}(X) = S$ if and only if $AT \oplus S = \mathbb{C}^m$, in which case X is unique and we denote it by $A^{(2)}_{T,S}$.

It is well known that the Moore–Penrose inverse and the weighted Moore–Penrose inverse $A^{\dagger}$, $A^{\dagger}_{M,N}$, as well as the Drazin and the group inverse A^D, $A^{\#}$ can be presented by a unified approach, as generalized inverses $A^{(2)}_{T,S}$ for appropriate choice of matrices T and S. For example, the next is valid for a rectangular matrix A [6]:

$$A^{\dagger} = A^{(2)}_{\mathcal{R}(A^*),\mathcal{N}(A^*)}, \quad A^{\dagger}_{M,N} = A^{(2)}_{\mathcal{R}(A^{\sharp}),\mathcal{N}(A^{\sharp})},$$

where M, N are positive definite matrices of appropriate orders and $A^{\sharp} = N^{-1}A^*M$. For a given square matrix A the next identities are satisfied [6, 24]:

$$A^{\mathrm{d}} = A^{(2)}_{\mathcal{R}(A^k),\mathcal{N}(A^k)}, \quad A^{\#} = A^{(2)}_{\mathcal{R}(A),\mathcal{N}(A)},$$

where $k = \mathrm{ind}(A)$.

A useful representation for $A^{(2)}_{T,S}$ based on the usage of the group inverse is presented in [289]. This representation, restated in the next proposition, gave a new computational aspect for $A^{(2)}_{T,S}$ inverse.

Proposition 2.3.2 ([289]) *Let $A \in \mathbb{C}^{m\times n}$ be of rank r, let T be a subspace of $\mathbb{C}^n$ of dimension $s \le r$, and let S be a subspace of $\mathbb{C}^m$ of dimension $m-s$. In addition, suppose that $R \in \mathbb{C}^{n\times m}$ satisfies $\mathcal{R}(R) = T$ and $\mathcal{N}(R) = S$. If A has $A^{(2)}_{T,S}$ then* $\mathrm{ind}(AR) = \mathrm{ind}(RA) = 1$ *and*

$$A^{(2)}_{T,S} = R(AR)^{\#} = (RA)^{\#}R.$$

A general integral representation for $A^{(2)}_{T,S}$ was introduced in [292].

Full-rank representation of $\{2\}$-inverses with prescribed range and null space is determined in the next proposition, which was originated in [208].

Proposition 2.3.3 ([208]) *Let $A \in \mathbb{C}_r^{m\times n}$, T be a subspace of $\mathbb{C}^n$ of dimension $s \le r$ and let S be a subspace of $\mathbb{C}^m$ of dimensions $m-s$. In addition, suppose that $R \in \mathbb{C}^{n\times m}$ satisfies $\mathcal{R}(R) = T$, $\mathcal{N}(R) = S$. Let R has an arbitrary full-rank decomposition, that is $R = FG$. If A has a $\{2\}$-inverse $A^{(2)}_{T,S}$, then:*

(1) *GAF is an invertible matrix;*
(2) $A^{(2)}_{T,S} = F(GAF)^{-1}G$.

Representation of outer inverses in the general form $A^{(2)}_{T,S} = R(AR)^{\#} = (RA)^{\#}R$ is widely exploited in scientific literature. This representation is used in the construction of iterative methods [144, 295, 296, 336]. Also, the full rank representation $A^{(2)}_{T,S} = F(GAF)^{-1}G$ is applicable in numerical calculations. For example, such a representation has been exploited to define the determinantal representation of $A^{(2)}_{T,S}$ inverse in [208] or the set $A\{2\}_s$ in [244]. Also, this representation has been used in the construction of the general successive matrix squaring algorithm for computing $A^{(2)}_{T,S}$ [219] or in the block representation of the set $A\{2\}_s$ [233].

On the other hand, the general representations of {2, 4} and {2, 3}-inverses of the form $(VA)^{\dagger}V$ and $U(AU)^{\dagger}$, respectively, are not widely exploited in the literature. Various representations of {2, 3} and {2, 4}-inverses with prescribed range and null space have been investigated in [20, 208, 328, 329].

Here we propose usage of GNN(A, B, D) design in computing {2, 3} and {2, 4} inverses. Main idea is to apply GNN(A, B, D) in solving matrix equations $GAX = G$ and $XAF = F$. According to Theorem 2.2.2, the general solution to $GAX = G$ produced by GNN(GA, I, G) is equal to

$$(GA)^{\dagger}G + V(0) - (GA)^{\dagger}GAV(0).$$

So, in the case $V(0) = 0$ or if GA is left invertible, the GNN(GA, I, G) design produces $(GA)^{\dagger}G \in A\{2, 4\}$.

According to Theorem 2.2.2, the general solution to $XAF = F$, given as the output of GNN(I, AF, F), is equal to

$$F(AF)^{\dagger} + V(0) - V(0)AF(AF)^{\dagger}.$$

So, in the case $V(0) = 0$ or if AF is right invertible, the GNN(I, AF, F) design produces $F(AF)^{\dagger} \in A\{2, 3\}$.

Example 2.3.1 Consider the matrices

$$A = \begin{bmatrix} -8 & 8 & -4 \\ 11 & 4 & -7 \\ 1 & -4 & 3 \\ 0 & 12 & -10 \\ 6 & 12 & -12 \end{bmatrix}, \quad D = \begin{bmatrix} -84 & -2252 & 484 & -1894 & -2778 \\ 2524 & -623 & -885 & 2278 & 1524 \\ 304 & 2897 & -701 & 2652 & 3750 \end{bmatrix}, \quad A_1 = DA$$

satisfying $\mathrm{rank}(A_1) = \mathrm{rank}(D)$ and the condition $A_1A_1^{\dagger}D = D$. Therefore, the matrix equation $A_1X = D$ is solvable. The linear GNN(A, B, D) model (2.2.2) defined by $B = I_3$, the scaling parameter $\gamma = 10^9$, solver `ode15s`, the time interval $[0, 10^{-5}]$, and the zero initial state $V(0) = 0$ generates the best approximate solution $X = A_1^{\dagger}D = (DA)^{\dagger}D$ of the matrix equation $A_1X = D$, given by

$$X = A_1^{\dagger} D = (DA)^{\dagger} D$$
$$= \begin{bmatrix} -0.0431089904 & 0.044951941 & 0.0082532071 & -0.0114583301 & 0.0151442382 \\ 0.0224358734 & -0.0048076485 & -0.0080128156 & 0.020833341 & 0.01442309565 \\ -0.0043269712 & -0.010977511 & 0.0039262879 & -0.013541657 & -0.017067285 \end{bmatrix}.$$

According to known representations of $\{2, 3\}$ and $\{2, 4\}$-inverses from [220], the matrix X is a $\{2, 4\}$-inverse of A. The matrix equation $A_1 X B = D$ is satisfied with the accuracy
$\|D - A_1 X B\| = 1.969 \cdot 10^{-13}$.

On the other hand, using the initial state

$$V(0) = \begin{bmatrix} 1 & 1 & 1 & 1 & 1 \\ 1 & 1 & 1 & 1 & 1 \\ 1 & 1 & 1 & 1 & 1 \end{bmatrix}$$

the generated approximation is equal to

$$X = A^{\dagger} D + V(0) - A^{\dagger} A V(0) = \begin{bmatrix} 0.3569 & 0.4450 & 0.4083 & 0.3885 & 0.4151 \\ 1.0224 & 0.9952 & 0.9920 & 1.0208 & 1.0144 \\ 1.1957 & 1.1890 & 1.2039 & 1.1865 & 1.1829 \end{bmatrix}.$$

2.3.1 GNN for Solving the Matrix Equation $GAX = G$ or $XAG = G$

The main property of the outer inverse $A^{(2)}_{\mathcal{R}(G),\mathcal{N}(G)}$ which is a basis for the dynamic state equation and later for the induced GNN is given in Lemma 2.3.1.

In order to define the dynamic state equation underlying in the corresponding recurrent neural network it is necessary to introduce Lemma 2.3.1 and several observations.

Lemma 2.3.1 *Let $A \in \mathbb{C}_r^{m\times n}$ be given and $G \in \mathbb{C}_s^{n\times m}$ be arbitrarily chosen matrix satisfying $0 < s \le r$. Assume that $X := A^{(2)}_{\mathcal{R}(G),\mathcal{N}(G)}$ exists. Then the matrix equations*

$$GAX = G, \quad XAG = G \tag{2.3.1}$$

are satisfied.

Proof The following known facts are used to verify the statements (cf. [266]):

$$AX = P_{A\mathcal{R}(G),\mathcal{N}(G)}, \quad XA = P_{\mathcal{R}(G),\left(A^*\mathcal{N}(G)^{\perp}\right)^{\perp}}.$$

Now, the first and the second statement in (2.3.1) follows, respectively, from the next known results (see, for example [6]): $P_{L,M} G = G$ if and only if $\mathcal{R}(G) \subseteq L$ and $G P_{L,M} = G$ if and only if $\mathcal{N}(G) \supseteq M$. □

The GNN model for solving the matrix equation $AXB = D$ was investigated in [231] and termed as GNN(A, B, D). Therefore, GNN(A, B, D) is based on the error matrix $E(t) = AV(t)B - D$. Therefore, GNN(GA, I, G) is defined for solving the matrix equations $GAX = G$, where $A \in \mathbb{R}_r^{m\times n}$ is given and $G \in \mathbb{R}_s^{n\times m}$, $0 < s \le r$, is appropriately chosen matrix. This model is applicable in generating the outer inverse $A^{(2)}_{\mathcal{R}(G),\mathcal{N}(G)}$.

The first matrix equation in (2.3.1) is more appropriate in the case $m \ge n$ and the second in the case $m < n$. The matrix equations in (2.3.1) define the error matrix $E(t)$ in one of the following two dual forms

$$E_{GA}(t) = GAV(t) - G, \quad E_{AG}(t) = V(t)AG - G, \tag{2.3.2}$$

where $V(t) \in \mathbb{R}^{n\times m}$ denotes the unknown time-varying matrix to be solved. Clearly, the time-varying matrix $V(t)$ corresponds to the outer inverse with prescribed range and null space $X := A^{(2)}_{\mathcal{R}(G),\mathcal{N}(G)}$. Our intention is to solve one of the equations included in (2.3.2) with respect to the unknown matrix $V(t)$ using the dynamic-system approach in conjunction with the symbolic data processing.

In this section we propose the LRNN(A, B, D) defined as follows:

$$\dot{V}(t) = \gamma(D - AV(t)B). \tag{2.3.3}$$

For this purpose, the generally adopted rule for solving $GAX = G$ or $XAG = G$ is to use one of the following two dual scalar-valued error functions, defined as the residual Frobenius norm

$$\varepsilon(t) = \begin{cases} \frac{\|GAV(t)-G\|_F^2}{2}, & if\ m \ge n, \\ \frac{\|V(t)AG-G\|_F^2}{2}, & if\ m < n. \end{cases} \tag{2.3.4}$$

A computational scheme for computing the minimum of $\varepsilon(V(t))$ is defined along the gradient descent direction of $\varepsilon(t)$. The derivative of $\varepsilon(t)$ with respect to V is equal to

$$\frac{\partial\varepsilon(t)}{\partial V} = \begin{cases} (GA)^{\mathrm{T}}\left(GAV(t) - G\right), & if\ m \ge n, \\ \left(V(t)AG - G\right)(AG)^{\mathrm{T}}, & if\ m < n. \end{cases} \tag{2.3.5}$$

According to the general GNN design (2.1.1), we obtain the following GNN(GA, I, G) dynamics:

$$\frac{\mathrm{d}V_{GA}(t)}{\mathrm{d}t} = -\gamma\,(GA)^{\mathrm{T}}\left(GAV_{GA}(t) - G\right), \quad if\ m \ge n, \tag{2.3.6}$$

and GNN(I, AG, G) dynamics

$$\frac{\mathrm{d}V_{AG}(t)}{\mathrm{d}t} = -\gamma\,\left(V_{AG}(t)AG - G\right)(AG)^{\mathrm{T}}, \quad if\ m < n. \tag{2.3.7}$$

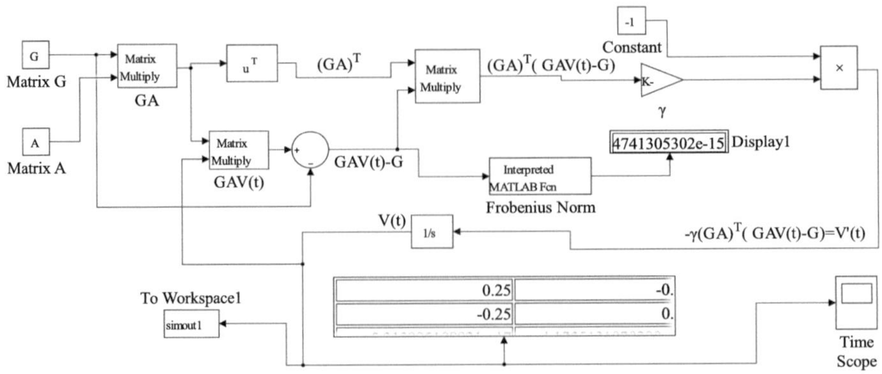

Fig. 2.14 Simulink implementation of GNN(GA, I, G)

The Simulink implementation of GNN(GA, I, G) is illustrated in Fig. 2.14.

Corollary 2.3.1 *Assume that the real matrices* $A \in \mathbb{R}_r^{m\times n}$, $G \in \mathbb{R}_s^{n\times m}$ *satisfy* $0 < s \le r$ *and* $\operatorname{rank}(GA) = \operatorname{rank}(G)$. *Then the following statements hold.*
(i) *The unknown matrix* $V(t)$ *of the model* GNN(GA, I, G) *given by (2.3.6) is convergent as* $t \to +\infty$ *and has the limit value*

$$\tilde{V}_{V(0)} = (GA)^{\dagger}G + V(0) - (GA)^{\dagger}GAV(0), \tag{2.3.8}$$

for every initial matrix $V(0) \in \mathbb{R}^{n\times m}$.
(ii) *In particular, the zero initial state* $V(0) = O$ *initiates* $\tilde{V}_O = (GA)^{\dagger}G = A^{(2,4)}_{\mathcal{N}(GA)^{\perp},\mathcal{N}(G)}$.

Proof **(i)** The assumption $\operatorname{rank}(GA) = \operatorname{rank}(G)$ implies the conditions $GA(GA)^{\dagger}G = G$ for solvability of the matrix equation $GAX = G$. In addition, according to (2.2.12), an arbitrary solution $\tilde{V}_{V(0)}$ of (2.3.6) is given by (2.3.8).
(ii) This part of the proof can be completed using representations of $\{2, 3\}$ and $\{2, 4\}$-inverses from [219]. □

Corollary 2.3.2 *Assume that the real matrices* $A \in \mathbb{R}_r^{m\times n}$, $G \in \mathbb{R}_s^{n\times m}$ *satisfy* $0 < s \le r$ *and* $\operatorname{rank}(AG) = \operatorname{rank}(G)$. *Then the following statements hold.*
(i) *The unknown matrix* $V(t)$ *of the dynamical model* GNN(I, AG, G) *defined by (2.3.7) is convergent when* $t \to +\infty$ *and has the limit value is equal to*

$$\tilde{V}_{V(0)} = G(AG)^{\dagger} + V(0) - V(0)AG(AG)^{\dagger}, \tag{2.3.9}$$

for every initial matrix $V(0) \in \mathbb{R}^{n\times m}$.
(ii) *In particular, the zero initial state* $V(0) = O$ *gives*

$$\tilde{V}_O = G(AG)^{\dagger} = A^{(2,3)}_{\mathcal{N}(G),\mathcal{N}(AG)^{\perp}}.$$

2.3.2 Particular Cases of GNN (GA, I, G) Model

In the particular case $G = A^{\mathrm{T}}$ we immediately derive known results concerning the usual inverse [272] when matrix A is nonsingular, as well as the Moore–Penrose inverse when matrix A is rectangular or rank-deficient [273].

The choice $G = A^{\sharp} = N^{-1}A^{\mathrm{T}}M$ produces the results corresponding to the weighted Moore–Penrose inverse $A^{\dagger}_{M,N}$ [290].

The case $G = A^l$, $l \geq \mathrm{ind}(A)$ yields corresponding GNN dynamic system for computing the Drazin inverse. But, in this case we have additional possibilities to generate the Drazin inverse using appropriate choice of the exponent m. Two different approaches in resolving this problem are described in [251, 252].

For a matrix A of index $\mathrm{ind}(A) = 1$ the Drazin inverse becomes the group inverse $A^{\#}$ of A.

The choice $\mathrm{rank}(G) = r = \mathrm{rank}(A)$ implies $\overline{V}_G \in A\{1, 2\}$.

2.4 GNN for Solving the Matrix Equation $AXA = A$

The GNN model for solving the matrix equation $AXA = A$ is based on the error matrix $E(t) = AV(t)A - A$. Then the dynamics is defined using the error function $\varepsilon(t) = \frac{1}{2}\|A - AV(t)A\|$. Such GNN model was investigated in [240]. In this case, the GNN(A, B, D) model reduces to the GNN(A, A, A) model, defined in [240] as the following GNN-MP model:

$$\dot{V}(t) = \gamma A^{\mathrm{T}}\mathcal{F}\left(A - AV(t)A\right)A^{\mathrm{T}} = \gamma A^{\mathrm{T}}\mathcal{F}\left(E(t)\right)A^{\mathrm{T}}, \quad V(0) \text{ arbitrary}, \quad (2.4.1)$$

where $\mathcal{F}$ denotes appropriately selected activation function.

2.4.1 Convergence Analysis of the GNN-MP Model

The following two statements are valid for the recurrent neural network models (2.4.1) with an arbitrary odd and monotonically increasing activation array function $\mathcal{F}(\cdot)$.

Theorem 2.4.1 ([240]) *Consider an arbitrary real matrix $A \in \mathbb{R}^{m\times n}$. If an odd and monotonically increasing function $f(\cdot)$ is utilized to define the array activation function $\mathcal{F}(\cdot)$ and the gain parameter γ satisfies $\gamma > 0$, then the state matrix $V(t) \in \mathbb{R}^{n\times m}$ of the GNN model (2.4.1) satisfies $AV(t)A \to A$ as $t \to \infty$, for an arbitrary initial state $V(0)$.*

Theorem 2.2.2 shows that $V(t)$ is convergent and determines its limit value. The limit value of $V(t)$ directed by the choice of $V(0)$ is denoted by $\tilde{V}_{V(0)}$.

Theorem 2.4.2 ([240]) *Under the assumptions of Theorem 2.4.1, the matrix of activation state variables $V(t)$ is convergent when $t \to +\infty$ and its limit value is*

$$\tilde{V}_{V(0)} = A^{\dagger} + V(0) - A^{\dagger}AV(0)AA^{\dagger}, \tag{2.4.2}$$

for an arbitrary initial matrix $V(0) \in \mathbb{R}^{n\times m}$.

Proof Denote by $V_1(t) = A^{\dagger}AV(t)AA^{\dagger}$, $V_2(t) = V(t) - V_1(t)$, and $E(t) = A - AV(t)A$. Since the identities $X^{\dagger}XX^{\mathrm{T}} = X^{\mathrm{T}} = X^{\mathrm{T}}XX^{\dagger}$ are valid for an arbitrary real matrix X, directly from the dynamic equation (2.4.1) it can be obtained

$$\begin{aligned}\dot{V}_1(t) &= \gamma A^{\dagger}A\dot{V}(t)AA^{\dagger} = \gamma A^{\dagger}AA^{\mathrm{T}}\mathcal{F}\left(E(t)\right)A^{\mathrm{T}}AA^{\dagger}\\ &= \gamma A^{\mathrm{T}}\mathcal{F}\left(E(t)\right)A^{\mathrm{T}}\\ &= \dot{V}(t),\end{aligned}$$

which implies

$$\frac{\mathrm{d}V_2(t)}{\mathrm{d}t} = 0 \Longrightarrow V_2(t) = V_2(0), \quad (t \geq 0).$$

Furthermore, according to Theorem 2.4.1, it follows $AV(t)A \to A$, which implies

$$V_1(t) = A^{\dagger}(AV(t)A)A^{\dagger} \to A^{\dagger}AA^{\dagger} = A^{\dagger}$$

as $t \to +\infty$. In that way, $V(t)$ is convergent and

$$V(t) = V_1(t) + V_2(t) \to A^{\dagger} + V_2(0) = \tilde{V}_{V(0)}.$$

Finally, the identities

$$V_2(0) = V(0) - V_1(0) = V(0) - A^{\dagger}AV(0)AA^{\dagger}$$

complete the proof of the theorem. □

Corollary 2.4.1 ([240]) *Limiting the values of the activation state matrix $V(t)$, we generate a set of inner inverses:*

$$\left\{\tilde{V}_{V(0)} \mid V(0) \in \mathbb{R}^{n\times m}\right\} = A\{1\}. \tag{2.4.3}$$

Proof The limit value $\tilde{V}$, given in (2.4.2), can be rewritten in the equivalent form

$$\tilde{V}_{V(0)} = A^{\dagger}AA^{\dagger} + V(0) - A^{\dagger}AV(0)AA^{\dagger},$$

which implies that $\tilde{V}_{V(0)}$ is an appearance of the general solution to the matrix equation $AXA = A$, where $V(0)$ can be an arbitrary matrix. The rest of the proof follows from Theorem 2.4.2 and [6, Corollary 1, P. 52], [266, Theorem 1.2.5]. □

The following theorem and corollary define the conditions which define the convergence of $V(t)$ to specific classes of generalized inverses or certain generalized inverses.

Theorem 2.4.3 ([240]) *Depending on the initial matrix $V(0)$, the limit matrix $\tilde{V}_{V(0)}$ satisfies*

1. $\tilde{V}_{V(0)} \in A\{1, 2\}$ *if and only if* $V(0)_{22} = V(0)_{21} A_{11} V(0)_{12}$;
2. $\tilde{V}_{V(0)} \in A\{1, 3\}$ *if and only if* $V(0)_{12} = 0$;
3. $\tilde{V}_{V(0)} \in A\{1, 4\}$ *if and only if* $V(0)_{21} = 0$;
4. $\tilde{V}_{V(0)} = A^{\dagger}$ *if and only if* $V(0)_{12} = 0$, $V(0)_{21} = 0$ *and* $V(0)_{22} = 0$.

Proof Consider the matrices A and $V(t)$ in the following block matrix form, with respect to the direct sum of subspaces:

$$\begin{aligned} A &= \begin{bmatrix} A_{11} & 0 \\ 0 & 0 \end{bmatrix} : \begin{bmatrix} \mathcal{R}(A^{\mathrm{T}}) \\ \mathcal{N}(A) \end{bmatrix} \to \begin{bmatrix} \mathcal{R}(A) \\ \mathcal{N}(A^{\mathrm{T}}) \end{bmatrix} \\ V(t) &= \begin{bmatrix} V(t)_{11} & V(t)_{12} \\ V(t)_{21} & V(t)_{22} \end{bmatrix} : \begin{bmatrix} \mathcal{R}(A) \\ \mathcal{N}(A^{\mathrm{T}}) \end{bmatrix} \to \begin{bmatrix} \mathcal{R}(A^{\mathrm{T}}) \\ \mathcal{N}(A) \end{bmatrix} \end{aligned} \tag{2.4.4}$$

Here A_{11} is a regular $r \times r$ matrix and $r = \mathrm{rank}(A)$. Since $AA^{\dagger}$ and $A^{\dagger}A$ are orthogonal projectors on $\mathcal{R}(A)$ and $\mathcal{R}(A^{\mathrm{T}})$, respectively, the limit matrix $\tilde{V}_{V(0)}$ has the form

$$\tilde{V}_{V(0)} = \begin{bmatrix} A_{11}^{-1} & V(0)_{12} \\ V(0)_{21} & V(0)_{22} \end{bmatrix} : \begin{bmatrix} \mathcal{R}(A) \\ \mathcal{N}(A^{\mathrm{T}}) \end{bmatrix} \to \begin{bmatrix} \mathcal{R}(A^{\mathrm{T}}) \\ \mathcal{N}(A) \end{bmatrix}. \tag{2.4.5}$$

Then the proof can be completed using the known properties of the generalized inverses of block matrices. □

Corollary 2.4.2 ([240]) *Under the assumptions of Theorem 2.4.1 we have*

1. $\tilde{V}_{V(0)} \in A\{1, 2, 3\}$ *if and only if* $\mathcal{N}(V(0)) \supseteq \mathcal{N}(A^{\mathrm{T}})$;
2. $\tilde{V}_{V(0)} \in A\{1, 2, 4\}$ *if and only if* $\mathcal{R}(V(0)) \subseteq \mathcal{R}(A^{\mathrm{T}})$;
3. $\tilde{V}_{V(0)} = A^{\dagger}$ *if and only if* $\mathcal{N}(V(0)) \supseteq \mathcal{N}(A^{\mathrm{T}})$ *and* $\mathcal{R}(V(0)) \subseteq \mathcal{R}(A^{\mathrm{T}})$.

Remark 2.4.1 The GNN-MP model (2.4.1) can be used in the computation of the Moore–Penrose inverse in the cases $V(0) = 0$ or $V(0) = \alpha A^{\mathrm{T}}$, where α is an arbitrary real parameter.

The Impact of Nonlinear Activation in the GNN-MP Model

The following statement holds when the power-sigmoid activation function is used in the GNN-MP model (2.4.1).

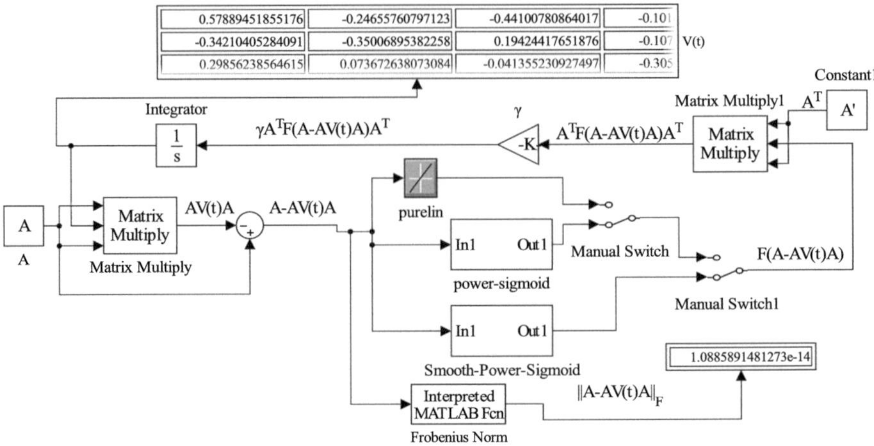

Fig. 2.15 Simulink implementation of the GNN-MP model

Corollary 2.4.3 ([240]) *If the conditions of Theorem 2.4.1 are satisfied, then the* (GNN-MP, $f_{\rm ps}$) *achieves superior convergence over the error range* $(-\infty, +\infty)$ *with respect to the linear case* (GNN-MP, $f_{\rm lin}$).

Proof The proof is similar to the proof of Theorem 3.4 in [282]. The only difference is in the notation. We use $W(t) = A\overline{V}(t)A = (w_{ij})$ instead of the notation $W = (A + \lambda I)^{l+1}\ \overline{V}(t) = (w_{ij})$ from [282, Theorem 3.4]. □

The following statement is valid for the GNN model (2.4.1) activated by the smooth power-sigmoid activation function. It can be verified generalizing results from [282, Theorem 3.4].

Corollary 2.4.4 ([240]) *If the conditions of Theorem 2.4.1 are satisfied, then the (GNN-MP,* $f_{\rm sps}$*) achieves superior convergence over the error range* $(-\infty, +\infty)$ *with respect to the linear case (GNN-MP,* $f_{\rm lin}$*).*

In order to simulate and test the behavior of our GNN models, we built two MATLAB Simulink models. Simulations were performed in *MATLAB* R2014b. The Simulink implementation of the GNN-MP model is presented in Fig. 2.15. The choice of different activation functions is implemented using two Manual Switch blocks.

2.5 GNN Model for Solving the Matrix Equation $A^kXA = A^k$

In this subsection, we construct a GNN model for solving the matrix equation $A^kXA = A^k$, $A \in \mathbb{R}^{n\times n}$, $k \geq \mathrm{ind}(A)$. The influence of various nonlinear activation functions on the defined GNN model is studied. In specific cases, the model is aimed at computing the Drazin inverse.

2.5.1 Definition of the GNN-D Model

The following matrix equation with respect to the unknown time-varying matrix $\Upsilon(t) \in \mathbb{R}^{n\times n}$ should be solved according to the matrix equations (1^k) and (5):

$$A^k \Upsilon(t) A = A^k, \ k \geq \operatorname{ind}(A). \tag{2.5.1}$$

Accordingly, the following matrix-valued error function appears:

$$E_D(t) = A^k - A^k \Upsilon(t) A, \ k \geq \operatorname{ind}(A). \tag{2.5.2}$$

In order to solve (2.5.2) for $\Upsilon(t)$ via dynamic-system approach, we can define a scalar-valued norm-based error function:

$$\varepsilon_D(t) = \varepsilon_D(V(t)) = \frac{1}{2}\|E(t)\|_F^2 = \frac{\|A^k - A^k V(t) A\|_F^2}{2}.$$

The descent direction of $\varepsilon_D(\Upsilon(t))$ along the negative gradient is defined in [240] by

$$-\frac{\partial \varepsilon_D(\Upsilon(t))}{\partial \Upsilon} = -\frac{\partial\left(\frac{1}{2}\|E_D(t)\|_F^2\right)}{\partial \Upsilon} = (A^k)^{\mathrm{T}}\left(A^k - A^k \Upsilon(t) A\right) A^{\mathrm{T}}. \tag{2.5.3}$$

Now, the following dynamic state equation for solving (2.5.1) can be obtained:

$$\frac{\mathrm{d}\Upsilon(t)}{\mathrm{d}t} = -\gamma \frac{\partial \varepsilon_D(\Upsilon(t))}{\partial \Upsilon} = \gamma (A^k)^{\mathrm{T}}\left(A^k - A^k \Upsilon(t) A\right) A^{\mathrm{T}}, \quad \Upsilon(0) = \Upsilon_0, \ k \geq \operatorname{ind}(A). \tag{2.5.4}$$

As in the previous case, the linear GNN model (2.5.4) can be improved by employing a nonlinear matrix-valued activation function array $\mathcal{F}(\cdot)$ into the following nonlinear GNN model:

$$\frac{\mathrm{d}\Upsilon(t)}{\mathrm{d}t} = \gamma (A^k)^{\mathrm{T}} \mathcal{F}\left(A^k - A^k \Upsilon(t) A)\right) A^{\mathrm{T}}, \quad \Upsilon(0) = \Upsilon_0. \tag{2.5.5}$$

The dynamic equation (2.5.5) will be simply termed the GNN-D model.

Remark 2.5.1 Despite the formal similarity, the GNN-MP model is not a particular case $k = 1$ of the GNN-D model, since the GNN-MP model is applicable to arbitrary rectangular real matrices, while the GNN-D model requires a square real matrix.

2.5.2 Convergence Analysis of the GNN-D Model

The following two statements are valid for the recurrent neural network model (2.5.5) with four types of activation functions.

Theorem 2.5.1 ([240]) *Consider a real square matrix $A \in \mathbb{R}^{n\times n}$. If an odd and monotonically increasing function $f(\cdot)$ is used to define the array activation function $\mathcal{F}(\cdot)$, then the state matrix $\Upsilon(t) \in \mathbb{R}^{n\times n}$ of the GNN model (2.5.5) satisfies $A^k\Upsilon(t)A \to A^k$ when $t \to +\infty$, for an arbitrary initial state matrix $\Upsilon(0)$.*

Proof Let $\overline{\Upsilon}(t) = \Upsilon(t) - A^{\mathrm{D}}$ be the difference between the matrix of state variables $\Upsilon(t)$ of the GNN model (2.5.5) and the Drazin inverse A^{D}. The substitution $\Upsilon(t) = \overline{\Upsilon}(t) + A^{\mathrm{D}}$ into the dynamic equation (2.5.5) implies (in accordance with $A^kA^{\mathrm{D}}A = A^k$)

$$\begin{aligned}\frac{\mathrm{d}\overline{\Upsilon}(t)}{\mathrm{d}t} = \frac{\mathrm{d}\Upsilon(t)}{\mathrm{d}t} &= \gamma(A^k)^{\mathrm{T}}\mathcal{F}\left(A^k - A^k\Upsilon(t)A\right)A^{\mathrm{T}}\\ &= \gamma(A^k)^{\mathrm{T}}\mathcal{F}\left(A^k - A^k\left(\overline{\Upsilon}(t) + A^{\mathrm{D}}\right)A\right)A^{\mathrm{T}}\\ &= -\gamma(A^k)^{\mathrm{T}}\mathcal{F}\left(A^k\overline{\Upsilon}(t)A\right)A^{\mathrm{T}}.\end{aligned} \tag{2.5.6}$$

The Lyapunov function candidate is defined similarly as in (2.2.7):

$$L(\overline{\Upsilon}(t), t) = \frac{1}{2}\left\|\overline{\Upsilon}(t)\right\|_F^2 = \frac{1}{2}\mathrm{Tr}\left(\overline{\Upsilon}(t)^{\mathrm{T}}\,\overline{\Upsilon}(t)\right). \tag{2.5.7}$$

Then the rest of the proof proceeds analogically as the proof of Theorem 2.2.1. □

Now we prove the main convergence theorem of the model (2.5.5).

Theorem 2.5.2 ([240]) *Under the assumptions of Theorem 2.5.1, $\Upsilon(t)$ is convergent when $t \to +\infty$ and has the limit value*

$$\tilde{\Upsilon}_{\Upsilon(0)} = (A^k)^{\dagger}A^kA^{\dagger} + V(0) - (A^k)^{\dagger}A^k\Upsilon(0)AA^{\dagger}, \tag{2.5.8}$$

for every initial matrix $\Upsilon(0) \in \mathbb{R}^{n\times n}$.

Proof Denote by $\Upsilon_1(t) = (A^k)^{\dagger}A^k\Upsilon(t)AA^{\dagger}$, $\Upsilon_2(t) = \Upsilon(t) - \Upsilon_1(t)$. An application of the Eq. (2.5.5) and a simplification arising from the basic properties of the Moore–Penrose inverse gives

$$\begin{aligned}\frac{\mathrm{d}\Upsilon_1(t)}{\mathrm{d}t} &= \gamma(A^k)^{\dagger}A^k(A^k)^{\mathrm{T}}\mathcal{F}\left(E_{\Upsilon}(t)\right)A^{\mathrm{T}}AA^{\dagger}\\ &= (A^k)^{\mathrm{T}}\mathcal{F}\left(E_{\Upsilon}(t)\right)A^{\mathrm{T}}\\ &= \frac{\mathrm{d}\Upsilon(t)}{\mathrm{d}t},\end{aligned}$$

which further implies

$$\frac{\mathrm{d}\Upsilon_2(t)}{\mathrm{d}t} = 0 \Longrightarrow \Upsilon_2(t) = \Upsilon_2(0), \quad (t \geq 0).$$

Furthermore, according to Theorem 2.5.1, it follows that

$$\Upsilon_1(t) = (A^k)^\dagger (A^k \Upsilon(t) A) A^\dagger \to (A^k)^\dagger A^k A^\dagger = \tilde{\Upsilon}_1, \; t \to +\infty.$$

Therefore, $\Upsilon(t)$ is convergent and $\Upsilon(t) = \Upsilon_1(t) + \Upsilon_2(t) \to \tilde{\Upsilon}_1 + \Upsilon_2(0) = \tilde{\Upsilon}_{\Upsilon(0)}$. Finally,

$$\Upsilon_2(0) = \Upsilon(0) - \Upsilon_1(0) = \Upsilon(0) - (A^k)^\dagger A^k \Upsilon(0) A A^\dagger$$

completes the proof of the theorem. □

Let us note that $(A^k)^\dagger A^k$ and $AA^\dagger$ are projectors onto $\mathcal{R}\left((A^k)^{\mathrm{T}}\right)$ and $\mathcal{R}(A)$ and, therefore, $\Upsilon(t)$ and its limit value $\tilde{\Upsilon}$ possess the following block matrix representations:

$$\begin{aligned}\Upsilon(t) &= \begin{bmatrix} \Upsilon(t)_{11} & \Upsilon(t)_{12} \\ \Upsilon(t)_{21} & \Upsilon(t)_{22} \end{bmatrix} : \begin{bmatrix} \mathcal{R}(A) \\ \mathcal{N}(A^{\mathrm{T}}) \end{bmatrix} \to \begin{bmatrix} \mathcal{R}((A^k)^{\mathrm{T}}) \\ \mathcal{N}(A^k) \end{bmatrix} \\ \tilde{\Upsilon}_{\Upsilon(0)} &= \begin{bmatrix} M & \Upsilon(0)_{12} \\ \Upsilon(0)_{21} & \Upsilon(0)_{22} \end{bmatrix} : \begin{bmatrix} \mathcal{R}(A) \\ \mathcal{N}(A^{\mathrm{T}}) \end{bmatrix} \to \begin{bmatrix} \mathcal{R}\left((A^k)^{\mathrm{T}}\right) \\ \mathcal{N}(A^k) \end{bmatrix},\end{aligned} \tag{2.5.9}$$

where M is the full-row rank $\mathbb{R}^{s \times r}$ matrix and $s = \operatorname{rank}(A) - \operatorname{ind}(A)$. The next lemma gives necessary and sufficient conditions which provide $\tilde{\Upsilon}_{\Upsilon(0)} = A^{\mathrm{D}}$.

Lemma 2.5.1 ([240]) *Under the assumptions of Theorem 2.5.2, it follows that* $\tilde{\Upsilon}_{\Upsilon(0)} = A^{\mathrm{D}}$ *if and only if* $\Upsilon(0) = A^{\mathrm{D}} + W$ *and the matrix* $W \in \mathbb{R}^{n \times n}$ *satisfies*

$$(A^k)^\dagger A^k W A A^\dagger = W. \tag{2.5.10}$$

Proof By replacing $\Upsilon(0) = A^{\mathrm{D}} + W$ into (2.5.8) and putting $\tilde{\Upsilon}_{\Upsilon(0)} = A^{\mathrm{D}}$, we conclude

$$A^{\mathrm{D}} = (A^k)^\dagger A^k A^\dagger + A^{\mathrm{D}} + W - (A^k)^\dagger A^k (A^{\mathrm{D}} + W) A A^\dagger.$$

Now using $A^k A^{\mathrm{D}} A = A^k$ and the previous equation, the statement of the lemma is evident. □

The result approved in Lemma 2.5.1 implies that $W = \Upsilon(0) - A^{\mathrm{D}}$ has the following block matrix form:

$$W = \begin{bmatrix} W_{11} & 0 \\ 0 & 0 \end{bmatrix} : \begin{bmatrix} \mathcal{R}(A) \\ \mathcal{N}(A^{\mathrm{T}}) \end{bmatrix} \to \begin{bmatrix} \mathcal{R}\left((A^k)^{\mathrm{T}}\right) \\ \mathcal{N}(A^k) \end{bmatrix},$$

where W_{11} is $s \times r$ matrix.

In the general case, it is difficult to effectively provide the matrix $\Upsilon(0)$ such that the condition (2.5.10) is satisfied. On the other hand, it would be interesting to see when these conditions are satisfied for $\Upsilon(0) = 0$, or equivalently, $W = -A^{\mathrm{D}}$. The more general case is done by Lemma 2.5.2.

Lemma 2.5.2 *Under the assumptions* $\mathcal{R}(\Upsilon(0))=\mathcal{R}(A^k)$ *and* $\mathcal{N}(\Upsilon(0))=\mathcal{N}(A^k)$ *we have that* $\tilde{\Upsilon}_{\Upsilon(0)}=A^{\mathrm{D}}$ *if and only if* $\mathcal{R}((A^k)^{\mathrm{T}}) = \mathcal{R}(A^k)$.

Proof Note that $W = \Upsilon(0) - A^{\mathrm{D}}$ implies $\mathcal{R}(W) = \mathcal{R}(A^k)$ and $\mathcal{N}(W) = \mathcal{N}(A^k)$. Furthermore, using $P_{\mathcal{R}((A^k)^{\mathrm{T}})} = (A^k)^{\dagger}A^k$ in conjunction with $P_{\mathcal{R}(A)} = AA^{\dagger}$ and Lemma 2.5.1, one gets that $\tilde{\Upsilon}_{\Upsilon(0)} = A^{\mathrm{D}}$ if and only if

$$P_{\mathcal{R}((A^k)^{\mathrm{T}})} W P_{\mathcal{R}(A)} = W. \tag{2.5.11}$$

Therefore, we have to prove that the condition (2.5.11) is satisfied if and only if $\mathcal{R}((A^k)^{\mathrm{T}}) = \mathcal{R}(A^k)$.

Assume that (2.5.11) is satisfied. Comparing the ranges of the left and right hand side of (2.5.11), we get $\mathcal{R}(A^k) \subseteq \mathcal{R}((A^k)^{\mathrm{T}})$. Now since A^k and $(A^k)^{\mathrm{T}}$ are of the same rank, one concludes that $\mathcal{R}(A^k) = \mathcal{R}((A^k)^{\mathrm{T}})$.

Suppose now $\mathcal{R}((A^k)^{\mathrm{T}}) = \mathcal{R}(A^k)$. It directly implies

$$P_{\mathcal{R}((A^k)^{\mathrm{T}})} W = P_{\mathcal{R}(A^k)} W = W$$

since $\mathcal{R}(W) = \mathcal{R}(A^k)$. Furthermore, using $\mathcal{R}((A^k)^{\mathrm{T}}) \oplus \mathcal{N}(A^k) = \mathbb{R}^n$ and $\mathcal{R}((A^k)) \oplus \mathcal{N}((A^k)^{\mathrm{T}}) = \mathbb{R}^n$ ($\oplus$ is the orthogonal direct sum of subspaces), we conclude that $\mathcal{N}(A^k) = \mathcal{N}((A^k)^{\mathrm{T}}) \supseteq \mathcal{N}(A)$. Combining the last inequality with $\mathcal{N}(W) = \mathcal{N}(A^k)$, one has $W P_{\mathcal{R}(A)} = W$ and hence

$$P_{\mathcal{R}((A^k)^{\mathrm{T}})} W P_{\mathcal{R}(A)} = W P_{\mathcal{R}(A)} = W.$$

This completes the proof of the lemma. □

Remark 2.5.2 The GNN model (2.5.5) can be changed by omitting the constant terms $(A^k)^{\mathrm{T}}$ and A^{T}. Then (2.5.5) becomes

$$\frac{\mathrm{d}\Upsilon(t)}{\mathrm{d}t} = \gamma\mathcal{F}\left(A^k - A^k\Upsilon(t)A)\right), \quad \Upsilon(0) = \Upsilon_0. \tag{2.5.12}$$

However, the global convergence will be lost.

The simplified model (2.5.12) leads to an analogy of the model given in [251].

2.6 Existence, Representations, and Computation of Matrix Generalized Inverses Using GNN Design

This section represents an interesting combination of an algebraic approach and dynamical system approach in computation of various classes of generalized inverses with prescribed range and/or null space. Namely, the algebraic approach gives some useful representations of generalized inverses, while the dynamical systems are defined in order to solve required matrix equations. Leading intention in [221] was to define appropriate numerical algorithms for computing generalized inverses

$$A^{(2)}_{T,S},\ A^{(1)}_{T,*},\ A^{(1)}_{*,S},\ A^{(1)}_{T,S},\ A^{(2)}_{T,*},\ A^{(2)}_{*,S},\ A^{(1,2)}_{T,*},\ A^{(1,2)}_{*,S},\ A^{(1,2)}_{T,S}$$

in both the time-varying and time-invariant cases. Since the dynamical systems used in [221] are GNN models, the convergence is ensured only in time-invariant case. But, GNN models can be applied in time-varying case, without theoretically verified convergence. For this purpose, we observed that the neural dynamic approach, as a powerful tool in solving matrix algebra problems, has very closed correlations with some algebraic problems. Further, the theoretical investigations initiated in [221] have been used in deriving some result about inner and outer inverses with prescribed range and/or null space in rings [119, 120]

Very useful theoretical tool is the full-rank representation of {2}-inverses with prescribed range and null space is determined in the Proposition 2.3.3, which originates from [208].

2.6.1 Existence and Representations of Generalized Inverses

The algorithms developed in this section develop methods for computing {2}- and {1, 2}-inverses of complex matrices with prescribed ranges and/or null spaces. The theoretical basis for the proposed methods and developed GNN models are presented in this subsection. Theorem 2.6.1 provides a theoretical basis for computing outer inverses with prescribed range space.

Theorem 2.6.1 ([221]) *Let $A \in \mathbb{C}^{m\times n}$ and $B \in \mathbb{C}^{n\times k}$.*
(a) *The following statements are equivalent:*

(i) *there exists a {2}-inverse X of A satisfying $\mathcal{R}(X) = \mathcal{R}(B)$, denoted by $A^{(2)}_{\mathcal{R}(B),*}$;*
(ii) *there exists $U \in \mathbb{C}^{k\times m}$ such that $BUAB = B$;*
(iii) *$\mathcal{N}(AB) = \mathcal{N}(B)$;*
(iv) *$\operatorname{rank}(AB) = \operatorname{rank}(B)$;*
(v) *$B(AB)^{(1)}AB = B$, for some (equivalently every) $(AB)^{(1)} \in (AB)\{1\}$.*

(b) *If the statements in* **(a)** *are true, then the set of all outer inverses with prescribed range $\mathcal{R}(B)$ is represented by*

$$\begin{aligned} A\{2\}_{\mathcal{R}(B),*} &= \left\{B(AB)^{(1)} \mid (AB)^{(1)} \in (AB)\{1\}\right\} \\ &= \{BU \mid U \in \mathbb{C}^{k\times m},\ BUAB = B\}. \end{aligned} \tag{2.6.1}$$

Moreover,

$$A\{2\}_{\mathcal{R}(B),*} = \left\{ B(AB)^{(1)} + BY\left(I_m - AB(AB)^{(1)}\right) \mid Y \in \mathbb{C}^{k\times m} \right\}, \qquad (2.6.2)$$

where $(AB)^{(1)} \in (AB)\{1\}$ *is an arbitrary but fixed.*

Proof **(a)** (i)⇒(ii). Let $X \in \mathbb{C}^{n\times m}$ such that $XAX = X$ and $\mathcal{R}(X) = \mathcal{R}(B)$. Then $X = BU$ and $B = XW$, for some $U \in \mathbb{C}^{k\times m}$ and $W \in \mathbb{C}^{m\times k}$, so $B = XW = XAXW = XAB = BUAB$.

(ii)⇒(iii). As we know, $\mathcal{N}(B) \subseteq \mathcal{N}(AB)$. On the other hand, based on $BUAB = B$ for some $U \in \mathbb{C}^{k\times m}$, it follows that $\mathcal{N}(AB) \subseteq \mathcal{N}(BUAB) = \mathcal{N}(B)$, and hence, $\mathcal{N}(AB) = \mathcal{N}(B)$.

(iii)⇒(v). Let $(AB)^{(1)}$ be an arbitrary $\{1\}$-inverse of AB. As $\mathcal{N}(AB) = \mathcal{N}(B)$ implies $B = VAB$, for some $V \in \mathbb{C}^{n\times m}$, it follows that

$$B = VAB = VAB(AB)^{(1)}AB = B(AB)^{(1)}AB.$$

(v)⇒(i). Let $B = B(AB)^{(1)}AB$, for some $(AB)^{(1)} \in (AB)\{1\}$, and set $X = B(AB)^{(1)}$. Then

$$XAX = B(AB)^{(1)}AB(AB)^{(1)} = B(AB)^{(1)} = X,$$

and by $X = B(AB)^{(1)}$ and $B = B(AB)^{(1)}AB = XAB$ it follows that X is a $\{2\}$-inverse of A which satisfies $\mathcal{R}(X) = \mathcal{R}(B)$.

(iii)⇒(v). This result is well-known.

(iii)⇔(iv). This result is well-known.

(b) From the proofs of (i)⇒(ii) and (iv)⇒(i), and the fact that $B = BUAB$ implies $U \in (AB)\{1\}$, it follows

$$\begin{aligned} A\{2\}_{\mathcal{R}(B),*} &\subseteq \{BU \mid U \in \mathbb{C}^{k\times m},\ BUAB = B\} \subseteq \left\{B(AB)^{(1)} \mid (AB)^{(1)} \in (AB)\{1\}\right\} \\ &\subseteq A\{2\}_{\mathcal{R}(B),*}, \end{aligned}$$

and hence, (2.6.1) holds.

According to Theorem 1 [6, Sect. 2] (or [266, Theorem 1.2.5]), the condition (v) ensures consistency of the matrix equation $BUAB = B$ and enables its general solution

$$\{U \in \mathbb{C}^{k\times m} \mid BUAB = B\} = \left\{B^{(1)}B(AB)^{(1)} + Y - B^{(1)}BYAB(AB)^{(1)} \mid Y \in \mathbb{C}^{k\times m}\right\},$$

whence we obtain

$$\begin{aligned} A\{2\}_{\mathcal{R}(B),*} &= \{BU \mid U \in \mathbb{C}^{k\times m},\ BUAB = B\} \\ &= \left\{ B(AB)^{(1)} + BY\left(I_m - AB(AB)^{(1)}\right) \mid Y \in \mathbb{C}^{k\times m} \right\}. \end{aligned}$$

This proves validity of (2.6.2). □

Remark 2.6.1 Five equivalent conditions for the existence and representations of the class of generalized inverses $A^{(2)}_{T,*}$ were given in [20, Theorem 1]. Theorem 2.6.1 gives two new and important conditions (i) and (v). These conditions are related with solvability of certain matrix equations. Further, the representations of generalized inverses $A^{(2)}_{T,*}$ were presented in [20, Theorem 2]. Theorem 2.6.1 gives two new and important representations: the second representation in (2.6.1) and the representation (2.6.2).

Theorem 2.6.2 provides a theoretical basis for computing outer inverses with prescribed kernel.

Theorem 2.6.2 ([221]) *Let $A \in \mathbb{C}^{m\times n}$ and $C \in \mathbb{C}^{l\times m}$.*
(a) *The following statements are equivalent:*

(i) *there exists a $\{2\}$-inverse X of A satisfying $\mathcal{N}(X) = \mathcal{N}(C)$, denoted by $A^{(2)}_{*,\mathcal{N}(C)}$;*
(ii) *there exists $V \in \mathbb{C}^{n\times l}$ such that $CAVC = C$;*
(iii) $\mathcal{R}(CA) = \mathcal{R}(C)$;
(iv) $\mathrm{rank}(CA) = \mathrm{rank}(C)$;
(v) $CA(CA)^{(1)}C = C$, *for some (equivalently every) $(CA)^{(1)} \in (CA)\{1\}$.*

(b) *If the statements in* **(a)** *are true, then the set of all outer inverses with prescribed null space $\mathcal{N}(C)$ is represented by*

$$\begin{aligned} A\{2\}_{*,\mathcal{N}(C)} &= \left\{(CA)^{(1)}C \mid (CA)^{(1)} \in (CA)\{1\}\right\} \\ &= \{VC \mid V \in \mathbb{C}^{n\times l},\ CAVC = C\}. \end{aligned} \tag{2.6.3}$$

Moreover,

$$A\{2\}_{*,\mathcal{N}(C)} = \left\{(CA)^{(1)}C + \left(I_l - (CA)^{(1)}CA\right)YC \mid Y \in \mathbb{C}^{n\times l}\right\}, \tag{2.6.4}$$

where $(CA)^{(1)}$ is an arbitrary fixed matrix from $(CA)\{1\}$.

Theorem 2.6.3 is a theoretical basis for computing a $\{2\}$-inverse with both prescribed range and null space.

Theorem 2.6.3 ([221]) *Let $A \in \mathbb{C}^{m\times n}$, $B \in \mathbb{C}^{n\times k}$ and $C \in \mathbb{C}^{l\times m}$.*
(a) *The following statements are equivalent:*

(i) *there exists a $\{2\}$-inverse X of A satisfying $\mathcal{R}(X) = \mathcal{R}(B)$ and $\mathcal{N}(X) = \mathcal{N}(C)$;*
(ii) *there exist $U \in \mathbb{C}^{k\times l}$ such that $BUCAB = B$ and $CABUC = C$;*
(iii) *there exist $U, V \in \mathbb{C}^{k\times l}$ such that $BUCAB = B$ and $CABVC = C$;*
(iv) *there exist $U \in \mathbb{C}^{k\times m}$ and $V \in \mathbb{C}^{n\times l}$ such that $BUAB = B$, $CAVC = C$ and $BU = VC$;*
(v) *there exist $U \in \mathbb{C}^{k\times m}$ and $V \in \mathbb{C}^{n\times l}$ such that $CABU = C$ and $VCAB = B$;*
(vi) $\mathcal{N}(CAB) = \mathcal{N}(B)$, $\mathcal{R}(CAB) = \mathcal{R}(C)$;
(vii) $\mathrm{rank}(CAB) = \mathrm{rank}(B) = \mathrm{rank}(C)$;

(viii) $B(CAB)^{(1)}CAB = B$ *and* $CAB(CAB)^{(1)}C = C$, *for some (equivalently every)* $(CAB)^{(1)} \in (CAB)\{1\}$.

(b) *If the statements in* **(a)** *are true, then the unique* $\{2\}$*-inverse of* A *with prescribed range* $\mathcal{R}(B)$ *and null space* $\mathcal{N}(C)$ *is represented by*

$$\begin{aligned} A^{(2)}_{\mathcal{R}(B),\mathcal{N}(C)} &= B(CAB)^{(1)}C \\ &= BUC, \end{aligned} \tag{2.6.5}$$

for arbitrary $(CAB)^{(1)} \in (CAB)\{1\}$ *and arbitrary* $U \in \mathbb{C}^{k\times l}$ *satisfying* $BUCAB = B$ *and* $CABUC = C$.

Proof **(a)** (i)⇒(ii). Let $X \in \mathbb{C}^{n\times m}$ be such that $XAX = X$, $\mathcal{R}(X) = \mathcal{R}(B)$ and $\mathcal{N}(X) = \mathcal{N}(C)$. Then there exists $U \in \mathbb{C}^{k\times l}$ such that $X = BUC$. Also, B and C satisfy $B = XW$ and $C = VX$, for some $W \in \mathbb{C}^{m\times k}$, $V \in \mathbb{C}^{l\times n}$. This further implies

$$B = XW = XAXW = XAB = BUCAB$$

and

$$C = VX = VXAX = CAX = CABUC.$$

(ii)⇒(vi). According to $CABUC = C$, for some $U \in \mathbb{C}^{k\times l}$, it follows that

$$\mathcal{R}(C) = \mathcal{R}(CABUC) \subseteq \mathcal{R}(CAB) \subseteq \mathcal{R}(C),$$

and thus, $\mathcal{R}(CAB) = \mathcal{R}(C)$. Further, by $B = BUCAB$, for some $U \in \mathbb{C}^{k\times l}$, it follows that

$$\mathcal{N}(B) \subseteq \mathcal{N}(CAB) \subseteq \mathcal{N}(BUCAB) = \mathcal{N}(B),$$

which yields $\mathcal{N}(CAB) = \mathcal{N}(B)$.

(vi)⇒(viii). Let $(CAB)^{(1)}$ be an arbitrary $\{1\}$-inverse of CAB. Since $\mathcal{R}(CAB) = \mathcal{R}(C)$ implies $C = CABW$, for some $W \in \mathbb{C}^{k\times m}$, it follows that

$$C = CABW = CAB(CAB)^{(1)}CABW = CAB(CAB)^{(1)}C.$$

Similarly, $\mathcal{N}(CAB) = \mathcal{N}(B)$ implies $B = VCAB$, for some $V \in \mathbb{C}^{n\times l}$ and

$$B = VCAB = VCAB(CAB)^{(1)}CAB = B(CAB)^{(1)}CAB.$$

(viii)⇒(i). Let $CAB(CAB)^{(1)}C = C$, for some $(CAB)^{(1)} \in (CAB)\{1\}$, and set $X = B(CAB)^{(1)}C$. Then

$$XAX = B(CAB)^{(1)}CAB(CAB)^{(1)}C = B(CAB)^{(1)}C = X$$

and by $X = B(CAB)^{(1)}C$, $B = B(CAB)^{(1)}CAB = XAB$ and $C = CAB(CAB)^{(1)}C = CAX$ it follows that X is a {2}-inverse of A which satisfies $\mathcal{R}(X) = \mathcal{R}(B)$, $\mathcal{N}(X) = \mathcal{N}(C)$.

(vi)⇔(vii). This statement follows from [266, Theorem 1.1.3, P. 3].

(ii)⇒(iii). This is evident.

(iii)⇒(ii). Let $U, V \in \mathbb{C}^{k\times l}$ be arbitrary matrices such that $BUCAB = B$ and $CABVC = C$. Then

$$BUC = BUCABVC = BVC,$$

whence

$$B = BUCAB = BVCAB \quad \text{and} \quad C = CABVC = CABUC.$$

Thus, (ii) holds.

(ii)⇒(iv). $U \in \mathbb{C}^{k\times l}$ such that $BUCAB = B$ and $CABUC = C$. Then

$$B = B(UC)AB, \quad C = CA(BU)C \quad \text{and} \quad B(UC) = (BU)C,$$

which means that (iv) is true.

(iv)⇒(v). Let $U \in \mathbb{C}^{k\times m}$ and $V \in \mathbb{C}^{n\times l}$ such that $BUAB = B$, $CAVC = C$ and $BU = VC$. Then

$$B = BUAB = VCAB \quad \text{and} \quad C = CAVC = CABU,$$

which confirms (v).

(v)⇒(iv). Let $U \in \mathbb{C}^{k\times m}$ and $V \in \mathbb{C}^{n\times l}$ such that $CABU = C$ and $VCAB = B$. Then

$$VC = VCABU = BU,$$

and

$$B = VCAB = BUAB \quad \text{and} \quad C = CABU = CAVC,$$

and hence, (iv) holds.

(iv)⇒(i). Let $U \in \mathbb{C}^{k\times m}$ and $V \in \mathbb{C}^{n\times l}$ such that $BUAB = B$, $CAVC = C$ and $BU = VC$, and set $X = BU = VC$. Then

$$XAX = BUABU = BU = X,$$

by $X = BU$ and $B = BUAB = XAB$ it follows that $\mathcal{R}(X) = \mathcal{R}(B)$, and by $C = CAVC = CAX$ it follows that $\mathcal{N}(X) = \mathcal{N}(C)$. Therefore, (i) is true.

(b) According to the proofs of (i)⇒(ii) and (iv)⇒(i), and the fact that $C = CABUC$ and $BUCAB = B$, for $U \in \mathbb{C}^{k\times l}$, implies $U \in (CAB)\{1\}$, it follows that

$$\begin{aligned} A^{(2)}_{\mathcal{R}(B),\mathcal{N}(C)} &= BUC, \\ &= B(CAB)^{(1)}C, \end{aligned}$$

and hence, (2.6.5) holds. □

Remark 2.6.2 It is worth mentioning that Drazin in [54] generalized the concept of the outer inverse with prescribed range and null space by introducing the concept of (b, c)-inverse in a semigroup. In the matrix case, these concepts can be defined as follows. Let $A \in \mathbb{C}^{m\times n}$, $X \in \mathbb{C}^{n\times m}$, $B \in \mathbb{C}^{n\times k}$, $C \in \mathbb{C}^{l\times m}$. Then, we call X a (B, C)-inverse of A if the following two relations hold:

$$XAB = B, \quad CAX = C \tag{2.6.6}$$

$$X = BU = VC, \quad \text{for some } U \in \mathbb{C}^{k\times m},\ V \in \mathbb{C}^{n\times l}. \tag{2.6.7}$$

It is easy to see that X is a (B, C)-inverse of A if and only if X is a $\{2\}$-inverse of A satisfying $\mathcal{R}(X) = \mathcal{R}(B)$ and $\mathcal{N}(X) = \mathcal{N}(C)$.

Theorem 2.6.4 is applicable for computing a $\{1\}$-inverse X of A satisfying $\mathcal{R}(X) \subseteq \mathcal{R}(B)$.

Theorem 2.6.4 ([221]) *Let $A \in \mathbb{C}^{m\times n}$ and $B \in \mathbb{C}^{n\times k}$.*
(a) *The following statements are equivalent:*

(i) *there exists an $\{1\}$-inverse X of A satisfying $\mathcal{R}(X) \subseteq \mathcal{R}(B)$;*
(ii) *there exists $U \in \mathbb{C}^{k\times m}$ such that $ABUA = A$;*
(iii) $\mathcal{R}(AB) = \mathcal{R}(A)$;
(iv) $AB(AB)^{(1)}A = A$, *for some (equivalently every)* $(AB)^{(1)} \in (AB)\{1\}$
(v) $\operatorname{rank}(AB) = \operatorname{rank}(A)$.

(b) *If the statements in* **(a)** *are true, then the set of all inner inverses of A whose range is contained in $\mathcal{R}(B)$ is represented by*

$$\begin{aligned} \{X \in A\{1\} \mid \mathcal{R}(X) \subseteq \mathcal{R}(B)\} &= \left\{B(AB)^{(1)} \;\middle|\; (AB)^{(1)} \in (AB)\{1\}\right\} \\ &= \{BU \mid U \in \mathbb{C}^{k\times m},\ ABUA = A\}. \end{aligned} \tag{2.6.8}$$

Moreover,

$$\{X \in A\{1\} \,|\, \mathcal{R}(X) \subseteq \mathcal{R}(B)\} = \left\{ B(AB)^{(1)}AA^{(1)} + BY - B(AB)^{(1)}ABYAA^{(1)} \;\middle|\; Y \in \mathbb{C}^{k\times m} \right\}, \tag{2.6.9}$$

where $(AB)^{(1)} \in (AB)\{1\}$ and $A^{(1)} \in A\{1\}$ are arbitrary but fixed.

Theorem 2.6.5 can be used for computing a $\{1\}$-inverse X of A satisfying $\mathcal{N}(C) \subseteq \mathcal{N}(X)$.

Theorem 2.6.5 ([221]) *Let $A \in \mathbb{C}^{m\times n}$ and $C \in \mathbb{C}^{l\times m}$.*
(a) *The following statements are equivalent:*

(i) *there exists an* $\{1\}$*-inverse* X *of* A *satisfying* $\mathcal{N}(C) \subseteq \mathcal{N}(X)$*;*
(ii) *there exists* $V \in \mathbb{C}^{n\times l}$ *such that* $AVCA = A$*;*
(iii) $\mathcal{N}(CA) = \mathcal{N}(A)$*;*
(iv) $A(CA)^{(1)}CA = A$*, for some (equivalently every)* $(CA)^{(1)} \in (CA)\{1\}$
(v) $\operatorname{rank}(CA) = \operatorname{rank}(A)$.

(b) *If the statements in* **(a)** *are true, then the set of all inner inverses of* A *whose null space is contained in* $\mathcal{N}(C)$ *is represented by*

$$\begin{aligned}\{X \in A\{1\} \mid \mathcal{N}(C) \subseteq \mathcal{N}(X)\} &= \left\{(CA)^{(1)}C \;\middle|\; (CA)^{(1)} \in (CA)\{1\}\right\} \\ &= \{VC \mid V \in \mathbb{C}^{n\times l},\ AVCA = A\}.\end{aligned} \tag{2.6.10}$$

Moreover,

$$\{X \in A\{1\} \mid \mathcal{N}(C) \subseteq \mathcal{N}(X)\} = \left\{A^{(1)}A(CA)^{(1)}C + YC - A^{(1)}AYCA(CA)^{(1)}C \;\middle|\; Y \in \mathbb{C}^{n\times l}\right\}, \tag{2.6.11}$$

where $(CA)^{(1)} \in (CA)\{1\}$ *and* $A^{(1)} \in A\{1\}$ *are arbitrary but fixed.*

Theorem 2.6.6 provides several equivalent conditions for the existence and representations for computing a $\{1, 2\}$-inverse with prescribed range.

Theorem 2.6.6 ([221]) *Let* $A \in \mathbb{C}^{m\times n}$ *and* $B \in \mathbb{C}^{n\times k}$.
(a) *The following statements are equivalent:*

(i) *there exists a* $\{1, 2\}$*-inverse* X *of* A *satisfying* $\mathcal{R}(X) = \mathcal{R}(B)$*, denoted by* $A^{(1,2)}_{\mathcal{R}(B),*}$*;*
(ii) *there exist* $U, V \in \mathbb{C}^{k\times m}$ *such that* $BUAB = B$ *and* $ABVA = A$*;*
(iii) *there exists* $W \in \mathbb{C}^{k\times m}$ *such that* $BWAB = B$ *and* $ABWA = A$*;*
(iv) $\mathcal{N}(AB) = \mathcal{N}(B)$ *and* $\mathcal{R}(AB) = \mathcal{R}(A)$*;*
(v) $\operatorname{rank}(AB) = \operatorname{rank}(A) = \operatorname{rank}(B)$*;*
(vi) $B(AB)^{(1)}AB = B$ *and* $AB(AB)^{(1)}A = A$*, for some (equivalently every)* $(AB)^{(1)} \in (AB)\{1\}$.

(b) *If the statements in* **(a)** *are true, then the set of all* $\{1, 2\}$*-inverses with prescribed range* $\mathcal{R}(B)$ *is represented by*

$$A\{1, 2\}_{\mathcal{R}(B),*} = A\{2\}_{\mathcal{R}(B),*} = \{X \in A\{1\} \mid \mathcal{R}(X) \subseteq \mathcal{R}(B)\}. \tag{2.6.12}$$

Theorem 2.6.7 provides several equivalent conditions for the existence and representations of $A^{(1,2)}_{*,\mathcal{N}(C)}$.

Theorem 2.6.7 ([221]) *Let* $A \in \mathbb{C}^{m\times n}$ *and* $C \in \mathbb{C}^{l\times m}$.
(a) *The following statements are equivalent:*

(i) *there exists a* $\{1, 2\}$*-inverse* X *of* A *satisfying* $\mathcal{N}(X) = \mathcal{N}(C)$*, denoted by* $A^{(1,2)}_{*,\mathcal{N}(C)}$*;*
(ii) *there exist* $U, V \in \mathbb{C}^{n\times l}$ *such that* $CAUC = C$ *and* $AVCA = A$*;*

(iii) *there exists* $W \in \mathbb{C}^{n\times l}$ *such that* $CAWC = C$ *and* $AWCA = A$;
(iv) $\mathcal{N}(CA) = \mathcal{N}(A)$ *and* $\mathcal{R}(CA) = \mathcal{R}(C)$;
(v) $\text{rank}(CA) = \text{rank}(A) = \text{rank}(C)$;
(vi) $CA(CA)^{(1)}C = C$ *and* $A(CA)^{(1)}CA = A$, *for some (equivalently every)* $(CA)^{(1)} \in (CA)\{1\}$.

(b) *If the statements in* **(a)** *are true, then the set of all* $\{1, 2\}$*-inverses with the range* $\mathcal{R}(B)$ *is given by*

$$A\{1, 2\}_{*,\mathcal{N}(C)} = A\{2\}_{*,\mathcal{N}(C)} = \{X \in A\{1\} \mid \mathcal{N}(C) \subseteq \mathcal{N}(X)\}. \tag{2.6.13}$$

Theorem 2.6.8 is a theoretical basis for computing a $\{1, 2\}$-inverse with the predefined range and null space.

Theorem 2.6.8 ([221]) *Let* $A \in \mathbb{C}^{m\times n}$, $B \in \mathbb{C}^{n\times k}$ *and* $C \in \mathbb{C}^{l\times m}$.
(a) *The following statements are equivalent:*

(i) *there exists a* $\{1, 2\}$*-inverse* X *of* A *satisfying* $\mathcal{R}(X) = \mathcal{R}(B)$ *and* $\mathcal{N}(X) = \mathcal{N}(C)$, *denoted by* $A^{(1,2)}_{\mathcal{R}(B),\mathcal{N}(C)}$;
(ii) *there exist* $U \in \mathbb{C}^{k\times m}$ *and* $V \in \mathbb{C}^{n\times l}$ *such that* $BUAB = B$, $ABUA = A$, $CAVC = C$ *and* $AVCA = A$;
(iii) $\mathcal{N}(AB) = \mathcal{N}(B)$, $\mathcal{R}(AB) = \mathcal{R}(A)$, $\mathcal{R}(CA) = \mathcal{R}(C)$ *and* $\mathcal{N}(CA) = \mathcal{N}(A)$;
(iv) $\text{rank}(AB) = \text{rank}(A) = \text{rank}(B)$, $\text{rank}(CA) = \text{rank}(A) = \text{rank}(C)$;
(v) $\text{rank}(CAB) = \text{rank}(C) = \text{rank}(B) = \text{rank}(A)$;
(vi) $B(AB)^{(1)}AB = B$, $AB(AB)^{(1)}A = A$, $CA(CA)^{(1)}C = C$ *and* $A(CA)^{(1)}CA = A$, *for some (equivalently every)* $(AB)^{(1)} \in (AB)\{1\}$ *and* $(CA)^{(1)} \in (CA)\{1\}$.

(b) *If the statements in* **(a)** *are true, then the unique* $\{1, 2\}$*-inverse of* A *with prescribed range* $\mathcal{R}(B)$ *and null space* $\mathcal{N}(C)$ *is represented by*

$$\begin{aligned} A^{(1,2)}_{\mathcal{R}(B),\mathcal{N}(C)} &= B(AB)^{(1)}A(CA)^{(1)}C \\ &= BUAVC \\ &= B(CAB)^{(1)}C, \end{aligned} \tag{2.6.14}$$

for arbitrary $(AB)^{(1)} \in (AB)\{1\}$, $(CA)^{(1)} \in (CA)\{1\}$ *and* $(CAB)^{(1)} \in (CAB)\{1\}$, *and arbitrary* $U \in \mathbb{C}^{k\times m}$ *and* $V \in \mathbb{C}^{n\times l}$ *satisfying* $BUAB = B$ *and* $CAVC = C$.

2.6.2 Algorithms and Simulink Implementation

The representations presented in Sect. 2.6.1 provide two different frameworks for computing generalized inverses. The first approach arises from the direct computation of various generalizations or certain variants of the Urquhart formula, derived in Sect. 2.6.1. The second approach enables computation of generalized inverses by means of solving certain matrix equations.

The approach based on solving matrix equation is used in this section. Simulation results confirm the efficiency of the ZNN and GNN approach in solving both time-varying and time-invariant linear matrix equations. We refer to Refs. [28, 346, 365] for further details. In the case of constant coefficient matrices A, B, C, it is possible to use the linear GNN(A, B, C) model

$$\dot{X} = -\gamma A^{\mathrm{T}}(AXB - C)B^{\mathrm{T}}. \tag{2.6.15}$$

The generalized nonlinearly activated GNN model (G-GNN model) is applicable in both time-varying and time-invariant case and possesses the form

$$\dot{X}(t) = -\gamma A(t)^{\mathrm{T}}\mathcal{F}\left(A(t)X(t)B(t) - C(t)\right) B(t)^{\mathrm{T}}, \tag{2.6.16}$$

where $\mathcal{F}(C) = (f(c_{kj}))$, wherein $f(\cdot)$ is an odd and monotonically increasing function.

Besides the linear activation function, $f(x) = x$, here we use the power-sigmoid activation function.

Theorem 2.6.1 provides not only criteria for the existence of an outer inverse $A(t)^{(2)}_{\mathcal{R}(B),*}$ with prescribed range, but also, it provides a method for computing such an inverse. Namely, the problem of computing a $\{2\}$-inverse X of A satisfying $\mathcal{R}(X) = \mathcal{R}(B)$ boils down to the problem of computing a solution to the matrix equation $BUAB = B$, where U is an unknown matrix taking values in $\mathbb{C}^{k\times m}$. If U is an arbitrary solution to this equation, then $X := BU$ is a $\{2\}$-inverse of A satisfying $\mathcal{R}(X) = \mathcal{R}(B)$.

Algorithm 2.6.1 Computing an outer inverse with prescribed range.

Input: Constant or time-varying matrices $A \in \mathbb{C}^{m\times n}$ and $B \in \mathbb{C}^{n\times k}$.
1: Verify $\operatorname{rank}(AB) = \operatorname{rank}(B)$.
If these conditions are satisfied then continue.
2: Solve the matrix equation $BUAB = B$ with respect to $U \in \mathbb{C}^{k\times m}$.
3: Return $X(t) = B(t)U(t) = A(t)^{(2)}_{\mathcal{R}(B),*}$.

The Simulink implementation of Algorithm 2.6.1 in the set of real time-varying matrices is based on the G-GNN model of the form (2.2.4). Since the goal is solving the matrix equation $B(t)U(t)A(t)B(t) = B(t)$, it is necessary to implement the model $GNN(BU, B, B)$. *MATLAB* Simulink implementation is presented in Fig. 2.16. The Simulink Scope and Display block denoted by $U(t)$ represents input signals corresponding to the solution $U(t)$ of the matrix equation $B(t)U(t)A(t)B(t) = B(t)$ with respect to the time t. The underlying G-GNN dynamics in the Simulink presented in Fig. 2.16 is

$$\dot{U}(t) = -\gamma B(t)^{\mathrm{T}}\mathcal{F}\left(B(t)U(t)A(t)B(t) - B(t)\right)\left(A(t)B(t)\right)^{\mathrm{T}}.$$

The Display block denoted by BU displays inputs signals corresponding to the solution $X(t) = B(t)U(t)$.

The block `Subsystem` implements the Power-sigmoid activation function and it is presented in Fig. 2.17.

Theorem 2.6.2 reduces the problem of computing a $\{2\}$-inverse X of A satisfying $\mathcal{N}(X) = \mathcal{N}(C)$ to the problem of computing a solution to the matrix equation $CAVC = C$, where V is an unknown matrix taking values in $\mathbb{C}^{n\times l}$. Then $X := A^{(2)}_{*,\mathcal{N}(C)} = VC$.

Algorithm 2.6.2 Computing an outer inverse with prescribed null space.

Input: Constant or time-varying matrices $A(t) \in \mathbb{C}^{m\times n}$ and $C(t) \in \mathbb{C}^{l\times m}$.

1: Verify $\text{rank}(CA) = \text{rank}(C)$.
   If these conditions are satisfied then continue.

2: Solve the matrix equation $CAV = C$ with respect to an unknown matrix $V \in \mathbb{C}^{n\times l}$.

3: Return $X = VC = A^{(2)}_{*,\mathcal{N}(C)}$.

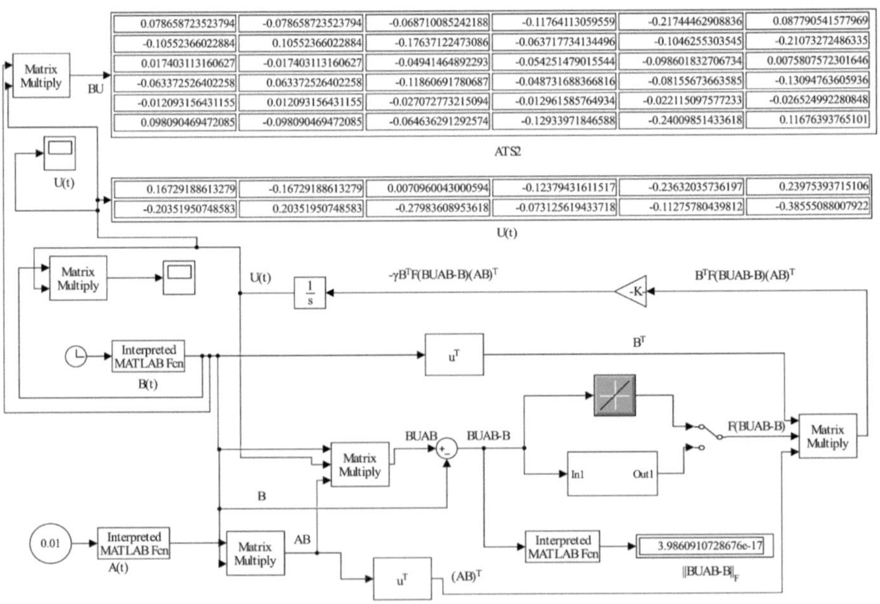

Fig. 2.16 Simulink implementation of the GNN model for computing $BUAB = B$, $X = BU \in A\{2\}_{\mathcal{R}(B),*}$

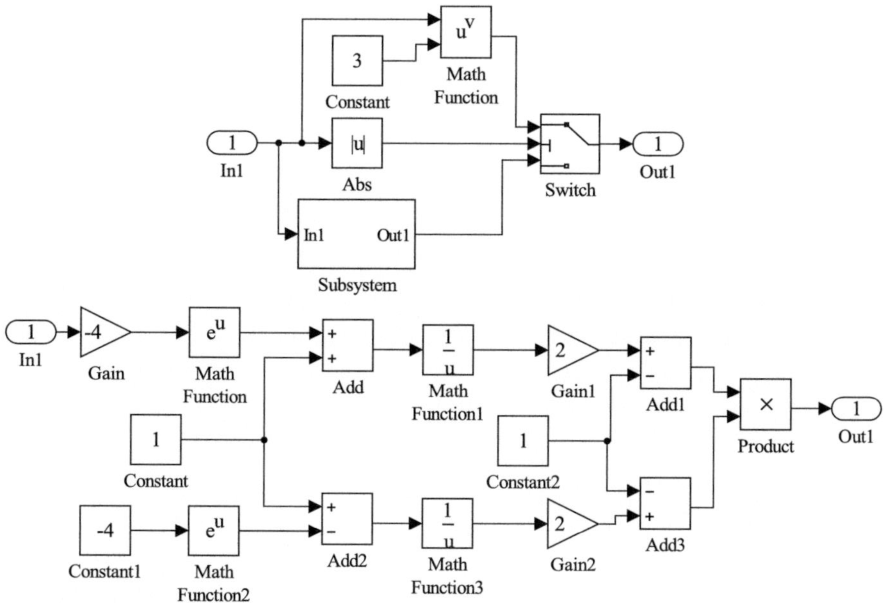

Fig. 2.17 Block for the implementation of the Power-sigmoid activation function (left) and its Subsystem (right)

The Simulink implementation of Algorithm 2.6.2 which is based on the G-GNN model for solving $C(t)A(t)V(t)C(t)=C(t)$ and computing $X(t) = V(t)C(t)$ is presented in Fig. 2.18.

The underlying G-GNN design for the Simulink presented in Fig. 2.18 is the GNN(CA, C, C) dynamics:

$$\dot{V}(t) = -\gamma(C(t)A(t))^{\mathrm{T}}\mathcal{F}\left(C(t)A(t)V(t)C(t) - C(t)\right)C(t)^{\mathrm{T}}.$$

The *Display* block denoted by $V(t)$ displays inputs signals corresponding to the solution $V(t)$ of the matrix equation $CAV(t)C = C$ with respect to simulation time. The block *Display* denoted by $ATS2$ displays inputs signals corresponding to the solution $X(t) = V(t)C(t)$.

Theorem 2.6.3 provides a powerful representation of a $\{2\}$-inverse X of A satisfying $\mathcal{R}(X) = \mathcal{R}(B)$ and $\mathcal{N}(X) = \mathcal{N}(C)$. Also, it suggests the following procedure for computing those generalized inverses. First, it is necessary to verify whether $\operatorname{rank}(CAB) = \operatorname{rank}(B) = \operatorname{rank}(C)$. If this is true, then by Theorem 2.6.3 it follows that the equations $BUCAB = B$ and $CABVC = C$ are solvable and have the same sets of solutions. We compute an arbitrary solution U of the equation $BUCAB = B$, and then $X = BUC$ is the desired $\{2\}$-inverse of A.

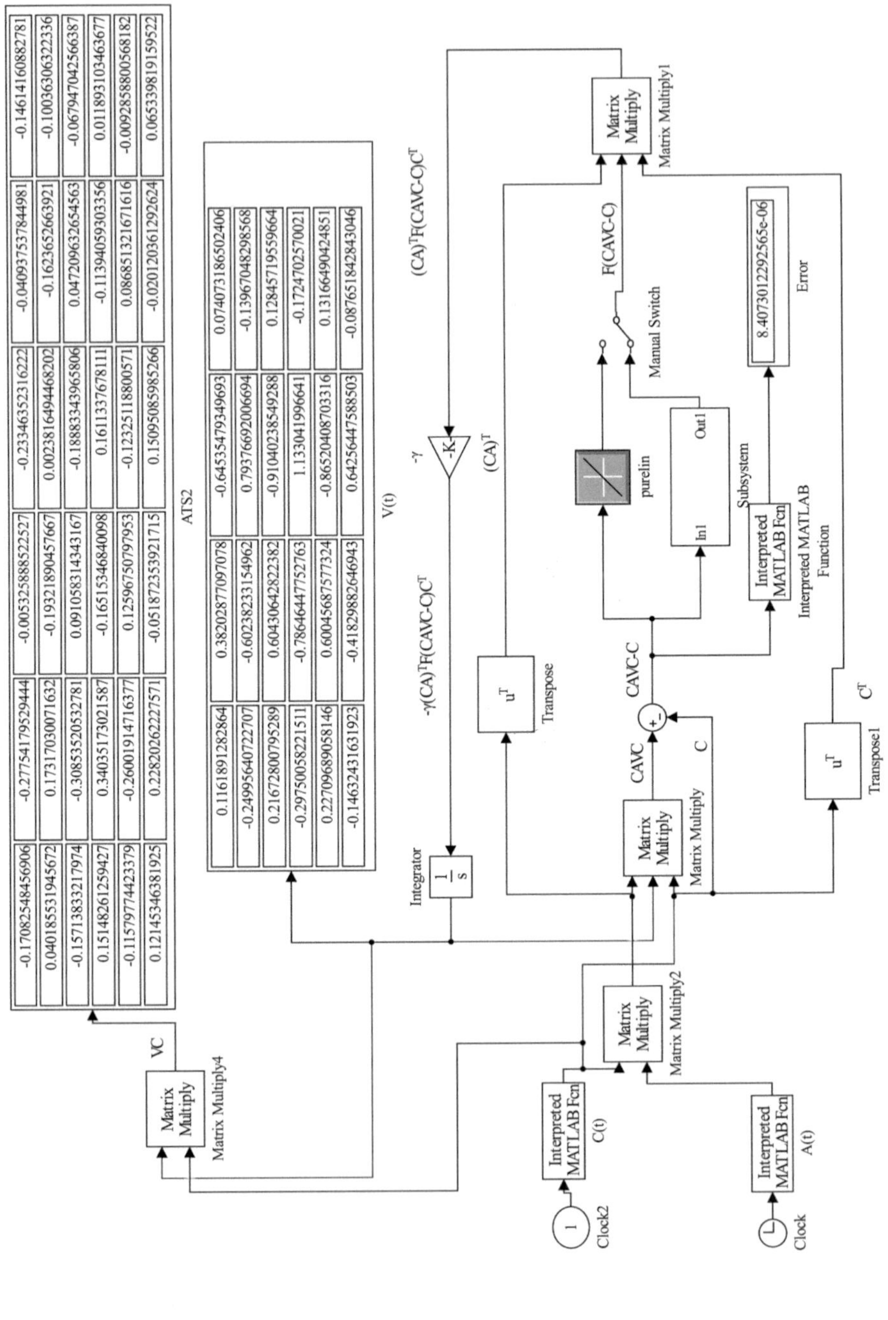

Fig. 2.18 Simulink implementation of the GNN model for computing $CAVC = C$, $X = VC \in A\{2\}_{*,N(C)}$

Algorithm 2.6.3 Computing a {2}-inverse with prescribed range and null space.

Input: Constant or time-varying matrices $A \in \mathbb{C}^{m\times n}$, $B \in \mathbb{C}^{n\times k}$ and $C \in \mathbb{C}^{l\times m}$.

1: Verify $\text{rank}(CAB) = \text{rank}(B) = \text{rank}(C)$.
   If these conditions are satisfied then continue.
2: Solve the matrix equation $BUCAB = B(t)$ with respect to an unknown matrix $U \in \mathbb{C}^{k\times m}$.
3: Return $X = BUC = A^{(2)}_{\mathcal{R}(B),\mathcal{N}(C)}$.

The Simulink implementation of the $GNN(B, CAB, B)$ model for solving $B(t)U(t)C(t)A(t)B(t) = B(t)$ and computing the outer inverse $X(t) = B(t)U(t)C(t)$, defined in Algorithm 2.6.3, is presented in Fig. 2.20. The underlying G-GNN model in Fig. 2.20 is

$$\dot{U}(t) = -\gamma B(t)^{\mathrm{T}}\mathcal{F}\left(B(t)U(t)C(t)A(t)B(t) - B(t)\right)\left(C(t)A(t)B(t)\right)^{\mathrm{T}}.$$

The implementation of the dual approach, based on the solution of $C(t)A(t)BV(t)C(t) = C(t)$ and generating the outer inverse $X(t) = B(t)V(t)C(t)$, is presented in Fig. 2.22. The underlying $GNN(CAB, C, C)$ model in Fig. 2.22 is

$$\dot{V}(t) = -\gamma(C(t)A(t)B(t))^{\mathrm{T}}\mathcal{F}\left(C(t)A(t)B(t)V(t)C(t) - C(t)\right)C(t)^{\mathrm{T}}.$$

Theorem 2.6.4 can be used in a similar way as Theorem 2.6.1: if the equation $ABUA = A$ is solvable and any its solution U is computed, then a {1}-inverse X of A satisfying $\mathcal{R}(X) \subseteq \mathcal{R}(B)$ is computed as $X = BU$.

Algorithm 2.6.4 Computing a {1}-inverse X of A satisfying $\mathcal{R}(X) \subseteq \mathcal{R}(B)$.

Input: Constant or time-varying matrices $A \in \mathbb{C}^{m\times n}$ and $B \in \mathbb{C}^{n\times k}$.

1: Check the condition $\text{rank}(AB) = \text{rank}(A)$.
   If this condition is satisfied then continue.
2: Solve the matrix equation $ABUA = A$ with respect to $U \in \mathbb{C}^{k\times m}$.
3: Return a {1}-inverse $X = BU$ of A satisfying $\mathcal{R}(X) \subseteq \mathcal{R}(B)$.

Similarly, Theorem 2.6.5 can be used for computing a {1}-inverse X of A satisfying $\mathcal{N}(C) \subseteq \mathcal{N}(X)$, as it is presented in Algorithm 2.6.5.

Algorithm 2.6.5 Computing a $\{1\}$-inverse X of A satisfying $\mathcal{N}(C) \subseteq \mathcal{N}(X)$.

Input: Constant or time-varying matrices $A \in \mathbb{C}^{m\times n}$ and $C \in \mathbb{C}^{l\times m}$.
1: Check the condition $\text{rank}(CA) = \text{rank}(A)$.
   If this condition is satisfied then continue.
2: Solve the matrix equation $AVCA = A$ with respect to $V \in \mathbb{C}^{n\times l}$.
3: Return a $\{1\}$-inverse $X = VC$ of A satisfying $\mathcal{N}(C) \subseteq \mathcal{N}(X)$.

An algorithm for computing a $\{1, 2\}$-inverse with prescribed range is based on Theorem 2.6.6. According to this theorem we first check the condition $\text{rank}(AB) = \text{rank}(A) = \text{rank}(B)$. If it is satisfied, then the equation $BUAB = B$ is solvable and we compute an arbitrary solution U to this equation, after which we compute a $\{2\}$-inverse X of A satisfying $\mathcal{R}(X) = \mathcal{R}(B)$ as $X = BU$. By Theorem 2.6.6, X is also a $\{1\}$-inverse of A. Algorithm 2.6.1 differs from Algorithm 2.6.6 only in the first step. Therefore, the implementation of Algorithm 2.6.6 uses the Simulink implementation of Algorithm 2.6.1 in the case when $\text{rank}(AB) = \text{rank}(A) = \text{rank}(B)$.

Algorithm 2.6.6 Computing a $\{1, 2\}$-inverse with prescribed range.

Input: Constant or time-varying matrices $A \in \mathbb{C}^{m\times n}$ and $B \in \mathbb{C}^{n\times k}$.
1: Check the condition $\text{rank}(AB) = \text{rank}(A) = \text{rank}(B)$.
   If these conditions are satisfied then continue.
2: If the previous condition is satisfied, then solve the matrix equation $BUAB = B$ with respect to an unknown matrix $U \in \mathbb{C}^{k\times m}$.
3: Return a $\{1, 2\}$-inverse $X = BU$ of A satisfying $\mathcal{R}(X) = \mathcal{R}(B)$.

Similarly, Theorem 2.6.7 provides an algorithm for computing $A^{(1,2)}_{*,\mathcal{N}(C)}$. The implementation of Algorithm 2.6.6 uses the Simulink implementation of Algorithm 2.6.1 in the case $\text{rank}(CA) = \text{rank}(C) = \text{rank}(A)$.

Algorithm 2.6.7 Computing a $\{1, 2\}$-inverse with prescribed null space.

Input: Constant or time-varying matrices $A(t) \in \mathbb{C}^{m\times n}$ and $C(t) \in \mathbb{C}^{l\times m}$.
1: Check the condition $\text{rank}(CA) = \text{rank}(A) = \text{rank}(C)$.
   If these conditions are satisfied then continue.
2: Solve the matrix equation $CAWC = C$ with respect to an unknown matrix $W \in \mathbb{C}^{l\times m}$.
3: Return a $\{1, 2\}$-inverse $X = VC$ of A satisfying $\mathcal{N}(X) = \mathcal{N}(C)$.

Theorem 2.6.8 suggests the following procedure for computing a $\{1, 2\}$-inverse X of A satisfying $\mathcal{R}(X) = \mathcal{R}(B)$ and $\mathcal{N}(X) = \mathcal{N}(C)$. First we check the condition $\text{rank}(CAB) = \text{rank}(B) = \text{rank}(C) = \text{rank}(A)$. If this is true, then the equations $BUAB = B$ and $CAVC = C$ are solvable, and we compute an arbitrary solution U to the first one and an arbitrary solution V of the second one. According to Theorem 2.6.8, $X = BUAVC$ is a $\{1,2\}$-inverse X of A with $\mathcal{R}(X) = \mathcal{R}(B)$ and $\mathcal{N}(X) = \mathcal{N}(C)$.

Algorithm 2.6.8 Computing a $\{1,2\}$-inverse with prescribed range and null space.

Input: Constant or time-varying matrices $A(t) \in \mathbb{C}^{m\times n}$, $B(t) \in \mathbb{C}^{n\times k}$ and $C(t) \in \mathbb{C}^{l\times m}$.

Input: Verify $\text{rank}(CAB) = \text{rank}(B) = \text{rank}(C) = \text{rank}(A)$.
If these conditions are satisfied then continue.

1: Solve the matrix equation $BUAB = B$ with respect to $U \in \mathbb{C}^{k\times m}$.

2: Solve the matrix equation $CAVC = C$ with respect to $V \in \mathbb{C}^{n\times l}$.

3: Return $X = BUAVC = A^{(1,2)}_{\mathcal{R}(B),\mathcal{N}(C)}$.

The Simulink implementation of Algorithm 2.6.8 based on the G-GNN models for solving $B(t)U(t)A(t)B(t) = B(t)$ and $C(t)A(t)V(t)C(t) = C(t)$ and computing $X(t) = B(t)U(t)A(t)V(t)C(t)$ is presented in Fig. 2.19. In this case, it is necessary to implement two parallel G-GNN models of the form

$$\dot{U}(t) = -\gamma B(t)^{\mathrm{T}}\mathcal{F}\left(B(t)U(t)A(t)B(t) - B(t)\right)\left(A(t)B(t)\right)^{\mathrm{T}},$$
$$\dot{V}(t) = -\gamma (C(t)A(t))^{\mathrm{T}}\mathcal{F}\left(C(t)A(t)V(t)C(t) - C(t)\right)C(t)^{\mathrm{T}}.$$

There is also an alternative way to compute a $\{1,2\}$-inverse X of A with $\mathcal{R}(X) = \mathcal{R}(B)$ and $\mathcal{N}(X) = \mathcal{N}(C)$. Namely, first we check whether $\text{rank}(CAB) = \text{rank}(B) = \text{rank}(C) = \text{rank}(A)$. If this is true, then by Theorem 2.6.8 it follows that there exists a $\{2\}$-inverse of A with prescribed range $\mathcal{R}(B)$ and null space $\mathcal{N}(C)$, and each such inverse is also a $\{1\}$-inverse of A. Therefore, to compute a $\{1,2\}$-inverse of A having the range $\mathcal{R}(B)$ and null space $\mathcal{N}(C)$ we have to compute a $\{2\}$-inverse X of A with $\mathcal{R}(X) = \mathcal{R}(B)$ and $\mathcal{N}(X) = \mathcal{N}(C)$ in exactly the same way as in Algorithm 2.6.3. In other words, we compute an arbitrary solution U to the equation $BUCAB = B$, and then $X = BUC$ is the desired $\{1,2\}$-inverse of A.

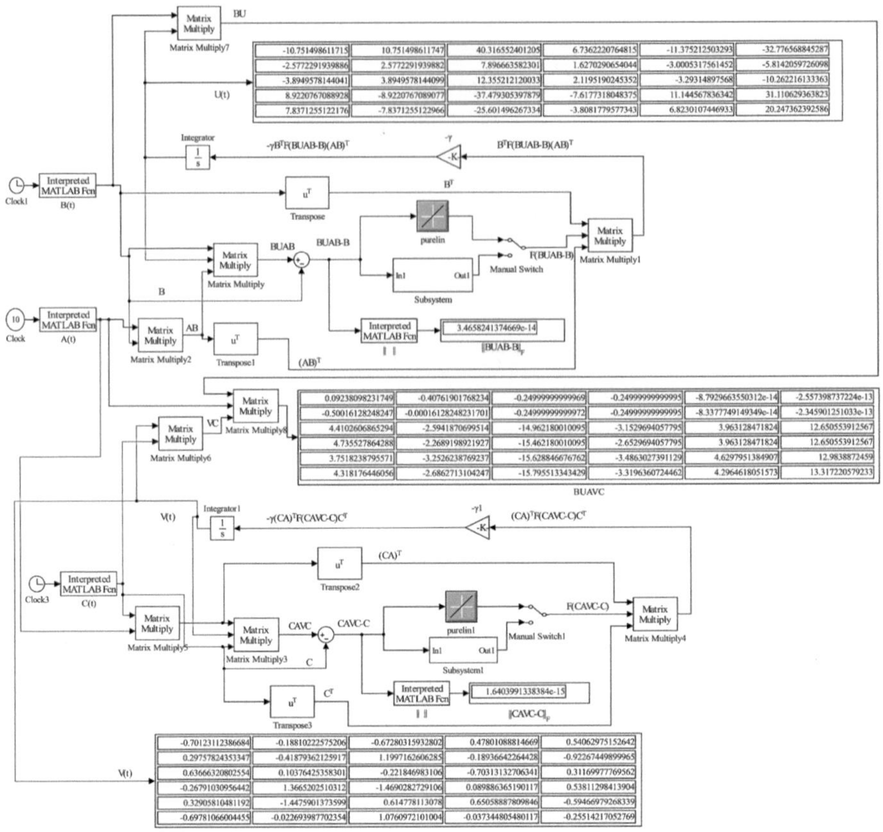

Fig. 2.19 Simulink implementation of Algorithm 2.6.9

Algorithm 2.6.9 An alternative computing of a $\{1, 2\}$-inverse with prescribed range and null space.

Input: Constant or time-varying matrices $A \in \mathbb{C}^{m\times n}$, $B \in \mathbb{C}^{n\times k}$ and $C \in \mathbb{C}^{l\times m}$.

Input: Verify $\operatorname{rank}(CAB) = \operatorname{rank}(B) = \operatorname{rank}(C) = \operatorname{rank}(A)$.
If these conditions are satisfied then continue.

1: Solve the matrix equation $BUCAB = B$ with respect to $U \in \mathbb{C}^{k\times m}$.

2: Return $X = BUC = A^{(1,2)}_{\mathcal{R}(B),\mathcal{N}(C)}$.

2.6.3 Examples on Various Outer and Inner Inverses

All numerical experiments are performed starting from the zero initial condition.

Example 2.6.1 Consider

$$A = \begin{bmatrix} 1 & -1 & 0 & 0 & 0 & 0 \\ -1 & 1 & 0 & 0 & 0 & 0 \\ -1 & -1 & 1 & -1 & 0 & 0 \\ -1 & -1 & -1 & 1 & 0 & 0 \\ -1 & -1 & -1 & 0 & 2 & -1 \\ -1 & -1 & 0 & -1 & -1 & 2 \end{bmatrix} \in \mathbb{R}_5^{6\times 6}$$

and

$$B = \begin{bmatrix} 0.793372 & 0.265655 \\ 0.140305 & 0.633824 \\ 0.329002 & 0.184927 \\ 0.141169569 & 0.427424 \\ 0.0468532 & 0.0979332 \\ 0.89494969 & 0.253673 \end{bmatrix} \in \mathbb{R}_2^{6\times 2},$$

$$C = \begin{bmatrix} 0.714297 & 0.734462 & 0.790305 & 1.1837035 & 0.850446 & 1.143219 \\ 0.596075 & 0.5652303 & 0.745458 & 1.011021 & 0.785712 & 1.013570 \\ 0.780387 & 0.931596 & 0.630581 & 1.23033 & 0.723199 & 1.0876717 \\ 0.298214 & 0.30235998 & 0.337657 & 0.496275 & 0.361875 & 0.482631 \end{bmatrix} \in \mathbb{R}_4^{4\times 6}.$$

(a) This part of the example illustrates results of Theorem 2.6.3 and it is based on the implementation of Algorithm 2.6.3. The matrices A, B, C satisfy $\text{rank}(B) = 2$, $\text{rank}(C) = 4$, $\text{rank}(CAB) = 2$. Since the conditions in (vii) of Theorem 2.6.3 are not satisfied, there is not an exact solution of the system of matrix equations $BUCAB = B$ and $CABUC = C$. The outer inverse $X = B(CAB)^{(1)}C$ can be computed using the RNN approach, as follows. The Simulink implementation of Algorithm 2.6.3, which is based on the G-GNN model for solving the matrix equation $B(t)U(t)C(t)A(t)B(t) = B(t)$, gives the result which is presented in Fig. 2.20. The display denoted by $U(t)$ denotes an approximate solution of the matrix equation $BU(t)CAB = B$. The time interval is $[0, 0.5]$, the solver is `ode15s`, the power sigmoid activation is selected and $\gamma = 10^6$.
Step 1. Solve the matrix equation $B(t)U(t)C(t)A(t)B(t) = B(t)$ with respect to $U(t)$ using an appropriate adaptation of the G-GNN approach developed in [28, 346] and restated in (2.2.4). In the particular case, the model becomes

$$\dot{U}(t) = -\gamma B(t)^{\mathrm{T}} \mathcal{F}\left(B(t)U(t)C(t)A(t)B(t) - B(t)\right)\left(C(t)A(t)B(t)\right)^{\mathrm{T}}. \quad (2.6.17)$$

The matrix B is of full-column rank, and it possesses the left inverse B_l^{-1}. Therefore, the matrix equation $BUCAB = B$ is equivalent to the equation $UCAB - I = 0$.

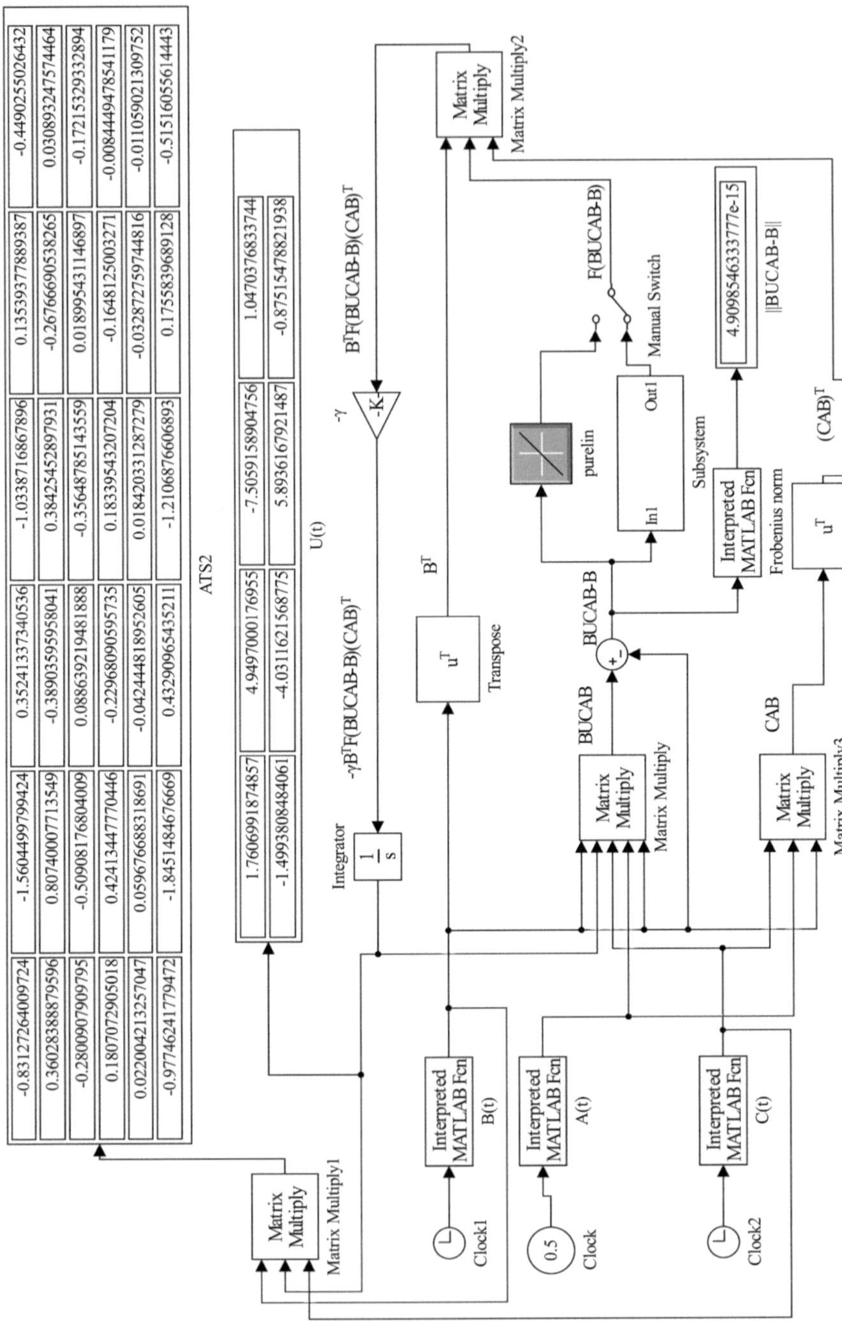

Fig. 2.20 G-GNN model for computing $B(t)U(t)C(t)A(t)B(t) = B(t)$, $X(t) = B(t)U(t)C(t)$

Then the G-GNN model (2.6.17) reduces to the well-known GNN model for computing the pseudoinverse of CAB. The GNN models for computing the pseudoinverse of rank deficient matrices were introduced and described in [273]. We further confirm the results derived in MATLAB Simulink by means of the programming package *Mathematica*. *Mathematica* gives

$$(CAB)^{\dagger} = \begin{bmatrix} 1.76069 & 4.94967 & -7.50589 & 1.04706 \\ -1.49938 & -4.03114 & 5.8936 & -0.875175 \end{bmatrix},$$

which coincides with the result displayed in $U(t)$ in Fig. 2.20.

Step 2. The matrix $X(t) = B(t)U(t)C(t)$ is showed in Fig. 2.20, in the display denoted by $ATS2$. The residual norm of X is equal to $\|XAX - X\|_2 = 6.5360016 * 10^{-15}$.

As a confirmation, *Mathematica* gives

$$X = B(CAB)^{\dagger}C$$
$$= \begin{bmatrix} -0.83127 & -1.56045 & 0.352412 & -1.03387 & 0.135393 & -0.449024 \\ 0.360282 & 0.807398 & -0.389035 & 0.384252 & -0.267666 & 0.0308925 \\ -0.28009 & -0.509081 & 0.0886387 & -0.356487 & 0.0189951 & -0.172153 \\ 0.180706 & 0.424133 & -0.22968 & 0.183394 & -0.164812 & -0.00844537 \\ 0.0220041 & 0.0596765 & -0.0424447 & 0.0184201 & -0.0328727 & -0.0110591 \\ -0.977459 & -1.84514 & 0.432908 & -1.21068 & 0.175583 & -0.515159 \end{bmatrix},$$

which coincides with the contents of the *Display* block denoted as ATS2 in Fig. 2.20. Further, the matrix $U = (CAB)^{\dagger}$ is an approximate solution of the matrix equations $CABUC = C$ and $BUCAB = B$. Also, $X = BUC$ is an approximate solution of the Eqs. (2.6.6), since

$$\|CABUC - C\| = \|CAX - C\| = 2.23452290 * 10^{-14},$$
$$\|BUCAB - B\| = \|XAB - B\| = 9.4574123 * 10^{-15}.$$

Therefore, the equations in (2.6.6) are satisfied. In addition, (2.6.7) is satisfied by the definition of X. Therefore, X is an approximate (B, C)-inverse of A.

Trajectories of the entries in the matrix $B(t)U(t)C(t)$ generated inside the time $[0, 5 * 10^{-2}]$ using $\gamma = 10^6$ and `ode15s` solver are presented in Fig. 2.21.

(b) Dual approach in Theorem 2.6.3, as well as in the implementation of Algorithm 2.6.3, is based on the solution of $C(t)A(t)V(t)C(t) = C(t)$ and the associated outer inverse $X_1(t) = B(t)V(t)C(t)$. The Simulink implementation of the G-GNN model which is based on the matrix equation $CABV(t)C = C$ and the matrix product $X_1(t) = BV(t)C$ gives the result which is presented in Fig. 2.22. The display denoted by $V(t)$ represents an approximate solution of the matrix equation $CABV(t)C = C$. The time interval is [0, 0.5], the solver is `ode15s`, the linear activation is selected and $\gamma = 10^{11}$.

Since the matrix C is right invertible, the matrix equation $CABV(t)C = C$ gives the dual form of the matrix equation for computing $(CAB)^\dagger$, that is $CABV(t) = I$.

Therefore, both X and X_1 are approximations of the same outer inverse of A, equal to $B(CAB)^\dagger C$. To that end, it can be verified that X and X_1 satisfy $\|X - X_1\| = 4.143699{*}10^{-11}$.

2.7 GNN for Dynamics Based on the Urquhart Formula

Characterizations and representations of outer inverses originated in [241] are based on symbolic computation of the matrix equation $BUCAB = B$ or $CUAB = C$ in the programming package Wolfram *Mathematica*.

Our motivation follows from Urquhart formula for computing {2}-inverses, given in Proposition [264].

Proposition 2.7.1 (Urquhart formula)
Let $A \in \mathbb{C}_r^{m\times n}$, $B \in \mathbb{C}^{n\times p}$, $C \in \mathbb{C}^{q\times m}$ *and* $X = B(CAB)^{(1)}C$, *where* $(CAB)^{(1)}$ *is a fixed but arbitrary element of* $(CAB)\{1\}$. *Then*
(1) $X \in A\{1\} \Longleftrightarrow \operatorname{rank}(CAB) = r$;
(2) $X \in A\{2\}_{\mathcal{R}(B)} \Longleftrightarrow \operatorname{rank}(CAB) = \operatorname{rank}(B)$;

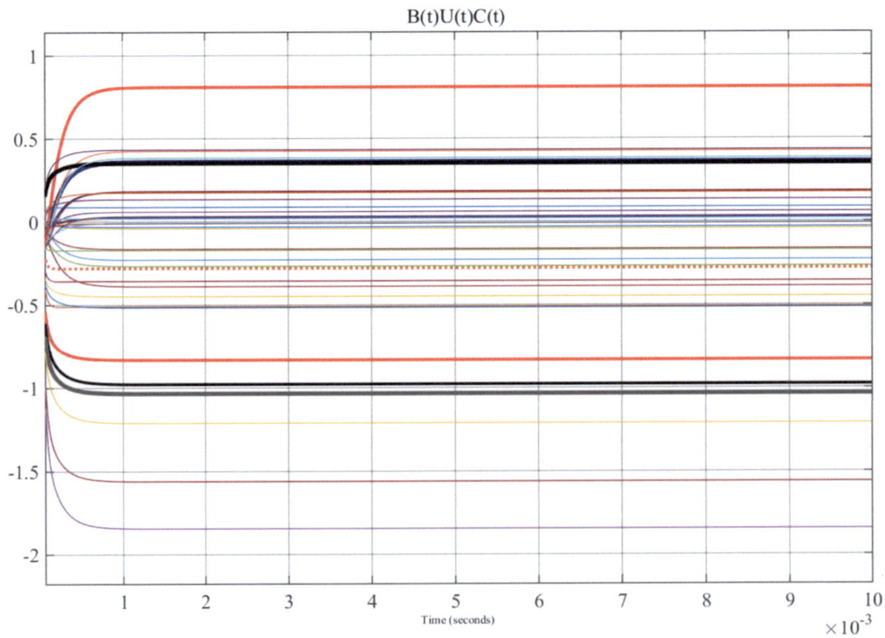

Fig. 2.21 Trajectories of elements of the matrix BUC

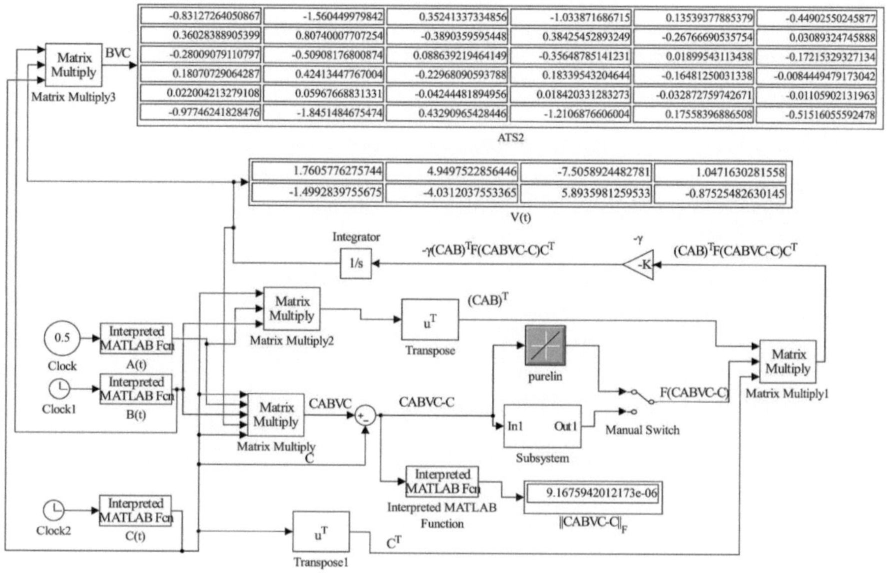

Fig. 2.22 Simulink implementation of the GNN model for computing $CABV(t)C = C$, $X_1 = BVC$

(3) $X \in A\{2\}_{\mathcal{N}(C)} \iff \text{rank}(CAB) = \text{rank}(C)$;
(4) $X = A^{(2)}_{\mathcal{R}(B),\mathcal{N}(C)} \iff \text{rank}(CAB) = \text{rank}(B) = \text{rank}(C)$;
(5) $X = A^{(1,2)}_{\mathcal{R}(B),\mathcal{N}(C)} \iff \text{rank}(CAB)=\text{rank}(B)=\text{rank}(C)=r$.

Solution $\tilde{V}_{V(0)}$ to the matrix equation $BXCAB = B$ can be derived by the GNN(B, CAB, B) model which is defined by the dynamics

$$\dot{V}(t) = B^{\mathrm{T}}\mathcal{F}(B - BV(t)CAB)(CAB)^{\mathrm{T}}. \tag{2.7.1}$$

The general solution to (2.7.1) satisfies $\lim_{t\to\infty} V(t) = \tilde{V}_{V(0)} \in (CAB)\{1\}$, i.e., belongs to the class of inner inverses of CAB. Then $X = B\tilde{V}C$ gives various representations of outer inverses according to Urguhart formula. The Urquhart formula was originated in [264] and later extended in [266, Theorem 1.3.3], [6, Theorem 13, P. 72] and [120].

Proposition 2.7.2 *Let $A \in \mathbb{R}^{m\times n}_r$, $B \in \mathbb{R}^{n\times p}$, $C \in \mathbb{R}^{q\times m}$ and consider the matrix expression $X := B\tilde{V}_{V(0)}C = B(CAB)^{(1)}C$, where $(CAB)^{(1)}$ is a fixed but arbitrary element of $(CAB)\{1\}$. Then*
(1) $X \in A\{1\}$ *if and only if* $\text{rank}(CAB) = r$;
(2) $X \in A\{2\}$ *and* $\mathcal{R}(X) = \mathcal{R}(B)$ *if and only if* $\text{rank}(CAB) = \text{rank}(B)$;
(3) $X \in A\{2\}$ *and* $\mathcal{N}(X) = \mathcal{N}(C)$ *if and only if* $\text{rank}(CAB) = \text{rank}(C)$;
(4) $X = A^{(2)}_{\mathcal{R}(B),\mathcal{N}(C)}$ *if and only if* $\text{rank}(CAB) = \text{rank}(B) = \text{rank}(C)$;
(5) $X = A^{(1,2)}_{\mathcal{R}(B),\mathcal{N}(C)}$ *if and only if* $\text{rank}(CAB) = \text{rank}(B) = \text{rank}(C) = r$.

According to the results presented in [241], it is possible to state the Algorithm 2.7.1 for computing $B(CAB)^{(1)}C$.

Algorithm 2.7.1 Computing outer inverse with prescribed range and/or null space.

Input: $A \in \mathbb{K}(\mathbf{x})^{m\times n}$, $B \in \mathbb{K}(\mathbf{x})^{n\times k}$, $C \in \mathbb{K}(\mathbf{x})^{l\times m}$.
1: Solve symbolically the LME
$BUCAB = B$ if $\text{rank}(CAB) = \text{rank}(B)$ or
$CABUC = C$ if $\text{rank}(CAB) = \text{rank}(C)$.
2: Compute $X := BUC$.
3: Return X.

The output of Algorithm 2.7.1 is defined in Corollary 2.7.1.

Corollary 2.7.1 *Let* $A \in \mathbb{K}(\mathbf{x})^{m\times n}$, $B \in \mathbb{K}(\mathbf{x})^{n\times k}$, $C \in \mathbb{K}(\mathbf{x})^{l\times m}$. *Then the output* X *of Algorithm 2.7.1 satisfies*
(1) $X = A^{(2)}_{\mathcal{R}(B),*} \iff \text{rank}(CAB) = \text{rank}(B)$;
(2) $X = A^{(2)}_{*,\mathcal{N}(C)} \iff \text{rank}(CAB) = \text{rank}(C)$;
(3) $X = A^{(2)}_{\mathcal{R}(B),\mathcal{N}(C)} \iff \text{rank}(CAB) = \text{rank}(B) = \text{rank}(C)$;
(4) $X = A^{(1,2)}_{\mathcal{R}(B),\mathcal{N}(C)} \iff \text{rank}(CAB) = \text{rank}(B) = \text{rank}(C) = \text{rank}(A)$.

The following application of GNN(A, B, D) is motivated by the results from [221] and the Urquhart formula.

Corollary 2.7.2 *Assume that given matrices* $A \in \mathbb{R}^{m\times n}_r$, $B \in \mathbb{R}^{n\times k}$, *and* $C \in \mathbb{R}^{l\times m}$ *satisfy* $\text{rank}(CAB) = \text{rank}(B) \iff B(CAB)^{\dagger}CAB = B$.
(i) *If an odd and monotonically increasing function* $f(\cdot)$ *is used to define the array activation function* $\mathcal{F}(\cdot)$, *then the state matrix* $V(t) \in \mathbb{R}^{n\times s}$ *of the GNN*(B, CAB, B) *model (2.7.1) satisfies* $BV(t)CAB \to B$ *as* $t \to +\infty$, *for an arbitrary initial state matrix* $V(0)$. *When* $t \to +\infty$, *the matrix* $V(t)$ *is convergent and its limiting value* $\tilde{V}_{V(0)}$ *satisfies*

$$\tilde{V}_{V(0)} = B^{\dagger}B(CAB)^{\dagger} + V(0) - B^{\dagger}BV(0)CAB(CAB)^{\dagger} \in (CAB)\{1\}, \quad (2.7.2)$$

for every initial state matrix $V(0) \in \mathbb{R}^{k\times l}$.
(ii) *The following statements are valid:*
(1) *The condition* $\text{rank}(CAB) = r$ *initiates*

$$X := B\tilde{V}_{V(0)}C \in A\{1\}.$$

(2) *The condition* $\text{rank}(CAB) = \text{rank}(B)$ *implies*

$$X := B\tilde{V}_{V(0)}C \in A\{2\}_{\mathcal{R}(B),*}.$$

(3) *If the conditions* $\text{rank}(CAB) = \text{rank}(B) = \text{rank}(C)$ *are satisfied, then*

$$X := B\tilde{V}_{V(0)}C = A^{(2)}_{\mathcal{R}(B),\mathcal{N}(C)}.$$

(4) *If the conditions* $\operatorname{rank}(CAB) = \operatorname{rank}(B) = \operatorname{rank}(C) = r$ *are satisfied, then*

$$X := B\tilde{V}_{V(0)}C = A^{(1,2)}_{\mathcal{R}(B),\mathcal{N}(C)}.$$

Corollary 2.7.3 *Assume that given matrices* $A \in \mathbb{R}^{m\times n}_r$, $B \in \mathbb{R}^{n\times k}$, *and* $C \in \mathbb{R}^{l\times m}$ *satisfy* $\operatorname{rank}(CAB) = \operatorname{rank}(C) \iff (CAB)(CAB)^{\dagger}C = C$.
(i) *If an odd and monotonically increasing function* $f(\cdot)$ *is used to define the array activation function* $\mathcal{F}(\cdot)$, *then the state matrix* $V(t) \in \mathbb{R}^{n\times s}$ *of the GNN(*CAB, C, C*) model (2.7.1) satisfies* $CABV(t)C \to C$ *as* $t \to +\infty$, *for an arbitrary initial state matrix* $V(0)$. *When* $t \to +\infty$, *the matrix* $V(t)$ *is convergent and its limiting value* $\tilde{V}_{V(0)}$ *satisfies*

$$\tilde{V}_{V(0)} = (CAB)^{\dagger}CC^{\dagger} + V(0) - (CAB)^{\dagger}CABV(0)CC^{\dagger} \in (CAB)\{1\}, \quad (2.7.3)$$

for every initial state matrix $V(0) \in \mathbb{R}^{k\times l}$.
(ii) *The following statements are valid:*
(1) *The condition* $\operatorname{rank}(CAB) = r$ *initiates*

$$X := B\tilde{V}_{V(0)}C \in A\{1\}.$$

(2) *The condition* $\operatorname{rank}(CAB) = \operatorname{rank}(C)$ *implies*

$$X := B\tilde{V}_{V(0)}C \in A\{2\}_{*,\mathcal{N}(C)}.$$

(3) *If the conditions* $\operatorname{rank}(CAB) = \operatorname{rank}(B) = \operatorname{rank}(C)$ *are satisfied, then*

$$X := B\tilde{V}_{V(0)}C = A^{(2)}_{\mathcal{R}(B),\mathcal{N}(C)}.$$

(4 *If the conditions* $\operatorname{rank}(CAB) = \operatorname{rank}(B) = \operatorname{rank}(C) = r$ *are satisfied, then*

$$X := B\tilde{V}_{V(0)}C = A^{(1,2)}_{\mathcal{R}(B),\mathcal{N}(C)}.$$

2.8 Dynamical Systems Arising from GNN Models

This section is aimed to some simplifications of the GNN models (2.3.6) and (2.3.7). Obtained dynamical systems do not follow true GNN design. They are simpler for numerical calculations. But, unfortunately, they lose global stability which is main characteristic of pure GNN models. Because of that, it will be necessary to modify defined simplified models in order to ensure global convergence again.

2.8.1 RNN Models for Computing Outer Inverses

The authors of [399] omitted the constant term $(GA)^{\mathrm{T}}$ (resp. $(AG)^{\mathrm{T}}$) from (2.3.6) (resp. (2.3.7)), and considered two dual linear RNNs, defined as follows:

$$\begin{cases} \frac{\mathrm{d}V_R(t)}{\mathrm{d}t} = -\gamma\ (GAV_R(t) - G), & V_R(0) = O,\ if\ m \geq n, \\ \frac{\mathrm{d}V_L(t)}{\mathrm{d}t} = -\gamma\ (V_L(t)AG - G), & V_L(0) = O,\ if\ m < n. \end{cases} \tag{2.8.1}$$

The dynamical evolution (2.8.1) will be termed as GNNATS2.

Practically, it is possible to consider two RNN models for computing outer inverses $A^{(2)}_{\mathcal{R}(G),\mathcal{N}(G)}$. More effective model in the case $m \geq n$ the dynamics GNNATS2 is termed as GNNATS2R≡ RNN(GA, I, G) and defined by

$$\frac{\mathrm{d}V_R(t)}{\mathrm{d}t} = -\gamma\ (GAV_R(t) - G)\,. \tag{2.8.2}$$

More effective model in the case $m < n$ is the dynamics GNNATS2, termed as GNNATS2L≡ RNN(I, AG, G) and defined by

$$\frac{\mathrm{d}V_L(t)}{\mathrm{d}t} = -\gamma\ (V_L(t)AG - G)\,. \tag{2.8.3}$$

The Simulink implementation of $GNNATS2R$ is illustrated in Fig. 2.23.

The application of the dynamic evolution design (2.8.1) assumes that the real parts of eigenvalues of GA or AG lie in the open right halfplane [399]:

$$\begin{cases} \sigma(GA) \subset \{z : \mathrm{Re}\,(z) \geq 0\}, & m \geq n, \\ \sigma(AG) \subset \{z : \mathrm{Re}\,(z) \geq 0\}, & m < n. \end{cases} \tag{2.8.4}$$

More precisely, the RNN(GA, I, G) model fails in the case when $\mathrm{Re}\,(\sigma(GA))$ contains negative values. Clearly, the model (2.8.1) is simpler than the models (2.3.6) and (2.3.7), but it loses global stability. Two approaches can be used to generate the solution in the case when (2.8.4) is not satisfied:
(i) An approach was proposed in [399], and it is based on the replacement of G by $G_0 = G(GAG)^{\mathrm{T}}G$ in the GNNATS2 evolution (2.8.1).
(ii) The second possibility is to use GNN(GA, I, G) instead of RNN(GA, I, G) dynamics.

Example 2.8.1 Consider the input matrix

$$A = \begin{bmatrix} 1 & 2 & 3 & 4 & 1 \\ 1 & 3 & 4 & 6 & 2 \\ 2 & 3 & 4 & 5 & 3 \\ 3 & 4 & 5 & 6 & 4 \\ 4 & 5 & 6 & 7 & 6 \\ 6 & 6 & 7 & 7 & 8 \end{bmatrix}$$

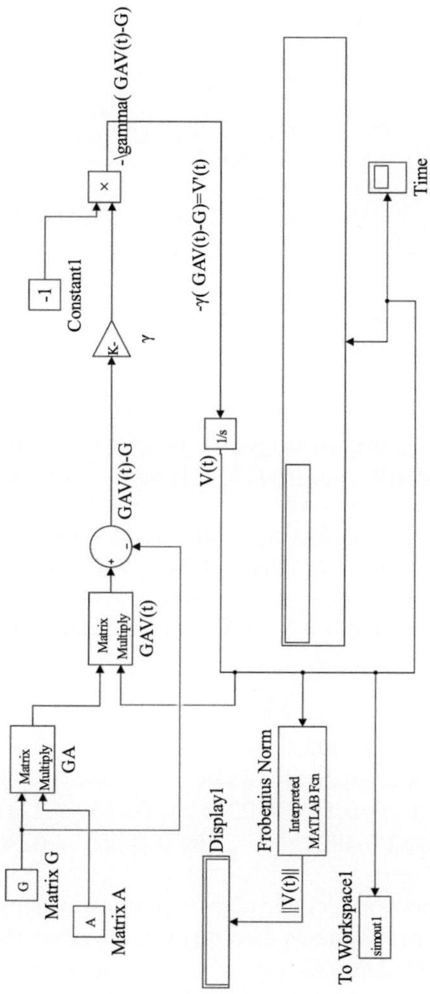

Fig. 2.23 Simulink implementation of (2.8.2)

and the joined matrix

$$G = \begin{bmatrix} 0\,0\,0\,0\,0\,0 \\ 1\,2\,1\,2\,1\,2 \\ 2\,3\,2\,3\,2\,3 \\ 3\,5\,3\,5\,3\,5 \\ 0\,1\,0\,1\,0\,1 \end{bmatrix}.$$

The model GNN(GA, I, G) with $\gamma = 10^8$ in the case $V(0) = O$ generates the result

$$X = (GA)^\dagger G$$
$$= \begin{bmatrix} -0.281213 & 0.230678 & -0.281213 & 0.230678 & -0.281213 & 0.230678 \\ -0.0760999 & 0.0719382 & -0.0760999 & 0.0719382 & -0.0760999 & 0.0719382 \\ 0.129013 & -0.0868014 & 0.129013 & -0.0868014 & 0.129013 & -0.0868014 \\ 0.334126 & -0.245541 & 0.334126 & -0.245541 & 0.334126 & -0.245541 \\ -0.32937 & 0.272295 & -0.32937 & 0.272295 & -0.32937 & 0.272295 \end{bmatrix}$$

within the time interval $[0, 10^{-5}]$. The initial state $V(0) = A^{\mathrm{T}}$ in $GNN(GA, I, G)$ under the same environment initiates the result

$$X = (GA)^\dagger G + A^{\mathrm{T}} - (GA)^\dagger GAA^{\mathrm{T}}$$
$$= \begin{bmatrix} -0.049346 & 0.0564804 & -0.327586 & 0.323424 & -0.466706 & 0.312128 \\ 0.0606421 & 0.328181 & -0.103448 & 0.126635 & -0.185493 & -0.239001 \\ 0.17063 & -0.400119 & 0.12069 & -0.0701546 & 0.0957194 & 0.209869 \\ 0.280618 & -0.128419 & 0.344828 & -0.266944 & 0.376932 & -0.34126 \\ -0.596908 & 0.357907 & -0.275862 & 0.165279 & -0.115339 & 0.293698 \end{bmatrix}.$$

On the other hand, the solution of $GNNATS2R$ under the zero initial state $V(0) = O$ is

$$A^{(2)}_{\mathcal{R}(G),\mathcal{N}(G)} = \begin{bmatrix} 0 & 0 & 0 & 0 & 0 & 0 \\ -0.1207 & 0.1092 & -0.1207 & 0.1092 & -0.1207 & 0.1092 \\ 0.3448 & -0.2644 & 0.3448 & -0.2644 & 0.3448 & -0.2644 \\ 0.2241 & -0.1552 & 0.2241 & -0.1552 & 0.2241 & -0.1552 \\ -0.5862 & 0.4828 & -0.5862 & 0.4828 & -0.5862 & 0.4828 \end{bmatrix}.$$

The recurrent neural network defined above is a linear dynamic system in matrix form. According to the linear systems theory [115], the closed-form solution of the state matrix can be described as follows:

$$\tilde{V}_G(t) = \begin{cases} \tilde{V}_R(t) := \mathrm{e}^{-\gamma GAt} V_R(0) + \gamma \mathrm{e}^{-\gamma GAt} \int_0^t \mathrm{e}^{\gamma GA\tau} G \, \mathrm{d}\tau, \; m \geq n, \\ \tilde{V}_L(t) := V_L(0)\mathrm{e}^{-\gamma AGt} + \gamma G \mathrm{e}^{-\gamma AGt} \int_0^t \mathrm{e}^{\gamma AG\tau} \, \mathrm{d}\tau, \; m < n. \end{cases} \tag{2.8.5}$$

To analyze the convergence and stability of a neural network, it is of major interest to know the eigenvalues of the matrix GA (or AG for the dual case). Using the principles from [399], it can be easily shown that

$$\lim_{t\to\infty} \mathrm{e}^{-\gamma GAt} = O. \tag{2.8.6}$$

if the matrix GA has only nonnegative eigenvalues. In dual case, $\lim_{t\to\infty} \mathrm{e}^{-\gamma AGt} = O$ if the spectrum of AG has only nonnegative eigenvalues. That fact and (2.8.5) imply the following representation for $\lim_{t\to\infty} V_G(t) = \overline{V}_G$:

$$\overline{V}_G \equiv \lim_{t\to\infty} V_G(t) = \begin{cases} \lim\limits_{t\to\infty} \gamma \mathrm{e}^{-\gamma GAt} \int_0^t \mathrm{e}^{\gamma GA\tau} G \,\mathrm{d}\tau, \ m \geq n, \\ \lim\limits_{t\to\infty} \gamma G \mathrm{e}^{-\gamma AGt} \int_0^t \mathrm{e}^{\gamma AG\tau} \,\mathrm{d}\tau, \ m < n. \end{cases} \tag{2.8.7}$$

Theorem 2.8.1 ([399]) *Let $A \in \mathbb{R}^{m\times n}$ be given matrix, $G \in \mathbb{R}_s^{n\times m}$ be arbitrary matrix satisfying $0 < s \leq r$, and $\sigma(GA) = \{\lambda_1, \lambda_2, \dots, \lambda_n\}$ be the spectrum of GA. Suppose that the condition*

$$\mathrm{Re}(\lambda_j) \geq 0, \ j = 1, 2, \dots, n \tag{2.8.8}$$

is satisfied. Then the limiting expression (2.8.7) produces the outer inverse $A^{(2)}_{\mathcal{R}(G),\mathcal{N}(G)}$, i.e.,

$$\overline{V}_{GA} = A^{(2)}_{\mathcal{R}(G),\mathcal{N}(G)}. \tag{2.8.9}$$

Proof Applying several elementary transformations and the basic property (2.3.1) of the outer inverse $A^{(2)}_{\mathcal{R}(G),\mathcal{N}(G)}$, it follows that

$$\begin{aligned} \overline{V}_{GA} &= \lim_{t\to\infty} \gamma \mathrm{e}^{-\beta GAt} \int_0^t \mathrm{e}^{\gamma GA\tau} G \,\mathrm{d}\tau \\ &= \left[\lim_{t\to\infty} \mathrm{e}^{-\gamma GAt} \int_0^t \mathrm{e}^{\gamma GA\tau} (\gamma GA) \,\mathrm{d}\tau \right] A^{(2)}_{\mathcal{R}(G),\mathcal{N}(G)} \\ &= \left[\lim_{t\to\infty} \mathrm{e}^{-\gamma GAt} \mathrm{e}^{\gamma GA\tau} \Big|_{\tau=0}^{\tau=t} \right] A^{(2)}_{\mathcal{R}(G),\mathcal{N}(G)} \\ &= \left[\lim_{t\to\infty} \mathrm{e}^{-\gamma GAt} \left(\mathrm{e}^{\gamma GAt} - I \right) \right] A^{(2)}_{\mathcal{R}(G),\mathcal{N}(G)} \\ &= \left(I - \lim_{t\to\infty} \mathrm{e}^{-\gamma GAt} \right) A^{(2)}_{\mathcal{R}(G),\mathcal{N}(G)}. \end{aligned}$$

Then (2.8.9) can be verified using (2.8.6). □

Remark 2.8.1 Analogous statement can be analogously verified when the outer inverse $A^{(2)}_{\mathcal{R}(G),\mathcal{N}(G)}$ is generated using dual equation $XAG = G$ in (2.3.1). This situation is useful in the case $m < n$.

According to Theorem 2.8.1, the application of the dynamic equation

$$\frac{\mathrm{d}V_G(t)}{\mathrm{d}t} = -\gamma\mathcal{H}\left(\nabla_G(A)\right) = \begin{cases} -\gamma\,\mathcal{H}\left(GAV_G(t) - G\right), & V_G(0) = O,\ if\ m \geq n, \\ -\gamma\,\mathcal{H}\left(V_G(t)AG - G\right), & V_G(0) = O,\ if\ m < n. \end{cases} \tag{2.8.10}$$

where $V_G(t)$ is a matrix of activation state variables, γ is a positive scaling constant, and O is an appropriate zero matrix. The convergence of dynamic equation (2.8.10) is conditioned by the restriction (2.8.4) on the spectrum of the matrix GA or AG:

$$\begin{cases} \sigma(GA) \subset \{z : Re(z) \geq 0\},\ m \geq n, \\ \sigma(AG) \subset \{z : Re(z) \geq 0\},\ m < n. \end{cases} \tag{2.8.11}$$

More precisely, the first RNN approach used in [399] fails in the case when $\mathrm{Re}\,(\sigma(GA))$ contains negative values.

The neural network used in our implementation is composed from a number of independent subnetworks, in the similar way as it has already been discussed in [272, 273]. Specifically, the number of subnetworks is m if $m \geq n$ or it is equal to n if $m < n$. The connection weight matrix W of the neurons is identical in each subnetwork and defined as

$$W = \begin{cases} -\gamma GA,\ m \geq n \\ -\gamma AG,\ m < n. \end{cases}$$

Note that the size of the connection weight matrix is $\min\{m, n\} \times \min\{m, n\}$ which usually has fewer connections than being defined otherwise. The threshold (input) matrix of the neuron array is $-\gamma G$.

Consider the case $m \geq n$. Let us denote by $v_j(t)$ (resp. g_j) the jth column vector of $V_G(t)$ (resp. G), for $j = 1, 2, \ldots, m$. The dynamics of the jth subnetwork can be expressed in the general form

$$\frac{\mathrm{d}v_j(t)}{\mathrm{d}t} = -\gamma\,\left(GAv_j(t) - g_j\right). \tag{2.8.12}$$

Each subnetwork exploits the same connection weight matrix equal to $W = -\gamma GA$ and $\gamma g_j = \{\gamma g_{1j}, \gamma g_{2j}, \ldots, \gamma g_{nj}\}$ is the biasing threshold vector for the jth subnetwork.

Elements v_{ij} of unknown matrix $V_G(t)$ are computed using

$$\dot{v}_{ij} = \frac{\mathrm{d}v_{ij}}{\mathrm{d}t} = \sum_{k=1}^{n} w_{ik}v_{kj} + \gamma g_{ji},\ i = 1, 2, \ldots, n;\ j = 1, 2, \ldots, m, \tag{2.8.13}$$

where w_{ij} are elements of W. It is important to mention that the column vector $v_j = \{v_{1j}, v_{2j}, \ldots, v_{nj}\}$ is output in the jth subnetwork.

2.8.2 *Simplified RNN for Computing the Drazin Inverse*

Particularly, a simplification of the GNN model which is applicable in computing the Drazin inverse A^{D} was proposed in [251]. This model can be derived by removing the constant term $\left(A^{k+1}\right)^{\mathrm{T}}$ in GNN(A^k, A, A^k), $k \geq \mathrm{ind}(A)$. We can rewrite the equation (1^k) in the form

$$A^{k+1}V_D(t) - A^k = 0, \tag{2.8.14}$$

where $m \geq \mathrm{ind}(A)$, and $V_D \in \mathbb{R}^{n\times n}$ denotes the unknown matrix to be solved which corresponds to the matrix A^{d}. To solve for $V_D(t)$ via dynamic system approach, we can define a scalar-valued norm based error function:

$$\varepsilon_D(t) = \frac{\|A^{k+1}V_D(t) - A^k\|_F^2}{2}, \quad k \geq \mathrm{ind}(A).$$

Note that minimal value $\varepsilon(\overline{t}) = 0$ of the residual-error function $\varepsilon(t)$ is achieved in a minimum point $\overline{V} = V(\overline{t})$ if and only if $V(\overline{t})$ is the exact solution of (2.8.14). A computational scheme could be designed to evolve along a descent direction of this error function $E(t)$, until the minimum point $\overline{V}$ is reached. The derivative of E with respect to $V \in \mathbb{R}^{n\times n}$ could simply be derived as [75, Chap. 5]

$$\frac{\partial \varepsilon_D(t)}{\partial V_D} = \left(A^{k+1}\right)^{\mathrm{T}}\left(A^{k+1}V_D(t) - A^k\right). \tag{2.8.15}$$

GNN for Computing Drazin Inverse with Restriction on Spectrum

A particular RNN for computing the Drazin inverse A^{D} was proposed in [251]. This model can be derived removing the first constant term in GNN(A^k, A, A^k), $k \geq \mathrm{ind}(A)$, and it is defined as the RNN(A^k, A, A^k) dynamics

$$\dot{V}_D(t) = -\gamma\left(A^{k+1}V_D(t) - A^k\right), \quad k \geq \mathrm{ind}(A), \quad V_D(0) = V_0. \tag{2.8.16}$$

Accordingly, an application of the model (2.8.16) is conditioned by

$$\mathrm{Re}\left(\lambda_j^{k+1}\right) \geq 0, \quad j = 1, \ldots, n, \tag{2.8.17}$$

where $\sigma(A^{k+1}) = \{\lambda_1^{k+1}, \ldots, \lambda_n^{k+1}\}$ is the spectrum of A^{k+1}, $k \geq \mathrm{ind}(A)$ and $\{\lambda_1, \ldots, \lambda_n\}$ is the spectrum of A [251]. One method to resolve the limitation (2.8.17) is proposed in [251], and it is based on the possibility to find an appropriate power k such that (2.8.17) holds. Another possibility to ensure the nonnegativity of the spectrum of the form (2.8.17) was proposed in [252], and it is based on the usage of the matrix $A^k \left(A^{2k+1}\right)^{\mathrm{T}} A^k$, $k = \mathrm{ind}(A)$.

One important case in (2.8.16) is $G = A^k,\ k \geq \mathrm{ind}(A)$, which defines a recurrent neural network for computing the Drazin inverse of a real square matrix in real time. This case was considered in [251]. The main goal of [251] was to find the first integer $k \geq 0$ such that

$$\sigma\left(A^{k+1}\right) \subset \{z : Re(z) \geq 0\}. \tag{2.8.18}$$

In [73] are found conditions ensuring that the nonzero spectrum of A^{k+1} lies in the open left/right half of the complex plane as an equivalent to the condition that the nonzero spectrum of A lies in the union of $k+1$ angular regions. Therefore, the authors in [73] found conditions for the arguments of eigenvalues. If the required conditions are not satisfied, the Drazin inverse cannot be generated using the neural network approach. A different approach was proposed in [251]: find corresponding conditions for the matrix exponent k.

Theorem 2.8.2 ([251]) *Let $A \in \mathbb{R}^{n\times n}$ be given matrix, $\sigma(A) = \{\lambda_1, \ldots, \lambda_n\}$ be the spectrum of A, and $k \geq \mathrm{ind}(A)$. Suppose that the condition (2.8.17) is satisfied. Then the limit of GNN(A^k, A, A^k), $k \geq \mathrm{ind}(A)$ as $t \to \infty$ produces the Drazin inverse A^d, i.e.,*

$$\overline{V} = A^D.$$

There are several cases, depending on eigenvalues of A, for selecting the parameter m which ensures that eigenvalues of the matrix A^{k+1} contain nonnegative real parts. These cases are discussed in Theorem 2.8.3. Before the main results, we present several supporting facts in Lemma 2.8.1 and some notations.

Lemma 2.8.1 ([251]) *Let $A \in \mathbb{R}^{n\times n}$ be given matrix, $\lambda_j \in \sigma(A)$ and k be a fixed positive integer. Then the condition $\mathrm{Re}\left(\lambda_j^k\right) \geq 0$ is ensured in the following cases:*

Case 1. $\lambda_j \in \mathbb{R}^+$.
Case 2. $\lambda_j \in \mathbb{R}$, *k is even.*
Case 3. λ_j *and k satisfy*

$$\varphi_j = \mathrm{Arg}\,\lambda_j = \pm\frac{\pi}{2},\quad k \in \{4l : l \in \mathbb{N}^+\}. \tag{2.8.19}$$

Case 4. λ_j *satisfies*

$$\lambda_j \in \mathbb{C} \cap \{z = r_j e^{\imath\varphi_j} : 0 < |\varphi_j| < \pi\} \tag{2.8.20}$$

and the parameter k satisfies

$$\frac{4s-1}{2}\frac{\pi}{|\varphi_j|} < k < \frac{4s+1}{2}\frac{\pi}{|\varphi_j|},\quad s \in \mathbb{N}_0, \tag{2.8.21}$$

Proof The proof in Cases 1-3 is evident. In Case 4 it follows

$$\lambda_j = a_j + \imath\, b_j = r_j \left(\cos(\varphi_j) + \imath \sin(\varphi_j)\right),$$

where

$$r_j = |\lambda_j| = \sqrt{a_j^2 + b_j^2},$$

$$\varphi_j = \begin{cases} -1, & a_j = 0, b_j = 0 \\ \frac{\pi}{2}, & a_j = 0, b_j > 0 \\ -\frac{\pi}{2}, & a_j = 0, b_j < 0 \\ \arctan\left(\frac{b_j}{a_j}\right), & a_j \neq 0. \end{cases}$$

The real part of

$$\lambda_j^k = r_j^k \left(\cos(k\,\varphi_j) + \imath \sin(k\,\varphi_j)\right)$$

is equal to

$$\mathrm{Re}\big(\lambda_j^k\big) = r_j^k \cos(k\varphi_j).$$

Therefore, $\mathrm{Re}\left(\lambda_j^k\right) > 0$ is satisfied in the case $\cos(k\,|\varphi_j|) > 0$, or equivalently

$$-\frac{\pi}{2} + 2s\pi < k\,|\varphi_j| < \frac{\pi}{2} + 2s\pi, \quad s \in \mathbb{N}_0.$$

Therefore, the condition

$$\frac{4s-1}{2}\frac{\pi}{|\varphi_j|} < k < \frac{4s+1}{2}\frac{\pi}{|\varphi_j|}, \quad s \in \mathbb{N}_0 \tag{2.8.22}$$

ensures $\mathrm{Re}(\lambda_j^k) \geq 0$ and (2.8.22) in conjunction with $\gamma > 0$ implies $\gamma\,\mathrm{Re}\left(\lambda_j^k\right) \geq 0$. Therefore, (2.8.21) must hold. The length of interval (2.8.21) which limits values for k is equal to $\frac{\pi}{|\varphi_j|}$ so that the condition

$$\frac{\pi}{|\varphi_j|} > 1 \tag{2.8.23}$$

must be satisfied. Therefore, necessity of (2.8.20) and (2.8.21) is verified. □

Theorem 2.8.3 ([251]) *Let $A \in \mathbb{R}^{n\times n}$ be given matrix of index $k = \mathrm{ind}(A)$ whose eigenvalues are contained in $\sigma(A) = \{\lambda_j,\ j = 1, \ldots, n\}$. Let us denote by M_2, M_3, M_4 sets of values of the parameter m defined by the cases* C2, C3, C4 *from Lemma 2.8.1. The recurrent neural network defined by (2.8.16) is globally stable and the steady-state matrix of the recurrent neural network is equal to A^{D}, i.e.,*

$$\lim_{t\to\infty} V(t) = A^{\mathrm{D}}$$

in the following cases:

Case 1. $\sigma(A) \subset \mathbb{R}^{+}$, $k \geq \operatorname{ind}(A)$.
Case 2. $\sigma(A) \subset \mathbb{R}$ *and* $\lambda_j < 0$, *for at least one index* j, m *satisfies* $k \geq \operatorname{ind}(A)$, $k+1$ *is even.*
Case 3. *The spectrum of* A *satisfies*

$$\sigma(A) \subset \mathbb{C} \cap \{z = r\, e^{\iota\theta} : 0 < |\theta| < \pi\}, \tag{2.8.24}$$

parameter m *satisfies the following condition (2.8.25)*

$$k+1 \geq \max\{\operatorname{ind}(A)+1, \min(M_2 \cap M_3 \cap M_4)\}. \tag{2.8.25}$$

Proof It is necessary to ensure existence of condition (2.8.18). Also, to ensure the existence of A^{d}, the condition $k \geq \operatorname{ind}(A)$ must be satisfied in all cases. The proof in Case 1 and Case 2 is obvious from Lemma 2.8.1, since $\sigma(A^{k+1}) = \{\lambda_j^{k+1},\ j = 1,\ldots,n\}$. In Case 1 and Case 2 the conditions (2.8.18) are satisfied. According to Lemma 2.8.1, the condition (2.8.24) ensures $\operatorname{Re}\left(\lambda_j^k\right) \geq 0$. Finally, (2.8.24) and (2.8.25) ensure (2.8.18). □

Examples of Simplified GNN for Computing the Drazin Inverse

In order to check the validity and performance of the neural network approach to the computation of the Drazin inverse some computer simulations have been done. Very good agreement with the theoretical considerations have been obtained. In all examples it is assumed that A^{d} denotes the exact Drazin inverse of A and A^{d} denotes its approximation derived by our NN approach. In the case $\operatorname{ind}(A) = 1$, by A^{g} we denote exact group inverse of A and A^{d} denotes its approximation.

Example 2.8.2 Find the group inverse matrix of the singular matrix

$$A = \begin{bmatrix} 1 & 0 & -1 \\ 0 & 1 & 0 \\ 0 & -1 & 0 \end{bmatrix}$$

with the index $\operatorname{ind}(A) = 1$. It is easy to check that the group inverse of matrix A has the form

$$A^{g} = \begin{bmatrix} 1 & -2 & -1 \\ 0 & 1 & 0 \\ 0 & -1 & 0 \end{bmatrix}.$$

The computer simulation has been done on the basis of the system of differential equations (2.8.1). Since $\operatorname{ind}(A) = 1$ and $\sigma(A) = \{1, 1, 0\}$, we can choose $k = 1$, so the weighted matrix $A^{k+1} = A^2$ has eigenvalues $\sigma(A^2) = \{1, 1, 0\}$. The learning rate γ was chosen as $\gamma = 10^7$. Therefore, the weighted matrix $-\gamma A^{k+1} = -10^7 \cdot A^2$ possesses the eigenvalues $\{-10^7, -10^7, 0\}$. Because all the eigenvalues are non-positive, neural network defined by (2.8.1) is globally stable. Convergence behavior of the network in time 10^{-6} seconds (shortly 10^{-6}s is illustrated in Fig. 2.24.

The resulting matrix

$$A^d = \begin{bmatrix} 1 & -1.9990 & -1.0000 \\ 0 & 1.0000 & 0 \\ 0 & -1.0000 & 0 \end{bmatrix}$$

has been obtained after 10^{-6}s, which is in good agreement with the exact solution. Figure 2.24 also shows that the states of neural network stabilize after approximately $5/(10000000) = 0.0000005$s.

Example 2.8.3 Consider the matrix

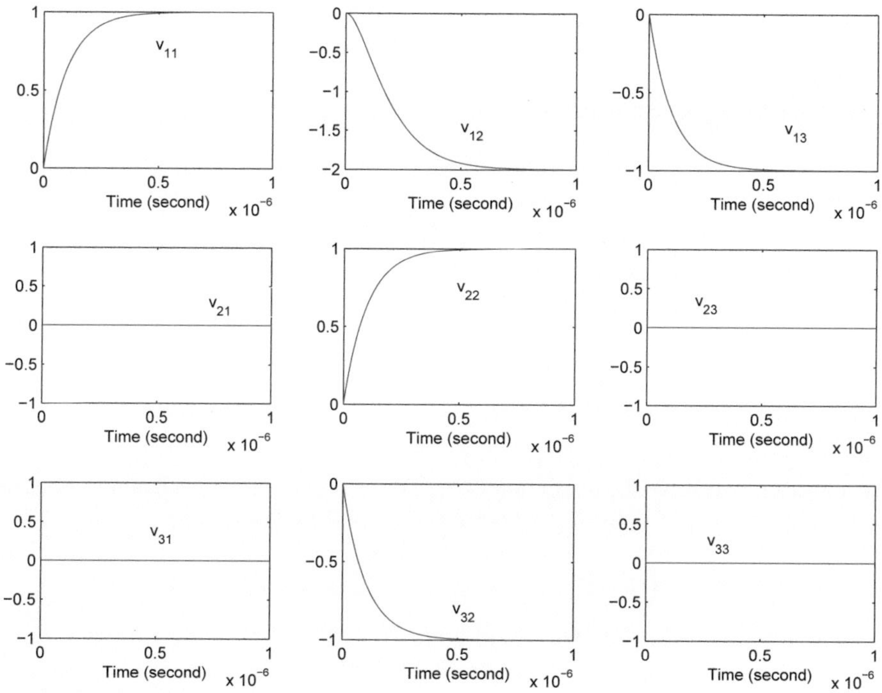

Fig. 2.24 Convergence behavior of the network in time 10^{-6}s

$$A = \begin{bmatrix} 2 & -1.6 & 5.6 & -5.6 & 0 & 5.6 \\ 0 & 1 & 6 & -6 & 0 & 6 \\ 0 & 0 & 4 & -4 & 0.1 & 3.9 \\ 0 & 0 & 0 & 0 & 0.1 & -0.1 \\ 0 & 0 & 0 & 0 & 0.1 & 0.1 \\ 0 & 0 & 0 & 0 & 0 & 0 \end{bmatrix}$$

with $\mathrm{ind}(A) = 3$. The Drazin inverse of A is equal to

$$A^{\mathrm{d}} = \begin{bmatrix} 0.5000 & 0.8000 & -1.9000 & 1.9000 & 0 & -1.9000 \\ 0 & 1.0000 & -1.5001 & 1.5001 & 0 & -1.5001 \\ 0 & 0 & 0.2500 & -0.2500 & 0 & 0.2500 \\ 0 & 0 & 0 & 0 & 0 & 0 \\ 0 & 0 & 0 & 0 & 0 & 0 \\ 0 & 0 & 0 & 0 & 0 & 0 \end{bmatrix}.$$

The eigenvalues of A are $\{4, 2, 1, 0, 0, 0\}$, so we can select $k = \mathrm{ind}(A) = 3$, because the matrix A^4 has all nonnegative eigenvalues. We choose $\gamma = 10^7$, so the connection weight matrix $W = -10^7 \cdot A^4$ has all non-positive eigenvalues. Trajectories for zero initial conditions are depicted in Fig. 2.25 and they show convergence behavior in time 10^{-7}s.

The matrix produced by our algorithm is

$$A^{d} = \begin{bmatrix} 0.5000 & 0.8000 & -1.9000 & 1.9000 & 0 & -1.9000 \\ 0 & 1.0000 & -1.5000 & 1.5000 & 0 & -1.5000 \\ 0 & 0 & 0.2500 & -0.2500 & 0 & 0.2500 \\ 0 & 0 & 0 & 0 & 0 & 0 \\ 0 & 0 & 0 & 0 & 0 & 0 \end{bmatrix}.$$

2.8.3 Simplified GNN Without Restrictions on Spectrum

An approach to resolve the restriction (2.8.11) was proposed in [399], and it is based on the replacement of G by $G_0 = G(GAG)^{\mathrm{T}}G$ in (2.8.2) or (2.8.3):

$$\begin{cases} \frac{\mathrm{d}V_{G_0}(t)}{\mathrm{d}t} = -\gamma\left(G_0 A V_{G_0}(t) - G_0\right), \; V_{G_0}(0) = V_0, \; if\ m \geq n, \\ \frac{\mathrm{d}V_{G_0}(t)}{\mathrm{d}t} - \gamma\left(V_{G_0}(t) A G_0 - G_0\right), \; V_{G_0}(0) = V_0, \; if\ m < n, \end{cases} \tag{2.8.26}$$

where $G_0 = G(GAG)^{\mathrm{T}}G$. But, this approach requires additional matrix multiplications during the computation of the matrix G_0 in (2.8.26). This approach eliminates the requirements (2.8.11), at the cost of increasing the number of matrix operations. Also, the numbers in G_0 grow, which could cause numerical instability.

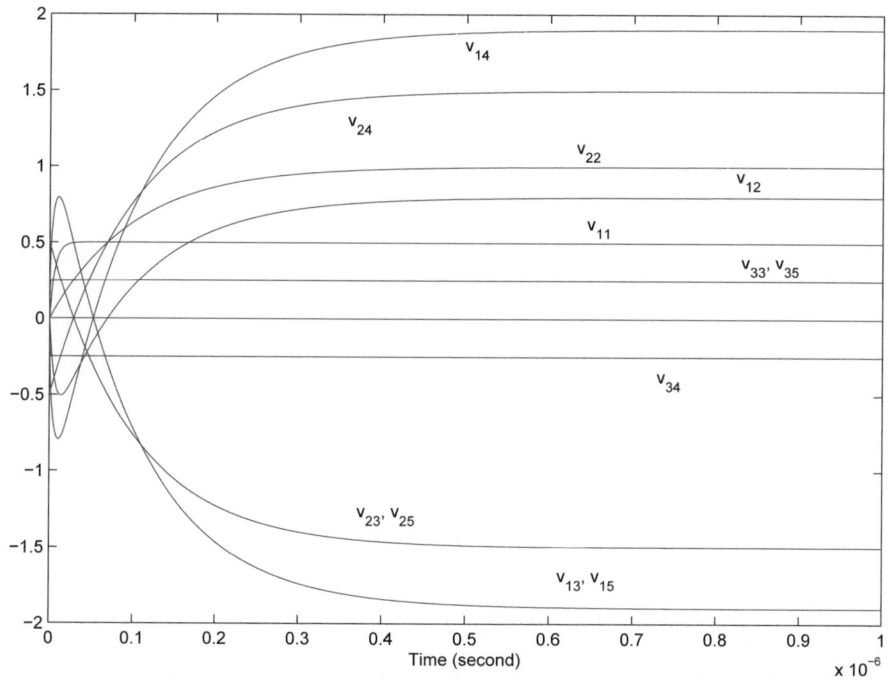

Fig. 2.25 Convergence behavior of the network in time 10^{-7}s

The RNN based on (2.8.10) is composed from a number of independent subnetworks, in the similar way as it has already been discussed in [272, 273]. Specifically, the number of subnetworks is m if $m \leq n$ or it is equal to n if $m > n$. These subnetworks can operate concurrently, so parallel and distributed processing can be achieved. In this way, the computational advantages over the existing sequential algorithms can be attained in real-time applications. The conditions which ensure stability of the proposed neural network are presented. The results corresponding to the most common generalized inverses are derived in particular cases.

Remark 2.8.2 Usage of the matrix product $G_0 = G(GAG)^{\mathrm{T}}G$ sometimes causes numerical instability. But, despite the disadvantage, the gradient based RNN and the dynamical state equation based on G_0 offer a solution of the problem and is capable to generate $A^{(2)}_{\mathcal{R}(G),\mathcal{N}(G)}$. The solution has its own drawbacks, it is charged by an increased number of matrix multiplications, but in some way overcomes the problem and generates an approximation of $A^{(2)}_{\mathcal{R}(G),\mathcal{N}(G)}$.

2.8.4 Globally Convergent Simplified GNN for Computing Drazin Inverse

The stability of the gradient-based neural networks for computing the regular matrix inverse as well as the Moore–Penrose inverse [272, 273] is warranted, as the matrix $M = A^{\mathrm{T}}A \in \mathbb{R}^{n\times n}$ possesses the spectrum $\sigma(M) = \{\lambda_1, \lambda_2, \ldots, \lambda_n\}$ which satisfies

$$\lambda_j \geq 0,\ j = 1, \ldots, n. \tag{2.8.27}$$

Regarding weighted Moore–Penrose inverse [291] the stability is guaranteed because the spectrum $\sigma(M) = \{\lambda_1, \ldots, \lambda_n\}$ of the matrix $M = A^{\sharp}A \in \mathbb{R}^{n\times n}$, $A^{\sharp} = N^{-1}A^{\mathrm{T}}M$, also satisfies (2.8.27).

In addition, the conditions that need to be applied to the spectrum of A^m guarantee the stability of the GNN initiated in [251, Lemma 1]. In the case of the Drazin inverse, the stability is guaranteed when the spectrum $\sigma(M) = \{\lambda_1, \lambda_2, \ldots, \lambda_n\}$ of the matrix $M = A^kA = A^{k+1} \in \mathbb{R}^{n\times n}$ fulfills

$$\mathrm{Re}\big(\lambda_j\big) \geq 0,\ j = 1, \ldots, n. \tag{2.8.28}$$

A major result of [251] was the algorithm, suggested in Lemma 4, for obtaining an suitable exponent k such that the power A^{m+1} satisfies (2.8.28). But, that algorithm may request an unbounded in advance exponent k. Now, we present an alternative approach developed in [252]. Basically, rather than using the matrix A^k, in which the parameter k is selected like in [251, Lemma 4], our current intention is to use the matrix $A^k\left(A^{2k+1}\right)^{\mathrm{T}}A^k$, $k = \mathrm{ind}(A)$. That matrix requires relatively great exponent A^{4k+1}, $k = \mathrm{ind}(A)$. However, this exponent is fixed and predefined in advance. Necessary details will be explained in the next section.

Neural Network Architecture of GNN for Computing the Drazin Inverse

Leading idea of this subsection is a further representation of the Drazin inverse A^{D} from [74].

Proposition 2.8.1 ([74]) *Assume* $A \in \mathbb{C}^{n\times n}$ *and* $k = \mathrm{ind}(A)$. *Consequently*

$$A^D = \int_0^\infty \exp\Big(-A^k\left(A^{2k+1}\right)^T A^{k+1}\tau\Big)\, A^k\left(A^{2k+1}\right)^T A^k\, \mathrm{d}\tau. \tag{2.8.29}$$

Remark 2.8.3 As we have seen in the previous chapter, the general property of GNNs for computing generalized inverses is centered on a predominant notion which necessitates to solve one specific matrix equation via dynamic-system method. Further, the dynamic-system method is determined using an error function (denoted by $E(t)$) which is based on matrix norm, therefore it has scalar values. Consequently, the suggested method attempts to identify the minimal value of the residual-norm function $E(t)$ utilizing the design schema $\mathrm{d}V(t)/\mathrm{d}t = -\gamma \partial E(t)/\partial V$. Hence, the main concept utilized in [251, 273, 290] is the same as the leading principle from [300]: using a matrix norm optimization problem as a tool to solve the generalized inverse problem.

As the main concepts, we will use the integral representations of the Drazin inverse in addition to the dynamic state equation which is outlined as a suitably determined error function $E(t)$ based on matrix norm. For simplicity, we utilize the substitution $G = A^k \left(A^{2k+1}\right)^{\mathrm{T}} A^k$, $k = \mathrm{ind}(A)$. It is clear that

$$GAA^{\mathrm{D}} = G$$

is fulfilled. Consequently, the subsequent matrix equation in relation to unknown state matrix $V(t)$ could be outlined:

$$GAV(t) - G = 0. \tag{2.8.30}$$

The norm based error function relating to (2.8.30) is defined as

$$\varepsilon(t) = \frac{\|GAV(t) - G\|_F^2}{2}. \tag{2.8.31}$$

Derivative $-\partial\varepsilon(t)/\partial V$ defines the gradient descent direction of $E(t)$is represented as (see, for example, [75, Chapter 5])

$$\frac{\partial\varepsilon(t)}{\partial V} = (GA)^{\mathrm{T}} \left(GAV(t) - G\right). \tag{2.8.32}$$

To make the equation simpler, the authors of [252] omitted the constant factor $(GA)^{\mathrm{T}}$ and the dynamics corresponding to the integral representation (2.8.29) can be represented

$$\begin{aligned} \frac{\mathrm{d}V(t)}{\mathrm{d}t} &= -\gamma G\left(AV(t) - I\right), \quad G = A^k \left(A^{2k+1}\right)^{\mathrm{T}} A^k, \quad k = \mathrm{ind}(A), \\ V(0) &= V_0. \end{aligned} \tag{2.8.33}$$

$V(t)$ denotes a matrix of activation states, $t \in [0, +\infty)$, $\gamma > 0$ denotes a scaling constant [272, 274].

The recurrent neural network (2.8.33) is a linear dynamic system in matrix form. According to the linear systems theory [115], the closed-form solution of the state matrix can be described as follows:

$$V(t) = \mathrm{e}^{-\gamma GAt} V(0) + \gamma \mathrm{e}^{-\gamma GAt} \int_0^t \mathrm{e}^{\gamma GA\tau} G \, \mathrm{d}\tau. \tag{2.8.34}$$

Remark 2.8.4 If the matrix A is positive definite (which means all its eigenvalues real and positive), then all the matrices A^m, $m \in \mathbb{N}$ are also positive definite.

Therefore the matrix $GA = A^k \left(A^{2k+1}\right)^{\mathrm{T}} A^{k+1}$, $k = \mathrm{ind}(A)$, is also positive definite in this case, so it has all eigenvalues real and positive. Since $\gamma > 0$, all eigenvalues of $-\gamma GA$ are real and negative, and according to the linear dynamic system theory, the recurrent neural network is asymptotically stable in the large. Therefore, the first matrix term in the right-hand side of (2.8.34) approaches the zero matrix of the same size as time approaches infinity, regardless of the initial states; i.e., for arbitrary $V(0)$

$$\lim_{t\to\infty} \exp(-\gamma GAt)\, V(0) = 0. \tag{2.8.35}$$

In the singular case, (2.8.35) is not true. Therefore, the steady-state matrix of the above recurrent neural network depends on the initial state matrix. Lemma 2.8.2 provides necessary conditions that (2.8.35) holds, even in the case when the matrix A is singular. This result is a generalization of Lemma 1 from [251] and can be proved in the same way.

Lemma 2.8.2 along with Proposition 2.8.2 provides necessary conditions that, for arbitrary $V(0)$, the first matrix expression in the right-hand side of (2.8.34) converges to the zero matrix as the time approaches infinity, i.e.,

$$\lim_{t\to\infty} \mathrm{e}^{-\gamma GAt}\, V(0) = 0. \tag{2.8.36}$$

Lemma 2.8.2 *Let $M \in \mathbb{R}^{n\times n}$ be given matrix, $\sigma(M) = \{\lambda_1, \lambda_2, \ldots, \lambda_n\}$ be the spectrum of M, and $t \in [0, +\infty)$. Assume that $\sigma(M)$ satisfies (2.8.28). Consequently, the subsequent limiting expression persists:*

$$\lim_{t\to\infty} \mathrm{e}^{-Mt} = 0. \tag{2.8.37}$$

Proof This result is a generalization of Lemma 1 from [251] and can be proved in the same way. □

Proposition 2.8.2 *The matrix $GA = A^k \left(A^{2k+1}\right)^T A^{k+1}$, $k = \mathrm{ind}(A)$, from (2.8.34) has a nonnegative spectrum, i.e., $\sigma(GA) = \{\lambda_1, \ldots, \lambda_n\}$ satisfies $\lambda_j \geq 0$, $j = 1, \ldots, n$, for any provided matrix $A \in \mathbb{R}^{n\times n}$.*

Proof The proof of this statement has been shown in the proof of Theorem 2.1 in [74]. □

Corollary 2.8.1 *The limit expression (2.8.36) is true for any matrix $A \in \mathbb{R}^{n\times n}$.*

At this point, Eqs. (2.8.34) and (2.8.36) suggest the following:

$$\overline{V} = \lim_{t\to\infty} V(t) = \lim_{t\to\infty} \gamma \mathrm{e}^{-\gamma GAt} \int_0^t \mathrm{e}^{\gamma GA\tau} G \,\mathrm{d}\tau. \tag{2.8.38}$$

On the basis of the matrix exponential definition and properties, we verify that the closed-form solution of $\overline{V}$ in (2.8.38) fulfills (1^k), (2), (5); also, we show the independence of the γ parameter. Consequently, (2.8.38) and

$$A^{\mathrm{D}} = \int_0^\infty \mathrm{e}^{-A^{k+1}\tau} A^k \, \mathrm{d}\tau. \tag{2.8.39}$$

produce the same result, A^{d}.

Theorem 2.8.4 ([252]) *Assume* $A \in \mathbb{R}^{n\times n}$ *is given and* $k = \mathrm{ind}(A)$. *The limiting expression (2.8.38) generates the Drazin inverse* A^D, *i.e.,*

$$\overline{V} = A^D.$$

Proof The elementary property (1^k) of the Drazin inverse suggests that the following holds

$$G = A^k \left(A^{2k+1}\right)^{\mathrm{T}} A^k = A^k \left(A^{2k+1}\right)^{\mathrm{T}} A^{k+1} A^{\mathrm{D}} = GAA^{\mathrm{D}},$$

which, in conjunction with (2.8.38), implies

$$\begin{aligned}
\overline{V} &= \lim_{t\to\infty} \gamma \exp\!\left(-\gamma A^k \left(A^{2k+1}\right)^{\mathrm{T}} A^{k+1} t\right) \int_0^t \exp\!\left(\gamma A^k \left(A^{2k+1}\right)^{\mathrm{T}} A^{k+1}\tau\right) A^k \left(A^{2k+1}\right)^{\mathrm{T}} A^k \, \mathrm{d}\tau \\
&= \lim_{t\to\infty} \exp(-\gamma G A t) \int_0^t \exp(\gamma G A \tau)\, \gamma G A A^{\mathrm{D}} \, \mathrm{d}\tau \\
&= \left[\lim_{t\to\infty} \exp(-\gamma G A t) \int_0^t \mathrm{d}(\exp(\gamma G A\, \tau))\right] A^{\mathrm{D}} \\
&= \left[\lim_{t\to\infty} \exp(-\gamma G A t) \exp(\gamma G A \tau) \Big|_{\tau=0}^{\tau=t}\right] A^{\mathrm{D}} \\
&= \left\{\lim_{t\to\infty} \exp(-\gamma G A t) \left[\exp(\gamma G A t) - I\right)\right]\right\} A^{\mathrm{D}} \\
&= \left[I - \lim_{t\to\infty} \exp(-\gamma G A t)\right] A^{\mathrm{D}}.
\end{aligned}$$

Applying Corollary 2.8.1, we get $\lim_{t\to\infty} \exp(-\gamma G A t) = 0$ and $\overline{V} = A^{\mathrm{D}}$. □

Setting zero initial states will force the first matrix expression in the right-hand side of the (2.8.34) to be zero, which can simplify the equation and accelerate the convergence. So, it is beneficial to use $V(0) = O$. In such case, the dynamic state equation of the recurrent neural network for generating the Drazin inverse could be outlined in the following way:

$$\dot{V}(t) = -\gamma G\left(AV(t) - I\right), \quad V(0) = O, \; G = A^k \left(A^{2k+1}\right)^{\mathrm{T}} A^k, \; k = \mathrm{ind}(A). \tag{2.8.40}$$

Before the examining the stability of the equilibrium state $\overline{V}$, we restate the next result from [1].

Lemma 2.8.3 ([1]) *If* P *and* Q *are two positive semidefinite matrices, the following trace inequality is valid:*

$$\mathrm{Tr}(PQ) \geq 0. \tag{2.8.41}$$

Theorem 2.8.5 ([252]) *The recurrent neural network described in (2.8.40) is stable in the sense of Lyapunov.*

Proof We utilize Lyapunov's approach (see [193, Chap. 2.9]) to verify the stability. The Lyapunov function candidate can be outlined in the following way:

$$\varepsilon(t) = \frac{\|GAV - G\|_F^2}{2} = \frac{\mathrm{Tr}\left((GAV - G)^{\mathrm{T}} (GAV - G)\right)}{2}. \tag{2.8.42}$$

The inequality $\varepsilon(t) \geq 0$ is evidently correct.

Using common matrix calculus (see, for example [75, Chap. 5]), applied on $E(t)$, one can obtain

$$\begin{aligned}
\frac{\partial \varepsilon(t)}{\partial V} &= \frac{1}{2}\frac{\partial \mathrm{Tr}[V^{\mathrm{T}}(GA)^{\mathrm{T}}GAV]}{\partial V} - \frac{1}{2}\frac{\partial \mathrm{Tr}[V^{\mathrm{T}}(GA)^{\mathrm{T}}G]}{\partial V} - \frac{1}{2}\frac{\partial \mathrm{Tr}[G^{\mathrm{T}}GAV]}{\partial V} \\
&= \frac{1}{2}\left\{\left[(GA)^{\mathrm{T}}GA\right] + \left[(GA)^{\mathrm{T}}GA\right]^{\mathrm{T}}\right\} V - (GA)^{\mathrm{T}}G \\
&= (GA)^{\mathrm{T}}GAV - (GA)^{\mathrm{T}}G \\
&= (GA)^{\mathrm{T}}(GAV - G).
\end{aligned}$$

Therefore, we obtain the time derivative of $E(t)$:

$$\begin{aligned}
\frac{\mathrm{d}E(t)}{\mathrm{d}t} &= \mathrm{Tr}\left(\left(\frac{\partial E}{\partial V}\right)^{\mathrm{T}} \frac{\mathrm{d}V}{\mathrm{d}t}\right) \\
&= \mathrm{Tr}\left((GAV - G)^{\mathrm{T}}\, GA\, (-\gamma\, (GAV - G))\right) \\
&= -\gamma \mathrm{Tr}\left(GA\, (GAV - G)\, (GAV - G)^{\mathrm{T}}\right).
\end{aligned} \tag{2.8.43}$$

Over the remainder of the proof, we use the Hermitian part of a square real matrix GA, defined as

$$H(GA) = \frac{1}{2}\left(GA + (GA)^{\mathrm{T}}\right).$$

As $(GAV - G)(GAV - G)^{\mathrm{T}}$ and $H(GA)$ are symmetric, following some matrix transformation it can be proven that

$$\frac{\mathrm{d}E(t)}{\mathrm{d}t} = -\gamma \mathrm{Tr}\left(H(GA)\, (GAV - G)\, (GAV - G)^{\mathrm{T}}\right).$$

In the spirit of the known property that all eigenvalues of a square matrix are equal to the eigenvalues of its transpose, the usage of Proposition 2.8.2 suggests $\lambda_j \geq 0$, for every $\lambda_j \in \sigma\,(H(GA))$. In addition, as $H(GA)$ is symmetric, therefore it is positive semidefinite. In the end, as both the matrices $H(GA)$ and $(GAV - G)(GAV - G)^{\mathrm{T}}$ are positive semidefinite, a further usage of Proposition 2.8.3 instantly indicates $\frac{\mathrm{d}\varepsilon(t)}{\mathrm{d}t} \leq 0$. Therefore, according to the Lyapunov stability theory, neural network (2.8.40) satisfies the property of stability. Thus, the verification is finalized. □

The configuration of a neural network is similar to what it has already been discussed in [272, 273]. It is composed from a number of independent subnetworks where each subnetwork is comprised of n neurons with linear activation function, which are organized in one layer. Each layer corresponds to a column vector of $V(t)$. We denote $v_j(t)$ (resp. a_j) as the jth column vector of $V(t)$ (resp. $A^k\left(A^{2k+1}\right)^{\mathrm{T}}A^k$), $j = 1, 2, \ldots, n$. The dynamic state equation of the jth subnetwork can be outlined in the following way:

$$\frac{\mathrm{d}v_j(t)}{\mathrm{d}t} = -\gamma\left(A^k\left(A^{2k+1}\right)^{\mathrm{T}}A^{k+1}v_j(t) - a_j\right). \tag{2.8.44}$$

The dynamics defined in (2.8.44) suggests that every subnetwork is basically similar to the recurrent neural network outlined in [272]. The connection weight matrix W is identical for each subnetwork, $W = -\gamma A^k\left(A^{2k+1}\right)^{\mathrm{T}}A^{k+1}$, while the biasing threshold vector for the jth subnetwork is $\gamma a_j = \{\gamma a_{1j}, \ldots, \gamma a_{nj}\}$. The matrix W (resp. V) elements are termed by w_{ij} (resp. v_{ij}).

Utilizing the relation (34) from [251], we can compute the elements of unknown matrix $V(t)$

$$\dot{v}_{ij} = \frac{\mathrm{d}v_{ij}}{\mathrm{d}t} = \sum_{k=1}^{n} v_{kj}w_{ik} + \gamma a_{ji},\ i, j = 1, \ldots, n. \tag{2.8.45}$$

This essentially means that the elements of the column $v_j = \{v_{1j}, \ldots, v_{nj}\}$ are generated in the jth subnetwork.

Because the connection weight matrix of the neurons in each subnetwork is identical, the proposed recurrent neural network can also be realized by single subnetwork with time-sharing threshold vectors. In each time slot, subnetwork biased by the corresponding threshold vector generates one column vector of the Drazin inverse. Therefore, the spatial complexity of the neural network can be reduced by a factor of n. As discussed in [381], the convergence rate of the linear recurrent neural network (2.8.1) is directly proportional to the smallest nonzero eigenvalue of the connection weight matrix. Specifically, the time needed for the recurrent neural network to converge is approximately $5/|\lambda_{\min}(W)|$, where $\lambda_{\min}(W)$ is the minimum nonzero eigenvalue of the connection weight matrix W. As will be shown through illustrative examples in Section 2.2.1, the recurrent neural network defined by the dynamic equation (2.8.1) is indeed able to generate the Drazin inverse of singular matrices at the projected convergence rate.

2.8.5 *Examples with Modified GNN Models*

Example 2.8.4 Consider

$$A = \begin{bmatrix} 1 & 2 & 3 & 4 & 1 \\ 1 & 3 & 4 & 6 & 2 \\ 2 & 3 & 4 & 5 & 3 \\ 3 & 4 & 5 & 6 & 4 \\ 4 & 5 & 6 & 7 & 6 \\ 6 & 6 & 7 & 7 & 8 \end{bmatrix}, \tag{2.8.46}$$

choose the following matrices $P \in \mathbb{R}^{5\times 2}$ and $Q \in \mathbb{R}^{2\times 6}$

$$P = \begin{bmatrix} 0 & 0 \\ 2 & 1 \\ 3 & 2 \\ 5 & 3 \\ 1 & 0 \end{bmatrix}, \quad Q = \begin{bmatrix} 0 & 1 & 0 & 1 & 0 & 1 \\ 1 & 0 & 1 & 0 & 1 & 0 \end{bmatrix} \tag{2.8.47}$$

and compute $G = PQ$.

The spectrum of the matrix GA is given by the set

$$\sigma(GA) = \{266.346716180717, 0.653283819282910, 0, 0, 0\}.$$

Using $G = PQ$ and $\gamma = 10^8$ we obtain the following approximation of $A^{(2)}_{\mathcal{R}(P),\mathcal{N}(Q)}$:

$$X = \begin{bmatrix} 0 & 0. & 0 & 0 & 0 & 0 \\ -0.1207 & 0.1092 & -0.1207 & 0.1092 & -0.1207 & 0.1092 \\ 0.3448 & 0.2644 & 0.3448 & -0.2644 & 0.3448 & -0.2644 \\ 0.2241 & -0.1552 & 0.2241 & -0.1552 & 0.2241 & -0.1552 \\ -0.5862 & 0.4828 & -0.5862 & 0.4828 & -0.5862 & 0.4828 \end{bmatrix}.$$

Corresponding exact $\{2\}$-inverse of A corresponding to G is defined by $P(QAP)^{-1}Q$ (see [208]) and it is equal to

$$A^{(2)}_{\mathcal{R}(P),\mathcal{N}(Q)} = \frac{1}{174}\begin{bmatrix} 0 & 0 & 0 & 0 & 0 & 0 \\ -21 & 19 & -21 & 19 & -21 & 19 \\ 60 & -46 & 60 & -46 & 60 & -46 \\ 39 & -27 & 39 & -27 & 39 & -27 \\ -102 & 84 & -102 & 84 & -102 & 84 \end{bmatrix}.$$

Convergence behavior of the network in time 10^{-7}s is graphically illustrated in Fig. 2.26.

Example 2.8.5 The goal of this example is to compute the Drazin inverse of the matrix

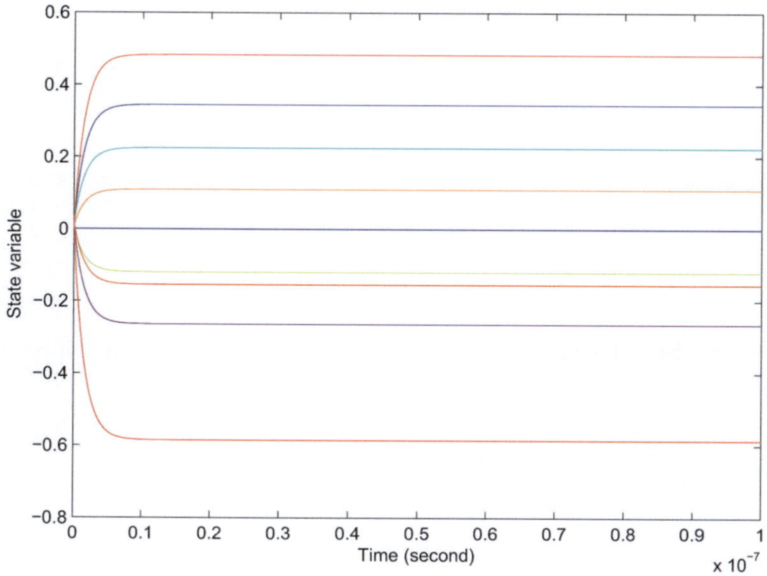

Fig. 2.26 Convergence behavior of the network in time 10^{-7}s for Example 2.8.4

$$A = \begin{bmatrix} 2 & 0.4 & 0 & 0 & 0 & 0 & 0 & 0 & 0 & 0 & 0 & 0 \\ -2 & 0.4 & 0 & 0 & 0 & 0 & 0 & 0 & 0 & 0 & 0 & 0 \\ -1 & -1 & 1 & -1 & 0 & 0 & 0 & 0 & -1 & 0 & 0 & 0 \\ -1 & -1 & -1 & 1 & 0 & 0 & 0 & 0 & 0 & 0 & 0 & 0 \\ 0 & 0 & 0 & 0 & 1 & 1 & -1 & -1 & 0 & 0 & -1 & 0 \\ 0 & 0 & 0 & 0 & 1 & 1 & -1 & -1 & 0 & 0 & 0 & 0 \\ 0 & 0 & 0 & -1 & -2 & 0.4 & 0 & 0 & 0 & 0 & 0 & 0 \\ 0 & 0 & 0 & 0 & 2 & 0.4 & 0 & 0 & 0 & 0 & 0 & 0 \\ 0 & -1 & 0 & 0 & 0 & 0 & 0 & 0 & 1 & -1 & -1 & -1 \\ 0 & 0 & 0 & 0 & 0 & 0 & 0 & 0 & -1 & 1 & -1 & -1 \\ 0 & 0 & 0 & 0 & 0 & 0 & 0 & 0 & 0 & 0 & 0.4 & -2 \\ 0 & 0 & 0 & 0 & 0 & 0 & 0 & 0 & 0 & 0 & 0.4 & 2 \end{bmatrix}$$

with $\mathrm{ind}(A) = 3$. Therefore, our first attempt is to use $G = A^3$. Since the eigenvalues of GA are equal to

$$\sigma(GA) = \{16.00 + 0.00i, 16.00 - 0.00i, 4.387, 0.717 - 2.458i, 0.717 + 2.458i, 0.717 - 2.458i\},$$

whose real parts are greater than zero, the choice $G = A^3$ is appropriate. Using $\gamma = 10^{10}$ we get the following approximation of the Drazin inverse with the precision 10^{-9} after 10^{-7}s:

$$A^d = \begin{bmatrix} 0.25 & -0.25 & 0. & 0. & 0. & 0. & 0. & 0. & 0. & 0. & 0. & 0. \\ 1.25 & 1.25 & 0. & 0. & 0. & 0. & 0. & 0. & 0. & 0. & 0. & 0. \\ -1.6640 & -0.9922 & 0.25 & -0.25 & 0. & 0. & 0. & 0. & -0.0625 & -0.0625 & 0. & 0.1563 \\ -1.1953 & -0.6797 & -0.25 & 0.25 & 0. & 0. & 0. & 0. & -0.0625 & 0.1875 & 0.6875 & 1.3437 \\ -2.7637 & -1.0449 & -1.875 & -1.25 & -1.25 & 1.25 & 1.25 & 1.25 & 1.4844 & 2.5781 & 3.3203 & 6.6406 \\ -2.7637 & -1.0449 & -1.875 & -1.25 & -1.25 & 1.25 & 1.25 & 1.25 & 1.4844 & 2.5781 & 4.5703 & 8.5156 \\ 14.1094 & 6.3008 & 6.625 & 3.375 & 5. & -3. & -5. & -5. & -4.1875 & -8.5 & -10.5078 & -22.4609 \\ -19.3242 & -8.5078 & -9.75 & -5.25 & -7.5 & 4.5 & 7.5 & 7.5 & 6.375 & 12.5625 & 15.9766 & 33.7891 \\ -0.625 & -0.3125 & 0. & 0. & 0. & 0. & 0. & 0. & 0.25 & -0.25 & -0.875 & -1.625 \\ -1.25 & -0.9375 & 0. & 0. & 0. & 0. & 0. & 0. & -0.25 & 0.25 & -0.875 & -1.625 \\ 0. & 0. & 0. & 0. & 0. & 0. & 0. & 0. & 0. & 0. & 1.25 & 1.25 \\ 0. & 0. & 0. & 0. & 0. & 0. & 0. & 0. & 0. & 0. & -0.25 & 0.25 \end{bmatrix}.$$

Convergence behavior of the network in time 10^{-7}s is graphically illustrated in Fig. 2.27.

Example 2.8.6 Consider the initial matrix A equal to

$$A = \begin{bmatrix} 1 & -1 & 0 & 0 & 0 & 0 \\ -1 & 1 & 0 & 0 & 0 & 0 \\ -1 & -1 & 1 & -1 & 0 & 0 \\ -1 & -1 & -1 & 1 & 0 & 0 \\ -1 & -1 & -1 & 0 & 2 & -1 \\ -1 & -1 & 0 & -1 & -1 & 2 \end{bmatrix}.$$

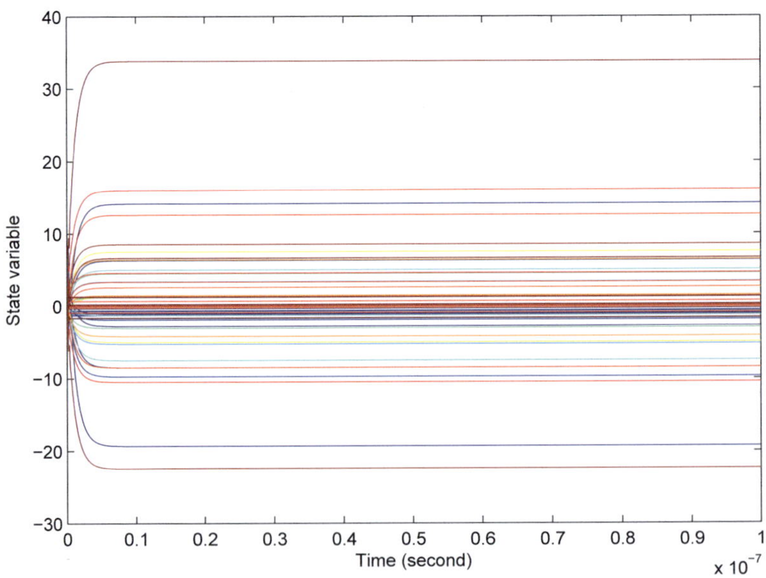

Fig. 2.27 Convergence behavior of the network in time 10^{-7}s for Example 2.8.5

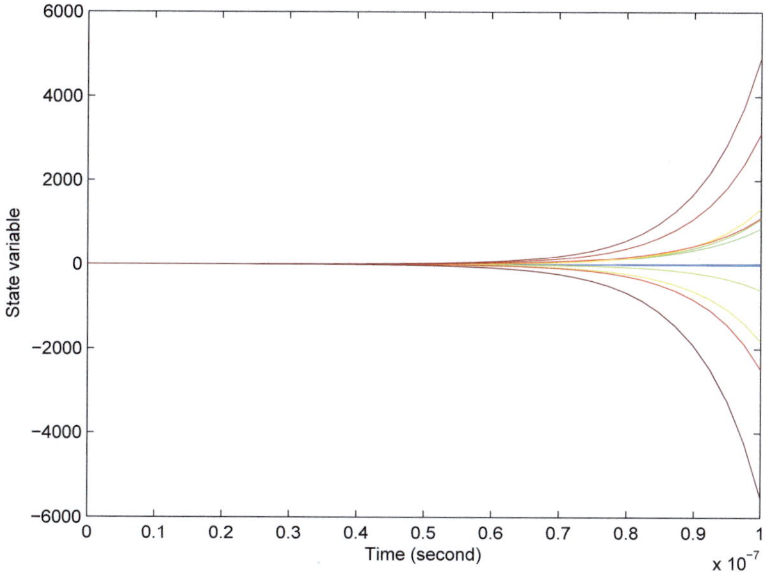

Fig. 2.28 Divergence of the network in time 10^{-7}s for Example 2.8.6

In the case $G = -A^{\mathrm{T}}$ the RNN does not provide a solution, since the matrix GA is negative semidefinite. Its eigenvalues are $\{-10.6056, -9.1231, -4.0000, -3.3944, -0.8769, 0.0000\}$. The divergence of neural network trajectories is illustrated in Fig. 2.28:

A useful representation of $A^{(2)}_{T,S}$, based on the usage of the group inverse, is presented in [289] and restated in Proposition 2.3.2. This representation, restated in the next proposition, gave a new computational aspect in relationship with the $A^{(2)}_{T,S}$ inverse.

Since $\mathrm{ind}(GA) = 1$, according to Proposition 2.3.2, $A^{(2)}_{\mathcal{R}(G),\mathcal{N}(G)}$ exists. But, in spite of its existence, $A^{(2)}_{\mathcal{R}(G),\mathcal{N}(G)}$ cannot be computed using the gradient based neural network approach. A possible solution is usage of the matrix $G = (-A^{\mathrm{T}})\left((-A^{\mathrm{T}})A(-A^{\mathrm{T}})\right)^{\mathrm{T}}(-A^{\mathrm{T}}) = A^{\mathrm{T}}\left(A^{\mathrm{T}}AA^{\mathrm{T}}\right)^{\mathrm{T}}A^{\mathrm{T}}$ instead of the matrix $G = -A^{\mathrm{T}}$. Such a choice initiates that the spectrum of GA contains elements with nonnegative real parts:

$$\sigma(GA) = 10^3\{1.1929, 0.7593, 0.0640, 0.0391, 0.0007, 0.0000\}.$$

Using $\gamma = 1000000000$ and $G = A^{\mathrm{T}}\left(A^{\mathrm{T}}AA^{\mathrm{T}}\right)^{\mathrm{T}}A^{\mathrm{T}}$ the RNN approach produces the Moore–Penrose inverse of A (Fig. 2.29):

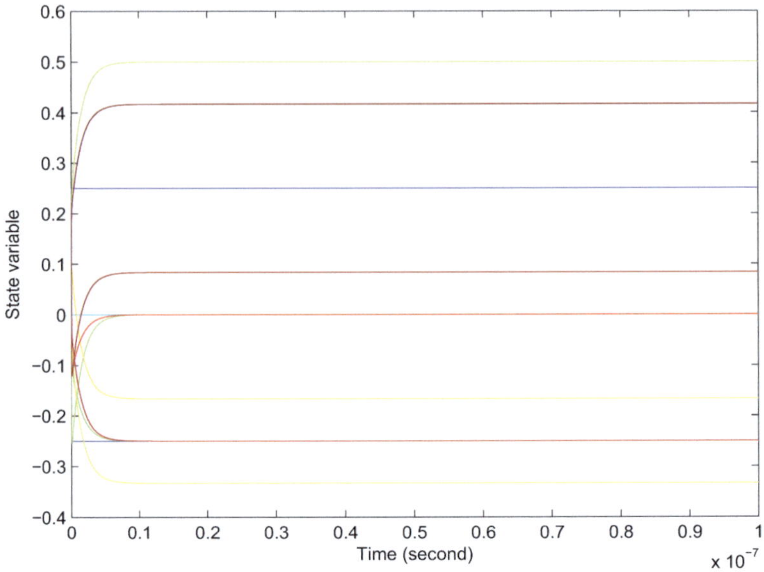

Fig. 2.29 Convergence behavior of the network in time 10^{-7}s for Example 2.8.6

$$A^{\dagger} = \begin{bmatrix} 0.2500 & -0.2500 & -0.2500 & -0.2500 & 0.0000 & 0.0000 \\ -0.2500 & 0.2500 & -0.2500 & -0.2500 & 0.0000 & 0.0000 \\ 0 & 0 & 0.5000 & 0.0000 & -0.2500 & -0.2500 \\ 0 & 0 & 0.0000 & 0.5000 & -0.2500 & -0.2500 \\ 0 & 0 & -0.1667 & -0.3333 & 0.4167 & 0.0833 \\ 0 & 0 & -0.3333 & -0.1667 & 0.0833 & 0.4167 \end{bmatrix}.$$

2.9 One Specific GNN Model for Computing the Moore–Penrose Inverse

This section describes one specific GNN model for computing the Moore–Penrose inverse which was defined in [231]. The development of the model starts from the fact that the Moore–Penrose inverse $X := A^{\dagger}$ is the unique inner inverse which additionally satisfies $\mathcal{R}(X) = \mathcal{R}(A^{\mathrm{T}})$ and $\mathcal{N}(X) = \mathcal{N}(A^{\mathrm{T}})$. Accordingly, there exists $V \in \mathbb{R}^{m\times n}$ such that $X = A^{\mathrm{T}}VA^{\mathrm{T}}$. Then the Penrose equation **(1)**, i.e., $AXA = A$, becomes

$$AA^{\mathrm{T}}VA^{\mathrm{T}}A = A. \tag{2.9.1}$$

The GNN model for solving the Eq. (2.9.1) is given by

$$\dot{V}(t) = \gamma AA^{\mathrm{T}}\mathcal{F}\left(A - AA^{\mathrm{T}}V(t)A^{\mathrm{T}}A\right)A^{\mathrm{T}}A. \tag{2.9.2}$$

Clearly, (2.9.2) is equal to $GNN(AA^{\mathrm{T}}, A^{\mathrm{T}}A, A)$ model. Theorem 2.9.1 shows the global convergence of the matrix $X(t) := A^{\mathrm{T}}V(t)A^{\mathrm{T}}$ to the Moore–Penrose inverse of A.

Theorem 2.9.1 ([231]) *Let $V(t)$ be the matrix of activation state variables of the $GNN(AA^{\mathrm{T}}, A^{\mathrm{T}}A, A)$ model (2.9.2). Then the matrix $X(t) = A^{\mathrm{T}}V(t)A^{\mathrm{T}}$ converges to the Moore–Penrose inverse $A^{\dagger}$ for every initial matrix $V(0)$.*

Proof Using the well-known properties of the Moore–Penrose inverse $A^{\dagger} = A^{\mathrm{T}}(AA^{\mathrm{T}})^{\dagger} = (A^{\mathrm{T}}A)^{\dagger}A^{\mathrm{T}}$ and $A^{\mathrm{T}} = A^{\mathrm{T}}AA^{\dagger} = A^{\dagger}AA^{\mathrm{T}}$, it is not difficult to verify

$$AA^{\mathrm{T}}(AA^{\mathrm{T}})^{\dagger}A(A^{\mathrm{T}}A)^{\dagger}A^{\mathrm{T}}A = A,$$

which implies that the conditions (2.2.11) are satisfied for solvability of the matrix equation $(AA^{\mathrm{T}})X(A^{\mathrm{T}}A) = A$. Consequently, the matrix equation (2.9.1) is solvable. According to Theorem 2.2.2, the limiting value of the activation state variables matrix $V(t)$ of the model $GNN(AA^{\mathrm{T}}, A^{\mathrm{T}}A, A)$ is given by

$$\tilde{V}_{V(0)} = (AA^{\mathrm{T}})^{\dagger}A(A^{\mathrm{T}}A)^{\dagger} + V(0) - (AA^{\mathrm{T}})^{\dagger}AA^{\mathrm{T}}V(0)A^{\mathrm{T}}A(A^{\mathrm{T}}A)^{\dagger}. \tag{2.9.3}$$

Therefore, $X(t) = A^{\mathrm{T}}V(t)A^{\mathrm{T}}$ converges to $\tilde{X} = A^{\mathrm{T}}\,\tilde{V}_{V(0)}A^{\mathrm{T}}$. Using again the same properties of the Moore–Penrose inverse, we obtain

$$\begin{aligned}
\tilde{X} &= A^{\mathrm{T}}\tilde{V}_{V(0)}A^{\mathrm{T}} \\
&= A^{\mathrm{T}}(AA^{\mathrm{T}})^{\dagger}A(A^{\mathrm{T}}A)^{\dagger}A^{\mathrm{T}} + A^{\mathrm{T}}V(0)A^{\mathrm{T}} - A^{\mathrm{T}}(AA^{\mathrm{T}})^{\dagger}AA^{\mathrm{T}}V(0)A^{\mathrm{T}}A(A^{\mathrm{T}}A)^{\dagger}A^{\mathrm{T}} \\
&= A^{\dagger}AA^{\dagger} + A^{\mathrm{T}}V(0)A^{\mathrm{T}} - A^{\dagger}AA^{\mathrm{T}}V(0)A^{\mathrm{T}}AA^{\dagger} \\
&= A^{\dagger} + A^{\mathrm{T}}V(0)A^{\mathrm{T}} - A^{\mathrm{T}}V(0)A^{\mathrm{T}} \\
&= A^{\dagger}.
\end{aligned}$$

This completes the proof of this theorem. □

Corollary 2.9.1 ([231]) *If $A \in \mathbb{R}^{m\times n}$ is an arbitrary real matrix, an odd and monotonically increasing function $f(\cdot)$ is used to define the array activation function $\mathcal{F}(\cdot)$ and the design parameter γ satisfies $\gamma > 0$, then the limiting values of the activation state matrix $V(t)$ of the model (2.9.2) satisfy*

$$\left\{A^{\mathrm{T}}\tilde{V}_{V(0)}A^{\mathrm{T}} \mid V(0) \in \mathbb{R}^{n\times m}\right\} = \left\{A^{\dagger}\right\}. \tag{2.9.4}$$

Example 2.9.1 The second example shows the behavior of GNN$(AA^{\mathrm{T}}, A^{\mathrm{T}}A, A)$ for computing the Moore–Penrose inverse. Consider the following matrix:

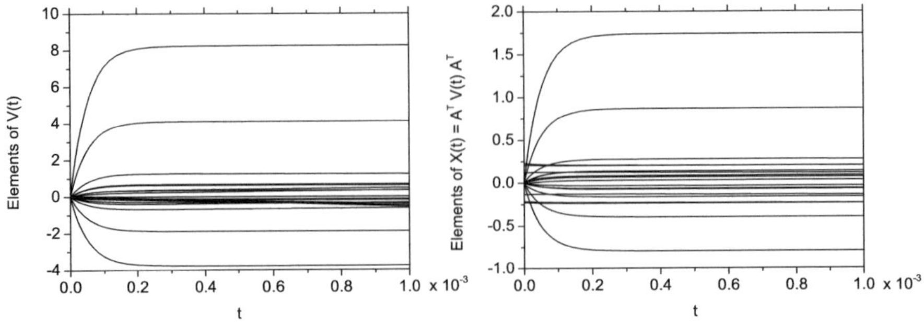

Fig. 2.30 Elementwise convergence trajectories of $V(t)$ in GNN(AA^{T}, $A^{\mathrm{T}}A$, A) with linear activation function (left) and the matrix $X(t) = A^{\mathrm{T}}V(t)A^{\mathrm{T}}$ in Example 2.9.1

$$A = \frac{1}{49}\begin{bmatrix} 45 & -1 & 22 & 3 & -49 \\ 0 & -8 & -2 & 18 & 4 \\ 0 & -4 & -1 & 9 & 2 \\ -49 & 5 & -33 & -4 & 47 \end{bmatrix}.$$

Its rank is equal to $r = \mathrm{rank}(A) = 3$. We use the GNN(AA^{T}, $A^{\mathrm{T}}A$, A) model with the following parameters: $\gamma = 10^7$ and $V(0) = 0$. Trajectories of the variable $V(t)$ are shown in Fig. 2.30 (left), while the trajectories of the matrix $X(t) = A^{\mathrm{T}}V(t)A^{\mathrm{T}}$, which converges to the Moore–Penrose inverse, are shown in Fig. 2.30 (right). In order to get better insight, some trajectories (leading to large magnitude limit values) are not shown in Fig. 2.30 (left).
Elementwise convergence trajectories of the state variable matrix $V(t)$ of the linear GNN(AA^{T}, $A^{\mathrm{T}}A$, A) (left) and the variable matrix $X(t) = A^{\mathrm{T}}V(t)A^{\mathrm{T}}$ converging to the Moore–Penrose inverse $A^{\dagger}$ (right) in Example 2.9.1. Some trajectories leading to large magnitude limit values are not shown on left figure. The range on both x and y axes in Fig. 2.30 is reduced in order to show the majority of the elements of $V(t)$ and $X(t)$ more clearer. Note that the exact value of the Moore–Penrose inverse is given by

$$A^{\dagger} = \begin{bmatrix} \frac{1337}{9476} & \frac{3101}{23690} & \frac{3101}{47380} & -\frac{2779}{9476} \\ \frac{240835}{208472} & -\frac{6419}{9476} & -\frac{6419}{18952} & \frac{233583}{208472} \\ -\frac{96145}{26059} & -\frac{5516}{11845} & -\frac{2758}{11845} & -\frac{94920}{26059} \\ \frac{68929}{104236} & \frac{43043}{23690} & \frac{43043}{47380} & \frac{53641}{104236} \\ -\frac{523271}{208472} & \frac{343}{9476} & \frac{343}{18952} & -\frac{395283}{208472} \end{bmatrix} \approx \begin{bmatrix} 0.14109 & 0.1309 & 0.06545 & -0.29327 \\ 1.15524 & -0.6775 & -0.3387 & 1.1204 \\ -3.6895 & -0.4657 & -0.2328 & -3.6425 \\ 0.66128 & 1.8169 & 0.9085 & 0.5146 \\ -2.510 & 0.036197 & 0.0181 & -1.896 \end{bmatrix}.$$

Figure 2.31 shows the residual norms $\|AX(t)A - A\|_F$ (left) and $\|X(t)AX(t) - X(t)\|_F$ (right), corresponding to the Penrose equations (1) and (2) respectively, for both linear and nonlinear activation functions.

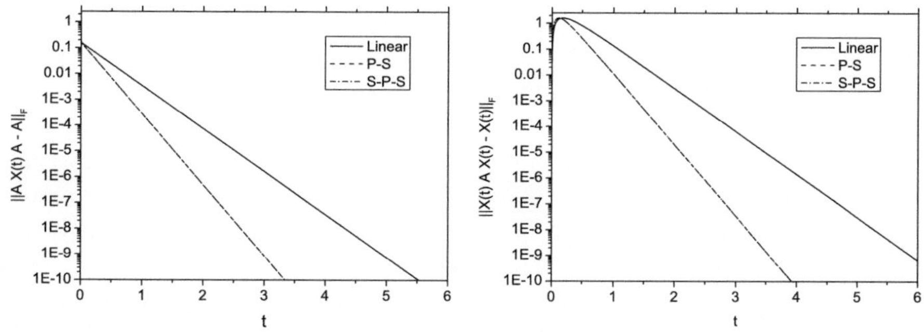

Fig. 2.31 Residual norms $\|AX(t)A - A\|_F$ (left) and $\|X(t)AX(t) - X(t)\|_F$ (right) of GNN(A, B, D) network with different activation functions in Example 2.9.1

According to Fig. 2.31, the nonlinear activation function still has an advantage in the convergence speed, comparing with the linear activation function. It is interesting that, in this case, the power-sigmoid and the smooth power-sigmoid activation functions give almost the same residual norms.

2.10 GNN Model for Solving Linear Systems

The parallel-processing computational GNN model (called as WNN model) for solving linear equations

$$Ax = b, \quad A \in \mathbb{R}^{n \times n}, \qquad x \in \mathbb{R}^n, \quad b \in \mathbb{R}^n, \tag{2.10.1}$$

where A is regular, and x is the unknown vector, was proposed in [275] and further investigated in [314]. A gradient neural network (GNN) for solving online a set of simultaneous linear equations was considered in [347]. In order to solve the linear system (2.10.1), it is necessary to consider the error function in vector form

$$e(t) = Av(t) - b.$$

and the scalar-valued error-monitoring goal function.

$$\varepsilon(t) = \frac{1}{2}\|Av(t) - b\|_F^2.$$

Now, the GNN dynamic is given as the dynamical system (2.10.11). An extension of WNN (called EWNN) by means of nonlinear activation functions was defined in [147], as follows:

$$\dot{v}(t) = -\gamma A^{\mathrm{T}} \mathcal{F}(Av(t) - b), \tag{2.10.2}$$

wherein $\mathcal{F}$ stands for one of two possible activation functions. A global converge to the unique theoretical solution $x^* = A^{-1}b$ was proved in [147]. Zhang et al. in [347] proved global exponential convergence for the GNN models solving $Ax = b$, as an addition to the previously presented asymptotical convergence.

In [231], Stanimirović and Petković avoided assumption of invertibility of the coefficient matrix A. Also, using GNN(A, B, D) design and developed theoretical results, some useful results about the convergence of a general system of linear equations were obtained. So, the general linear system

$$Ax = b, \quad A \in \mathbb{R}_r^{m\times n}, \qquad x \in \mathbb{R}^n, \quad b \in \mathbb{R}^m, \tag{2.10.3}$$

without the regularity constraint, was considered in [231]. Moreover, in [231], the authors considered even more general system $AX = B_1$ where $X \in \mathbb{R}^{n\times t}$ is the unknown matrix, while $B_1 \in \mathbb{R}^{m\times t}$ is the right-hand side coefficient matrix. Practically, each column of the righ-hand side coeffficient matrix B_1 defines a linear system of the form (2.10.3). For this purpose, it suffices to consider the specialization of the general GNN model to systems of linear equations. This specialization will be termed as GEWNN and defined as follows:

$$\dot{V}(t) = \gamma A^{\mathrm{T}}\mathcal{F}(B_1 - AV(t)),\ A \in \mathbb{R}_r^{m\times n},\ B_1 \in \mathbb{R}^{m\times l},\ \ V(t) \in \mathbb{R}^{n\times l}. \tag{2.10.4}$$

It is observable that the reduction GNN(A, I_l, B_1) of GNN(A, B, D) produces GEWNN defined in (2.10.4). Then the following convergence results follow from Theorems 2.2.1 and 2.2.2. According to these results, the convergence of GEWNN model is global.

Corollary 2.10.1 *Assume that $A \in \mathbb{R}_r^{m\times n}$ and $B_1 \in \mathbb{R}^{m\times l}$ satisfy*

$$AA^{(1)}B_1 = B_1, \tag{2.10.5}$$

for some inner inverse $A^{(1)}$. If an odd and monotonically increasing array activation function $\mathcal{F}(\cdot)$ based on an elementwise function $f(\cdot)$ is used, then the neural state matrix $V(t) \in \mathbb{R}^{n\times l}$ of the $GNN(A, I_l, B_1)$ model (2.10.4) converges to the solution of the matrix equation $AX = B_1$, i.e., $AV(t) \to B_1$ as $t \to +\infty$, for an arbitrary initial state matrix $V(0)$.

Corollary 2.10.2 *Assume that $A \in \mathbb{R}_r^{m\times n}$ and $B_1 \in \mathbb{R}^{m\times l}$ satisfy*

$$AA^{\dagger}B_1 = B_1. \tag{2.10.6}$$

Then the activation state variables matrix $V(t)$ of the model $GNN(A, I_l, B_1)$, defined by (2.10.4), is convergent as $t \to +\infty$ and has the equilibrium state

$$V(t) \to \tilde{V}_{V(0)} = A^{\dagger}B_1 + \left(I_n - A^{\dagger}A\right)V(0), \tag{2.10.7}$$

for every initial state matrix $V(0) \in \mathbb{R}^{n\times t}$.

In the case $l = 1$ the matrix $B_1 \in \mathbb{R}^{m\times l}$ becomes a vector, the system $AX = B_1$ reduces to the linear system $Ax = b$ and, accordingly, the model $GNN(A, I_l, B_1)$ becomes corresponding GEWNN model $GNN(A, I_1, b)$, defined as

$$\dot{v}(t) = \gamma A^{\mathrm{T}}\mathcal{F}(b - Av(t)),\ A \in \mathbb{R}_r^{m\times n},\ b \in \mathbb{R}^m,\ v \in \mathbb{R}^n. \tag{2.10.8}$$

The conditions in Corollary 2.10.1 are in the linear system case reduced to $AA^{(1)}b = b$. This condition is equivalent with the solvability of the system $Ax = b$. In such case, according to Corollary 2.10.2, the GEWNN model (2.10.8) converges to the general solution [231]

$$v(t) \to \tilde{v}_{v(0)} = A^{\dagger}b + \left(I_n - A^{\dagger}A\right)v(0) \tag{2.10.9}$$

of $Ax = b$, defined by the initial state $v(0)$. The minimum 2-norm solution $x^* = A^{\dagger}b$ can be obtained by the zero initial state $v(0) = 0$.

A fast convergent gradient neural network (FCGNN) model for solving system of linear equations (SoLE) was proposed and analyzed in [314]. The design of the FCGNN model is based on a nonlinear activation function which enables the better convergence speed. The convergence upper bound of the FCGNN model was estimated and tested in numerical examples. Fact convergence in the FCGNN model is achieved by the following activation function:

$$\mathcal{F}(\cdot) = \mathbf{Sgn}^r(\cdot) + \mathbf{Sgn}^{1/r}(\cdot) \tag{2.10.10}$$

where the parameter $r \in (0, 1)$ and the function $\mathbf{Sgn}^r(\cdot) : \mathbb{R}^n \to \mathbb{R}^n$ denotes an activation-function vector mapping, with each element $sgn^r(\cdot)$ defined as follows:

$$\mathbf{Sgn}^r(u) = \begin{cases} |u|^r, & \text{if } u > 0, \\ 0, & \text{if } u = 0, \\ -|u|^r, & \text{if } u < 0, \end{cases}$$

Example 2.10.1 Consider the linear system $Ax = b$ defined by

$$A = \begin{bmatrix} 59 & 58 & 120 & 64 & 84 \\ 62 & 76 & 86 & 39 & 81 \\ 106 & 120 & 182 & 111 & 125 \\ 139 & 180 & 176 & 96 & 158 \end{bmatrix},\quad b = \begin{bmatrix} 1513 \\ 1360 \\ 2416 \\ 2839 \end{bmatrix}.$$

The GNN(A, I_1, b) model

$$\dot{v}(t) = -\gamma A^{\mathrm{T}}(Av(t) - b), \tag{2.10.11}$$

in which $v(t)$ is the neural state vector and γ is the scaling parameter, can be used to compute the solution of the induced system $Ax = b$. The initial vector $v(0) =$

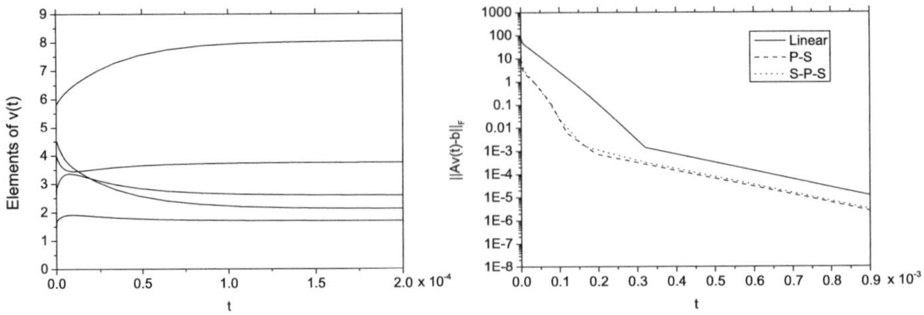

Fig. 2.32 Elementwise convergence trajectories of the GNN(A, I_1, b) with linear activation (left) and the trajectory of the error norm for different activation functions (right) in Example 2.10.1

$\left[1, 2, 3, 4, 5\right]^{\mathrm{T}}$ produces the solution

$$x = A^{\dagger}b + \left(I - A^{\dagger}A\right)v(0) = \begin{bmatrix} 1.7002 \\ 2.5904 \\ 3.7511 \\ 2.1211 \\ 8.0544 \end{bmatrix}.$$

The gain parameter of the model is $\gamma = 100$. Figure 2.32 shows the elementwise trajectories of $v(t)$ (left) for linear activation, and the residual norms $\|b - Av(t)\|_F$ (right) when different activation functions are used. All three activation functions assure the convergence, while nonlinear activations again provide a slightly faster convergence.

2.11 Dynamical Systems for Computing DMP Inverse

The aim of this section is to investigate the GNN dynamics for computing the DMP inverse $A^{\mathrm{d},\dagger} = A^{\mathrm{d}}AA^{\dagger}$ from [173]. Leading idea in this section is to show applicability of GNN design and corresponding modification in approximating composite generalized inverses. Such generalized inverses are defined as matrix expressions involving some known generalized inverses. According to the general GNN dynamics, the linear GNN model for solving $AXB = D$ is defined as [231]

$$\dot{V}(t) = -\gamma \frac{\partial \varepsilon(V(t))}{\partial V} = -\gamma A^{\mathrm{T}}(AV(t)B - D)B^{\mathrm{T}}. \tag{2.11.1}$$

Particularly, the GNN for generating solutions to the matrix equation $GAX = G$ is given by the dynamical system (2.3.6), given as

$$\dot{V}(t) = -\gamma\,(GA)^{\mathrm{T}}\,(GAV(t) - G)\,,\quad V(0) = V_0.$$

The authors of [399] defined the following simplified dynamical evolution for solving the matrix equation $GAX = G$:

$$\dot{V}(t) = -\gamma\,\,(GAV(t) - G)\,,\quad V(0) = O. \tag{2.11.2}$$

The dynamical evolution (2.11.2) will be termed as GNNATS2(GA, I, G).

The range space and null space of the DMP inverse are defined as follows:

$$\mathcal{R}(A^{\mathrm{d},\dagger}) = \mathcal{R}(A^k),\qquad \mathcal{N}(A^{\mathrm{d},\dagger}) = \mathcal{N}(A^kA^\dagger).$$

Several additional properties of the DMP inverse are developed in Lemma 2.11.1 from [173] in order to define an appropriate error monitoring function $E(t)$ and corresponding GNN model.

Lemma 2.11.1 *Let $A \in \mathbb{C}^{n\times n}$ satisfy* $\mathrm{ind}(A) = k$. *Then the DMP inverse $A^{d,\dagger}$ satisfies the following properties:*

$$A^kA^\dagger AA^{d,\dagger} = A^kA^{d,\dagger} = A^kA^\dagger \tag{2.11.3}$$
$$A^{d,\dagger}AA^* = A^dAA^* \tag{2.11.4}$$
$$A^kA^{d,\dagger}A = A^k \tag{2.11.5}$$

Proof The property (2.11.3) can be verified using the following transformations:

$$A^kA^\dagger AA^{\mathrm{d},\dagger} = A^kA^\dagger AA^{\mathrm{d}}AA^\dagger = [A^{k-1}(AA^\dagger A)A]A^{\mathrm{d}}A^\dagger = A^{k+1}A^{\mathrm{d}}A^\dagger = A^kA^\dagger.$$

The property (2.11.4) follows from $A^{\mathrm{d},\dagger}A = A^{\mathrm{d}}A$ in conjunction with $A^\dagger AA^* = A^*$, while the property (2.11.5) can be verified directly. □

In view of (2.11.3), the dynamical system of the form (2.11.2) for computing $A^{\mathrm{d},\dagger}$ can be defined by dynamics GNNATS2($A^kA^\dagger A, I, A^kA^\dagger$), which produces

$$\dot{V}(t) = -\gamma\,\left(A^kA^\dagger AV(t) - A^kA^\dagger\right),\quad V(0) = \mathbf{O}. \tag{2.11.6}$$

Corollary 2.11.1 *Let $A \in \mathbb{R}^{n\times n}$ be of index k and the spectrum $\sigma(A^k) = \{\lambda_1, \ldots, \lambda_n\}$ of A^k satisfies (2.8.8):* $\mathrm{Re}\big(\lambda_j\big) \geq 0,\ j = 1, \ldots, n$ *Then the DMP inverse of A is generated by the limiting expression*

$$\lim_{t\to\infty} V(t) = \lim_{t\to\infty} \gamma\exp(-\gamma A^kt)\int_0^t \exp(\gamma A^k\tau)A^kA^\dagger\,\mathrm{d}\tau = A^{d,\dagger},$$

where $V(t)$ is the activation state matrix of the evolution design (2.11.6).

Proof In view of $A^k A^\dagger A = A^k$ and $\mathrm{ind}(A) = k$, it follows that

$$\mathrm{rank}\left(\left(A^k A^\dagger A\right)^2\right) = \mathrm{rank}(A^{2k}) = \mathrm{rank}(A^k) = \mathrm{rank}\left(A^k A^\dagger A\right).$$

So, $\mathrm{ind}(A^k A^\dagger A) = 1$, and the existence of $A^{(2)}_{\mathcal{R}(A^k A^\dagger),\mathcal{N}(A^k A^\dagger)} = (A^k A^\dagger A)_g A^k A^\dagger$ is guaranteed. Also, $\sigma(A^k A^\dagger A) = \sigma(A^k)$. An application of [399, Theorem 1] in the case $G = A^k A^\dagger$ in conjunction with the simplification $GA = A^k A^\dagger A = A^k$ initiates $\lim_{t\to\infty} V(t) = A^{(2)}_{\mathcal{R}(A^k A^\dagger),\mathcal{N}(A^k A^\dagger)}$. □

Remark 2.11.1 It is important to mention that the symbolic implementation of GNNATS2(A^k, I, $A^k A^\dagger$) in the package Mathematica is based on finding the exact solution of (2.11.6) by means of its standard function `DSolve`. Also, the implementation of GNNATS2(A^k, I, $A^k A^\dagger$) in MATLAB is developed by appropriate Simulink for solving (2.11.6) by means of the integrator `Ode15s`.

Example 2.11.1 Consider the matrix Consider

$$A = \begin{bmatrix} 1 & -1 & 0 & 0 & 0 & 0 \\ -1 & 1 & 0 & 0 & 0 & 0 \\ -1 & -1 & 1 & -1 & 0 & 0 \\ -1 & -1 & -1 & 1 & 0 & 0 \\ -1 & -1 & -1 & 0 & 2 & -1 \\ -1 & -1 & 0 & -1 & -1 & 2 \end{bmatrix} \tag{2.11.7}$$

satisfying $\mathrm{rank}(A) = 5$, $\mathrm{rank}(A^2) = \mathrm{rank}(A^3) = 4$. Therefore, $k = \mathrm{ind}(A) = 2$ and $r_k = 4$. The DMP inverse of A in the error-free form is given by

$$A^{\mathrm{d},\dagger} = A^2(A^5)^\dagger A^3 A^\dagger = \begin{bmatrix} \frac{1}{4} & -\frac{1}{4} & 0 & 0 & 0 & 0 \\ -\frac{1}{4} & \frac{1}{4} & 0 & 0 & 0 & 0 \\ 0 & 0 & \frac{1}{4} & -\frac{1}{4} & 0 & 0 \\ 0 & 0 & -\frac{1}{4} & \frac{1}{4} & 0 & 0 \\ 0 & 0 & -\frac{5}{12} & -\frac{7}{12} & \frac{2}{3} & \frac{1}{3} \\ 0 & 0 & -\frac{7}{12} & -\frac{5}{12} & \frac{1}{3} & \frac{2}{3} \end{bmatrix}. \tag{2.11.8}$$

The condition (2.8.8) is satisfied for A^2. Then the implementation of $GNNATS2(A^2, I, A^2 A^\dagger)$ in the package *Mathematica* produces the following result in symbolic form:

$$X_{t,\gamma} = \begin{bmatrix} \frac{1}{4}-\frac{1}{4}e^{-4t\gamma} & \frac{1}{4}\left(-1+e^{-4t\gamma}\right) & 0 & 0 & 0 & 0 \\ \frac{1}{4}\left(-1+e^{-4t\gamma}\right) & \frac{1}{4}-\frac{1}{4}e^{-4t\gamma} & 0 & 0 & 0 & 0 \\ 0 & 0 & \frac{1}{4}-\frac{1}{4}e^{-4t\gamma} & \frac{1}{4}\left(-1+e^{-4t\gamma}\right) & 0 & 0 \\ 0 & 0 & \frac{1}{4}\left(-1+e^{-4t\gamma}\right) & \frac{1}{4}-\frac{1}{4}e^{-4t\gamma} & 0 & 0 \\ 0 & 0 & \frac{1}{12}e^{-9t\gamma}\left(2-3e^{5t\gamma}+6e^{8t\gamma}-5e^{9t\gamma}\right) & \frac{1}{12}e^{-9t\gamma}\left(-2+3e^{5t\gamma}+6e^{8t\gamma}-7e^{9t\gamma}\right) & \frac{1}{6}\left(4-e^{-9t\gamma}-3e^{-t\gamma}\right) & \frac{1}{6}\left(2+e^{-9t\gamma}-3e^{-t\gamma}\right) \\ 0 & 0 & \frac{1}{12}e^{-9t\gamma}\left(-2+3e^{5t\gamma}+6e^{8t\gamma}-7e^{9t\gamma}\right) & \frac{1}{12}e^{-9t\gamma}\left(2-3e^{5t\gamma}+6e^{8t\gamma}-5e^{9t\gamma}\right) & \frac{1}{6}\left(2+e^{-9t\gamma}-3e^{-t\gamma}\right) & \frac{1}{6}\left(4-e^{-9t\gamma}-3e^{-t\gamma}\right) \end{bmatrix}.$$

Finally, the limit $\lim_{t\to\infty} X_{t,10^3}$ produces the exact DMP inverse as in (2.11.8).

The elementwise trajectories of $X_{t,\gamma}$ in the case $\gamma = 10^6$ during the time $[0, 2 * 10^{-6}]$s (left) and $[0, 10^{-5}]$s (right) are presented in Fig. 2.33. The architecture of GNNATS2(A^2, I, $A^2A^\dagger$) is presented in the corresponding *MATLAB* Simulink implementation as in Fig. 2.34.

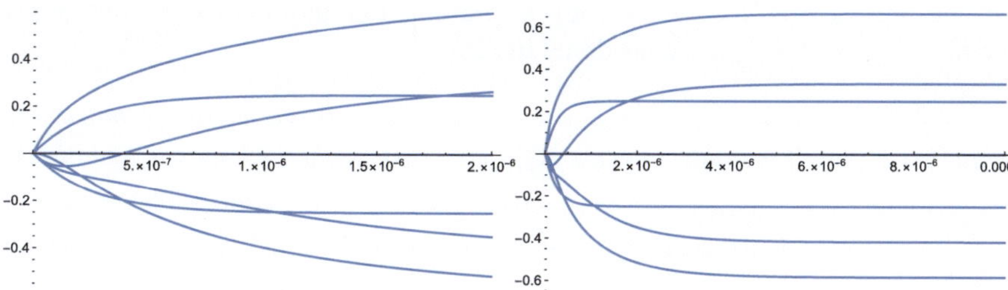

Fig. 2.33 Elementwise trajectories of $X_{t,10^6}$ in the time interval $[0, 2 * 10^{-6}]$s (left) and $[0, 10^{-5}]$s (right)

2.12 Convergence of Gradient-Based Dynamical Systems and Their Implementation

Convergence of gradient-based dynamical systems and their implementation are considered in Sect. 2.12. The objective of this section is dynamical systems arising from the GNN dynamics. Global convergence is lost, but the results of convergence are outer inverses. Some conditions for the establishment of the global convergence are investigated. Also, *MATLAB Simulink* implementation of considered dynamics is presented and tested in terms of expected convergence results. Numerical comparison between GNN and modified GNN is given.

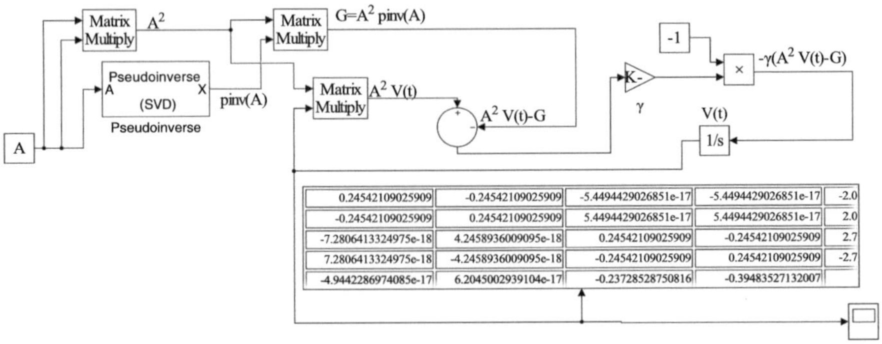

Fig. 2.34 Simulink implementation of GNNATS2$(A^2, I, A^2A^{\dagger})$

2.12.1 GNN Evolution for Generating Outer Inverses

The main property of $A^{(2)}_{\mathcal{R}(G),\mathcal{N}(G)}$, which is a basis for the dynamic state equation and later for the induced GNNs, is given in Lemma 2.3.1. Lemma 2.3.1 follows immediately from known facts $P_{L,M}G = G$ if and only if $\mathcal{R}(G) \subseteq L$ and $GP_{L,M} = G$ if and only if $\mathcal{N}(G) \supseteq M$ [6, 266]. Representation and approximations of the generalized inverse $A^{(2)}_{T,S}$ are considered in [287].

Lemma 2.12.1 *Let $A \in \mathbb{C}_r^{m\times n}$ be given and $G \in \mathbb{C}_s^{n\times m}$ be an arbitrary matrix satisfying $0 < s \le r$. If the outer inverse $A^{(2)}_{\mathcal{R}(G),\mathcal{N}(G)}$ exists, then the matrix equations*

$$GAA^{(2)}_{\mathcal{R}(G),\mathcal{N}(G)} = G, \quad A^{(2)}_{\mathcal{R}(G),\mathcal{N}(G)}AG = G \tag{2.12.1}$$

are satisfied.

It is observable that GNN(GA, I, G) (resp. GNN(I, AG, G)) are appropriate for solving $GAX = G$ (resp. $XAG = G$) in (2.12.1). That model is applicable for computing the outer inverse $A^{(2)}_{\mathcal{R}(G),\mathcal{N}(G)}$.

According to the general GNN(A, B, D) design (2.11.1), the linear GNN (GA, I, G) dynamics is defined by the dynamics

$$\frac{\mathrm{d}V_{GA}(t)}{\mathrm{d}t} = -\gamma\,(GA)^{\mathrm{T}}\,(GAV_{GA}(t) - G)\,, \quad if\ m \ge n, \tag{2.12.2}$$

as well as GNN(I, AG, G) dynamics:

$$\frac{\mathrm{d}V_{AG}(t)}{\mathrm{d}t} = -\gamma\,\,(V_{AG}(t)AG - G)\,(AG)^{\mathrm{T}}, \quad if\ m < n. \tag{2.12.3}$$

The Simulink implementation of GNN(GA, I, G) is illustrated in Fig. 2.14.

Initial research is related to the dynamics GNN(I, I, I) which is aimed to the online calculation of the ordinary matrix inverse A^{-1} [272, 364]. The global exponential convergence of gradient neural network GNN(I, I, I) in the case when A is nonsingular as well as its global stability when A is singular was verified in [363]. The general GNN(GA, I, G) (resp. GNN(I, AG, G)) evolutions are applicable for general class of outer inverses. In this way, they comprise many previously defined GNN models. Two dual improved GNN (IGNN) nonlinearly activated models for computing the Moore–Penrose inverse of a full-rank matrix were developed and investigated in [170]. These IGNN models are defined on the basis of GNN$(A^{\mathrm{T}}A, I, A^{\mathrm{T}})$ (resp. GNN$(I, AA^{\mathrm{T}}, A^{\mathrm{T}})$). A comparison between of GNN models with corresponding ZNN models was presented in [370, 374].

Specific GNN models which are not subject to GNN(GA, I, G) scheme were proposed in [282]. These dynamics are aimed to calculating the Drazin inverse of $A \in \mathbb{R}^{n\times n}$ and arise from the Tikhonov regularization principle [71]. One of these models possesses the form GNN$((A + \lambda I)^{l+1}, I, A^l)$, where $l \geq \operatorname{ind}(A)$.

2.12.2 RNN Models Arising from GNN Models

In the paper [399], authors considered the similar models to (2.12.3) and (2.12.2) but they omitted the constant terms $(GA)^{\mathrm{T}}$ and $(AG)^{\mathrm{T}}$ from the design equation. That resulted in two dual linear dynamical systems for approximating $A^{(2)}_{\mathcal{R}(G),\mathcal{N}(G)}$:

$$\frac{\mathrm{d}V_R(t)}{\mathrm{d}t} = -\gamma\ (GAV_R(t) - G)\,, \tag{2.12.4}$$

$$\frac{\mathrm{d}V_L(t)}{\mathrm{d}t} = -\gamma\ (V_L(t)AG - G)\,. \tag{2.12.5}$$

In the case $m < n$, the more effective is the model (2.12.5) denoted by RNN (I, AG, G), while otherwise, the more effective is (2.12.4), denoted by RNN (GA, I, G). The Simulink implementation of RNN(GA, I, G) is illustrated in Fig. 2.23.

The application of the dynamic evolution design RNN(GA, I, G) assumes that the real parts of eigenvalues of GA or AG lie in the open right complex halfplane, i.e., $\sigma(GA) \subset \{z : \operatorname{Re}(z) \geq 0\}$ [399]. Similarly, the evolution of RNN(I, AG, G) requires the constraint $\sigma(AG) \subset \{z : \operatorname{Re}(z) \geq 0\}$. Clearly, the models (2.12.4) and (2.12.5) are simpler than the models (2.12.2) and (2.12.3), but they lose global stability.

The recurrent neural network defined above is a linear dynamic system in matrix form. According to the linear systems theory [115], the closed-form solution of the state matrix can be described as follows:

$$\tilde{V}_G(t) = \begin{cases} \tilde{V}_R(t) := \mathrm{e}^{-\gamma GAt} V_R(0) + \gamma \mathrm{e}^{-\gamma GAt} \int_0^t \mathrm{e}^{\gamma GA\tau} G\, \mathrm{d}\tau, & m \geq n, \\ \tilde{V}_L(t) := V_L(0)\mathrm{e}^{-\gamma AGt} + \gamma G\mathrm{e}^{-\gamma AGt} \int_0^t \mathrm{e}^{\gamma AG\tau}\, \mathrm{d}\tau, & m < n. \end{cases} \tag{2.12.6}$$

To analyze the convergence and stability of a neural network, it is of major interest to know the eigenvalues of the matrix GA (or AG for the dual case). Using the principles from [399], it can be easily shown that the term $\lim_{t\to\infty} \mathrm{e}^{-\gamma GAt}$ vanishes if the matrix GA has nonnegative eigenvalues, i.e.,

$$\lim_{t\to\infty} \mathrm{e}^{-\gamma GAt} = O. \tag{2.12.7}$$

That fact and Eq. (2.12.6) imply the following representation for $\lim_{t\to\infty} V_G(t) = \overline{V}_G$:

$$\overline{V}_G = \begin{cases} \lim\limits_{t\to\infty} \gamma \mathrm{e}^{-\gamma GAt} \int_0^t \mathrm{e}^{\gamma GA\tau} G \, \mathrm{d}\tau, \; m \geq n, \\ \lim\limits_{t\to\infty} \gamma G \mathrm{e}^{-\gamma AGt} \int_0^t \mathrm{e}^{\gamma AG\tau} \, \mathrm{d}\tau, \; m < n, \end{cases} \tag{2.12.8}$$

The following theorem shows precise conditions when RNN(GA, I, G) and RNN(I, AG, G) are convergent, as well as the limit value of the state matrices $V_L(t)$ and $V_R(t)$ when $t \to +\infty$.

Theorem 2.12.1 ([399]) *Let $A \in \mathbb{R}^{m\times n}$ and $G \in \mathbb{R}_s^{n\times m}$ be chosen matrices satisfying $0 < s \leq r$. Let $\sigma(GA) = \{\lambda_1, \ldots, \lambda_n\}$ be the spectrum of GA. Suppose that the condition*

$$\mathrm{Re}\big(\lambda_j\big) \geq 0, \; j = 1, \ldots, n \tag{2.12.9}$$

is satisfied. Then the state variables matrix $V_R(t)$ converges to

$$\overline{V}_R = \lim_{t\to\infty} V_R(t) = A^{(2)}_{\mathcal{R}(G),\mathcal{N}(G)}. \tag{2.12.10}$$

An RNN model for computing the Drazin inverse A^{D} was proposed in [251]. This model can be derived removing the first constant term in RNN(A^k, A, A^k), $k \geq \mathrm{ind}(A)$.

Remark 2.12.1 In the general case, it is possible to use an appropriate replacement to generate the solution in the case the restrictions (2.12.9) on the spectrum of $\sigma(GA)$ or $\sigma(AG)$ are not satisfied: The approach used in [399] exploits the replacement of G by $G_0 = G(GAG)^{\mathrm{T}}G$ in the RNN(GA, I, G) evolution, i.e., it assumes usage of the RNN(G_0A, I, G_0) dynamics. The replacement of G by G_0 is approved by the identity $A^{(2)}_{\mathcal{R}(G),\mathcal{N}(G)} = A^{(2)}_{\mathcal{R}(G_0),\mathcal{N}(G_0)}$.

Our main intention is to find analytical expressions of matrices $V_{GA}(t)$ (resp. $V_{AG}(t)$) in GNN(GA, I, G) (resp. GNN(I, AG, G)). Thanks to such knowledge, we will be able to find the exact solution and follow the work of many GNN evolutions. In addition, the limit values as $t \to \infty$ will give accurate convergence results. Analogously, analytical expressions of matrices $V_R(t)$ (resp. $V_L(t)$) in RNN(GA, I, G) (resp. RNN(I, AG, G)) are investigated as well as $\lim\limits_{t\to\infty} V_R(t)$ and $\lim\limits_{t\to\infty} V_L(t)$). These results represent the essence of our research and are obtained in Sect. 2.12.3.

2.12.3 Further Convergence Properties

This section presents main results of the paper. Our first goal is to determine possible equilibrium, i.e., fixed or limit points of the previously mentioned dynamical systems. The most important result is the discovery that the final convergence results are given in the form of matrix inversions involving outer generalized inverses.

Lemma 2.12.2 gives an auxiliary result which will be frequently used in subsequent research.

Lemma 2.12.2 *Assume that* $M \in \mathbb{C}^{n\times n}$, $\operatorname{ind}(M) = 1$ *and all eigenvalues* λ_i, $i = 1, \dots, n$, *of* M *have nonnegative real part. Then* $\mathrm{e}^{-tM} \to I - M^{\#}M = P_{N(M)}$ *as* $t \to +\infty$.

Proof The condition $\operatorname{ind}(M) = 1$ implies that the Jordan matrix J of M in the Jordan normal form $M = QJQ^{-1}$ is diagonal, i.e., $J = \operatorname{diag}(\lambda_1, \lambda_2, \dots, \lambda_n)$. Assume that $r = \operatorname{rank} M$ and $\lambda_{r+1} = \lambda_{r+2} = \cdots = \lambda_n = 0$. The matrix exponential e^{-tM} can be written by

$$\begin{aligned}
\mathrm{e}^{-tM} &= \mathrm{e}^{-tQJQ^{-1}} = Q\mathrm{e}^{-tJ}Q^{-1} \\
&= Q \operatorname{diag}\left(\mathrm{e}^{-\lambda_1 t}, \mathrm{e}^{-\lambda_2 t}, \dots, \mathrm{e}^{-\lambda_r t}, \underbrace{1, 1, \dots, 1}_{n-r}\right) Q^{-1}.
\end{aligned}$$

Since $\operatorname{Re}(\lambda_i) \geq 0$, it means that $\mathrm{e}^{-\lambda_i t} \to 0$ when $t \to +\infty$, for every $i = 1, 2, \dots, r$. Therefore,

$$\begin{aligned}
\mathrm{e}^{-tM} &\to Q \operatorname{diag}\left(\underbrace{0, 0, \dots, 0}_{r}, \underbrace{1, 1, \dots, 1}_{n-r}\right) Q^{-1} \\
&= I - M^{\#}M = P_{N(M)}
\end{aligned}$$

as $t \to +\infty$. This completes the proof of the lemma. □

Convergence of GNN-Based Dynamics

The following theorem shows the convergence properties of the model $RNN(GA, I, G)$ introduced by (2.12.2).

Theorem 2.12.2 ([253]) *Let* $A \in \mathbb{R}^{m\times n}$ *be a given matrix, let* $G \in \mathbb{R}_s^{n\times m}$ *be an arbitrary matrix satisfying* $0 < s \leq r$ *and* $\operatorname{ind}(GA) = \operatorname{ind}(AG) = 1$. *Assume that the spectrum* $\sigma(GA) = \{\lambda_1, \dots, \lambda_n\}$ *satisfies the constraints (2.8.8):*

$$\operatorname{Re}(\lambda_j) \geq 0, \; j = 1, \dots, n.$$

Then the following statements are valid:

(a) *The exact solution to the $RNN(GA, I, G)$ evolution (2.12.4) is equal to*

$$V_R(t) := \mathcal{V}_R(G, t, \gamma, V_R(0)) = \mathrm{e}^{-\gamma t\, GA} V_R(0) + \left(I - \mathrm{e}^{-\gamma t\, GA}\right) A^{(2)}_{\mathcal{R}(G),\mathcal{N}(G)}. \tag{2.12.11}$$

(b) *If the initial approximation is the zero matrix: $V_R(0) = O$, then the exact solution to $RNN(GA, I, G)$ is*

$$V_R(t) := \mathcal{V}_R(G, t, \gamma, O) = \left(I - \mathrm{e}^{-\gamma t\, GA}\right) A^{(2)}_{\mathcal{R}(G),\mathcal{N}(G)}. \tag{2.12.12}$$

(c) *If $t > 0$ and $\gamma > 0$, then the limit of (2.12.11) is equal to*

$$\begin{aligned}\lim_{t\to\infty} V_R(t) := \tilde{\mathcal{V}}_R(V_R(0)) &= \lim_{t\to\infty} \mathcal{V}_R(G, t, \gamma, V_R(0)) = \lim_{\gamma\to\infty} \mathcal{V}_R(G, t, \gamma, V_R(0))\\ &= A^{(2)}_{\mathcal{R}(G),\mathcal{N}(G)} + \left(I - A^{(2)}_{\mathcal{R}(G),\mathcal{N}(G)} A\right) V_R(0).\end{aligned} \tag{2.12.13}$$

(d) *If $t > 0$ and $\gamma > 0$, then the limits of (2.12.11) with $V_R(0) = O$ are equal to*

$$\begin{aligned}\lim_{t\to\infty} V_R(t) := \tilde{\mathcal{V}}_R(O) &= \lim_{t\to\infty} \mathcal{V}_R(G, t, \gamma, O) = \lim_{\gamma\to\infty} \mathcal{V}_R(G, t, \gamma, O)\\ &= A^{(2)}_{\mathcal{R}(G),\mathcal{N}(G)} = G(GAG)^{\dagger} G.\end{aligned} \tag{2.12.14}$$

Proof The assumption $\mathrm{ind}(GA) = \mathrm{ind}(AG) = 1$ initiates the existence of $A^{(2)}_{\mathcal{R}(G),\mathcal{N}(G)}$ as well as its representation $A^{(2)}_{\mathcal{R}(G),\mathcal{N}(G)} = (GA)^{\#}G = G(AG)^{\#}$. Further, according to the linear dynamical systems theory [115], the closed-form solution of the state matrix $V(t)$ of RNN(GA, I, G) evolution (2.12.4) is equal to

$$\mathcal{V}_R(G, t, \gamma, V_R(0)) = \mathrm{e}^{-\gamma t\, GA} V_R(0) + \gamma \mathrm{e}^{-\gamma t\, GA} \int_0^t \mathrm{e}^{\gamma\, GA\tau} G\, \mathrm{d}\tau. \tag{2.12.15}$$

(a) The replacement $G = GAA^{(2)}_{\mathcal{R}(G),\mathcal{N}(G)}$ leads to

$$\begin{aligned}\gamma \mathrm{e}^{-\gamma t\, GA} \int_0^t \mathrm{e}^{\gamma\, GA\tau} G\, \mathrm{d}\tau &= \left[\mathrm{e}^{-\gamma t\, GA} \int_0^t \mathrm{e}^{\gamma\, GA\tau} (\gamma\, GA)\, \mathrm{d}\tau\right] A^{(2)}_{\mathcal{R}(G),\mathcal{N}(G)}\\ &= \left[\mathrm{e}^{-\gamma t GA} \int_0^t \mathrm{d}\left(\mathrm{e}^{\gamma GA\tau}\right)\right] A^{(2)}_{\mathcal{R}(G),\mathcal{N}(G)}.\end{aligned}$$

Further, using (2.12.1), one can obtain

$$\begin{aligned}\mathcal{V}_R(G,t,\gamma,V_R(0)) &= e^{-\gamma t\,GA}V_R(0)+\left[e^{-\gamma tGA}\,e^{\gamma GA\tau}\Big|_{\tau=0}^{\tau=t}\right]A^{(2)}_{\mathcal{R}(G),\mathcal{N}(G)}\\ &= e^{-\gamma t\,GA}V_R(0)+\left[\,e^{-\gamma tGA}\left(e^{\gamma tGA}-I\right)\right]A^{(2)}_{\mathcal{R}(G),\mathcal{N}(G)}.\\ &= e^{-\gamma t\,GA}V_R(0)+\left(I-e^{-\gamma tGA}\right)A^{(2)}_{\mathcal{R}(G),\mathcal{N}(G)}.\end{aligned}$$

This part of the proof is completed.

(b) This part of the proof is implied by the part **(a)**.

(c) Denote $X := A^{(2)}_{\mathcal{R}(G),\mathcal{N}(G)}$. By applying Lemma 2.12.2 on $M = GA$ and using $X = (GA)^{\#}G$, we obtain

$$e^{-\gamma tGA} \to I-(GA)^{\#}GA = I - XA \tag{2.12.16}$$

when $t,\gamma > 0$ and either $t \to +\infty$ or $\gamma \to +\infty$. The limit of (2.12.11) is now given by

$$\begin{aligned}\tilde{\mathcal{V}}_R(G,V_R(0)) &= \lim_{t\to\infty}\mathcal{V}_R(G,t,\gamma,V_R(0)) = \lim_{\gamma\to\infty}\mathcal{V}_R(G,t,\gamma,V_R(0))\\ &= \left[\lim_{t\to\infty}e^{-\gamma t\,GA}\right]V_R(0)+\left(I-\lim_{t\to\infty}e^{-\gamma t\,GA}\right)A^{(2)}_{\mathcal{R}(G),\mathcal{N}(G)}\\ &= (I-XA)V_R(0)+XAX = (I-XA)V_R(0)+X.\end{aligned}$$

So, the statement **(c)** is verified.

The statement **(d)** follows from **(c)**. This finally completes this proof. □

Remark 2.12.2 The statement **(d)** of Theorem 2.12.2 is valid in the general case when $\mathcal{R}(V_R(0)) \subset \mathcal{N}(G)$. That remark can be of help in choosing the initial matrix $V_R(0)$.

In the similar way, we can also prove the following theorem, which deals with the convergence properties of RNN(I, AG, G).

Theorem 2.12.3 ([253]) *Let* $A \in \mathbb{R}^{m\times n}$ *and* $G \in \mathbb{R}^{n\times m}_s$ *satisfy* $0 < s \le r$ *and* $\operatorname{ind}(AG) = 1$. *Assume that the spectrum* $\sigma(AG) = \{\lambda_1,\ldots,\lambda_m\}$ *satisfies the constraints*

$$\operatorname{Re}(\lambda_j) \ge 0,\; j = 1,\ldots,m. \tag{2.12.17}$$

Then the following assertions hold.

(a) *The exact solution to the the* $RNN(I, AG, G)$ *evolution (2.12.4) is equal to*

$$V_L(t) := \mathcal{V}_L(G,t,\gamma,V_L(0)) = V_L(0)e^{-\gamma t\,GA}+A^{(2)}_{\mathcal{R}(G),\mathcal{N}(G)}\left(I-e^{-\gamma t\,AG}\right). \tag{2.12.18}$$

(b) *If the initial approximation $V_L(0)$ is the zero matrix O, then the exact solution of $RNN(I, AG, G)$ is*

$$V_L(t) := \mathcal{V}_L(G, t, \gamma, O) = A^{(2)}_{\mathcal{R}(G),\mathcal{N}(G)}\left(I - e^{-\gamma t GA}\right). \tag{2.12.19}$$

(c) *If $t > 0$ and $\gamma > 0$, then the limits of (2.12.18) are equal to*

$$\begin{aligned}\lim_{t\to\infty} V_L(t) := \tilde{\mathcal{V}}_L(V_L(0)) &= \lim_{t\to\infty} \mathcal{V}_L(G, t, \gamma, V_L(0)) = \lim_{\gamma\to\infty} \mathcal{V}_L(G, t, \gamma, V_L(0)) \\ &= A^{(2)}_{\mathcal{R}(G),\mathcal{N}(G)} + V_L(0)\left(I - A^{(2)}_{\mathcal{R}(G),\mathcal{N}(G)}A\right).\end{aligned} \tag{2.12.20}$$

(d) *If $t > 0$ and $\gamma > 0$, then the limits of $V_G(t, \gamma)$ with $V_L(0) = O$ are equal to*

$$\begin{aligned}\lim_{t\to\infty} V_L(t) := \tilde{\mathcal{V}}_L(O) &= \lim_{t\to\infty} \mathcal{V}_L(G, t, \gamma, O) = \lim_{\gamma\to\infty} \mathcal{V}_L(G, t, \gamma, O) \\ &= A^{(2)}_{\mathcal{R}(G),\mathcal{N}(G)} = G(GAG)^{\dagger}G.\end{aligned} \tag{2.12.21}$$

Convergence of GNN Dynamics

The convergence of GNN(GA, I, G) in the case when A, G are constant matrices in the limiting case $t \to \infty$ follows from (2.2.12).

Corollary 2.12.1 *Assume that the $A \in \mathbb{R}^{m\times n}_r$, $G \in \mathbb{R}^{n\times m}_s$ satisfy $0 < s \leq r$ and* $\text{ind}(GA) = 1$. *The next statements are valid.*

(i) *The activation state variables matrix $V_G(t)$ of GNN(GA, I, G), defined by (2.12.2), converges as $t \to +\infty$ to the limit value*

$$\tilde{V}_{GA}(V_{GA}(0)) = (GA)^{\dagger}G + V_{GA}(0) - (GA)^{\dagger}GAV_{GA}(0), \tag{2.12.22}$$

for arbitrary $V_{GA}(0) \in \mathbb{R}^{n\times m}$.

(ii) *Particularly, $V_{GA}(0) = O$ leads to*

$$\tilde{V}_{GA}(O) = (GA)^{\dagger}G.$$

Proof The proof follows from Theorem 2.2.2 and $(GA)^{\dagger}G = A^{(2,4)}_{\mathcal{R}(GA)^{\mathrm{T}},\mathcal{N}(G)}$. □

The convergence of symbolic GNN(GA, I, G) dynamics is investigated in Theorem 2.12.4. The methods and the results are similar to the previous cases.

Theorem 2.12.4 ([253]) *Let $A \in \mathbb{R}^{m\times n}$ and $G \in \mathbb{R}^{n\times m}_s$ satisfy $0 < s \leq r$ and* $\text{ind}(GA) = 1$. *The following statements are valid.*

(a) *The exact solution to GNN(GA, I, G) dynamics (2.12.2) is equal to*

$$\begin{aligned} V_{GA}(t) &:= \mathcal{V}_{GA}(G, t, \gamma, V_{GA}(0)) \\ &= \mathrm{e}^{-\gamma t (GA)^{\mathrm{T}} GA} V_{GA}(0) + \left(I - \mathrm{e}^{-\gamma t (GA)^{\mathrm{T}} GA}\right) A^{(2,4)}_{\mathcal{R}(GA)^{\mathrm{T}}, \mathcal{N}(G)}. \end{aligned} \tag{2.12.23}$$

(b) *The exact solution to GNN(GA, I, G) with the zero initial stage $V_{GA}(0) = O$ is equal to*

$$V_{GA}(t) := \mathcal{V}_{GA}(G, t, \gamma, O) = \left(I - \mathrm{e}^{-\gamma t (GA)^{\mathrm{T}} GA}\right) A^{(2,4)}_{\mathcal{R}((GA)^{\mathrm{T}}), \mathcal{N}(G)}. \tag{2.12.24}$$

(c) *The limit of the exact solution (2.12.23) is given by*

$$\begin{aligned} \lim_{t\to\infty} V_{GA}(t) &:= (\tilde{V}_{GA})(V_{GA}(0)) = \tilde{\mathcal{V}}_{GA}(G, t, \gamma, V_{GA}(0)) \\ &= \lim_{t\to\infty} \mathcal{V}_{GA}(G, t, \gamma, V_{GA}(0)) = \lim_{\gamma\to\infty} \mathcal{V}_{GA}(G, t, \gamma, V_{GA}(0)) \\ &= A^{(2,4)}_{\mathcal{R}(GA)^{\mathrm{T}}, \mathcal{N}(G)} + \left(I - A^{(2,4)}_{\mathcal{R}(GA)^{\mathrm{T}}, \mathcal{N}(G)} A\right) V_{GA}(0). \end{aligned} \tag{2.12.25}$$

(d) *The exact solution (2.12.25) to GNN(GA, I, G) with the zero initial stage $V_{GA}(0) = O$ converges to*

$$\begin{aligned} \lim_{t\to\infty} V_{GA}(t) &:= (\tilde{V}_{GA})(O) = \tilde{\mathcal{V}}_{GA}(G, t, \gamma, O) \\ &= \lim_{t\to\infty} \mathcal{V}_{GA}(G, t, \gamma, O) = \lim_{\gamma\to\infty} \mathcal{V}_{GA}(G, t, \gamma, O) \\ &= (GA)^{\dagger} G = A^{(2,4)}_{\mathcal{R}(GA)^{\mathrm{T}}, \mathcal{N}(G)}. \end{aligned} \tag{2.12.26}$$

Proof **(a)** Using known results from [115], the closed-form solution to GNN (GA, I, G) is equal to

$$\begin{aligned} V_{GA}(t) &= \mathcal{V}_{GA}(G, t, \gamma, V_{GA}(0)) \\ &= \mathrm{e}^{-\gamma t (GA)^{\mathrm{T}} GA} V_{GA}(0) + \gamma \mathrm{e}^{-\gamma t (GA)^{\mathrm{T}} GA} \int_0^t \mathrm{e}^{\gamma (GA)^{\mathrm{T}} GA \tau} (GA)^{\mathrm{T}} G \, \mathrm{d}\tau. \end{aligned} \tag{2.12.27}$$

Since $\operatorname{ind}(GA) = 1$, it follows that $\operatorname{rank}(GA) = \operatorname{rank}(G)$, which further implies

$$GA(GA)^{\dagger} G = G.$$

Now, it follows that

$$\begin{aligned}
&\gamma \mathrm{e}^{-\gamma t\,(GA)^{\mathrm{T}}GA} \int_0^t \mathrm{e}^{\gamma (GA)^{\mathrm{T}}GA\tau} (GA)^{\mathrm{T}} G \, \mathrm{d}\tau \\
&= \left[\mathrm{e}^{-\gamma t\,(GA)^{\mathrm{T}}GA} \int_0^t \mathrm{e}^{\gamma (GA)^{\mathrm{T}}GA\tau} (\gamma (GA)^{\mathrm{T}} GA) \, \mathrm{d}\tau \right] (GA)^{\dagger} G \\
&= \left[\mathrm{e}^{-\gamma t\,(GA)^{\mathrm{T}}GA} \int_0^t \mathrm{d} \left(\mathrm{e}^{\gamma (GA)^{\mathrm{T}}GA\tau} \right) \right] (GA)^{\dagger} G.
\end{aligned}$$

Further, using several elementary integral transformations, we conclude

$$\begin{aligned}
&\mathcal{V}_{GA}(G, t, \gamma, V_{GA}(0)) \\
&= \mathrm{e}^{-\gamma t (GA)^{\mathrm{T}}GA} V_{GA}(0) + \left[\mathrm{e}^{-\gamma t\,(GA)^{\mathrm{T}}GA} \left(\mathrm{e}^{\gamma\,(GA)^{\mathrm{T}}GA\tau} \Big|_{\tau=0}^{\tau=t} \right) \right] (GA)^{\dagger} G \\
&= \mathrm{e}^{-\gamma t (GA)^{\mathrm{T}}GA} V_{GA}(0) + \left[\mathrm{e}^{-\gamma t\,(GA)^{\mathrm{T}}GA} \left(\mathrm{e}^{\gamma t\,(GA)^{\mathrm{T}}GA} - I \right) \right] (GA)^{\dagger} G.
\end{aligned}$$

Now, the proof of this part can be easily completed using $A^{(2,4)}_{\mathcal{R}(GA)^{\mathrm{T}}, \mathcal{N}(G)} = (GA)^{\dagger} G$.

(b) The zero initial state $V_{GA}(0) = O$ vanishes the first term of (2.12.27), and the proof follows from the part **(a)**.

(c) Since the matrix $(GA)^{\mathrm{T}} GA$ is of index 1 and has nonnegative real parts of all eigenvalues, an application of Lemma 2.12.2 implies

$$\begin{aligned}
\lim_{t\to\infty} \mathrm{e}^{-\gamma t (GA)^{\mathrm{T}}GA} &= I - \left((GA)^{\mathrm{T}} GA\right)^{\#} (GA)^{\mathrm{T}} GA \\
&= I - A^{(2)}_{\mathcal{R}((GA)^{\mathrm{T}}G), \mathcal{N}((GA)^{\mathrm{T}}G)} A \\
&= I - A^{(2)}_{\mathcal{R}((GA)^{\mathrm{T}}), \mathcal{N}(G)} A.
\end{aligned}$$

Using known representation of outer inverses $A^{(2)}_{\mathcal{R}(B), \mathcal{N}(C)} = B\,(CAB)^{\dagger}\, C$, one obtains

$$A^{(2)}_{\mathcal{R}((GA)^{\mathrm{T}}), \mathcal{N}(G)} = (GA)^{\mathrm{T}} \left(GA(GA)^{\mathrm{T}}\right)^{\dagger} G,$$

which implies $A^{(2)}_{\mathcal{R}((GA)^{\mathrm{T}}), \mathcal{N}(G)} \in A\{4\}$. Therefore,

$$\lim_{t\to\infty} \mathrm{e}^{-\gamma t (GA)^{\mathrm{T}}GA} = I - A^{(2,4)}_{\mathcal{R}(GA)^{\mathrm{T}}, \mathcal{N}(G)} A = I - (GA)^{\dagger} GA.$$

Now, the limit of (2.12.23) is equal to

$$(\tilde{V}_{GA})_{V_{GA}(0)} = \left(I - A^{(2,4)}_{\mathcal{R}(GA)^{\mathrm{T}}, \mathcal{N}(G)} A\right) V_{GA}(0) + A^{(2,4)}_{\mathcal{R}(GA)^{\mathrm{T}}, \mathcal{N}(G)} A A^{(2,4)}_{\mathcal{R}(GA)^{\mathrm{T}}, \mathcal{N}(G)}.$$

The proof can be easily completed.

(d) The proof follows from the part **(c)** and (2.12.23). □

The convergence of GNN(I, AG, G) in the case when A, G are constant matrices was investigated [253].

Corollary 2.12.2 ([253]) *Suppose that* $A \in \mathbb{R}_r^{m\times n}$ *and* $G \in \mathbb{R}_s^{n\times m}$ *satisfy* $0 < s \leq r$ *and and* $\operatorname{ind}(AG) = 1$. *The following assertions are true.*

(i) *The activation state variables matrix* $V_{AG}(t)$ *of GNN(I, AG, G) dynamics, defined by (2.12.3), converges as* $t \to +\infty$ *to the limit*

$$\begin{aligned}\tilde{V}_{AG}(V_{AG}(0)) &= \lim_{t\to+\infty} \mathcal{V}_{AG}(G, t, \gamma, V_{AG}(0)) \\ &= G(AG)^{\dagger} + V_{AG}(0) - V_{AG}(0)AG(AG)^{\dagger}, \end{aligned} \tag{2.12.28}$$

for every $V_{AG}(0) \in \mathbb{R}^{n\times m}$.

(ii) *Specifically, for* $V_{AG}(0) = O$, *it follows that*

$$\tilde{V}_{AG}(O) = G(AG)^{\dagger} = A^{(2,3)}_{\mathcal{N}(G),\mathcal{N}(AG)^{\perp}}.$$

In Theorem 2.12.5 we investigate the exact solution and convergence of symbolic GNN(I, AG, G) dynamics.

Theorem 2.12.5 ([253]) *Suppose that* $A \in \mathbb{R}^{m\times n}$ *and* $G \in \mathbb{R}_s^{n\times m}$ *satisfy* $0 < s \leq r$ *and* $\operatorname{ind}(AG) = 1$. *Then the next statements are true.*

(a) *The exact solution to GNN(I, AG, G) dynamics (2.12.3) is equal to*

$$\begin{aligned}V_{AG}(t) &:= \mathcal{V}_{AG}(G, t, \gamma, V_{AG}(0)) \\ &= \mathrm{e}^{-\gamma t AG(AG)^{\mathrm{T}}} V_{AG}(0) + A^{(2,3)}_{\mathcal{R}(G),\mathcal{N}(AG)^{\mathrm{T}}}\left(I - \mathrm{e}^{-\gamma t\, AG(AG)^{\mathrm{T}}}\right). \end{aligned} \tag{2.12.29}$$

(b) *The exact solution to GNN(I, AG, G) dynamics with the zero initial stage* $V_{AG}(0) = O$ *is defined by*

$$V_{AG}(t) := \mathcal{V}_{AG}(G, t, \gamma, O) = G(AG)^{\dagger}\left(I - \mathrm{e}^{-\gamma t\, AG(AG)^{\mathrm{T}}}\right). \tag{2.12.30}$$

(c) *The exact solution (2.12.3) to GNN(I, AG, G) converges to*

$$\begin{aligned}\lim_{t\to\infty} V_{AG}(t) &:= (\tilde{V}_{AG})(V_{AG}(0)) = \tilde{\mathcal{V}}_{AG}(G, t, \gamma, V_{AG}(0)) \\ &= \lim_{t\to\infty} \mathcal{V}_{AG}(G, t, \gamma, V_{AG}(0)) = \lim_{\gamma\to\infty} \mathcal{V}_{AG}(G, t, \gamma, V_{AG}(0)) \\ &= A^{(2,3)}_{\mathcal{R}(G),\mathcal{N}((AG)^{\mathrm{T}})} + \left(I - AA^{(2,3)}_{\mathcal{R}(G),\mathcal{N}((AG)^{\mathrm{T}})}\right) V_{AG}(0). \end{aligned} \tag{2.12.31}$$

(d) *The exact solution (2.12.31) to GNN(I, AG, G) with the zero initial stage $V_G(0) = O$ converges to*

$$\begin{aligned}\lim_{t\to\infty} V_{AG}(t) &:= (\tilde{V}_{AG})(O) = \tilde{\mathcal{V}}_{AG}(G, t, \gamma, O)\\ &= \lim_{t\to\infty} \mathcal{V}_{AG}(G, t, \gamma, O) = \lim_{\gamma\to\infty} \mathcal{V}_{GA}(G, t, \gamma, O) \\ &= G(AG)^{\dagger} = A^{(2,3)}_{\mathcal{R}(G),\mathcal{N}((AG)^{\mathrm{T}})}.\end{aligned} \tag{2.12.32}$$

Proof The closed-form solution of the state matrix $V(t)$ of the GNN(I, AGI, G) dynamics is equal to

$$V_G(t) = V_G(0)\mathrm{e}^{-\gamma t AG(AG)^{\mathrm{T}}} + \gamma G(AG)^{\mathrm{T}}\mathrm{e}^{-\gamma t AG(AG)^{\mathrm{T}}} \int_0^t \mathrm{e}^{\gamma AG(AG)^{\mathrm{T}}\tau}\, \mathrm{d}\tau. \tag{2.12.33}$$

□

Remark 2.12.3 It is important to mention that the parts **(a)** and **(b)** of Theorems 2.12.2, 2.12.3, 2.12.4, 2.12.5 discover convergence results of considered dynamical systems in each time instant. Such analysis has not been investigated so far. The parts **(c)** and **(d)** of Theorems 2.12.2, 2.12.3, 2.12.4, 2.12.5 consider convergence results of considered dynamical systems in the limit as $t \to \infty$. Also, one important remark is that the stated dynamical systems are in the essence systems of linear equations, for which the Duhamel formula provides an explicit solution. But, the main results are established using transformations on immediately obtained explicit solutions by means of tools available in linear algebra and generalized inverses theory. For example, only the result (2.12.15) of Theorem 2.12.2 is obtained using the Duhamel formula. All results are obtained using further transformations of this result.

Summary review of exact solutions of GNN and GNN-based models as well as limits of these solutions are presented in Table 2.2.

Exact solutions of GNN(A, I, I) and GNN(A, I, I) are discovered in [261]. The integer k in Table 2.2 means ind(A).

2.12.4 *Illustrative Examples*

Theoretical results obtained in previous sections are verified using software developed in package *Mathematica*, and described in [247]. Also, the *MATLAB Simulink* implementations of proposed dynamical evolutions are used. The exact solution of the model $\mathcal{M}$ represents its trajectory, and will be denoted by $\mathfrak{T}(\mathcal{M})$.

Example 2.12.1 Consider

Table 2.2 Exact solutions and limits of GNN and GNN-based models

Model	Exact solution	Limit
$RNN(GA, I, G)$	$\mathrm{e}^{-\gamma t\, GA} V_R(0) + \left(I - \mathrm{e}^{-\gamma t\, GA}\right) A^{(2)}_{\mathcal{R}(G),\mathcal{N}(G)}$	$A^{(2)}_{\mathcal{R}(G),\mathcal{N}(G)} + \left(I - A^{(2)}_{\mathcal{R}(G),\mathcal{N}(G)} A\right) V_R(0)$
$RNN(I, AG, G)$	$V_L(0)\mathrm{e}^{-\gamma t\, GA} + A^{(2)}_{\mathcal{R}(G),\mathcal{N}(G)} \left(I - \mathrm{e}^{-\gamma t\, AG}\right)$	$A^{(2)}_{\mathcal{R}(G),\mathcal{N}(G)} + V_L(0)\left(I - AA^{(2)}_{\mathcal{R}(G),\mathcal{N}(G)}\right)$
$GNN(GA, I, G)$	$\mathrm{e}^{-\gamma t (GA)^{\mathrm{T}} GA} V_{GA}(0) + \left(I - \mathrm{e}^{-\gamma t\, (GA)^{\mathrm{T}} GA}\right) A^{(2,4)}_{\mathcal{R}((GA)^{\mathrm{T}}),\mathcal{N}(G)}$	$A^{(2,4)}_{\mathcal{R}((GA)^{\mathrm{T}}),\mathcal{N}(G)} + \left(I - A^{(2,4)}_{\mathcal{R}((GA)^{\mathrm{T}}),\mathcal{N}(G)} A\right) V_{GA}(0)$
$GNN(I, AG, G)$	$\mathrm{e}^{-\gamma t AG (AG)^{\mathrm{T}}} V_{AG}(0) + A^{(2,3)}_{\mathcal{R}(G),\mathcal{N}(AG)^{\mathrm{T}}} \left(I - \mathrm{e}^{-\gamma t\, AG(AG)^{\mathrm{T}}}\right)$	$A^{(2,3)}_{\mathcal{R}(G),\mathcal{N}((AG)^{\mathrm{T}})} + V_{AG}(0)\left(I - AA^{(2,3)}_{\mathcal{R}(G),\mathcal{N}((AG)^{\mathrm{T}})}\right)$
$GNN(A, I, I)$	$\mathrm{e}^{-\gamma t A} V_A(0) + \left(I - \mathrm{e}^{-\gamma t\, A}\right) A^{-1}$	A^{-1}
$GNN(I, A, I)$	$V_A(0)\mathrm{e}^{-\gamma t A} + A^{-1}\left(I - \mathrm{e}^{-\gamma t\, A}\right)$	A^{-1}
$RNN(A, I, I)$	$\mathrm{e}^{-\gamma t\, A} V_R(0) + \left(I - \mathrm{e}^{-\gamma t\, A}\right) A^{-1}$	A^{-1}
$RNN(I, A, I)$	$V_L(0)\mathrm{e}^{-\gamma t\, A} + A^{-1}\left(I - \mathrm{e}^{-\gamma t\, A}\right)$	A^{-1}
$RNN(A^{\mathrm{T}}A, I, A^{\mathrm{T}})$	$\mathrm{e}^{-\gamma t\, A^{\mathrm{T}}A} V_R(0) + \left(I - \mathrm{e}^{-\gamma t\, A^{\mathrm{T}}A}\right) A^{\dagger}$	$A^{\dagger} + \left(I - A^{\dagger}A\right) V_R(0)$
$RNN(I, AA^{\mathrm{T}}, A^{\mathrm{T}})$	$V_L(0)\mathrm{e}^{-\gamma t\, A^{\mathrm{T}}A} + A^{\dagger}\left(I - \mathrm{e}^{-\gamma t\, AA^{\mathrm{T}}}\right)$	$A^{\dagger} + V_L(0)\left(I - AA^{\dagger}\right)$
$GNN(A^{\mathrm{T}}A, I, A^{\mathrm{T}})$	$\mathrm{e}^{-\gamma t (A^{\mathrm{T}}A)^{\mathrm{T}} A^{\mathrm{T}}A} V_{A^{\mathrm{T}}A}(0) + \left(I - \mathrm{e}^{-\gamma t\, (A^{\mathrm{T}}A)^{\mathrm{T}} A^{\mathrm{T}}A}\right) A^{\dagger}$	$A^{\dagger} + \left(I - A^{\dagger}A\right) V_{A^{\mathrm{T}}A}(0)$
$GNN(I, AA^{\mathrm{T}}, A^{\mathrm{T}})$	$\mathrm{e}^{-\gamma t AA^{\mathrm{T}} (AA^{\mathrm{T}})^{\mathrm{T}}} V_{AA^{\mathrm{T}}}(0) + A^{\dagger}\left(I - \mathrm{e}^{-\gamma t\, AA^{\mathrm{T}}(AA^{\mathrm{T}})^{\mathrm{T}}}\right)$	$A^{\dagger} + \left(I - AA^{\dagger}\right) V_{AA^{\mathrm{T}}}(0)$
$RNN(A^{k+1}, I, A^k)$	$\mathrm{e}^{-\gamma t\, A^{k+1}} V_R(0) + \left(I - \mathrm{e}^{-\gamma t\, A^{k+1}}\right) A^{\mathrm{d}}$	$A^{\mathrm{d}} + \left(I - A^{\mathrm{d}}A\right) V_R(0)$
$RNN(I, A^{k+1}, A^k)$	$V_L(0)\mathrm{e}^{-\gamma t\, A^{k+1}} + A^{\mathrm{d}}\left(I - \mathrm{e}^{-\gamma t\, A^{k+1}}\right)$	$A^{\mathrm{d}} + V_L(0)\left(I - AA^{\mathrm{d}}\right)$
$GNN(A^{k+1}, I, A^k)$	$\mathrm{e}^{-\gamma t (A^{k+1})^{\mathrm{T}} A^{k+1}} V_{A^{k+1}}(0) + \left(I - \mathrm{e}^{-\gamma t\, (A^{k+1})^{\mathrm{T}} A^{k+1}}\right) (A^{k+1})^{\dagger} A^k$	$(A^{k+1})^{\dagger} A^k + \left(I - (A^{k+1})^{\dagger} A^{k+1}\right) V_{A^{k+1}}(0)$
$GNN(I, A^{k+1}, A^k)$	$\mathrm{e}^{-\gamma t A^{k+1} (A^{k+1})^{\mathrm{T}}} V_{A^{k+1}}(0) + (A^{k+1})^{\dagger} A^k \left(I - \mathrm{e}^{-\gamma t\, A^{k+1}(A^{k+1})^{\mathrm{T}}}\right)$	$A^k (A^{k+1})^{\dagger} + \left(I - A^{k+1}(A^{k+1})^{\dagger}\right) V_{A^{k+1}}(0)$

$$
A = \begin{bmatrix} 1 & 2 & 3 & 4 & 1 \\ 1 & 3 & 4 & 6 & 2 \\ 2 & 3 & 4 & 5 & 3 \\ 3 & 4 & 5 & 6 & 4 \\ 4 & 5 & 6 & 7 & 6 \\ 6 & 6 & 7 & 7 & 8 \end{bmatrix}, \quad G = \begin{bmatrix} 0 & 0 & 0 & 0 & 0 & 0 \\ 1 & 2 & 1 & 2 & 1 & 2 \\ 2 & 3 & 2 & 3 & 2 & 3 \\ 3 & 5 & 3 & 5 & 3 & 5 \\ 0 & 1 & 0 & 1 & 0 & 1 \end{bmatrix}.
$$

(A) According to Theorem 2.12.4, the model GNN(GA, I, G) in the case $V_{GA}(0) = O$ generates the trajectory

$$\begin{aligned}\mathfrak{T}(GNN(GA, I, G)) :&= \mathcal{V}_{GA}(G, t, \gamma, O)\\ &= \left(I - \mathrm{e}^{-\gamma t\,(GA)^{\mathrm{T}}GA}\right) A^{(2,4)}_{\mathcal{R}(GA)^{\mathrm{T}},\mathcal{N}(G)}\\ &= \left(I - \mathrm{e}^{-\gamma t\,(GA)^{\mathrm{T}}GA}\right) (GA)^{\dagger}G.\end{aligned}$$

The limit of this expression is

$$(\tilde{V}_{GA})(O) := \lim_{t\to\infty} (\mathfrak{T}(GNN(GA, I, G))) = (GA)^{\dagger}G$$
$$= \begin{bmatrix} -\frac{473}{1682} & \frac{194}{841} & -\frac{473}{1682} & \frac{194}{841} & -\frac{473}{1682} & \frac{194}{841}\\ -\frac{64}{841} & \frac{121}{1682} & -\frac{64}{841} & \frac{121}{1682} & -\frac{64}{841} & \frac{121}{1682}\\ \frac{217}{1682} & -\frac{73}{841} & \frac{217}{1682} & -\frac{73}{841} & \frac{217}{1682} & -\frac{73}{841}\\ \frac{281}{841} & -\frac{413}{1682} & \frac{281}{841} & -\frac{413}{1682} & \frac{281}{841} & -\frac{413}{1682}\\ -\frac{277}{841} & \frac{229}{841} & -\frac{277}{841} & \frac{229}{841} & -\frac{277}{841} & \frac{229}{841} \end{bmatrix},$$

for arbitrary value of the gain parameter $\gamma > 0$.
MATLAB Simulink implementation of GNN(GA, I, G), presented in Fig. 2.35, generates the following approximation of $(GA)^{\dagger}G$ within the time interval $[0, 10^{-5}]$ (seconds):

$$X = \begin{bmatrix} -0.2812 & 0.2307 & -0.2812 & 0.2307 & -0.2812 & 0.2307\\ -0.0761 & 0.0719 & -0.0761 & 0.0719 & -0.0761 & 0.0719\\ 0.1290 & -0.0868 & 0.1290 & -0.0868 & 0.1290 & -0.0868\\ 0.3341 & -0.2455 & 0.3341 & -0.2455 & 0.3341 & -0.2455\\ -0.3294 & 0.2723 & -0.3294 & 0.2723 & -0.3294 & 0.2723 \end{bmatrix}.$$

The initial state $V_{GA}(0) = A^{\mathrm{T}}$ in GNN(GA, I, G) under the same environment initiates the result

$$\mathfrak{T}(GNN(GA, I, G)) = \mathcal{V}_{GA}(G, t, \gamma, A^{\mathrm{T}}) = (GA)^{\dagger}G + A^{\mathrm{T}} - (GA)^{\dagger}GAA^{\mathrm{T}}$$
$$= \begin{bmatrix} -\frac{83}{1682} & \frac{95}{1682} & -\frac{19}{58} & \frac{272}{841} & -\frac{785}{1682} & \frac{525}{1682}\\ \frac{51}{841} & \frac{276}{841} & -\frac{3}{29} & \frac{213}{1682} & -\frac{156}{841} & -\frac{201}{841}\\ \frac{287}{1682} & -\frac{673}{1682} & \frac{7}{58} & -\frac{59}{841} & \frac{161}{1682} & \frac{353}{1682}\\ \frac{236}{841} & -\frac{108}{841} & \frac{10}{29} & -\frac{449}{1682} & \frac{317}{841} & -\frac{287}{841}\\ -\frac{502}{841} & \frac{301}{841} & -\frac{8}{29} & \frac{139}{841} & -\frac{97}{841} & \frac{247}{841} \end{bmatrix}.$$

The limit of $\mathfrak{T}(GNN(GA, I, G))$ is

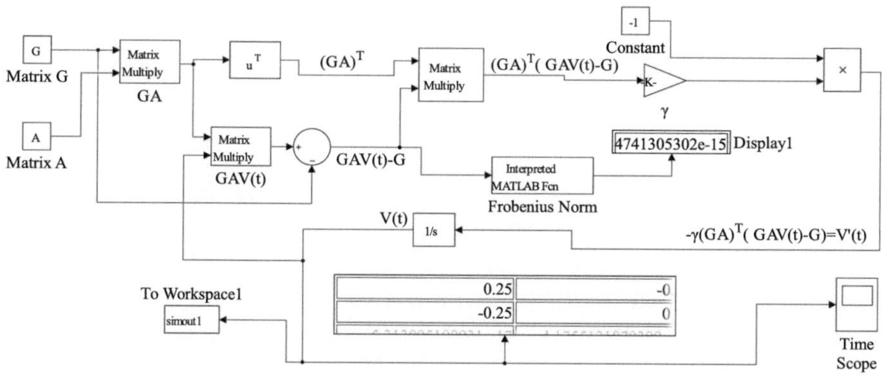

Fig. 2.35 Simulink implementation of GNN(GA, I, G)

$$\lim_{t\to\infty}(\mathfrak{T}(GNN(GA,I,G))) = (\tilde{V}_{GA})(A^{\mathrm{T}}) = (GA)^{\dagger}G$$

$$= \begin{bmatrix} -\frac{83}{1682} & \frac{95}{1682} & -\frac{19}{58} & \frac{272}{841} & -\frac{785}{1682} & \frac{525}{1682} \\ \frac{51}{841} & \frac{276}{841} & -\frac{3}{29} & \frac{213}{1682} & -\frac{156}{841} & -\frac{201}{841} \\ \frac{287}{1682} & -\frac{673}{1682} & \frac{7}{58} & -\frac{59}{841} & \frac{161}{1682} & \frac{353}{1682} \\ \frac{236}{841} & -\frac{108}{841} & \frac{10}{29} & -\frac{449}{1682} & \frac{317}{841} & -\frac{287}{841} \\ -\frac{502}{841} & \frac{301}{841} & -\frac{8}{29} & \frac{139}{841} & -\frac{97}{841} & \frac{247}{841} \end{bmatrix}.$$

MATLAB Simulink implementation of GNN(GA, I, G) generates the following approximation of $(GA)^{\dagger}G$ with $\gamma = 10^8$ and within the time interval $[0, 10^{-5}]$ (seconds):

$$X = (GA)^{\dagger}G + A^{\mathrm{T}} - (GA)^{\dagger}GAA^{\mathrm{T}}$$

$$= \begin{bmatrix} -0.0493 & 0.0565 & -0.3276 & 0.3234 & -0.4667 & 0.3121 \\ 0.0606 & 0.3282 & -0.1034 & 0.1266 & -0.1855 & -0.2390 \\ 0.1706 & -0.4001 & 0.1207 & -0.0702 & 0.0957 & 0.2099 \\ 0.2806 & -0.1284 & 0.3448 & -0.2669 & 0.376932 & -0.3413 \\ -0.5969 & 0.3579 & -0.2759 & 0.1653 & -0.1153 & 0.2937 \end{bmatrix}.$$

(B) On the other hand, the solution generated by the *MATLAB Simulink* implementation of RNN(GA, I, G), presented in Fig. 2.36, with $V_R(0) = O$ is equal to

$$\mathfrak{T}(RNN(GA,I,G)) = \mathcal{V}_R(G,t,\gamma,O) = \left(I - \mathrm{e}^{-\gamma t\, GA}\right) A^{(2)}_{\mathcal{R}(G),\mathcal{N}(G)}.$$

The limit of $\mathfrak{T}(RNN(GA, I, G))$ is equal to

$$\lim_{t\to\infty}(\mathfrak{T}(RNN(GA,I,G))) = \tilde{\mathcal{V}}_R(O) = A^{(2)}_{\mathcal{R}(G),\mathcal{N}(G)} = G(GAG)^{\dagger}G$$

$$= \begin{bmatrix} 0 & 0 & 0 & 0 & 0 & 0 \\ -\frac{7}{58} & \frac{19}{174} & -\frac{7}{58} & \frac{19}{174} & -\frac{7}{58} & \frac{19}{174} \\ \frac{10}{29} & -\frac{23}{87} & \frac{10}{29} & -\frac{23}{87} & \frac{10}{29} & -\frac{23}{87} \\ \frac{13}{58} & -\frac{9}{58} & \frac{13}{58} & -\frac{9}{58} & \frac{13}{58} & -\frac{9}{58} \\ -\frac{17}{29} & \frac{14}{29} & -\frac{17}{29} & \frac{14}{29} & -\frac{17}{29} & \frac{14}{29} \end{bmatrix}.$$

Simulink implementation of RNN(GA, I, G) with $V_R(0) = O$ approximates $A^{(2)}_{\mathcal{R}(G),\mathcal{N}(G)}$ by the matrix

$$\begin{bmatrix} 0 & 0 & 0 & 0 & 0 & 0 \\ -0.1207 & 0.1092 & -0.1207 & 0.1092 & -0.1207 & 0.1092 \\ 0.3448 & -0.2644 & 0.3448 & -0.2644 & 0.3448 & -0.2644 \\ 0.2241 & -0.1552 & 0.2241 & -0.1552 & 0.2241 & -0.1552 \\ -0.5862 & 0.4828 & -0.5862 & 0.4828 & -0.5862 & 0.4828 \end{bmatrix}.$$

The trajectory of RNN(GA, I, G) with $V_R(0) = A^{\mathrm{T}}$ is

$$\mathfrak{T}(RNN(GA,I,G)) = \mathcal{V}_R(GA,t,\gamma,A^{\mathrm{T}}) = \mathrm{e}^{-\gamma t\, GA}A^{\mathrm{T}} + \left(I - \mathrm{e}^{-\gamma t\, GA}\right)A^{(2)}_{\mathcal{R}(G),\mathcal{N}(G)}.$$

The limit of $\mathfrak{T}(RNN(GA, I, G))$ is equal to

$$\lim_{t\to\infty}(\mathfrak{T}(RNN(GA,I,G))) = \tilde{\mathcal{V}}_R(A^{\mathrm{T}}) = A^{(2)}_{\mathcal{R}(G),\mathcal{N}(G)} + \left(I - A^{(2)}_{\mathcal{R}(G),\mathcal{N}(G)}A\right)A^{\mathrm{T}}$$

$$= \begin{bmatrix} 1 & 1 & 2 & 3 & 4 & 6 \\ -\frac{13}{58} & \frac{1}{174} & -\frac{113}{174} & -\frac{91}{174} & -\frac{205}{174} & -\frac{257}{174} \\ \frac{31}{29} & \frac{40}{87} & \frac{178}{87} & \frac{188}{87} & \frac{326}{87} & \frac{421}{87} \\ -\frac{9}{58} & -\frac{31}{58} & -\frac{35}{58} & -\frac{79}{58} & -\frac{83}{58} & -\frac{153}{58} \\ -\frac{44}{29} & -\frac{13}{29} & -\frac{68}{29} & -\frac{64}{29} & -\frac{119}{29} & -\frac{139}{29} \end{bmatrix}.$$

Finally, the Simulink implementation of RNN(GA, I, G) under the nonzero initial state $V_R(0) = A^{\mathrm{T}}$ produces the following approximation of $(I - A^{(2)}_{\mathcal{R}(G),\mathcal{N}(G)}A)A^{\mathrm{T}} + A^{(2)}_{\mathcal{R}(G),\mathcal{N}(G)}$:

$$X = \begin{bmatrix} 1.0000 & 1.0000 & 2.0000 & 3.0000 & 4.0000 & 6.0000 \\ -0.2241 & 0.0057 & -0.6494 & -0.5230 & -1.1782 & -1.4770 \\ 1.0690 & 0.4598 & 2.0460 & 2.1609 & 3.7471 & 4.8391 \\ -0.1552 & -0.5345 & -0.6034 & -1.3621 & -1.4310 & -2.6379 \\ -1.5172 & -0.4483 & -2.3448 & -2.2069 & -4.1034 & -4.7931 \end{bmatrix}.$$

Example 2.12.2 This example is aimed to the input matrix from [247]

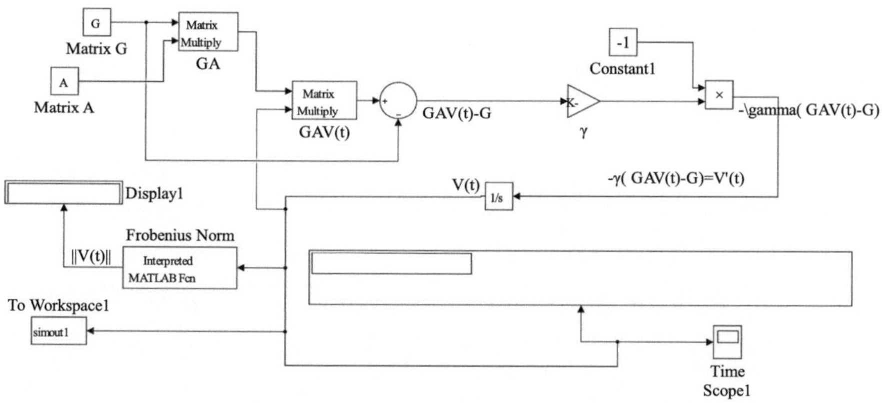

Fig. 2.36 Simulink implementation of RNN(GA, I, G)

$$A = \begin{bmatrix} 1 & -1 & 0 & 0 & 0 & 0 \\ -1 & 1 & 0 & 0 & 0 & 0 \\ -1 & -1 & 1 & -1 & 0 & 0 \\ -1 & -1 & -1 & 1 & 0 & 0 \\ -1 & -1 & -1 & 0 & 2 & -1 \\ -1 & -1 & 0 & -1 & -1 & 2 \end{bmatrix}$$

and choose various appropriate matrices G:

(A) The model $RNN(A^3, I, A^2)$ generated using $G = A^2$ in RNN(GA, I, G) under the zero initial state $V_R(0) = O$ was considered in [247]. It can be mentioned that the solution

$$V_R^D = \mathcal{V}_R(A^2, t, \gamma, O) = \mathfrak{T}\left(RNN(A^3, I, A^2)\right) = \left(I - \mathrm{e}^{-\gamma t A^3}\right) A^{\mathrm{d}}$$

can be generated by the *Mathematica* expression:

```
G=A.A; AD = G.PseudoInverse[G.A.G].G;
(IdentityMatrix[6] - MatrixExp[-\[Gamma]*t*G.A]).AD // Simplify
```

which is equal to

$$\begin{bmatrix} \frac{1}{4}\left(1-e^{-8t\gamma}\right) & \frac{1}{4}\left(-1+e^{-8t\gamma}\right) & 0 & 0 & 0 & 0 \\ \frac{1}{4}\left(-1+e^{-8t\gamma}\right) & \frac{1}{4}\left(1-e^{-8t\gamma}\right) & 0 & 0 & 0 & 0 \\ 0 & 0 & \frac{1}{4}\left(1-e^{-8t\gamma}\right) & \frac{1}{4}\left(-1+e^{-8t\gamma}\right) & 0 & 0 \\ 0 & 0 & \frac{1}{4}\left(-1+e^{-8t\gamma}\right) & \frac{1}{4}\left(1-e^{-8t\gamma}\right) & 0 & 0 \\ 0 & 0 & \frac{1}{12}\left(-5+2e^{-27t\gamma}-3e^{-8t\gamma}+6e^{-t\gamma}\right) & \frac{1}{12}\left(-7-2e^{-27t\gamma}+3e^{-8t\gamma}+6e^{-t\gamma}\right) & \frac{1}{6}\left(4-e^{-27t\gamma}-3e^{-t\gamma}\right) & \frac{1}{6}\left(2+e^{-27t\gamma}-3e^{-t\gamma}\right) \\ 0 & 0 & \frac{1}{12}\left(-7-2e^{-27t\gamma}+3e^{-8t\gamma}+6e^{-t\gamma}\right) & \frac{1}{12}\left(-5+2e^{-27t\gamma}-3e^{-8t\gamma}+6e^{-t\gamma}\right) & \frac{1}{6}\left(2+e^{-27t\gamma}-3e^{-t\gamma}\right) & \frac{1}{6}\left(4-e^{-27t\gamma}-3e^{-t\gamma}\right) \end{bmatrix}.$$

The limit of this expression gives A^{d} as follows:

$$\lim_{t\to\infty}\left(\mathfrak{T}(RNN(A^3, I, A^2))\right) = \tilde{\mathcal{V}}_R(O) = A^{\mathrm{d}} = \begin{bmatrix} \frac{1}{4} & -\frac{1}{4} & 0 & 0 & 0 & 0 \\ -\frac{1}{4} & \frac{1}{4} & 0 & 0 & 0 & 0 \\ 0 & 0 & \frac{1}{4} & -\frac{1}{4} & 0 & 0 \\ 0 & 0 & -\frac{1}{4} & \frac{1}{4} & 0 & 0 \\ 0 & 0 & -\frac{5}{12} & -\frac{7}{12} & \frac{2}{3} & \frac{1}{3} \\ 0 & 0 & -\frac{7}{12} & -\frac{5}{12} & \frac{1}{3} & \frac{2}{3} \end{bmatrix}.$$

Note that A^{d} can be computed using $A^{\mathrm{d}} = A(A^3)^{\dagger}A$.
The *Simulink* approximation of $\lim_{t\to\infty}\left(\mathfrak{T}(RNN(A^3, I, A^2))\right) = A^{\mathrm{d}}$ generated within the time $[0, 10^{-6}]$ (seconds) and using $\gamma = 10^8$ is given by

$$\begin{bmatrix} 0.25 & -0.25 & 0. & 0. & 0. & 0. \\ -0.25 & 0.25 & 0. & 0. & 0. & 0. \\ 0. & 0. & 0.25 & -0.25 & 0. & 0. \\ 0. & 0. & -0.25 & 0.25 & 0. & 0. \\ 3.298e{-48} & 3.301e{-48} & -0.416667 & -0.583333 & 0.666667 & 0.333333 \\ 2.949e{-48} & 2.953e{-48} & -0.583333 & -0.416667 & 0.333333 & 0.666667 \end{bmatrix}.$$

The state trajectories of the elements $(V_R^D)_{ij}$ under the parameters $\gamma \to 10^8$, $t \in [0, 10^{-6}]$ (seconds) are displayed in Fig. 2.37.

(B) On the other hand, the trajectory $\mathfrak{T}(GNN(A^3, I, A^2))$ of the original GNN evolution GNN(A^3, I, A^2) under the zero initial state $V_{GA}(0) = O$, given by $\mathcal{V}_{GA}(A^2, t, \gamma, O) = \left(I - \mathrm{e}^{-\gamma t\,(A^3)^{\mathrm{T}}A^3}\right)(A^3)^{\dagger}A^2$, can be traced by the *Mathematica* expression

```
G=A.A; AD = G.PseudoInverse[G.A.G].G;
(IdentityMatrix[6]-MatrixExp[-\[Gamma] *t*Transpose[G.A].G.A]).
        PseudoInverse[G.A].G // Simplify
```

The limit $(\tilde{V}_{GA})(O) := \lim_{t\to\infty}\mathcal{V}_{GA}(A^2, t, \gamma, O)$ can be achieved using the standard *Mathematica* function `Limit` as follows: $\tilde{V}_{GA}(O) =$ `Limit`

$[\mathcal{V}_{GA}(G, t, \gamma, O)/.\gamma \to 10^3, t \to \infty]$, which produces

$$(\tilde{V}_{GA})(O) = (A^3)^\dagger A^2 = \begin{bmatrix} \frac{1}{4} & -\frac{1}{4} & 0 & 0 & 0 & 0 \\ -\frac{1}{4} & \frac{1}{4} & 0 & 0 & 0 & 0 \\ 0 & 0 & \frac{1}{2} & 0 & -\frac{1}{4} & -\frac{1}{4} \\ 0 & 0 & 0 & \frac{1}{2} & -\frac{1}{4} & -\frac{1}{4} \\ 0 & 0 & -\frac{1}{6} & -\frac{1}{3} & \frac{5}{12} & \frac{1}{12} \\ 0 & 0 & -\frac{1}{3} & -\frac{1}{6} & \frac{1}{12} & \frac{5}{12} \end{bmatrix}.$$

Another possibility is to achieve the same result is usage of the limit as $\gamma \to \infty$ in finite time. It is possible to use the *Mathematica* command `Limit` to find $(\tilde{V}_{GA})(O)$:

$$\tilde{V}_{GA}(O) = \text{Limit}[\mathcal{V}_{GA}(G, t, \gamma, O)/.\gamma \to \infty, t \to 10^{-8}].$$

The Simulink implementation of $GNN(A^3, I, A^2)$ produces the following approximation of $(A^3)^\dagger A^2$ within the time $[0, 10^{-6}]$ and using $\gamma = 10^8$:

$$\begin{bmatrix} 0.25 & -0.25 & 0. & 0. & 0. & 0. \\ -0.25 & 0.25 & 0. & 0. & 0. & 0. \\ 0. & 0. & 0.500 & 7.976e-09 & -0.25000 & -0.250001 \\ 0. & 0. & -1.226e-06. & 0.500000 & -0.250001 & -0.250001 \\ 0. & 0. & -0.166667 & -0.333333 & 0.416667 & 0.0833333 \\ 0. & 0. & -0.333333 & -0.166667 & 0.0833333 & 0.416667 \end{bmatrix}.$$

(C) The expression $V_{A^T}(t) = \mathfrak{T}(RNN(A^TA, I, A^T))$ with $V_{A^T}(0) = O$ is the "symbolic Simulink" for computing $A^\dagger$. The limit expression $\tilde{V}_{A^T}$ =`Limit` $[V_{A^T}(t)/.\gamma \to 10^3, t \to \infty]$ gives the exact Moore–Penrose inverse of A under the assumption $V_{A^T}(0) = O$:

$$\tilde{V}_{A^T}(O) = A^\dagger = \begin{bmatrix} \frac{1}{4} & -\frac{1}{4} & -\frac{1}{4} & -\frac{1}{4} & 0 & 0 \\ -\frac{1}{4} & \frac{1}{4} & -\frac{1}{4} & -\frac{1}{4} & 0 & 0 \\ 0 & 0 & \frac{1}{2} & 0 & -\frac{1}{4} & -\frac{1}{4} \\ 0 & 0 & 0 & \frac{1}{2} & -\frac{1}{4} & -\frac{1}{4} \\ 0 & 0 & -\frac{1}{6} & -\frac{1}{3} & \frac{5}{12} & \frac{1}{12} \\ 0 & 0 & -\frac{1}{3} & -\frac{1}{6} & \frac{1}{12} & \frac{5}{12} \end{bmatrix}.$$

The limit of $V_{A^TA}(t) := \mathfrak{T}(GNN(A^TA, I, A^T))$ under the nonzero initial state $V_{A^TA}(0) = A^T$ generates $A^\dagger + \left(I - A^\dagger A\right) A^T = A^\dagger$, which is equal to

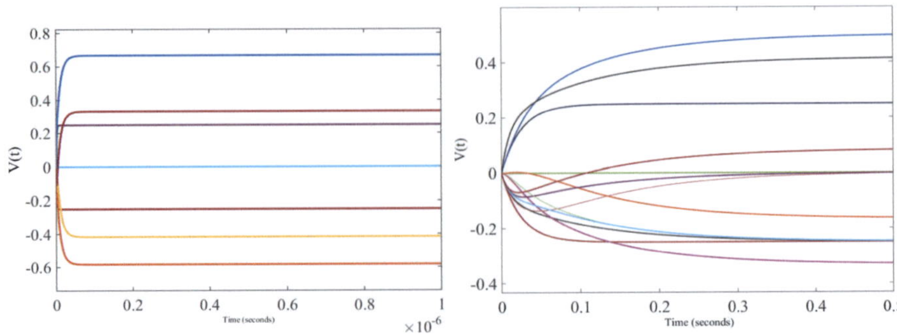

Fig. 2.37 The state trajectories of $(V_R^D)_{ij}$ with $\gamma = 10^8$, $t \in [0, 10^{-6}]s$ (left) and $\gamma = 10$, $t \in [0, 0.5]s$ (right) in Example 2.12.2

$$\begin{bmatrix} \frac{1}{4} & -\frac{1}{4} & -\frac{1}{4} & -\frac{1}{4} & 0 & 0 \\ -\frac{1}{4} & \frac{1}{4} & -\frac{1}{4} & -\frac{1}{4} & 0 & 0 \\ 0 & 0 & \frac{1}{2} & 0 & -\frac{1}{4} & -\frac{1}{4} \\ 0 & 0 & 0 & \frac{1}{2} & -\frac{1}{4} & -\frac{1}{4} \\ 0 & 0 & -\frac{1}{6} & -\frac{1}{3} & \frac{5}{12} & \frac{1}{12} \\ 0 & 0 & -\frac{1}{3} & -\frac{1}{6} & \frac{1}{12} & \frac{5}{12} \end{bmatrix}$$

The Simulink implementation of GNN$(A^{\mathrm{T}}A, I, A^{\mathrm{T}})$ and RNN$(A^{\mathrm{T}}A, I, A^{\mathrm{T}})$ with the initial state equal to A^{T} generate approximations of $A^{\dagger} + \left(I - A^{\dagger}A\right) A^{\mathrm{T}} = A^{\dagger}$.

Example 2.12.3 Consider the matrix

$$A = \begin{bmatrix} 1 & 2 & 3 & 4 & 1 \\ 1 & 3 & 4 & 6 & 2 \\ 2 & 3 & 4 & 5 & 3 \\ 3 & 4 & 5 & 6 & 4 \\ 4 & 5 & 6 & 7 & 6 \\ 6 & 6 & 7 & 7 & 8 \end{bmatrix}, \tag{2.12.1}$$

choose the following matrices $P \in \mathbb{R}^{5\times 2}$ and $Q \in \mathbb{R}^{2\times 6}$

$$P = \begin{bmatrix} 0 & 0 \\ 2 & 1 \\ 3 & 2 \\ 5 & 3 \\ 1 & 0 \end{bmatrix}, \quad Q = \begin{bmatrix} 0 & 1 & 0 & 1 & 0 & 1 \\ 1 & 0 & 1 & 0 & 1 & 0 \end{bmatrix} \tag{2.12.2}$$

and $G = PQ$. The trajectory $\mathfrak{T}(GNN(GA, I, G))$ in the case $V_{GA}(0) = O$ is given by the function

$$\mathcal{V}_{GA}(G,t,\gamma,O) := \left(I - e^{-\gamma t\,(GA)^{\mathrm{T}}GA}\right) A^{(2,4)}_{\mathcal{R}((GA)^{\mathrm{T}}),\mathcal{N}(G)} = \left(I - e^{-\gamma t\,(GA)^{\mathrm{T}}GA}\right)(GA)^{\dagger}G.$$

Also, $\lim_{t\to\infty} \mathfrak{T}(GNN(GA, I, G)) = (GA)^{\dagger}G$, which can be confirmed by

$$(\tilde{V}_{GA})(O) = \lim_{t\to+\infty,\gamma\to 10^3} \mathcal{V}_{GA}(G,t,\gamma,O)$$

$$= \begin{bmatrix} -\frac{473}{1682} & \frac{194}{841} & -\frac{473}{1682} & \frac{194}{841} & -\frac{473}{1682} & \frac{194}{841} \\ -\frac{64}{841} & \frac{121}{1682} & -\frac{64}{841} & \frac{121}{1682} & -\frac{64}{841} & \frac{121}{1682} \\ \frac{217}{1682} & -\frac{73}{841} & \frac{217}{1682} & -\frac{73}{841} & \frac{217}{1682} & -\frac{73}{841} \\ \frac{281}{841} & -\frac{413}{1682} & \frac{281}{841} & -\frac{413}{1682} & \frac{281}{841} & -\frac{413}{1682} \\ -\frac{277}{841} & \frac{229}{841} & -\frac{277}{841} & \frac{229}{841} & -\frac{277}{841} & \frac{229}{841} \end{bmatrix}.$$

The spectrum of GA contains only nonnegative real values:

$$\sigma(GA) = \{266.346716180717, 0.653283819282910, 0., 0., 0.\}.$$

The exact solution to RNN$(GA, I, G))$ is given by the function $\mathfrak{T}(RNN(GA, I, G))$. In the case $V_R(0) = O$, it is known that $\mathfrak{T}(RNN(GA, I, G)) := \mathcal{V}_R(G, t, \gamma, V_R(0) = \left(I - e^{-\gamma t\,GA}\right) A^{(2)}_{\mathcal{R}(G),\mathcal{N}(G)}$. Also, $\lim_{t\to\infty} \mathfrak{T}(RNN(GA, I, G)) = A^{(2)}_{\mathcal{R}(G),\mathcal{N}(G)}$, which can be confirmed by

$$\tilde{\mathcal{V}}_R(O) = \lim_{t\to+\infty,\gamma=10^3} \mathcal{V}_R(G,t,\gamma,O) = \begin{bmatrix} 0 & 0 & 0 & 0 & 0 & 0 \\ -\frac{7}{58} & \frac{19}{174} & -\frac{7}{58} & \frac{19}{174} & -\frac{7}{58} & \frac{19}{174} \\ \frac{10}{29} & -\frac{23}{87} & \frac{10}{29} & -\frac{23}{87} & \frac{10}{29} & -\frac{23}{87} \\ \frac{13}{58} & -\frac{9}{58} & \frac{13}{58} & -\frac{9}{58} & \frac{13}{58} & -\frac{9}{58} \\ -\frac{17}{29} & \frac{14}{29} & -\frac{17}{29} & \frac{14}{29} & -\frac{17}{29} & \frac{14}{29} \end{bmatrix}.$$

In order to verify $\tilde{\mathcal{V}}_R(O)$, it is possible to use the rank factorization of $A^{(2)}_{\mathcal{R}(P),\mathcal{N}(Q)} = P(QAP)^{-1}Q$ (see [208]). The state trajectories of the entries $(\tilde{\mathcal{V}}_{GA}(O))_{ij}$ and $(\tilde{\mathcal{V}}_R(O))_{ij}$ under the selection $\gamma \to 10^8$, $t \in [0, 10^{-6}]$ are illustrated in Fig. 2.38.

Example 2.12.4 The tridiagonal square test matrix

$$B_n = \begin{bmatrix} 1 & -1 & & & & \\ -1 & 2 & -1 & & \mathbf{0} & \\ & -1 & 2 & -1 & & \\ & & \ddots & \ddots & \ddots & \\ & \mathbf{0} & & -1 & 2 & -1 \\ & & & & -1 & 1 \end{bmatrix}$$

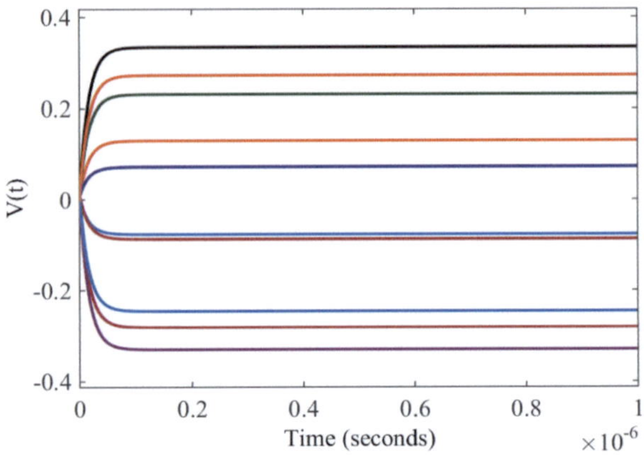

Fig. 2.38 The state trajectories of $(\tilde{\mathcal{V}}_{GA}(O))_{ij}$ with $\gamma = 10^8$, $t \in [0, 10^{-6}]$ in Example 2.12.3

of the order n of rank $\text{rank}(B_n) = n - 1$ is considered. In the case $n = 6$ let us choose $G_6 = B_6^{\mathrm{T}}$. Now, the expression $\mathfrak{T}(RNN(B_6^{\mathrm{T}} B_6, I, B_6^{\mathrm{T}})) := \mathcal{V}_R(B_6^{\mathrm{T}}, t, \gamma, O)$ in the zero initial state generates the exact Moore–Penrose inverse of B_6 using the limit

$$\tilde{V}_{B_6^{\mathrm{T}} B_6} = \lim_{t \to +\infty, \gamma = 10^3} \mathcal{V}_R(B_6^{\mathrm{T}}, t, \gamma, O) = A^{\dagger} = \begin{bmatrix} \frac{55}{36} & \frac{25}{36} & \frac{1}{36} & -\frac{17}{36} & -\frac{29}{36} & -\frac{35}{36} \\ \frac{25}{36} & \frac{31}{36} & \frac{7}{36} & -\frac{11}{36} & -\frac{23}{36} & -\frac{29}{36} \\ \frac{1}{36} & \frac{7}{36} & \frac{19}{36} & \frac{1}{36} & -\frac{11}{36} & -\frac{17}{36} \\ -\frac{17}{36} & -\frac{11}{36} & \frac{1}{36} & \frac{19}{36} & \frac{7}{36} & \frac{1}{36} \\ -\frac{29}{36} & -\frac{23}{36} & -\frac{11}{36} & \frac{7}{36} & \frac{31}{36} & \frac{25}{36} \\ -\frac{35}{36} & -\frac{29}{36} & -\frac{17}{36} & \frac{1}{36} & \frac{25}{36} & \frac{55}{36} \end{bmatrix}.$$

The elementwise trajectories of $(V_{B_6^{\mathrm{T}} B_6})_{ij}$ under the configuration $\gamma \to 10^{10}$, $t \in [0, 10^{-7}]s$ are displayed in Fig. 2.39. Clear and fast convergence to exact equilibrium points of each trajectory is observable.
Since $\text{ind}(A) = 1$, the exact solution $V_{B_6^2}$ to $\text{RNN}(B_6^2, I, B_6)$ can be used in deriving the exact group inverse of B_6:

$$\tilde{V}_{B_6^{\mathrm{T}} B_6} = \lim_{t \to +\infty, \gamma = 10^3} \mathcal{V}_R(B_6^{\mathrm{T}}, t, \gamma, O) = B_6^{\#} = B_6^{\dagger}.$$

2.13 Application of Gradient Optimization Methods in Improving GNN Dynamics

Main goal is to solve the equation $AXB = D$ and apply its particular cases in computing generalized inverses in real time improving the GNN model developed in [231] and denoted by $\text{GNN}(A, B, D)$. Our motivation is to improve the

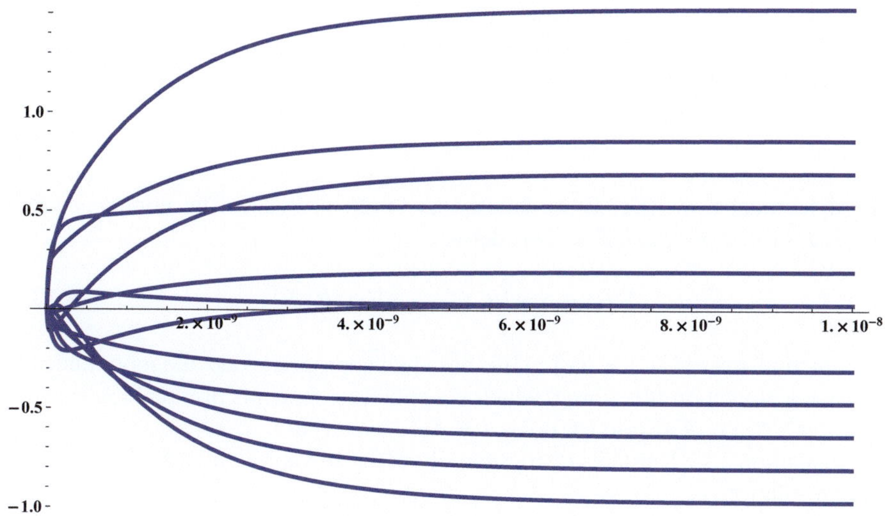

Fig. 2.39 The state trajectories of $(\mathcal{V}_R(B_6^{\mathrm{T}}, t, \gamma, O))_{ij}$ under the choice $\gamma \to 10^{10}$, $t \in [0, 10^{-7}]$ in Example 2.12.4

GNN(A, B, D) and develop the novel gradient-based GNN (GGNN) design, termed as GGNN(A, B, D) utilizing a novel type of dynamical system. The proposed GGNN model is defined evolving the standard GNN dynamics along the gradient of the standard error matrix. The convergence analysis reveals the global asymptotic convergence of GGNN(A, B, D) without restrictions, while the output belongs to the set of general solutions to the matrix equation $AXB = D$.

2.13.1 Motivation

The GNN used to solve the general linear matrix equation $AXB = D$ is defined over the error matrix $E(t) = D - AV(t)B$, where $t \in [0, +\infty)$ is the time and $V(t)$ is an unknown state-variable matrix that approximates the unknown matrix X in $AXB = D$. The goal function is $\varepsilon(t) = ||D - AV(t)B||_F^2/2$, where $\|\cdot\|_F = \sqrt{\sum_{ij} a_{ij}^2}$ denotes the Frobenius norm of a matrix. The gradient of $\varepsilon(t)$ is equal to

$$\frac{\partial \varepsilon(t)}{\partial V} = \nabla \varepsilon = \frac{1}{2}\frac{\partial ||D - AV(t)B||_F^2}{\partial V} = -A^{\mathrm{T}}(D - AV(t)B)B^{\mathrm{T}}.$$

The GNN evolutionary design is defined by the dynamic system

$$\dot{V}(t) = \frac{\mathrm{d}V(t)}{\mathrm{d}t} = -\gamma \frac{\partial \varepsilon(t)}{\partial V}, \quad V(0) = V_0, \tag{2.13.1}$$

where $\gamma > 0$ is real parameter used to speed up the convergence and $\dot{V}(t)$ denotes the time derivative of $V(t)$. Thus, the linear GNN aimed at solving $AXB = D$ is given by the following dynamics:

$$\dot{V}(t) = \gamma A^{\mathrm{T}}(D - AV(t)B)B^{\mathrm{T}}. \tag{2.13.2}$$

The dynamical flow (2.13.2) is denoted as GNN(A, B, D). The nonlinear GNN (A, B, D) for solving $AXB = D$ is defined by

$$\dot{V}(t) = \gamma A^{\mathrm{T}}\mathcal{F}\,(D - AV(t)B)\,B^{\mathrm{T}}. \tag{2.13.3}$$

The function array $\mathcal{F}(C) = \mathcal{F}([c_{ij}])$ is based on appropriate odd and monotonically increasing activation function f which is applied to elements of a real matrix $C = (c_{ij}) \in \mathbb{R}^{m\times n}$, i.e., $\mathcal{F}(C) = [f(c_{ij})]$, $i = 1, \ldots, m$, $j = 1, \ldots, n,$.

Proposition 2.13.1 restates restrictions about the solvability of $AXB = D$ and its general solution.

Proposition 2.13.1 ([6, 266]) *If $A \in \mathbb{R}^{m\times n}$, $B \in \mathbb{R}^{p\times q}$ and $D \in \mathbb{R}^{m\times q}$ then fulfillment of the condition*

$$AA^{\dagger}DB^{\dagger}B = D \tag{2.13.4}$$

is necessary and sufficient for solvability of the linear matrix equation $AXB = D$. In this case, the set of all solutions is given by

$$X = \left\{A^{\dagger}DB^{\dagger} + Y - A^{\dagger}AYBB^{\dagger} \mid Y \in \mathbb{R}^{n\times p}\right\}. \tag{2.13.5}$$

The following results from [231] describe the conditions of convergence and the limit of the unknown matrix $V(t)$ from (2.13.3) as $t \to +\infty$.

Proposition 2.13.2 ([231]) *Suppose the matrices $A \in \mathbb{R}^{m\times n}$, $B \in \mathbb{R}^{p\times q}$ and $D \in \mathbb{R}^{m\times q}$ satisfy (2.13.4). Then the unknown matrix $V(t)$ from (2.13.3) converges as $t \to +\infty$ with the equilibrium state*

$$V(t) \to \tilde{V} = A^{\dagger}DB^{\dagger} + V(0) - A^{\dagger}AV(0)BB^{\dagger} \tag{2.13.6}$$

for any initial state variable matrix $V(0) \in \mathbb{R}^{n\times p}$.

The research in [249] investigated various ZNN models based on optimization methods. The goal of current research is to develop the GNN model based on the gradient $E_G(t)$ of $\|E(t)\|_F^2$ instead of the original goal function $E(t)$.

2.13.2 Motivation and Derivation of GGNN Models

The standard GNN design (2.13.2) solves the GLME $AXB = D$ under the constraint (2.13.4). Our goal is to resolve this restriction and propose dynamic evolutions based on the error functions that tend to zero without restrictions.

Our goal is to define the GNN design for solving the GLME $AXB = D$ based on the error function

$$E_G(t) := \nabla\varepsilon(t) = A^{\mathrm{T}}\left(D - AV(t)B\right)B^{\mathrm{T}} = A^{\mathrm{T}}E(t)B^{\mathrm{T}}. \tag{2.13.7}$$

According to known results from nonlinear unconstrained optimization [187], the equilibria points of $\varepsilon(t)$ in (2.13.7) satisfy

$$E_G(t) := \nabla\varepsilon(t) = 0.$$

We continue investigation aimed to modification of ZNN design based on optimization methods which was investigated in [250]. More precisely, we will develop the GNN model based on the error function $E_G(t)$ instead of the error function $E(t)$. In this way, new neural dynamics is aimed to force the gradient E_G to zero instead of the standard goal function $E(t)$. It is reasonable to call such GNN design as gradient-based GNN (GGNN shortly).

Proposition 2.13.3 gives conditions for solvability of the matrix equations $E(t) = 0$ and $E_G(t) = 0$ and general solutions to these systems.

Proposition 2.13.3 ([250]) *Consider arbitrary matrices $A \in \mathbb{R}^{m\times n}$, $B \in \mathbb{R}^{k\times h}$ and $D \in \mathbb{R}^{m\times h}$. The next statements are true.*
(a) *The equation $E(t) = 0$ is solvable if and only if (2.13.4) is satisfied and the general solution to $E(t) = 0$ is given by (2.13.5).*
(b) *The equation $E_G(t) = 0$ is always solvable and its general solution coincides with (2.13.5).*

Proof (a) This part of the proof follows from known results on the solvability and general solution to the matrix equation $AXB = D$ of generalized inverses [6, P. 52, Theorem 1], and its application to the matrix equation $E(t) = 0 \Longleftrightarrow AV(t)B = D$.
(b) According to [6, P. 52, Theorem 1], the matrix equation

$$E_G(t) = 0 \Longleftrightarrow A^{\mathrm{T}}AVBB^{\mathrm{T}} = A^{\mathrm{T}}DB^{\mathrm{T}}$$

is consistent if and only if

$$A^{\mathrm{T}}A\left(A^{\mathrm{T}}A\right)^{\dagger}A^{\mathrm{T}}DB^{\mathrm{T}}\left(BB^{\mathrm{T}}\right)^{\dagger}BB^{\mathrm{T}} = A^{\mathrm{T}}DB^{\mathrm{T}} \tag{2.13.8}$$

is satisfied. Indeed, applying the properties $(A^{\mathrm{T}}A)^{\dagger}A^{\mathrm{T}} = A^{\dagger}$, $B^{\mathrm{T}}(BB^{\mathrm{T}})^{\dagger} = B^{\dagger}$ and $A^{\mathrm{T}}AA^{\dagger} = A^{\mathrm{T}}$, $B^{\dagger}BB^{\mathrm{T}} = B^{\mathrm{T}}$ of the Moore–Penrose inverse [6, 266], it can be obtained

$$A^{\mathrm{T}}A\left(A^{\mathrm{T}}A\right)^{\dagger}A^{\mathrm{T}}DB^{\mathrm{T}}\left(BB^{\mathrm{T}}\right)^{\dagger}BB^{\mathrm{T}}=A^{\mathrm{T}}AA^{\dagger}DB^{\dagger}BB^{\mathrm{T}}=A^{\mathrm{T}}DB^{\mathrm{T}},$$

which is a confirmation of (2.13.8).

In addition, based on [6, P. 52, Theorem 1] the general solution $V(t)$ to $E_G(t)=0$ is

$$\begin{aligned}V&=\left(A^{\mathrm{T}}A\right)^{\dagger}A^{\mathrm{T}}DB^{\mathrm{T}}\left(BB^{\mathrm{T}}\right)^{\dagger}+Y-\left(A^{\mathrm{T}}A\right)^{\dagger}A^{\mathrm{T}}AYBB^{\mathrm{T}}\left(BB^{\mathrm{T}}\right)^{\dagger}\\&=A^{\dagger}DB^{\dagger}+Y-A^{\dagger}AYBB^{\dagger},\end{aligned}\tag{2.13.9}$$

which coincides with (2.13.5). □

In this way, the matrix equation $E(t)=0$ is solvable under the condition (2.13.4), while the equation $E_G(t)=0$ is unconditionally consistent. In addition, the general solutions to equations $E(t)=0$ and $E_G(t)=0$ are identical (see [249]).

The next step is to develop the GGNN dynamics using the error matrix $E_G(t)$ and the objective function $\varepsilon_G=||E_G||_F^2/2$. The gradient of ε_G is equal to

$$\frac{\partial\varepsilon_G(V(t))}{\partial V}=\frac{\partial||A^{\mathrm{T}}(D-AV(t)B)B^{\mathrm{T}}||_F^2}{\partial V}=-A^{\mathrm{T}}A\left(A^{\mathrm{T}}(D-AV(t)B)B^{\mathrm{T}}\right)BB^{\mathrm{T}}.$$

The dynamical GGNN system is obtained by applying the GNN evolution along the gradient of $\varepsilon_G(V(t))$ based on $E_G(t)$, as follows:

$$\begin{aligned}\dot V(t)&=-\gamma\frac{\partial\varepsilon_G}{\partial V}\\&=\gamma A^{\mathrm{T}}A\left(A^{\mathrm{T}}\left(D-AV(t)B\right)B^{\mathrm{T}}\right)BB^{\mathrm{T}}.\end{aligned}\tag{2.13.10}$$

The nonlinear GGNN dynamics is defined as

$$\dot V(t)=\gamma A^{\mathrm{T}}A\mathcal{F}(A^{\mathrm{T}}\left(D-AV(t)B\right)B^{\mathrm{T}})BB^{\mathrm{T}},\tag{2.13.11}$$

in which $\mathcal{F}(C)=\mathcal{F}([c_{ij}])$ denotes elementwise application of an odd and monotonically increasing function $f(\cdot)$, as the GNN(A,B,D) model (2.13.3). The evolution (2.13.11) is termed as GGNN(A,B,D) and applicable to constant coefficient matrices A,B,D. Three activation functions $f(\cdot)$ are used in numerical experiments for testing GGNN(A,B,D):

1. Linear function

$$f_{lin}(x)=x;\tag{2.13.12}$$

2. Power-sigmoid activation function

$$f_{ps}(x,\rho,\varrho)=\begin{cases}x^{\rho} & \text{if } |x|\geq 1\\ \frac{1+e^{-\varrho}}{1-e^{-\varrho}}\cdot\frac{1+e^{-\varrho x}}{1-e^{-\varrho x}} & \text{if } |x|<1\end{cases}\tag{2.13.13}$$

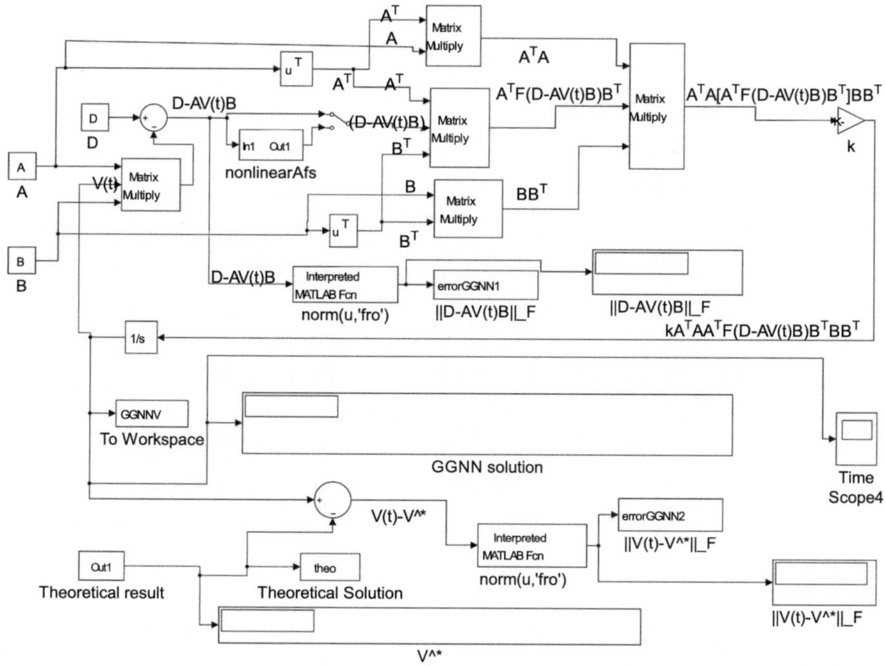

Fig. 2.40 Simulink implementation of GGNN(A, B, D) evolution (2.13.11)

where $\varrho > 2$ and $\rho \geq 3$ is odd integer.
3. Smooth power-sigmoid function

$$f_{sps}(x, \rho, \varrho) = \frac{1}{2}x^{\rho} + \frac{1+e^{-\varrho}}{1-e^{-\varrho}} \cdot \frac{1+e^{-\varrho x}}{1-e^{-\varrho x}}, \tag{2.13.14}$$

where $\varrho > 2$ and $\rho \geq 3$ is an odd integer.

Figure 2.40 represents the Simulink implementation of GGNN(A, B, D) dynamics (2.13.11).

On the other hand, the GZNN model defined upon the Zhangian matrix $E_G(t)$ is defined in [250] by the general evolution design

$$\dot{E}_G(t) = \frac{\mathrm{d}E_G(t)}{\mathrm{d}t} = -\gamma\mathcal{F}(E_G(t)). \tag{2.13.15}$$

Convergence Analysis of GGNN Dynamics

In this section, we will analyze convergence properties of GGNN model given by dynamics (2.13.11).

Theorem 2.13.1 ([250]) *Consider matrices $A \in \mathbb{R}^{m\times n}$, $B \in \mathbb{R}^{p\times q}$ and $D \in \mathbb{R}^{m\times q}$. If an odd and monotonically increasing array activation function $\mathcal{F}(\cdot)$ based on an elementwise function $f(\cdot)$ is used, then the activation state matrix $V(t) \in \mathbb{R}^{n\times p}$ of the* GGNN(A, B, D) *model (2.13.11) asymptotically converges to the solution of the matrix equation $AXB = D$, i.e., $A^{\mathrm{T}}AV(t)BB^{\mathrm{T}} \to A^{\mathrm{T}}DB^{\mathrm{T}}$ as $t \to +\infty$, for an arbitrary initial state matrix $V(0)$.*

Proof From the statement (b) of Proposition 2.13.3, the solvability of $A^{\mathrm{T}}AVBB^{\mathrm{T}} = A^{\mathrm{T}}DB^{\mathrm{T}}$ is ensured. The substitution $V(t) = \bar{V}(t) + A^{\dagger}DB^{\dagger}$ transforms the dynamics (2.13.11) into

$$\begin{aligned}\frac{\mathrm{d}\bar{V}(t)}{\mathrm{d}t} = \frac{\mathrm{d}V(t)}{\mathrm{d}t} &= \gamma A^{\mathrm{T}}A\mathcal{F}\left(A^{\mathrm{T}}\left(D - AV(t)B\right)B^{\mathrm{T}}\right)BB^{\mathrm{T}} \\ &= \gamma A^{\mathrm{T}}A\mathcal{F}\left(A^{\mathrm{T}}\left(D - A\left(\bar{V}(t) + A^{\dagger}DB^{\dagger}\right)B\right)B^{\mathrm{T}}\right)BB^{\mathrm{T}} \\ &= \gamma A^{\mathrm{T}}A\mathcal{F}\left(A^{\mathrm{T}}\left(D - A\bar{V}(t)B - AA^{\dagger}DB^{\dagger}B\right)B^{\mathrm{T}}\right)BB^{\mathrm{T}} \\ &\overset{(2.13.4)}{=} \gamma A^{\mathrm{T}}A\mathcal{F}\left(A^{\mathrm{T}}\left(D - A\bar{V}(t)B - D\right)B^{\mathrm{T}}\right)BB^{\mathrm{T}} \\ &= -\gamma A^{\mathrm{T}}A\mathcal{F}\left(A^{\mathrm{T}}A\bar{V}(t)BB^{\mathrm{T}}\right)BB^{\mathrm{T}}.\end{aligned} \tag{2.13.16}$$

Lyapunov function candidate which measures the convergence performance is defined by

$$\mathcal{L}\left(\bar{V}(t), t\right) = \frac{1}{2}||\bar{V}(t)||_F^2 = \frac{1}{2}\mathrm{Tr}\left(\bar{V}(t)^{\mathrm{T}}\bar{V}(t)\right). \tag{2.13.17}$$

The conclusion is $\mathcal{L}(\bar{V}(t), t) \geq 0$. According to (2.13.17), assuming (2.13.16) and using $\mathrm{dTr}(X^{\mathrm{T}}X) = 2\mathrm{Tr}(X^{\mathrm{T}}\mathrm{d}X)$ in conjunction with basic properties of the matrix trace function, one can express the time derivative of $\mathcal{L}(\bar{V}(t), t)$ as in the following

$$\begin{aligned}\frac{\mathrm{d}\mathcal{L}(\bar{V}(t), t)}{\mathrm{d}t} &= \frac{1}{2}\frac{\mathrm{d}\,\mathrm{Tr}\left(\bar{V}(t)^{\mathrm{T}}\bar{V}(t)\right)}{\mathrm{d}t} \\ &= \frac{1}{2}\cdot 2\cdot \mathrm{Tr}\left(\bar{V}(t)^{\mathrm{T}}\frac{\mathrm{d}\bar{V}(t)}{\mathrm{d}t}\right) \\ &= \mathrm{Tr}\left[\bar{V}(t)^{\mathrm{T}}\left(-\gamma A^{\mathrm{T}}A\mathcal{F}\left(A^{\mathrm{T}}A\bar{V}(t)BB^{\mathrm{T}}\right)BB^{\mathrm{T}}\right)\right] \\ &= -\gamma\mathrm{Tr}\left[\bar{V}(t)^{\mathrm{T}}A^{\mathrm{T}}A\mathcal{F}\left(A^{\mathrm{T}}A\bar{V}(t)BB^{\mathrm{T}}\right)BB^{\mathrm{T}}\right] \\ &= -\gamma\mathrm{Tr}\left[BB^{\mathrm{T}}\bar{V}(t)^{\mathrm{T}}A^{\mathrm{T}}A\mathcal{F}\left(A^{\mathrm{T}}A\bar{V}(t)BB^{\mathrm{T}}\right)\right] \\ &= -\gamma\mathrm{Tr}\left[\left(A^{\mathrm{T}}A\bar{V}(t)BB^{\mathrm{T}}\right)^{\mathrm{T}}\mathcal{F}\left(A^{\mathrm{T}}A\bar{V}(t)BB^{\mathrm{T}}\right)\right].\end{aligned} \tag{2.13.18}$$

Using the notation $W(t) = A^{\mathrm{T}}A\bar{V}(t)BB^{\mathrm{T}}$, the time derivative $\frac{\mathrm{d}\mathcal{L}(\bar{V}(t),t)}{\mathrm{d}t}$ is expressed in more readable form

$$\frac{\mathrm{d}\mathcal{L}(\bar{V}(t),t)}{\mathrm{d}t} = -\gamma \mathrm{Tr}\left[(W^{\mathrm{T}}\mathcal{F}(W))\right].$$

Since the scalar-valued function $f(\cdot)$ is an odd and monotonically increasing, it follows

$$\begin{aligned}\frac{\mathrm{d}L(\bar{V}(t),t)}{\mathrm{d}t} &= -\gamma \mathrm{Tr}\left[(W^{\mathrm{T}}\mathcal{F}(W))\right]\\ &= -\gamma\sum_{i=1}^{m}\sum_{j=1}^{n} w_{ij} f(w_{ij}) \begin{cases} < 0 \text{ if } W(t) := A^{\mathrm{T}}A\bar{V}(t)BB^{\mathrm{T}} \neq 0\\ = 0 \text{ if } W(t) := A^{\mathrm{T}}A\bar{V}(t)BB^{\mathrm{T}} = 0,\end{cases}\end{aligned} \tag{2.13.19}$$

which implies

$$\frac{\mathrm{d}\mathcal{L}(\bar{V}(t),t)}{\mathrm{d}t} \begin{cases} < 0 \text{ if } W(t) \neq 0\\ = 0 \text{ if } W(t) = 0.\end{cases} \tag{2.13.20}$$

Observing the identity

$$\begin{aligned} W(t) &= A^{\mathrm{T}}A\bar{V}(t)BB^{\mathrm{T}}\\ &= A^{\mathrm{T}}A\left(V(t) - A^{\dagger}DB^{\dagger}\right)BB^{\mathrm{T}}\\ &= A^{\mathrm{T}}AV(t)BB^{\mathrm{T}} - A^{\mathrm{T}}DB^{\mathrm{T}}\\ &= A^{\mathrm{T}}\left(AV(t)B - D\right)B^{\mathrm{T}},\end{aligned}$$

and using the Lyapunov stability theory, we conclude that $W(t) := A^{\mathrm{T}}(AV(t)B - D)\,B^{\mathrm{T}}$ globally converges to the zero matrix, from arbitrary initial value $V(0)$. □

Theorem 2.13.2 *The activation state variables matrix $V(t)$ of the model* GGNN (A, B, D)*, defined by (2.13.11), is convergent as $t \to +\infty$ and its equilibrium state is*

$$V(t) \to \tilde{V}(t) = A^{\dagger}DB^{\dagger} + V(0) - A^{\dagger}AV(0)BB^{\dagger} \tag{2.13.21}$$

for every initial state matrix $V(0) \in \mathbb{R}^{n\times p}$.

Proof From (2.13.11), the matrix $V_1(t) = (A^{\mathrm{T}}A)^{\dagger}A^{\mathrm{T}}AV(t)BB^{\mathrm{T}}(BB^{\mathrm{T}})^{\dagger}$ satisfies

$$\begin{aligned}\frac{\mathrm{d}V_1(t)}{\mathrm{d}t} &= (A^{\mathrm{T}}A)^{\dagger}A^{\mathrm{T}}A\frac{\mathrm{d}V(t)}{\mathrm{d}t}BB^{\mathrm{T}}(BB^{\mathrm{T}})^{\dagger}\\ &= \gamma(A^{\mathrm{T}}A)^{\dagger}A^{\mathrm{T}}A\left[A^{\mathrm{T}}A\left(A^{\mathrm{T}}(D - AV(t)B)B^{\mathrm{T}}\right)BB^{\mathrm{T}}\right]BB^{\mathrm{T}}(BB^{\mathrm{T}})^{\dagger}.\end{aligned}$$

According to the basic properties of the Moore–Penrose inverse [266], it follows

$$(BB^{\mathrm{T}})^{\mathrm{T}}BB^{\mathrm{T}}(BB^{\mathrm{T}})^{\dagger} = (BB^{\mathrm{T}})^{\mathrm{T}} = BB^{\mathrm{T}}, \quad (A^{\mathrm{T}}A)^{\dagger}A^{\mathrm{T}}A(A^{\mathrm{T}}A)^{\mathrm{T}} = (A^{\mathrm{T}}A)^{\mathrm{T}} = A^{\mathrm{T}}A$$

which further implies

$$\begin{aligned}\frac{\mathrm{d}V_1(t)}{\mathrm{d}t} &= \gamma A^{\mathrm{T}}A\left(A^{\mathrm{T}}(D - AV(t)B)B^{\mathrm{T}}\right)BB^{\mathrm{T}}\\ &= \frac{\mathrm{d}V(t)}{\mathrm{d}t}.\end{aligned}$$

Consequently, $V_2(t) = V(t) - V_1(t)$ satisfies $\frac{\mathrm{d}V_2(t)}{\mathrm{d}t} = \frac{\mathrm{d}V(t)}{\mathrm{d}t} - \frac{\mathrm{d}V_1(t)}{\mathrm{d}t} = 0$, which implies

$$\begin{aligned}V_2(t) &= V_2(0)\\ &= V(0) - V_1(0)\\ &= V(0) - (A^{\mathrm{T}}A)^{\dagger}A^{\mathrm{T}}AV(0)BB^{\mathrm{T}}(BB^{\mathrm{T}})^{\dagger}\\ &= V(0) - A^{\dagger}AV(0)BB^{\dagger}, \quad t \geq 0.\end{aligned} \tag{2.13.22}$$

Furthermore, from Theorem 2.13.1, $A^{\mathrm{T}}AV(t)BB^{\mathrm{T}} \to A^{\mathrm{T}}DB^{\mathrm{T}}$ and $V_1(t)$ converges to

$$\begin{aligned}V_1(t) &= (A^{\mathrm{T}}A)^{\dagger}A^{\mathrm{T}}AV(t)BB^{\mathrm{T}}(BB^{\mathrm{T}})^{\dagger} \to (A^{\mathrm{T}}A)^{\dagger}A^{\mathrm{T}}DB^{\mathrm{T}}(BB^{\mathrm{T}})^{\dagger}\\ &= A^{\dagger}DB^{\dagger}\end{aligned}$$

as $t \to +\infty$. Therefore, $V(t) = V_1(t) + V_2(t)$ converges to the equilibrium state

$$\tilde{V}(t) = A^{\dagger}DB^{\dagger} + V_2(t) = A^{\dagger}DB^{\dagger} + V(0) - A^{\dagger}AV(0)BB^{\dagger}.$$

The proof is finished. □

2.13.3 Numerical Experiments on GNN and GGNN Dynamics

Numerical examples in this section are represented based on the Simulink implementation of GGNN formula presented in Fig. 2.40.

The parameter γ, initial state $V(0)$, and parameters ρ and ϱ of the nonlinear activation functions (2.13.13) and (2.13.14) are entered directly in the model, while matrices A, B, and D are defined from the workspace. It is assumed $\rho = \varrho = 3$ in all examples. The `ode15s` differential equation solver is used in configuration parameters. In all examples V^* will denote the theoretical solution.

The blocks *powersig, smoothpowersig* and *transpmult* include previously defined codes `powersig` and `smoothpowersig` in Sect. 2.2.

Example 2.13.1 Let us consider the idempotent matrix A

$$A = \begin{bmatrix} 1 & 0 & 1 & 1\\ 0 & 1 & 1 & 2\\ 0 & 0 & 0 & 0\\ 0 & 0 & 0 & 0\end{bmatrix}$$

of $\text{rank}(A) = 2$ and with the theoretical Moore–Penrose inverse

$$V^* = A^\dagger = \frac{1}{3}\begin{bmatrix} 2 & -1 & 0 & 0 \\ -1 & 1 & 0 & 0 \\ 1 & 0 & 0 & 0 \\ 0 & 1 & 0 & 0 \end{bmatrix}.$$

The matrix equation corresponding to the Moore–Penrose inverse is $A^{\mathrm{T}}AX = A^{\mathrm{T}}$ [266], which implies the error function $E(t) = A^{\mathrm{T}}(I - AX)$. Corresponding GNN model is defined by $\text{GNN}(A^{\mathrm{T}}A, I_4, A^{\mathrm{T}})$, where I_4 denotes the identity 4×4 matrix. The constraint (2.13.4) reduces to the condition $AA^\dagger A^{\mathrm{T}} = A^{\mathrm{T}}$ which is not satisfied. The input parameters of $\text{GNN}(A^{\mathrm{T}}A, I_4, A^{\mathrm{T}})$ are $\gamma = 10^8$, $V(0) = O_4$, where O_4 denotes the zero 4×4 matrix. Corresponding $\text{GGNN}((A^{\mathrm{T}}A)^2, I, A^{\mathrm{T}}AA^{\mathrm{T}})$ design is based on the error matrix $E_G(t) = A^{\mathrm{T}}AA^{\mathrm{T}}(I - AV)$. The Simulink implementation of $\text{GGNN}(A, B, D)$ from Fig. 2.40 and the Simulink implementation of $\text{GNN}(A, B, D)$ from [231]) and explained in Sect. 2.2 export in this case the graphical results presented in Figs. 2.41 and 2.42 which display the behavior of the norms $||E_G(t)||_F = ||A^{\mathrm{T}}AA^{\mathrm{T}}(I - AV(t))||_F$ and $||V(t) - V^*||_F$ respectively. It is observable that the norms generated by the application of GGNN formula vanish faster to zero against corresponding norms in the GNN model. Graphs in presented figures strengthen the fast convergence of the GGNN dynamical system and the important role which can include the application of this specific model (2.13.11) to problems that require the computation of the Moore–Penrose inverse.

Example 2.13.2 Let us consider the matrices

$$A = \begin{bmatrix} -8 & 8 & -4 \\ 11 & 4 & -7 \\ 1 & -4 & 3 \\ 0 & 12 & -10 \\ 6 & 12 & -12 \end{bmatrix}, \quad B = \begin{bmatrix} 1 & 0 & 0 \\ 0 & 1 & 0 \\ 0 & 0 & 1 \\ 0 & 0 & 0 \end{bmatrix}, \quad D = \begin{bmatrix} -84 & 2524 & 304 \\ -2252 & -623 & 2897 \\ 484 & -885 & -701 \\ -1894 & 2278 & 2652 \\ -2778 & 1524 & 3750 \end{bmatrix}.$$

The exact minimum norm least squares solution is

$$V^* = A^\dagger D B^\dagger = \begin{bmatrix} -\frac{7409}{65} & -\frac{9564}{65} & \frac{8953}{65} & 0 \\ -\frac{968}{13} & \frac{1770}{13} & \frac{1402}{13} & 0 \\ \frac{6503}{65} & -\frac{4187}{65} & -\frac{8826}{65} & 0 \end{bmatrix}.$$

Ranks of input matrices are equal to $r = \text{rank}(A) = 2$, $\text{rank}(D) = 2$ and $\text{rank}(B) = 3$. The constraint (2.13.4) is satisfied in this case. The linear $\text{GGNN}(A, B, D)$ formula (2.13.11) is applied to solve the matrix equation $AXB = D$. The gain parameter of the model is $\gamma = 10^9$, $V(0) = 0$ and the stopping time is $t = 0.00001$, which gives

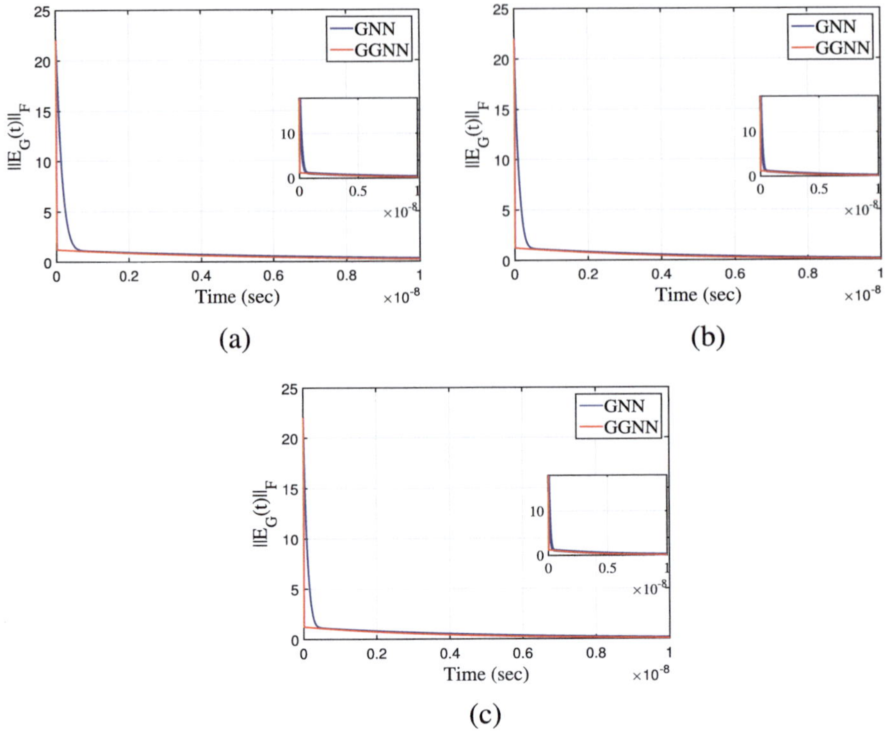

Fig. 2.41 **a** Linear activation. **b** Power-sigmoid activation. **c** Smooth power-sigmoid activation. $\|E_G(t)\|_F$ in GGNN$((A^{\mathrm{T}}A)^2, I, A^{\mathrm{T}}AA^{\mathrm{T}})$ against GNN$(A^{\mathrm{T}}A, I_4, A^{\mathrm{T}})$ in Example 2.13.1

$$X = \begin{bmatrix} -113.9846 & -147.1385 & 137.7385 & 0 \\ -74.4615 & 136.1538 & 107.8462 & 0 \\ 100.0462 & -64.4154 & -135.7846 & 0 \end{bmatrix} \approx A^{\dagger}DB^{\dagger}.$$

The norm of the error matrix E_G of both models GNN and GGNN under linear and nonlinear AFs is shown in Fig. 2.43a–c. The power-sigmoid and smooth power-sigmoid activation functions show superiority in the convergence speed, comparing with the linear activation. On each graph in Fig. 2.43a–c, Frobenius norm $\|E_G(t)\|_F$ of the error matrix $E_G(t)$ in the GGNN formula vanishes faster to zero than in the GNN model. Moreover, on each graph in Fig. 2.44a–c, Frobenius norm $\|E(t)\|_F$ in the GGNN formula vanishes faster to zero than in the GNN model. which strengthens the fact that the proposed dynamical system (2.13.11) initiates accelerated convergence against (2.13.3).

All graphs involved in Figs. 2.43 and 2.44 confirm applicability of the proposed GGNN design compared to the traditional GNN design even if the constraint (2.13.4) holds.

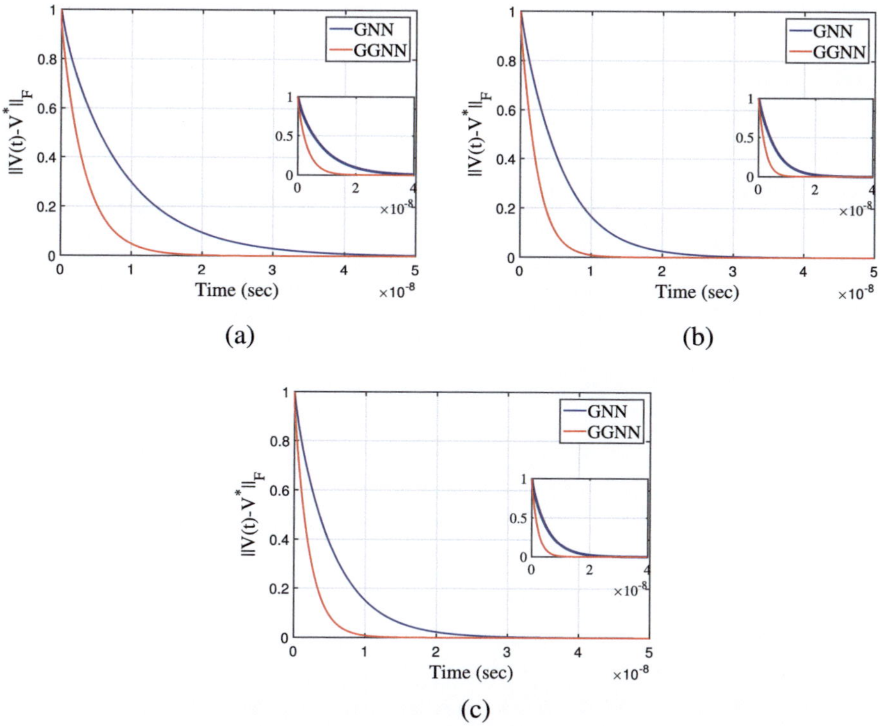

Fig. 2.42 **a** Linear activation. **b** Power-sigmoid activation. **c** Smooth power-sigmoid activation. $\|V(t) - V^*\|_F$ in GGNN($(A^{\mathrm{T}}A)^2$, I, $A^{\mathrm{T}}AA^{\mathrm{T}}$) against GNN($A^{\mathrm{T}}A$, I_4, A^{T}) in Example 2.13.1

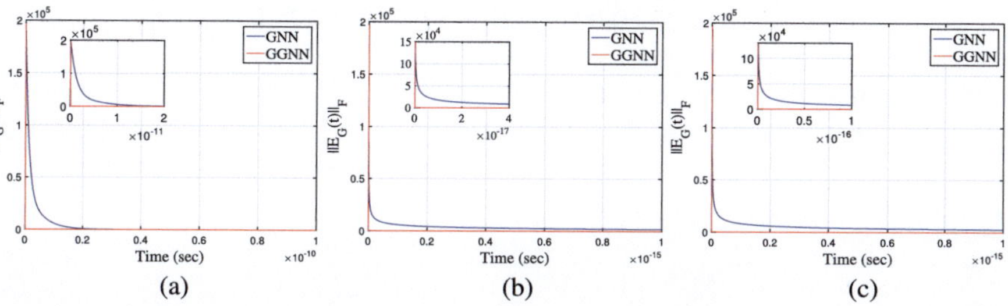

Fig. 2.43 **a** Linear activation. **b** Power-sigmoid activation. **c** Smooth power-sigmoid activation. $\|E_G(t)\|_F$ in GGNN(A, B, D) against GNN(A, B, D) in Example 2.13.2

Example 2.13.3 Let us explore behavior of GNN and GGNN dynamics for computing the Moore–Penrose inverse of the matrix

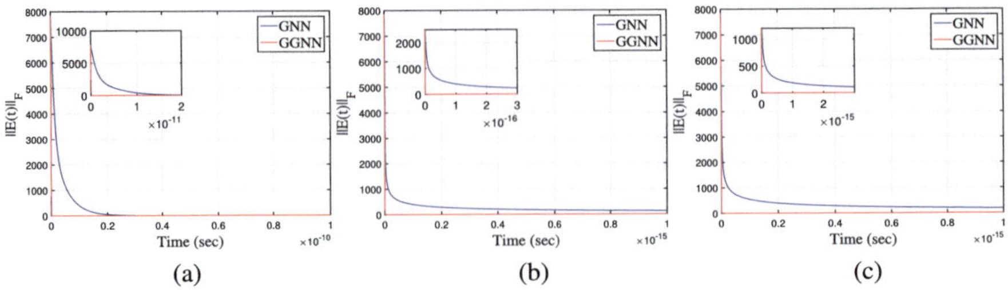

Fig. 2.44 **a** Linear activation. **b** Power-sigmoid activation. **c** Smooth power-sigmoid activation. $\|E(t)\|_F$ in GGNN(A, B, D) against GNN(A, B, D) in Example 2.13.2

$$A = \begin{bmatrix} 9 & 3 & -3 \\ -1 & 1 & 0 \\ 4 & 7 & 2 \\ 2 & 4 & -4 \\ 13 & 5 & 8 \end{bmatrix}.$$

The Moore–Penrose inverse of A is equal to

$$A^{\dagger} = \begin{bmatrix} \frac{9908}{127779} & -\frac{18037}{766674} & -\frac{6874}{127779} & -\frac{2663}{383337} & \frac{29941}{766674} \\ -\frac{5690}{127779} & \frac{14426}{383337} & \frac{16741}{127779} & \frac{25130}{383337} & -\frac{6392}{383337} \\ -\frac{3517}{42593} & \frac{1979}{255558} & \frac{1073}{42593} & -\frac{7373}{127779} & \frac{15049}{255558} \end{bmatrix}$$
$$\approx \begin{bmatrix} 0.0775 & -0.0235 & -0.0538 & -0.0069 & 0.0390 \\ -0.0445 & 0.0376 & 0.1310 & 0.0655 & -0.0167 \\ -0.0826 & 0.0077 & 0.0252 & -0.0577 & 0.0589 \end{bmatrix}.$$

Rank of the input matrix is equal to $r = \text{rank}(A) = 3$. Consequently, the matrix A is left invertible and satisfies $A^{\dagger}A = I$. The error matrix $E(t) = I - VA$ initiates the GNN(I, A, I) dynamics for computing $A^{\dagger}$. The gradient-based error matrix

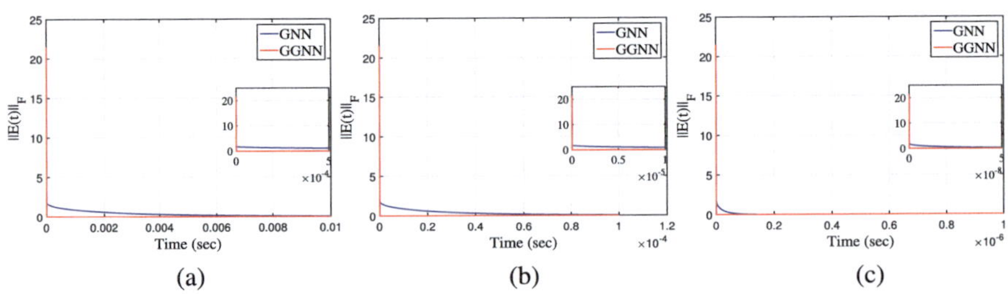

Fig. 2.45 **a** $\gamma = 10$, $t \in [0, 10^{-2}]$. **b** $\gamma = 10^3$, $t \in [0, 10^{-3}]$. **c** $\gamma = 10^6$, $t \in [0, 10^{-6}]$. $\|E(t)\|_F$ for different γ in GGNN$(I, AA^{\mathrm{T}}, A^{\mathrm{T}})$ against GNN(I, A, I) in Example 2.13.3

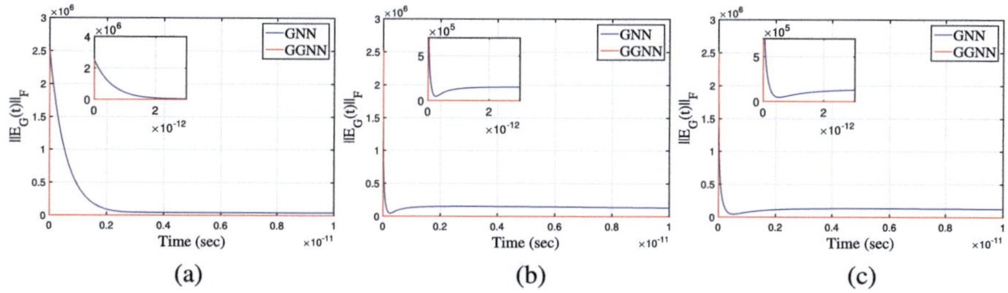

Fig. 2.46 **a** Linear activation. **b** Power-sigmoid activation. **c** Smooth power-sigmoid activation. $\|E_G(t)\|_F$ in GGNN$(I, AA^{\mathrm{T}}, A^{\mathrm{T}})$ against GNN(I, A, I) in Example 2.13.3

$$E_G(t) = (I - V(t)A)\,A^{\mathrm{T}}.$$

initiates the GGNN$(I, AA^{\mathrm{T}}, A^{\mathrm{T}})$ design.

The gain parameter of the model is $\gamma = 100$, initial state is $V(0) = 0$ with stop time $t = 0.00001$.

The Frobenius norm of the error matrix $E(t)$ generated inside the linear GNN and GGNN models for different values $\gamma = 10^2, \gamma = 10^3, \gamma = 10^6$ are shown in Fig. 2.45a–c. Graphs involved in these figures confirm increase in the convergence speed which is caused by the increase of the gain parameter γ. Because of that, considered time intervals are $[0, 10^{-2}]$, $[0, 10^{-3}]$ and $[0, 10^{-6}]$, respectively. In all three scenarios, a faster convergence of the GGNN model is observable compared to the GNN design. The values of the norm $\|E_G\|_F$ generated by both models GNN and GGNN with linear and two nonlinear activation functions are shown in Fig. 2.46a–c. Like the conclusion in the previous example, the perception is that GGNN converges faster compared to the GNN model.

In addition, graphs involved in Fig. 2.46b and c, corresponding to Power-sigmoid and Smooth power-sigmoid AF, respectively, show a certain level of instability in convergence as well as an increase in values $\|E_G(t)\|_F$.

Chapter 3
Zeroing Neural Network (ZNN)

Many gradient-related methods has been reported in the scientific literature on the solutions of algebraic equations and many optimization problems before the proposal of ZNN approach, i.e., zero-finding problems [29, 271, 363, 366]. Nevertheless, gradient descent methods may fail to work well when exploited to the online solution of dynamic problems with time-varying coefficients, which is intrinsically due to the lacking of the tracking time-varying coefficients. Consequently, any method designed intrinsically for computing the static problem can no longer guarantee the decrease of the performance index of a time-varying problem in view of the variability of coefficients. Hence, the task is possibly leading to a failure with a large residual error. For example, in [347, 364, 372] it is observed, investigated and analyzed that the residual error of gradient-based neural network (GNN) for solving a time-varying problem can not be eliminated and remains at a relative high level. In [108, 109, 112] it was shown that, when exploited to solve a time-varying problem, any traditional method that does not exploit the time-derivatives of time-varying coefficients can not converge to the theoretical solution.

3.1 Review of Basic ZNN Models

In [353], Zhang et al. present a recurrent neural network for solving the time-varying Sylvester equation, in order to solve a time-varying problem in an error-free manner which is depicted in an implicit dynamical system and can be deemed as the seminal work on ZNN.

The time-varying matrix inversion problem is depicted in the form of

$$A(t)X(t) = I, \tag{3.1.1}$$

P. Stanimirović et al., *Generalized Matrix Inversion: A Machine Learning Approach*,
https://doi.org/10.1007/978-3-032-01493-1_3

where $A(t) \in \mathbb{R}^{(n\times n)}$ is a smooth matrix with its derivative assumed to be known, $I \in \mathbb{R}^{(n\times n)}$ is the identity matrix and $X(t) \in \mathbb{R}^{(n\times n)}$ is the unknown matrix to be obtained. The convergence and stability of the corresponding ZNN model for solving the time-varying matrix inversion was analysed in [381].

The core in the design of ZNN model is definition of an appropriate error function $E(t)$, which is evidently different from the performance index of gradient-related methods. Following that, the ZNN design formula comes to enforce the corresponding $E(t)$ to converge to zero:

$$\dot{E}(t) = -\gamma \mathcal{F}(E(t)), \tag{3.1.2}$$

where $\gamma > 0$ and $\mathcal{F}(\cdot)$ is a matrix array of activation function $f(\cdot)$. In [84], the ZNN design formula is implemented for vector-valued time-varying problems, e.g., the system of linear equation $A(t)x(t) = b(t)$, with $A(t) \in \mathbb{R}^{(m\times n)}$, $x(t) \in \mathbb{R}^n$ and $b(t) \in \mathbb{R}^m$, and, as a consequence, the error function can be designed as $e(t) = A(t)x(t) - b(t)$. Even for the scalar-valued time-varying problems [331, 332, 355, 368], e.g., the time-varying 4th root finding problem $x^4(t) = \alpha(t)$, with $\alpha(t) \in \mathbb{R}$ and $x(t) \in \mathbb{R}$, the error function can be designed as $e(t) = x^4(t) - \alpha(t)$. Note that the matrix-valued (or vector-valued) error function is a decoupled system and thus its (ij)th (or ith) subsystem, i.e., the scale-valued dynamical system $\dot{e}(t) = -\gamma\phi(e(t))$, can be used to analyze the corresponding convergence and stability. In essence, any ZNN model for solving any time-varying problem can be deemed as an equivalently expansion of the ZNN design formula (see Table 3.1).

A brief review on the design of continuous-time ZNN models for various problems solving is presented in [85] and some new variations have taken place in [318].

The basic problems in the research of neural networks are convergence and stability. In general, there are three ways for proving the convergence of ZNN models,

Table 3.1 Continuous-time ZNN models constructed for solving time-varying problems

Dynamic problem	Error function	ZNN model	References
4th root $x^4(t) = \alpha(t)$	$e(t) = x^4(t) - \alpha(t)$	$\dot{x}(t) = \frac{\dot{\alpha}(t)-\gamma\phi(x^4(t)-\alpha(t))}{4x^3(t)}$	[368]
Linear system $A(t)x(t) = b(t)$	$e(t) = A(t)x(t) - b(t)$	$A(t)\dot{x}(t) = -\dot{A}(t)x(t) + \dot{b}(t) - \gamma\Phi(A(t)x(t) - b(t))$	[367]
Matrix inversion $A(t)X(t) = I$	$E(t) = A(t)X(t) - I$	$A(t)\dot{X}(t) = -\dot{A}(t)X(t) - \gamma\Phi(A(t)X(t) - I)$	[381]
Matrix square root $X^2(t) = A(t)$	$E(t) = X^2(t) - A(t)$	$X(t)\dot{X}(t) + \dot{X}(t)X(t) = -\gamma\Phi(X^2(t) - A(t)) - \dot{A}(t)$	[354, 373]
Nonlinear equations $f(x(t), t) = 0$	$e(t) = f(x(t), t)$	$\dot{x}(t) = -J^{-1}(x(t), t)(\gamma\Phi(f(x(t), t)) + \frac{\partial f(x(t),t)}{\partial t})$	[375]

i.e., proof based on Lyapunov theory [83, 109, 190, 357, 358], ordinary differential equation (ODE) [84, 109, 381], or Laplace transform [110, 111, 113, 153, 364].

3.1.1 Complex-Valued ZNN Models

Over the past few years, different ZNN models have received considerable studies in many scientific and engineering fields, which successfully tackles the estimation error problem in the real domain. In comparison to these ZNN models defined in the real domain, the research on complex-valued ZNN models shows advantage over conventional real-valued neural networks in complex problems solving. The first proposition of a complex-valued ZNN model capable to solve the time-varying matrix-inversion problems in complex domain was made in [358]. Note that it uses only linear activation functions for ensuring the global convergence of the neural network. As mentioned earlier, many nonlinear function can be used to accelerate the convergence speed of ZNN model, which inspires Li et al. to explore nonlinear complex-valued activation functions to accelerate the convergence of ZNN with guaranteed global convergence in [129]. They find two classes of activation functions to achieve the global convergence of the complex-valued ZNN for solving the complex-valued time-varying Sylvester equation and then accelerate it to finite-time convergence. By exploiting different techniques to compute the time-varying complex-valued pseudoinversion problem Liao et al. proposed five ZNN models in [126], which is further generalized to complex-valued matrix inversion in [150]. Qiao et al. presented two complex-valued ZNN models for computing the Drazin inverse, which is accelerated to finite-time with proven upper bounds of the convergence time in [203]. Also, online solution of complex-valued systems of linear equations was investigated in the complex domain via a GNN model in [316].

3.1.2 Noise-Tolerant ZNN Models

For solving time-varying problems, many computational models such as the conventional ZNN models usually assume that the solving task is free of noises or that the noise reduction operation has been completed before the computation [27]. Point out that time is valuable for the online computation of time-varying problems and the preprocessing for denoising may consume extra time. In that case, the online computation is no longer required. In order to avoid that an integration-enhanced ZNN design formula is proposed in [113] for time-varying matrix inversion, which leverages the integration-control technique in control theoretical and is able to handle simultaneously the noises. In [153], such a noise-tolerant ZNN design formula is used to construct continuous-time ZNN model with nonlinear activation function exploited for accelerating the noise-tolerant model. Authors in [110] analyze and design different ZNN models via a systematic approach from the control perspective for solving time-varying problems. The essence of the noise-tolerant ZNN models is to leverage the error integration information to eliminate the constant bias errors. The challenge in this approach is how to determine the suitable activation functions for accelerating its convergence.

3.2 Solving Time-Varying Inversion by ZNN Neurodynamics

Solving the time-varying problem effectively is important problem in scientific and engineering circles. Many of the neural-dynamic approaches are efficient for static problems, but less efficient for most of the time-varying problems. The base of Zhang dynamics (ZD) is indefinite Zhang function (ZF), also called Zhangian. ZF is error-monitoring function, and indefinite means that function can be positive, zero, negative or even lower-unbounded. Also, ZF can be matrix or vector-valued. So, ZF is not common error or energy function (such is used in gradient-based dynamics). In this work, we will consider three different class of ZF, scalar-valued ZFs, vector-valued ZFs and matrix-valued ZFs. Based on them, we get ZD models for solving time-varying (scalar/vector/matrix-valued) problems. After that, we make simulations in MATLAB Simulink with examples to substantiate the efficacy of the proposed ZD models.

In this section we will consider three different class of ZF, scalar-valued ZFs, vector-valued ZFs and matrix-valued ZFs. Based on them, we will develop ZD models for solving time-varying (scalar/vector/matrix-valued) problems. After that, we make simulations in MATLAB Simulink with examples to substantiate the efficacy of the proposed ZD models.

3.2.1 *Time-Varying Reciprocal*

In this chapter, we present 4 different error functions (ZF) and adequate ZD models for solving time-varying reciprocal problem. First, we consider:

$$f(x(t), t) = a(t)x(t) - 1 = 0 \in \mathbb{R}, \quad t \in [0, +\infty), \tag{3.2.1}$$

where $a(t) \neq 0 \in \mathbb{R}$ is a smoothly time-varying scalar, and $\dot{a}(t)$ its time derivative. We are looking for $x(t) \in \mathbb{R}$ such that (3.2.1) holds true.

After we define indefinite ZF (as error-monitoring function), which we denote with $e(t)$, we choose its time-derivative $\dot{e}(t)$ to force $e(t)$ to converge to zero:

$$\dot{e}(t) = \mathrm{d}e(t)/\mathrm{d}t = -\gamma\phi(e(t)), \tag{3.2.2}$$

where $\gamma > 0$ is parameter and ϕ is activation function. There exist many activation functions such as linear, sigmoid, power-sigmoid, hyperbolic sine etc. In this study we consider linear activation function $\phi(e(t)) = e(t)$, which gives us exponential convergence.

Theorem 3.2.1 *The ZF* $e(t)$ *that solves* (3.2.2) *globally and exponentially converges to zero with rate* $\gamma > 0$, *starting from an initial error* $e(0) \in \mathbb{R}$.

Proof From calculus, we know analytical solution of (3.2.2) for initial condition $e(0)$:

$$e(t) = e(0)e^{-\gamma t}. \tag{3.2.3}$$

So, starting from any $e(0)$, $e(t)$ globally and exponentially converges to 0 with rate γ, as time t tends to infinity. □

Now, we define 4 ZFs for computing inverse of time-varying scalars, as follows:

$$e(t) = x(t) - \frac{1}{a(t)} \tag{3.2.4}$$

$$e(t) = a(t) - \frac{1}{x(t)} \tag{3.2.5}$$

$$e(t) = a(t)x(t) - 1 \tag{3.2.6}$$

$$e(t) = \frac{1}{a(t)x(t)} - 1. \tag{3.2.7}$$

Every single ZF in combination with (3.2.2) gives different ZD model. Below, we derive Zhang model for each ZF.

Combining ZF (3.2.4) with (3.2.2) in linear case gives

$$\dot{e}(t) = \dot{x}(t) + \frac{1}{a^2(t)}\dot{a}(t) = -\gamma e(t) = -\gamma \left(x(t) - \frac{1}{a(t)} \right).$$

Further, multiplying both sides by $a^2(t) \neq 0$ we obtain

$$a^2(t)\dot{x}(t) = -\dot{a}(t) - \gamma \left(x(t)a^2(t) - a(t) \right),$$

and the final model

$$\dot{x}(t) = (1 - a^2(t))\dot{x}(t) - \dot{a}(t) - \gamma \left(x(t)a^2(t) - a(t) \right) \tag{3.2.8}$$

Secondly, (3.2.5) and (3.2.2) lead to

$$\dot{a}(t) + \frac{1}{x^2(t)}\dot{x}(t) = -\gamma \left(a(t) - \frac{1}{x(t)} \right)$$

and further

$$\dot{x}(t) = -\dot{a}(t)x^2(t) - \gamma(x^2(t)a(t) - x(t)). \tag{3.2.9}$$

ZF (3.2.6) in combination with (3.2.2) gives

$$\dot{a}(t)x(t) + a(t)\dot{x}(t) = \dot{e}(t) = -\gamma(a(t)x(t) - 1),$$

which implies

$$a(t)\dot{x}(t) = -\dot{a}(t)x(t) - \gamma(a(t)x(t) - 1)$$

and produces the final model

$$\dot{x}(t) = (1 - a(t))\dot{x}(t) - \dot{a}(t)x(t) - \gamma(a(t)x(t) - 1). \tag{3.2.10}$$

Lastly, ZF (3.2.7) and (3.2.2) lead to

$$-\frac{1}{a^2(t)x^2(t)}(\dot{a}(t)x(t) + a(t)\dot{x}(t)) = \dot{e}(t) = -\gamma\left(\frac{1}{a(t)x(t)} - 1\right),$$

which gives

$$a(t)\dot{x}(t) = -\dot{a}(t)x(t) + \gamma(a(t)x(t) - a^2(t)x^2(t))$$

and

$$\dot{x}(t) = (1 - a(t))\dot{x}(t) - \dot{a}(t)x(t) + \gamma(a(t)x(t) - a^2(t)x^2(t)). \tag{3.2.11}$$

Convergence Properties of the Proposed ZD Models

In this section we will discuss about convergence properties of described ZD models for time-varying reciprocal finding.

Proposition 3.2.1 *Consider a smoothly time-varying scalar* $a(t) \neq 0 \in \mathbb{R}$ *involved in time-varying reciprocal problem* (3.2.1). *Starting from randomly-generated initial state* $x(0) \neq 0 \in \mathbb{R}$ *which has the same sign as* $a(0)$*, the neural state* $x(t)$ *of ZD model* (3.2.8) *derived from ZF* (3.2.4) *exponentially converges to the theoretical time-varying reciprocal* $x^*(t)$ *of* $a(t)$*, i.e.,* $a^{-1}(t)$*.*

Proof To prove this Proposition we use Lyapunov method. First, define the Lyapunov function candidate as

$$V(x(t), t) = \frac{1}{2}\left(x(t) - \frac{1}{a(t)}\right)^2 \geq 0$$

then $V(x(t), t) = 0$ for any $x(t) = \frac{1}{a(t)}$, $V(x(t), t) > 0$ for any $x(t) \neq \frac{1}{a(t)}$. We take time derivative which leads to

$$\dot{V}(x(t),t) = \frac{\mathrm{d}V(x(t),t)}{\mathrm{d}t} = \frac{\mathrm{d}\left(\frac{1}{2}\left(x(t)-\frac{1}{a(t)}\right)^2\right)}{\mathrm{d}t} = \left(x(t)-\frac{1}{a(t)}\right)\left(x(t)-\frac{1}{a(t)}\right)'$$
$$= \left(x(t)-\frac{1}{a(t)}\right)\left(\dot{x}(t)+\frac{1}{a^2(t)}\dot{a}(t)\right) = -\gamma\left(x(t)-\frac{1}{a(t)}\right)^2$$
$$= -2\gamma V(x(t),t).$$

Since $V(x(t),t) \geq 0$, we have that $V'(x(t),t) = -2\gamma V(x(t),t) \leq 0$. All of three conditions of Lyapunov exponential stability theorem are satisfied, so $e(t)$ exponentially converges to 0. Furthermore, $V'(x(t),t) = -2\gamma V(x(t),t)$ initiates

$$V'(x(t),t) = \frac{\mathrm{d}V(x(t),t)}{\mathrm{d}t} = -2\gamma V(x(t),t).$$

Now, it is necessary to solve the differential equation

$$\frac{\mathrm{d}V(x(t),t)}{V(x(t),t)} = -2\gamma\mathrm{d}t.$$

If we integrate above equation we get

$$\ln V(x(t),t) = -2\gamma t + \ln c, \quad c \in \mathbb{R}.$$

Then applying exp on both side of equation:

$$V(x(t),t) = e^{-2\gamma t}e^{\ln c} = ce^{-2\gamma t}$$

For $t = 0$, it follows $V(x(0),0) = ce^0 = c$. Accordingly,

$$V(x(t),t) = V(x(0),0)e^{-2\gamma t}$$

$$V(x(t),t) = \frac{1}{2}\left(x(t)-\frac{1}{a(t)}\right)^2 = \frac{1}{2}\left(x(0)-\frac{1}{a(0)}\right)^2 e^{-2\gamma t}$$

$$\left|x(t)-\frac{1}{a(t)}\right| = \left|x(0)-\frac{1}{a(0)}\right| e^{-\gamma t}$$

Now, we see that for randomly-generated initial state $x(0) \neq 0 \in \mathbb{R}$, which has the same sign as $a(0)$, the neural state $x(t)$ exponentially converges to the theoretical time-varying reciprocal $x^*(t)$ of $a(t)$, i.e., $a^{-1}(t)$. □

For the rest of ZFs results are analogous, and we will just give a statement of proposition for that Zhang functions.

Proposition 3.2.2 *Consider a smoothly time-varying scalar* $a(t) \neq 0 \in \mathbb{R}$ *involved in the time-varying reciprocal problem* (3.2.1). *Starting from randomly-generated initial state* $x(0) \neq 0 \in \mathbb{R}$ *which has the same sign as* $a(0)$, *the neural state* $x(t)$ *of ZD model* (3.2.9) *derived from ZF* (3.2.5) *converges to the theoretical time-varying reciprocal* $x^*(t)$ *of* $a(t)$, *i.e.,* $a^{-1}(t)$, *with the error defined in ZF* (3.2.5) *exponentially convergent to 0.*

Proposition 3.2.3 *Consider a smoothly time-varying scalar* $a(t) \neq 0 \in \mathbb{R}$ *involved in time-varying reciprocal problem* (3.2.1). *Starting from randomly-generated initial state* $x(0) \neq 0 \in \mathbb{R}$ *which has the same sign as* $a(0)$, *the neural state* $x(t)$ *of ZD model* (3.2.10) *derived from ZF* (3.2.6) *converges to the theoretical time-varying reciprocal* $x^*(t)$ *of* $a(t)$, *i.e.,* $a^{-1}(t)$, *with the error defined in ZF* (3.2.6) *exponentially convergent to 0.*

Proposition 3.2.4 *Consider a smoothly time-varying scalar* $a(t) \neq 0 \in \mathbb{R}$ *involved in time-varying reciprocal problem* (3.2.1). *Starting from randomly-generated initial state* $x(0) \neq 0 \in \mathbb{R}$ *which has the same sign as* $a(0)$, *the neural state* $x(t)$ *of ZD model* (3.2.11) *derived from ZF* (3.2.7) *converges to the theoretical time-varying reciprocal* $x^*(t)$ *of* $a(t)$, *i.e.,* $a^{-1}(t)$, *with the error defined in ZF* (3.2.2) *exponentially convergent to 0.*

MATLAB Simulink Modeling for Time-Varying Reciprocal

We will present MATLAB Simulink model for each previously discussed Zhang dynamics for time-varying reciprocal and give an example.

–First, we consider Zhang function given with (3.2.4) and its Zhang model given with (3.2.8). Implementation of (3.2.8) in MATLAB Simulink presented in Fig. 3.1.

Now, we using Zhang function (3.2.5) we already get Zhang dynamics model (3.2.9). Implementation of (3.2.9) in MATLAB Simulink is given in Fig. 3.2.

Next, we consider Zhang function (3.2.6) with its Zhang dynamics model given by (3.2.10). MATLAB Simulink model for this is presented in Fig. 3.3.

Finally, we observed last Zhang function (3.2.7) with Zhang dynamics (3.2.11). MATLAB Simulink model for (3.2.11) is given in Fig. 3.4.

Example 3.2.1 Let us consider time-varying reciprocal problem if $a(t)$ is given with $a(t) = \sin(3t) + 2$:

$$f(x(t), t) = (\sin(3t) + 2)x(t) - 1$$

We will use $\gamma = 10$ in all tested Zhang dynamics models. Theoretical initial reciprocal value is $x^*(0) = \frac{1}{a(0)} = \frac{1}{2}$. Initial state $x(0)$ is randomly generated in MATLAB inside the interval (0.4, 0.6) applying the standard function `rand`. We create a vector x_0 of 10 random values, and use `rand` function to generate values from a uniform distribution in the open interval (0.4, 0.6). The model has been repeated 10 times with fixed starting point x_0 to generate time-varying reciprocal $x(t)$ and computational errors $e(t)$ in Fig. 3.5 (Figs. 3.6, 3.7 and 3.8).

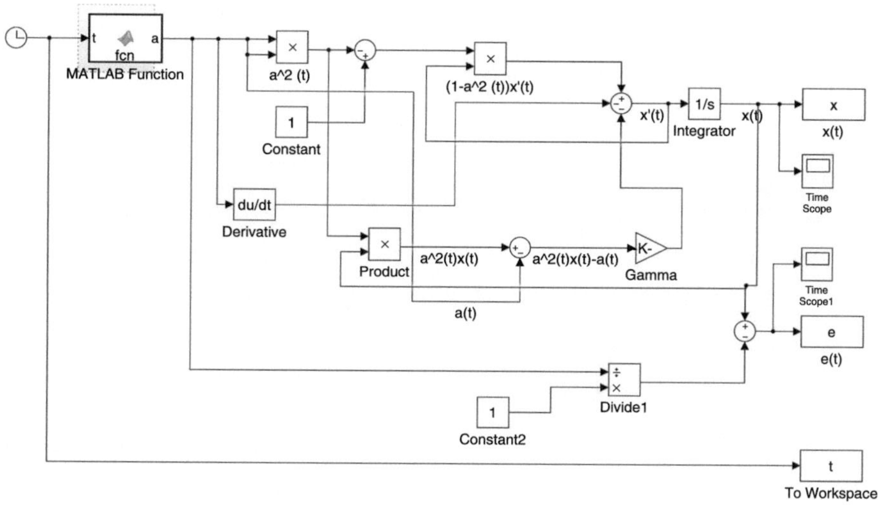

Fig. 3.1 MATLAB Simulink model for the error function (3.2.4) and dynamics (3.2.8)

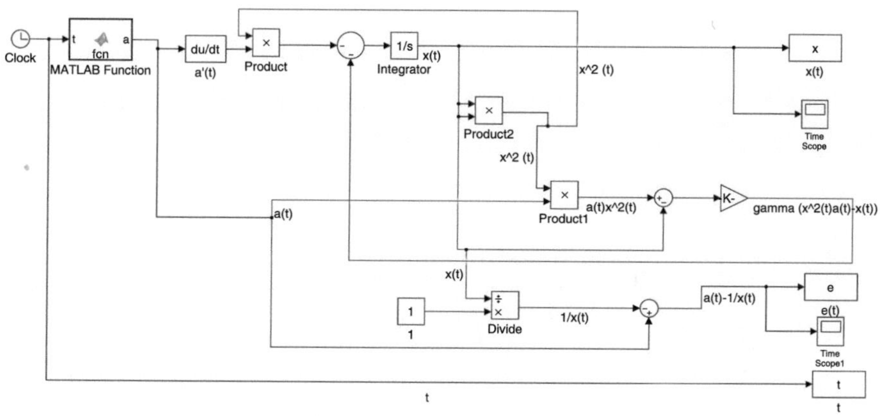

Fig. 3.2 MATLAB Simulink model for the error function (3.2.5) and dynamics (3.2.9)

3.2.2 *Time-Varying Vector Inverse*

Solving system of linear equation is widely spread problem in science, applied in many area, like steps for optimization, signal processing, electromagnetic systems, and robots inverse kinematics. Many algorithms are proposed to solve this problem, but they are not so effective for solving the system of time-varying linear equations

$$A(t)\boldsymbol{x}(t) = \boldsymbol{b}(t) \in \mathbb{R}^m, \quad t \in [0, +\infty), \tag{3.2.12}$$

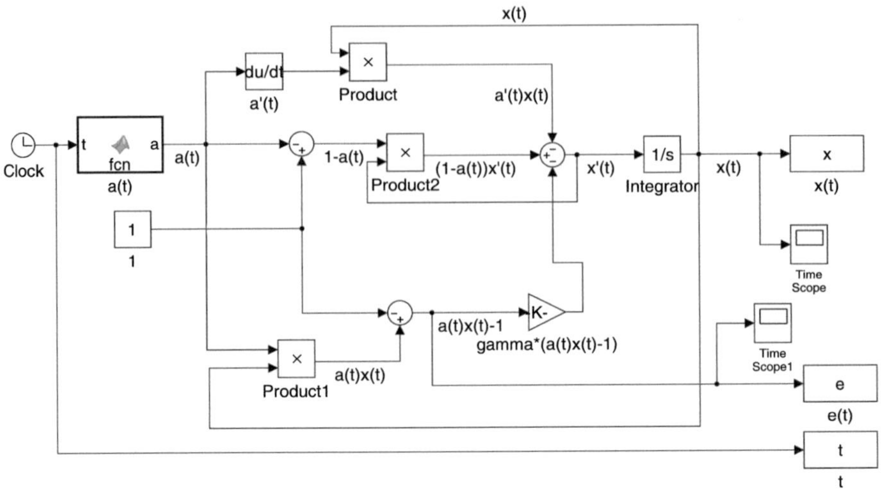

Fig. 3.3 MATLAB Simulink model for the error function (3.2.6) and dynamics (3.2.10)

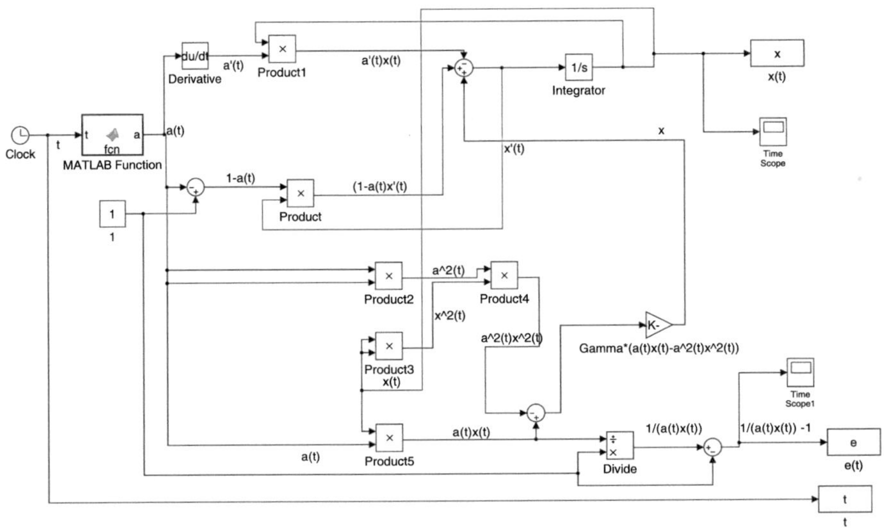

Fig. 3.4 MATLAB Simulink model for the error function (3.2.7) and dynamics (3.2.11)

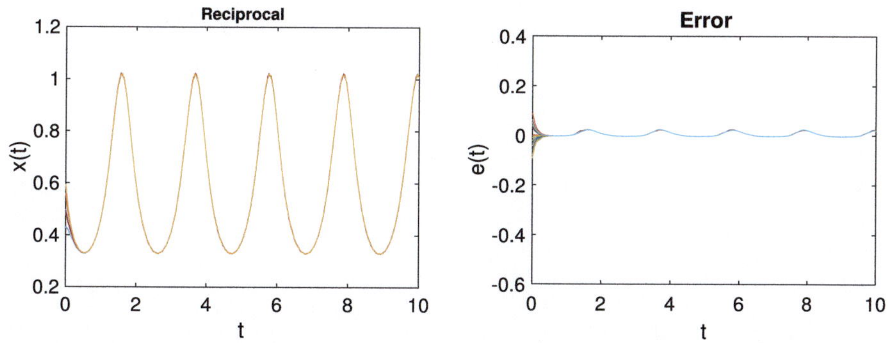

(a) State trajectory (b) Computational error

Fig. 3.5 State trajectory $x(t)$ and computational error $e(t)$ for Zhang function $e(t) = x(t) - \frac{1}{\sin(3t)+2}$

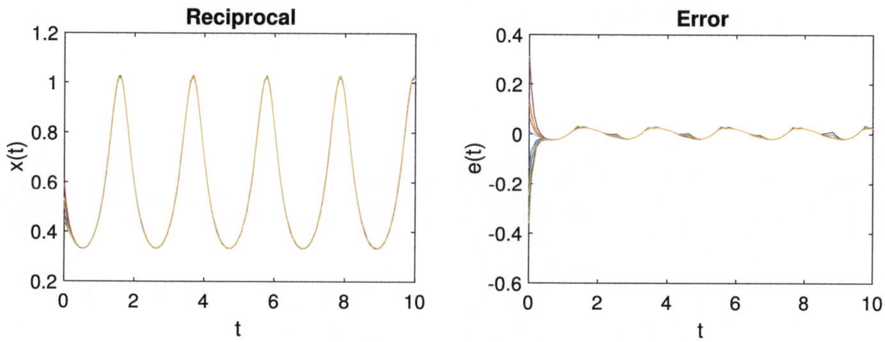

(a) State trajectory (b) Computational error

Fig. 3.6 State trajectory $x(t)$ and computational error $e(t)$ for Zhang function $e(t) = \sin(3t) + 2 - \frac{1}{x(t)}$

where $A(t) \in \mathbb{R}^{m\times n}$ is known coefficient matrix, $\boldsymbol{b}(t) \in \mathbb{R}$ coefficient vector and $\boldsymbol{x}(t) \in \mathbb{R}^m$ is unknown vector, which we want to find.

We define two error-monitoring Zhang functions and develop Zhang dynamics (ZD) model for them. Notice that (3.2.12) is presented in matrixâŁ“vector form, unlike scalar form in previously chapter. Vector-valued ZF and its time derivative we denote with $\boldsymbol{e}(t)$ and $\dot{\boldsymbol{e}}(t)$. ZD design formula (3.2.2) is generalized in the vector-form

$$\dot{\boldsymbol{e}}(t) = \frac{\mathrm{d}\boldsymbol{e}(t)}{\mathrm{d}t} = -\gamma \boldsymbol{e}(t), \tag{3.2.13}$$

where $\gamma \in \mathbb{R}$ is a parameter.

Now we introduce first ZF as vector-valued error function:

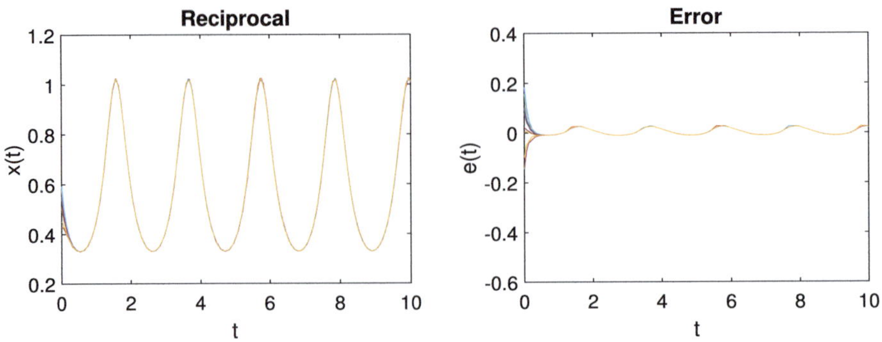

(a) State trajectory (b) Computational error

Fig. 3.7 State trajectory $x(t)$ and computational error $e(t)$ for Zhang function $e(t) = (\sin(3t) + 2)x(t) - 1$

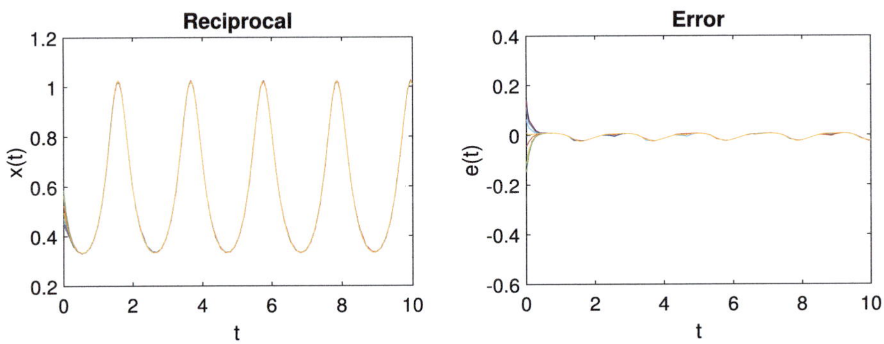

(a) State trajectory (b) Computational error

Fig. 3.8 State trajectory $x(t)$ and computational error $e(t)$ for Zhang function $e(t) = \frac{1}{(\sin(3t)+2)x(t)} - 1$

$$\boldsymbol{e}(t) = A(t)\boldsymbol{x}(t) - \boldsymbol{b}(t) \in \mathbb{R}^n. \tag{3.2.14}$$

Zhang dynamics initiated by this Zhang function is

$$\dot{\boldsymbol{e}}(t) = -\gamma \boldsymbol{e}(t) \Longleftrightarrow -\gamma(A(t)\boldsymbol{x}(t) - \boldsymbol{b}(t)) = \dot{A}(t)\boldsymbol{x}(t) + A(t)\dot{\boldsymbol{x}}(t) - \dot{\boldsymbol{b}}(t),$$

which implies

$$A(t)\dot{\boldsymbol{x}}(t) = \dot{\boldsymbol{b}}(t) - \dot{A}(t)\boldsymbol{x}(t) - \gamma(A(t)\boldsymbol{x}(t) - \boldsymbol{b}(t)).$$

We transform this equation in the form (Fig. 3.9)

$$\dot{\boldsymbol{x}}(t) = (I - A(t))\dot{\boldsymbol{x}}(t) + \dot{\boldsymbol{b}}(t) - \dot{A}(t)\boldsymbol{x}(t) - \gamma(A(t)\boldsymbol{x}(t) - \boldsymbol{b}(t)). \tag{3.2.15}$$

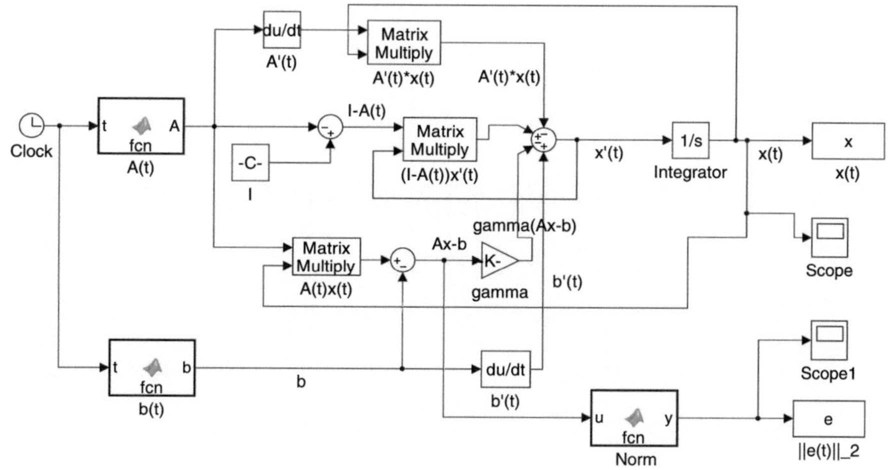

Fig. 3.9 MATLAB Simulink model for the ZF (3.2.14) and dynamics (3.2.15)

Second Zhang function is defined as

$$\boldsymbol{e}(t) = \boldsymbol{x}(t) - A^{-1}(t)\boldsymbol{b}(t) \tag{3.2.16}$$

Proposition 3.2.5 gives the time derivative of the inverse matrix.

Proposition 3.2.5 *The time derivative of the time-varying matrix inverse $A^{-1}(t)$ is formulated as*

$$\dot{A}^{-1}(t) = \frac{\mathrm{d}A^{-1}(t)}{\mathrm{d}t} = -A^{-1}(t)\dot{A}(t)A^{-1}(t).$$

Proposition 3.2.5 initiates the ZD model

$$\dot{\boldsymbol{e}}(t) = -\gamma(\boldsymbol{x}(t) - A^{-1}(t)\boldsymbol{b}(t)) = \dot{\boldsymbol{x}}(t) - A^{-1}(t)\dot{A}(t)A^{-1}(t)\boldsymbol{b}(t) - A^{-1}(t)\dot{\boldsymbol{b}}(t)$$

Multiplying both sides from left by $A(t)$ we obtain (Fig. 3.10)

$$-\gamma(A(t)\boldsymbol{x}(t) - \boldsymbol{b}(t)) = A(t)\dot{\boldsymbol{x}}(t) + \dot{A}(t)A^{-1}(t)\boldsymbol{b}(t) - A^{-1}(t)\boldsymbol{b}(t)$$

$$\dot{\boldsymbol{x}}(t) = (I - A(t))\dot{\boldsymbol{x}}(t) - \dot{A}(t)A^{-1}(t)\boldsymbol{b}(t) - \gamma(A(t)\boldsymbol{x}(t) - \boldsymbol{b}(t)) + \dot{\boldsymbol{b}}(t). \tag{3.2.17}$$

Convergence Properties

In this subsection we present convergence properties of the proposed ZD models for time-varying vector inverse.

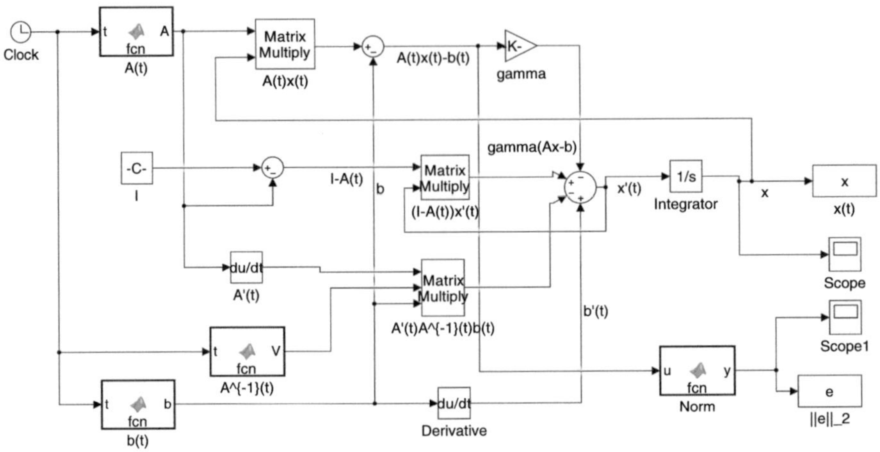

Fig. 3.10 MATLAB Simulink model for the ZF (3.2.16) and dynamics (3.2.17)

Theorem 3.2.2 *Given a smoothly time-varying nonsingular coefficient matrix $A(t) \in \mathbb{R}^{n\times n}$ and a smoothly time-varying vector $\boldsymbol{b}(t) \in \mathbb{R}^n$ in (3.2.12), the state vector $\boldsymbol{x}(t) \in \mathbb{R}^n$ of ZD model (3.2.15), starting from a randomly-generated initial state $\boldsymbol{x}(0)$ converges globally and exponentially to the theoretical time-varying solution $\boldsymbol{x}^*(t)$ of (3.2.12) with time.*

Proof Let $\tilde{\boldsymbol{x}}(t) = \boldsymbol{x}(t) - \boldsymbol{x}^*(t)$ denote the difference between the time-varying solution of ZD model (3.2.15) and the theoretical time-varying solution of (3.2.17). Then:

$$\boldsymbol{x}(t) = \tilde{\boldsymbol{x}}(t) + \boldsymbol{x}^*(t).$$

Substitution of the last identity into (3.2.15) leads to

$$A(t)\boldsymbol{x}^*(t) - \boldsymbol{b}(t) = \boldsymbol{0}$$

$$A(t)\dot{\tilde{\boldsymbol{x}}}(t) = -\dot{A}(t)\tilde{\boldsymbol{x}}(t) - \gamma A(t)\tilde{\boldsymbol{x}}(t).$$

Now,

$$\boldsymbol{e}(t) = A(t)\boldsymbol{x}(t) - \boldsymbol{b}(t) = A(t)(\tilde{\boldsymbol{x}}(t) + \boldsymbol{x}^*(t)) - \boldsymbol{b}(t) = A(t)\tilde{\boldsymbol{x}}(t)$$

and above equation can be rewritten equivalently as

$$\dot{\boldsymbol{e}}(t) = -\gamma \boldsymbol{e}(t)$$

or in elementwise form

$$\dot{e}_i = -\gamma e_i \in \mathbb{R}.$$

Finally, we define a Lyapunov function $v = e_i^2/2 \geq 0$ for last ith subsystem. So, $v > 0$ for any $e_i \neq 0$, and $v = 0$ only for $e_i = 0$. This gives us that Lyapunov function is positive-definite. Now we consider its time derivative:

$$v' = \frac{\mathrm{d}v}{\mathrm{d}t} = e_i \dot{e}_i = -\gamma e_i e_i \leq 0.$$

Therefore, v' is negative-definite. From Lyapunov theory, e_i globally converges to zero for any $i \in 1, 2, \ldots, n$.

Since $\boldsymbol{e}(t) = A(t)\tilde{\boldsymbol{x}}(t)$ and A is invertible, we have that $\lim_{t\to\infty} \tilde{\boldsymbol{x}}(t) = 0 = \lim_{t\to\infty} \boldsymbol{x}(t) - \boldsymbol{x}^*(t)$. This means that the state vector $\boldsymbol{x}(t)$ of ZD model (3.2.15) converges globally to the theoretical time-varying solution $\boldsymbol{x}^*(t) \in \mathbb{R}^n$.

We already know analytical solution of differential equation for $e(t)$:

$$\boldsymbol{e}(t) = \boldsymbol{e}(0)exp(-\gamma t).$$

If we take a second norm of both sides, we get:

$$||\boldsymbol{e}(t)||_2 = ||\boldsymbol{e}(0)||_2 exp(-\gamma t)$$

$$\lim_{t\to\infty} ||\boldsymbol{e}(t)||_2 = \lim_{t\to\infty} ||A(t)\tilde{\boldsymbol{x}}(t)||_2 = \lim_{t\to\infty} ||\boldsymbol{e}(0)||_2 exp(-\gamma t) = 0$$

Let $\alpha > 0$ be minimum eigenvalue of $A^{\mathrm{T}}(t)A(t)$, then

$$||A(t)\tilde{\boldsymbol{x}}(t)||_2 = \tilde{\boldsymbol{x}}(t)^{\mathrm{T}} A^{\mathrm{T}}(t)A(t)\tilde{\boldsymbol{x}}(t) \geq \alpha ||\tilde{\boldsymbol{x}}(t)||_2$$

so starting from any randomly-generated initial state $\boldsymbol{x}(0)$, the state vector $\boldsymbol{x}(t)$ of ZD model (3.2.15) converges exponentially to the theoretical time-varying solution $\boldsymbol{x}^*(t)$ of (3.2.12)). □

Theorem 3.2.3 *Given a smoothly time-varying nonsingular coefficient matrix* $A(t) \in \mathbb{R}^{n\times n}$ *and a smoothly time-varying vector* $\boldsymbol{b}(t) \in \mathbb{R}^n$ *in* (3.2.12)*, the state vector* $\boldsymbol{x}(t) \in \mathbb{R}^n$ *of ZD model* (3.2.17)*, starting from a randomly-generated initial state* $\boldsymbol{x}(0)$ *converges globally and exponentially to the theoretical time-varying solution* $\boldsymbol{x}^*(t)$ *of* (3.2.12) *with rate* $\gamma > 0$*, as time* t *goes on.*

Numerical Examples

We investigated before two ZD models of two ZF. Now, we give example where we are solving the system of time-varying linear equations to see efficacy of ZD models.

Example 3.2.2 We consider $A(t)$ and $\boldsymbol{b}(t)$ defined as $A(t) = \begin{bmatrix} \sin t & -\cos t \\ \cos t & \sin t \end{bmatrix}$, $\boldsymbol{b}(t) = \begin{bmatrix} \sin t + 1 \\ \cos t \end{bmatrix}$ Now, the theoretical time-varying solution $\boldsymbol{x}^*(t)$ of (3.2.12) is given as

$$\boldsymbol{x}^*(t) = A^{-1}(t)\boldsymbol{b}(t) = \begin{bmatrix} \sin t & \cos t \\ -\cos t & \sin t \end{bmatrix} \begin{bmatrix} \sin t + 1 \\ \cos t \end{bmatrix} = \begin{bmatrix} \sin t + 1 \\ -\cos t \end{bmatrix}.$$

Simulation results are shown in Figs. 3.11, 3.12, 3.13 and 3.14.

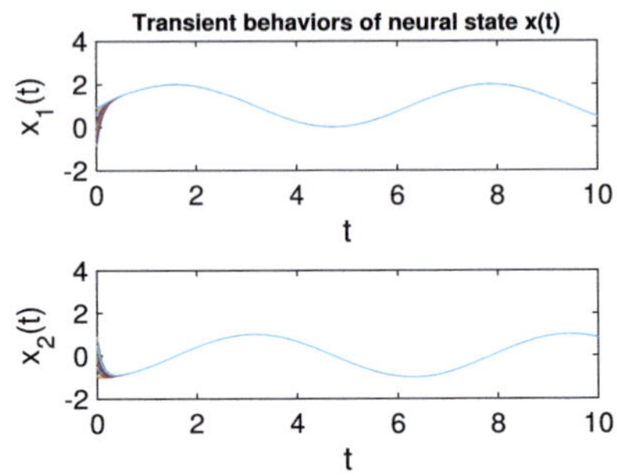

Fig. 3.11 State trajectory $x(t)$ for ZD (3.2.15)

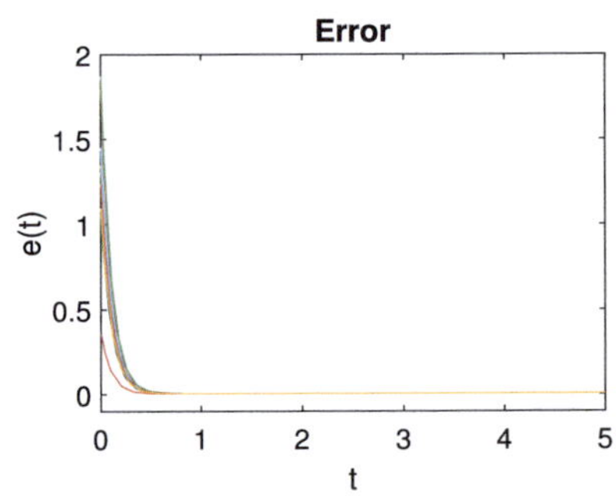

Fig. 3.12 Computational error $e(t)$ for ZD (3.2.15)

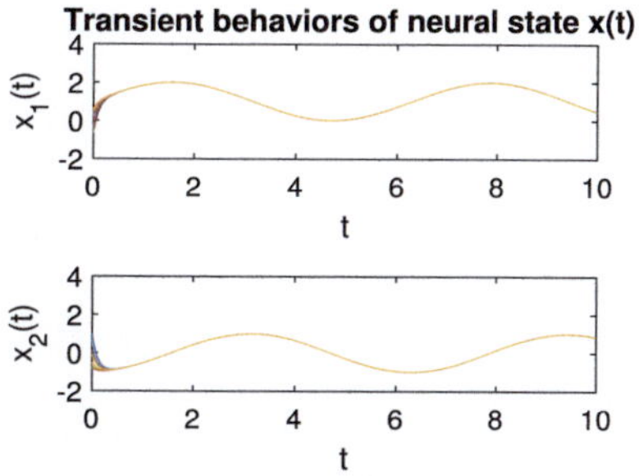

Fig. 3.13 State trajectory $x(t)$ for ZD (3.2.17)

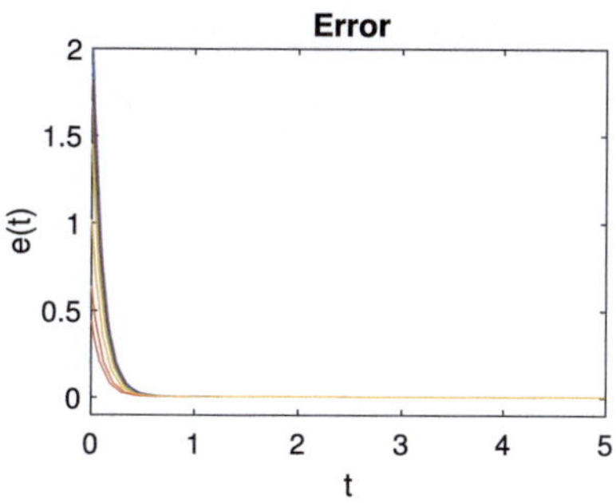

Fig. 3.14 Computational error $e(t)$ for ZD (3.2.17)

3.2.3 Time-Varying Matrix Inverse

In this chapter, we introduce Zhang functions and appropriate Zhang dynamics model for solving problem of time-varying matrix inversion:

$$A(t)X(t) - I = 0 \in \mathbb{R}^{n\times n} \tag{3.2.18}$$

where $A(t) \in \mathbb{R}^{n\times n}$ is smoothly time-varying nonsingular coefficient matrix and $I \in \mathbb{R}^{n\times n}$ identity matrix. Our goal is to obtain time-varying matrix $X(t) \in \mathbb{R}^{n\times n}$. Problem (3.2.18) is in matrix form, $E(t)$ denotes the Zhang function in the matrix form and its time derivative with $\dot{E}(t)$. Now, the standard ZD design formula is presented in the matrix form as follows:

$$\dot{E}(t) = \frac{\mathrm{d}E(t)}{\mathrm{d}t} = -\gamma E(t) \tag{3.2.19}$$

where $\gamma \in \mathbb{R}$ is the gain parameter, as defined before.

We define different ZFs for solving TV matrix inversion and developing corresponding ZDs. In this case, six ZFs are available, because matrix multiplication is not commutative:

$$E(t) = A(t)X(t) - I \tag{3.2.20}$$

$$E(t) = X(t)A(t) - I \tag{3.2.21}$$

$$E(t) = A^{-1}(t) - X(t) \tag{3.2.22}$$

$$E(t) = A(t) - X^{-1}(t) \tag{3.2.23}$$

$$E(t) = (A(t)X(t))^{-1} - I \tag{3.2.24}$$

$$E(t) = (X(t)A(t))^{-1} - I \tag{3.2.25}$$

Now, for different Zhang functions we construct Zhang dynamics model for time-varying matrix inversion.

Utilizing (3.2.19) and ZF (3.2.20) and taking derivative, it can be obtained

$$\dot{E}(t) = \frac{\mathrm{d}(A(t)X(t) - I)}{\mathrm{d}t} = \dot{A}(t)X(t) + A(t)\dot{X}(t) = -\gamma(A(t)X(t) - I)$$

and later the final model

$$\dot{X}(t) = (I - A(t))\dot{X}(t) - \gamma(A(t)X(t) - I) - \dot{A}(t)X(t). \tag{3.2.26}$$

Usage of (3.2.19) and (3.2.21) implies

$$\dot{E}(t) = \frac{\mathrm{d}(X(t)A(t) - I)}{\mathrm{d}t} = \dot{X}(t)A(t) + X(t)\dot{A}(t) = -\gamma(X(t)A(t) - I),$$

or equivalently

$$\dot{X}(t)A(t) = -X(t)\dot{A}(t) - \gamma(X(t)A(t) - I),$$

which gives the final dynamics

$$\dot{X}(t) = \dot{X}(t)(I - A(t)) - \gamma(X(t)A(t) - I) - X(t)\dot{A}(t) \tag{3.2.27}$$

Let us consider (3.2.19), (3.2.22) and Proposition 3.2.5, which leads to

$$\dot{E}(t) = \frac{\mathrm{d}(A^{-1}(t) - X(t))}{\mathrm{d}t} = -A^{-1}(t)\dot{A}(t)A^{-1}(t) - \dot{X}(t) = -\gamma(A^{-1}(t) - X(t))$$

Now, we multiply both sides of last equation by matrix $A(t)$ from both sides and then add $\dot{X}(t)$, which leads to

$$-\gamma(A(t) - A(t)X(t)A(t)) = -\dot{A}(t) - A(t)\dot{X}(t)A(t)$$

and

$$\begin{aligned}\dot{X}(t) &= \dot{X}(t) + A(t)\dot{X}(t)A(t) + \dot{A}(t) - \gamma\left(A(t) - A(t)X(t)A(t)\right)\\ &= \dot{X}(t) + \dot{A}(t) - \gamma\left(I - A(t)X(t)\right)A(t) + A(t)\dot{X}(t)A(t)\\ &= \dot{X}(t) + \dot{A}(t) + \left(\gamma\left(A(t)X(t) - I\right) + A(t)\dot{X}(t)\right)A(t).\end{aligned}$$

Finally, the ZD model based on the ZF (3.2.22) is defined by

$$\dot{X}(t) = \dot{X}(t) + \dot{A}(t) + (A(t)\dot{X}(t) + \gamma(A(t)X(t) - I)A(t). \tag{3.2.28}$$

Consider now (3.2.19) based on (3.2.23). Computing the time derivative of $E(t)$ and expanding the ZD dynamics, it can be obtained

$$\dot{E}(t) = \frac{\mathrm{d}(A(t) - X^{-1}(t))}{\mathrm{d}t} = \dot{A}(t) + X^{-1}(t)\dot{X}(t)X^{-1}(t) = -\gamma(A(t) - X^{-1}(t)).$$

We multiply both sides of the last equation by $X(t)$ and obtain

$$X(t)\dot{A}(t)X(t) + \dot{X}(t) = -\gamma\left(X(t)A(t)X(t) - X(t)\right),$$

which yields

$$\dot{X}(t) = -X(t)\dot{A}(t)X(t) - \gamma X(t)(A(t)X(t) - I). \tag{3.2.29}$$

Zhang function (3.2.24) in conjunction with (3.2.19) implies

$$\begin{aligned}\dot{E}(t) &= \frac{\mathrm{d}((A(t)X(t))^{-1} - I)}{\mathrm{d}t}\\ &= -(A(t)X(t))^{-1}\left(\dot{A}(t)X(t) + A(t)\dot{X}(t)\right)(A(t)X(t))^{-1}\\ &= -\gamma E(t) = -\gamma\left((A(t)X(t))^{-1} - I\right).\end{aligned}$$

Both sides of above equation are multiplied by $A(t)X(t)$ from left and right, which produces

$$-\gamma(I - A(t)X(t))A(t)X(t) = -\dot{A}(t)X(t) - A(t)\dot{X}(t).$$

Hence, the final design is defined by

$$\begin{aligned}\dot{X}(t) &= \dot{X}(t) - A(t)\dot{X}(t) - \dot{A}(t)X(t) + \gamma(I - A(t)X(t))A(t)X(t)\\ &= (I - A(t))\dot{X}(t) - \gamma(I - A(t)X(t))A(t)X(t) - \dot{A}(t)X(t).\end{aligned} \tag{3.2.30}$$

According to (3.2.19), the Zhang function (3.2.25) satisfies

$$\begin{aligned}\dot{E}(t) &= \frac{\mathrm{d}((X(t)A(t))^{-1} - I)}{\mathrm{d}t} \\ &= -(X(t)A(t))^{-1}\left(\dot{X}(t)A(t) + X(t)\dot{A}(t)\right)(X(t)A(t))^{-1} \\ &= -\gamma E(t) = -\gamma\left((X(t)A(t))^{-1} - I\right).\end{aligned}$$

Analogous to the previous case, we multiply both sides by $X(t)A(t)$, which produces

$$-\gamma X(t)A(t)(I - X(t)A(t)) = -\dot{X}(t)A(t) - X(t)\dot{A}(t)$$

and finally

$$\dot{X}(t) = \dot{X}(t)(I - A(t)) - X(t)\dot{A}(t) - \gamma X(t)A(t)(X(t)A(t) - I). \qquad (3.2.31)$$

Convergence Properties of the Proposed ZD Models

In this part of study work, we will discuss about convergence properties of the proposed ZD models for time-varying matrix inversion finding.

Theorem 3.2.4 *Let us consider a smoothly time-varying nonsingular matrix* $A(t) \in \mathbb{R}^{n\times n}$ *in* (3.2.12). *Starting from an initial state* $X(0) \in \mathbb{R}^{n\times n}$, *the state matrix* $X(t)$ *of ZD model* (3.2.22) *derived from ZF* (3.2.14) *globally and exponentially converges to the theoretical time-varying inverse* $A^{-1}(t)$ *of matrix* $A(t)$.

Proof From ZD design formula $\dot{E}(t) = -\gamma E(t)$, we have a set of $n \times n$ differential equations:

$$\dot{e}_{ij} = -\gamma e_{ij}(t), \qquad (3.2.32)$$

for any $i, j \in \{1, 2, \ldots, n\}$. First, we define a Lyapunov function candidate $v_{ij} = e_{ij}^2(t)/2 \geq 0$ with its time derivative:

$$v_{ij}'(t) = \frac{\mathrm{d}v_{ij}(t)}{\mathrm{d}t} = e_{ij}(t)\dot{e}_{ij}(t) = -\gamma e_{ij}^2(t) \leq 0,$$

which guarantees the negative-definiteness of v_{ij}'.

By Lyapunov theory, equilibrium point $e_{ij} = 0$ of (3.2.28) is globally asymptotically stable, so the matrix-valued error function $E(t) = [e_{ij}(t)] \in \mathbb{R}^{n\times n}$ is globally convergent to zero. As we have it $E(t) = A^{-1}(t) - X(t) \Rightarrow X(t) = A^{-1}(t) - E(t)$.

$$\lim_{t\to\infty} E(t) = 0 \Rightarrow \lim_{t\to\infty} X(t) = A^{-1}(t)$$

This is a proof of global convergence.

We already know that analytical solution of (3.2.28) is $e_{ij}(t) = e^{-\gamma t}e_{ij}(0)$, so error function $E(t) \in \mathbb{R}^{n\times n}$ can be expressed as

$$E(t) = \begin{bmatrix} e_{11}(0) & e_{12}(0) & \cdots & e_{1n}(0) \\ e_{21}(0) & e_{22}(0) & \cdots & e_{2n}(0) \\ \vdots & \vdots & \ddots & \vdots \\ e_{n1}(0) & e_{n2}(0) & \cdots & e_{nn}(0) \end{bmatrix} e^{-\gamma t} = E(0)e^{-\gamma t}.$$

That is $E(t) = A^{-1}(t) - X(t) = E(0)e^{-\gamma t}$, which means that $X(t)$ exponentially converges to $A^{-1}(t)$ with rate $\gamma > 0$, so the proof is completed. □

For the rest od ZD models, results are analogous and we will just give a statement of proposition for them.

Proposition 3.2.6 *Consider a smoothly time-varying nonsingular matrix* $A(t) \in \mathbb{R}^{n\times n}$ *in* (3.2.12). *Starting from an initial state* $X(0) \in \mathbb{R}^{n\times n}$ *that is close enough to the initial theoretical inverse* $A^{-1}(0)$, *the state matrix* $X(t)$ *of ZD model* (3.2.23) *derived from ZF* (3.2.15), *the state matrix* $X(t)$ *of ZD model* (3.2.24) *derived from ZF* (3.2.18), *and the state matrix* $X(t)$ *of ZD model* (3.2.25) *derived from ZF* (3.2.19) *exponentially converge to the theoretical time-varying inverse* $A^{-1}(t)$ *of matrix* $A(t)$.

Proposition 3.2.7 *Consider a smoothly time-varying nonsingular matrix* $A(t) \in \mathbb{R}^{n\times n}$ *in* (3.2.12). *Starting from an initial state* $X(0) \in \mathbb{R}^{n\times n}$, *the state matrix* $X(t)$ *of ZD model* (3.2.20) *derived from ZF (16) and the state matrix* $X(t)$*of ZD model* (3.2.21) *derived from ZF* (3.2.17) *globally and exponentially converge to the theoretical time-varying inverse* $A^{-1}(t)$ *of matrix* $A(t)$.

MATLAB Simulink Models

In this section, we test MATLAB Simulink model for previously discussed Zhang dynamics models for computing time-varying matrix inverses (Figs. 3.15, 3.16, 3.17, 3.18, 3.19 and 3.20).

Example 3.2.3 Let us consider time-varying matrix inverse problem if $A(t)$ is given with:

$$A(t) = \begin{bmatrix} \sin t & -\cos t \\ \cos t & \sin t \end{bmatrix}$$

and its theoretical time-varying inverse is

$$A^{-1}(t) = \begin{bmatrix} \sin t & \cos t \\ -\cos t & \sin t \end{bmatrix}.$$

We present solution of proposed Zhang dynamics model with theoretical time-varying matrix inverse, each element X_{ij} separately. In this example, we use $\gamma = 10$ and initial state equal to the identity matrix I. Also, we present residual errors for each model with different values of $\gamma = 10$, 100 and 1000. Simulation results for each of six ZD models are shown in following (Figs. 3.21, 3.22, 3.23, 3.24, 3.25, 3.26, 3.27, 3.28, 3.29, 3.30, 3.31 and 3.32).

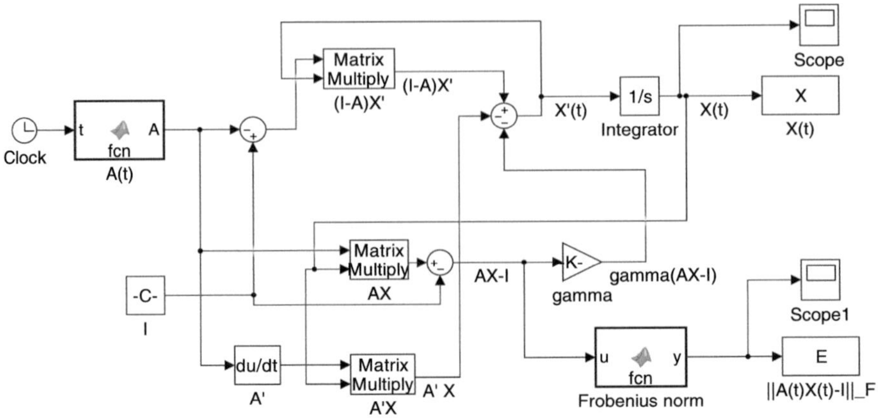

Fig. 3.15 MATLAB Simulink model for (3.2.20) and (3.2.26)

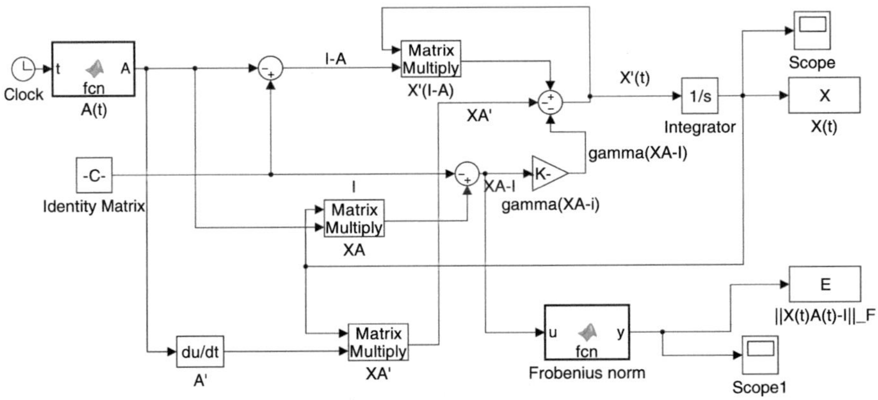

Fig. 3.16 MATLAB Simulink model for (3.2.21) and (3.2.27)

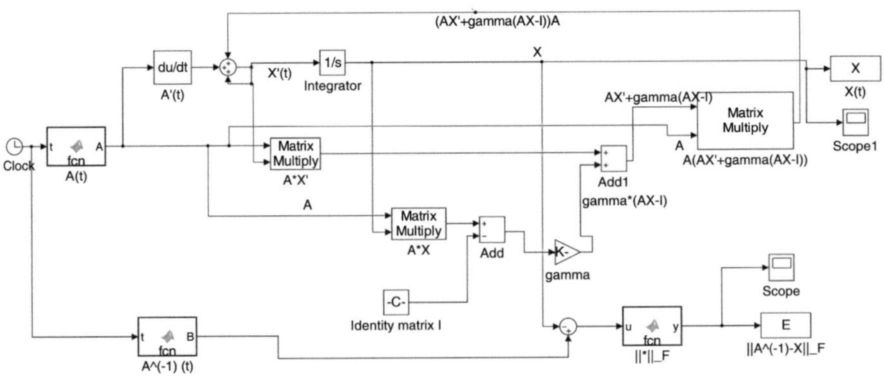

Fig. 3.17 MATLAB Simulink model for (3.2.22) and (3.2.28)

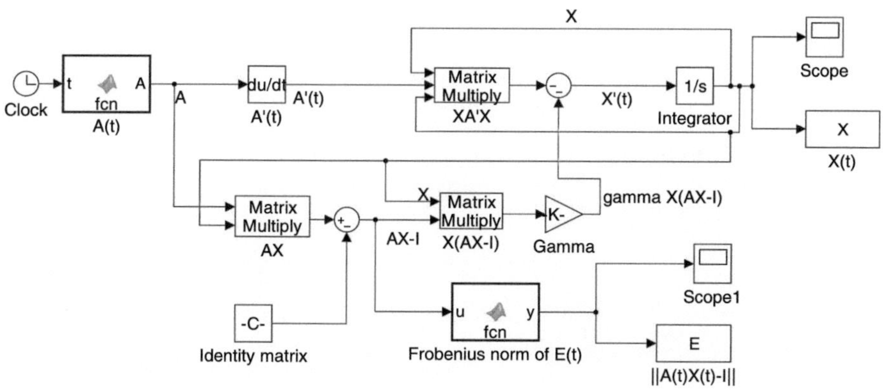

Fig. 3.18 MATLAB Simulink model for (3.2.23) and (3.2.29)

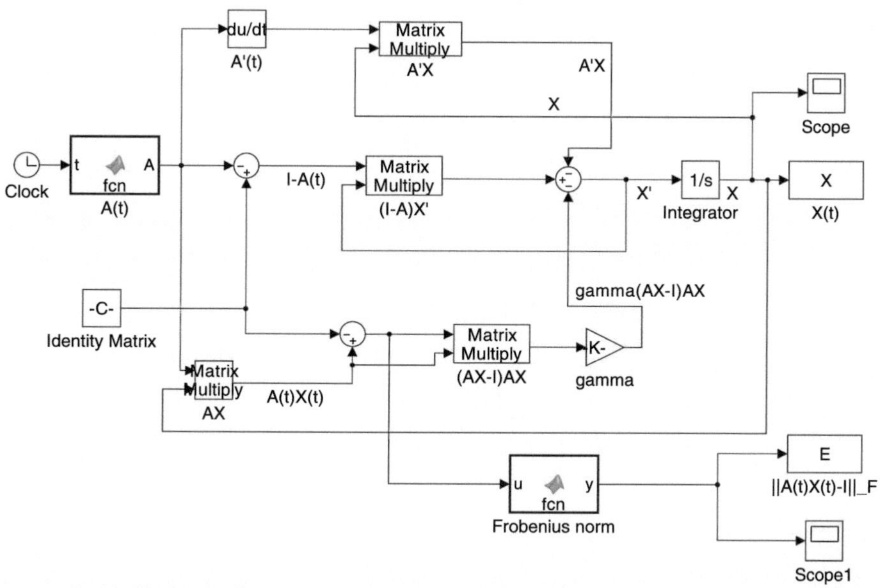

Fig. 3.19 MATLAB Simulink model for (3.2.24) and (3.2.30)

Nonlinear ZNN Model for Time-Varying Matrix Inverse

Nonlinear dynamics of (3.2.26) includes nonlinear activation $\mathcal{F}$, as follows:

$$\dot{X}(t) = (I - A(t))\dot{X}(t) - \gamma\mathcal{F}(A(t)X(t) - I) - \dot{A}(t)X(t). \qquad (3.2.33)$$

Simulink aimed to implementation of nonlinear dynamics (3.2.33) is presented in Fig. 3.33.

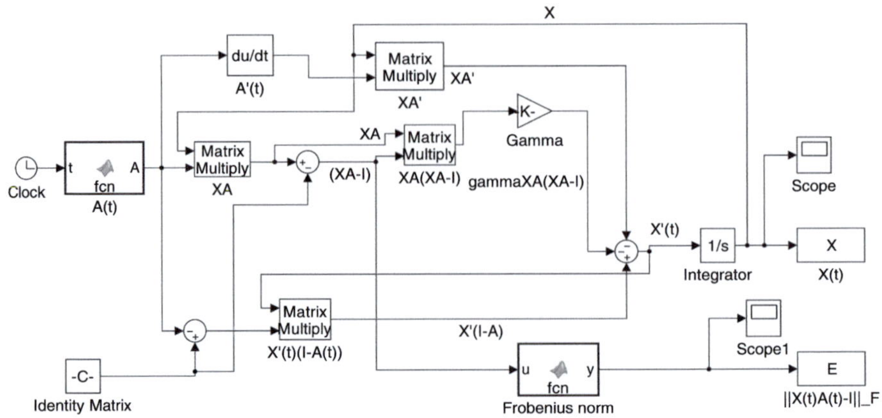

Fig. 3.20 MATLAB Simulink model for (3.2.25) and (3.2.31)

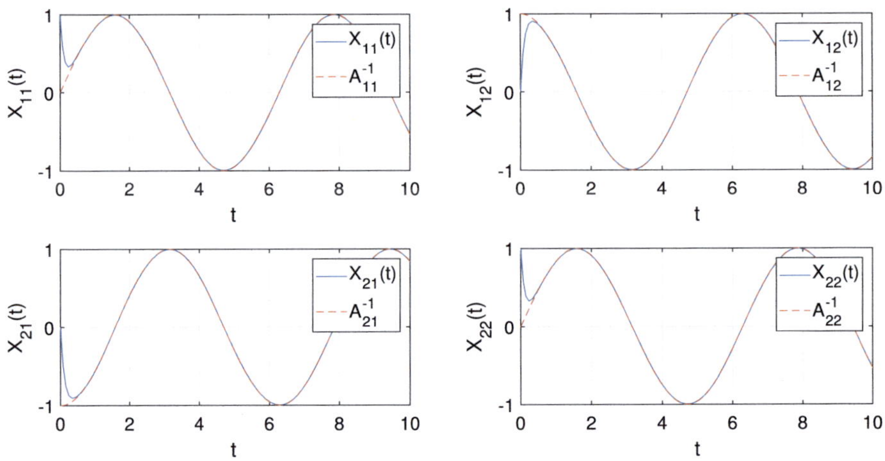

Fig. 3.21 State trajectory $x(t)$ for ZD model (3.2.28)

Fig. 3.22 Computational errors $e(t)$ of ZD model (26) using $\gamma = 10$, 100 and 1000

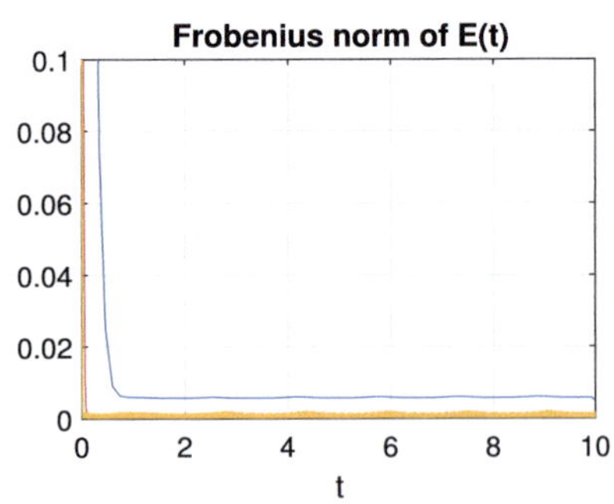

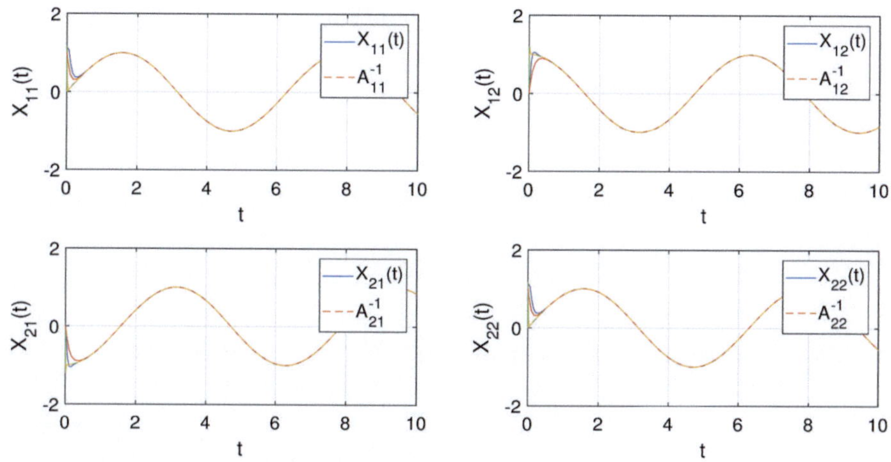

Fig. 3.23 State trajectory $x(t)$ for ZD model (3.2.29)

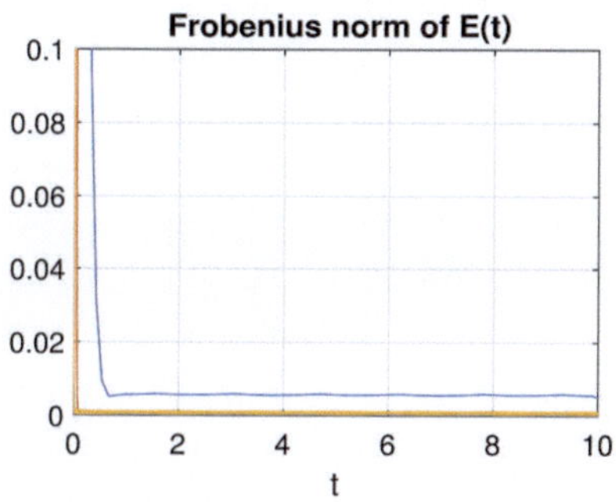

Fig. 3.24 Computational errors $e(t)$ of ZD model (3.2.29) using $\gamma = 10$, 100 and 1000

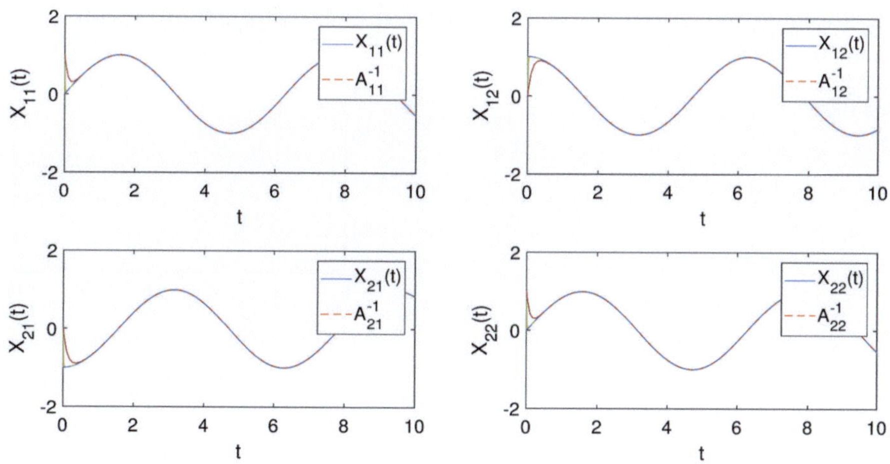

Fig. 3.25 State trajectory $x(t)$ for ZD model (3.2.26)

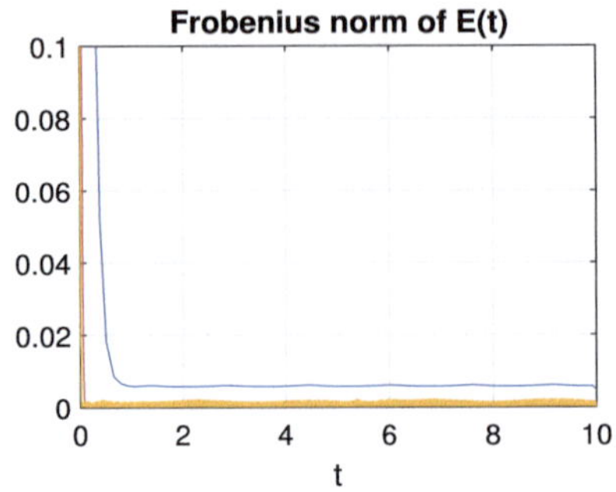

Fig. 3.26 Computational errors $e(t)$ of ZD model (3.2.26) using $\gamma = 10$, 100 and 1000

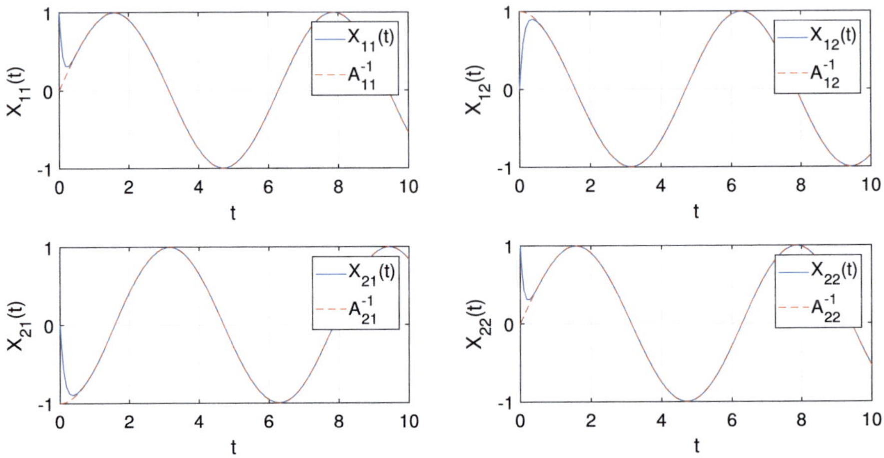

Fig. 3.27 State trajectory $x(t)$ for ZD model (3.2.27)

Fig. 3.28 Computational errors $e(t)$ of ZD model (3.2.27) using $\gamma = 10$, 100 and 1000

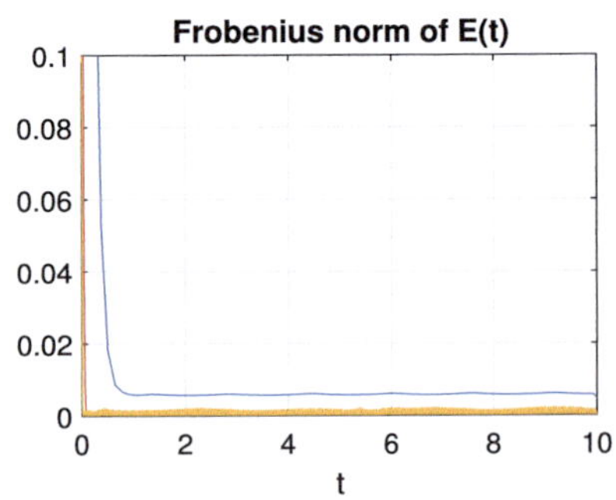

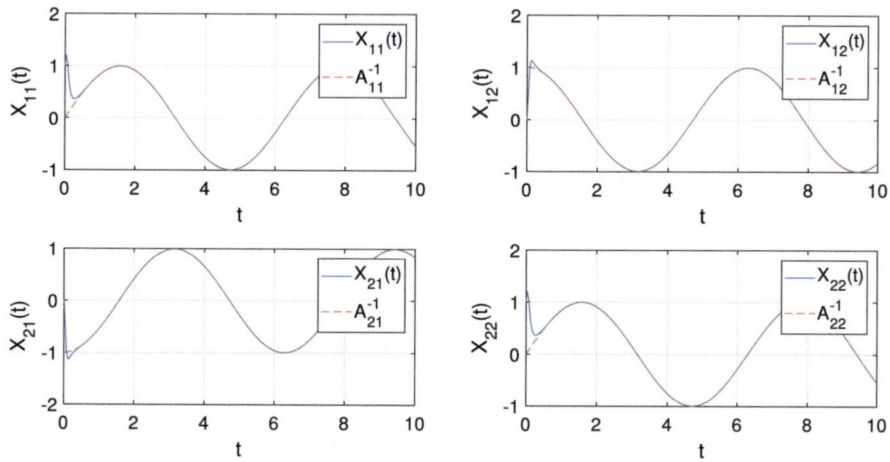

Fig. 3.29 State trajectory $x(t)$ for ZD model (3.2.30)

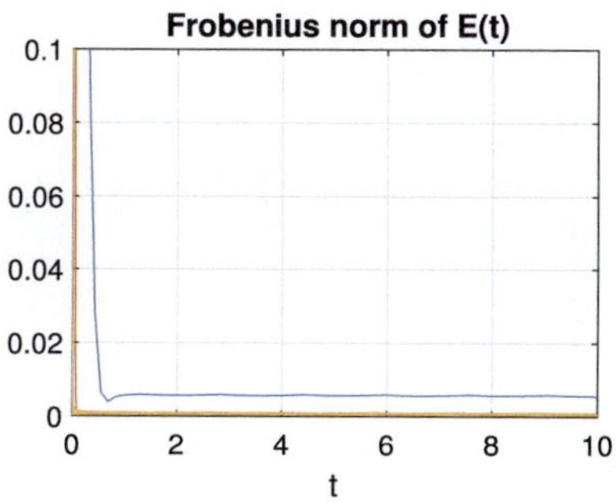

Fig. 3.30 Computational errors $e(t)$ of ZD model (3.2.30) using $\gamma = 10$, 100 and 1000

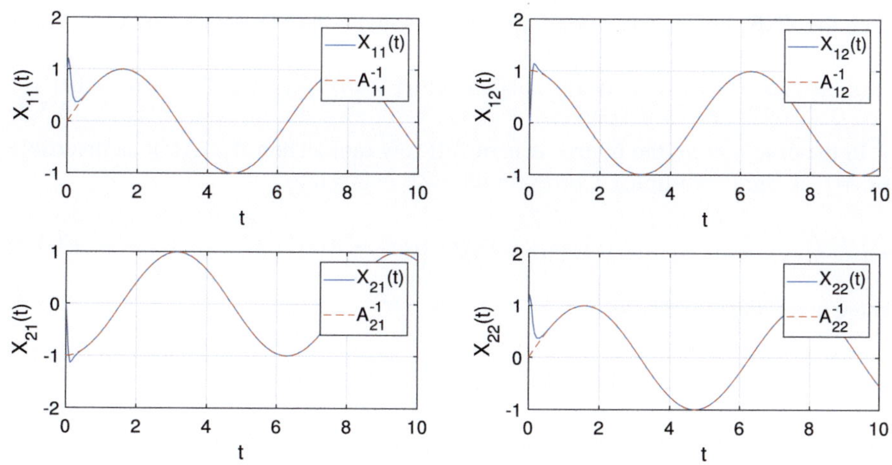

Fig. 3.31 State trajectory $x(t)$ for ZD model (3.2.31)

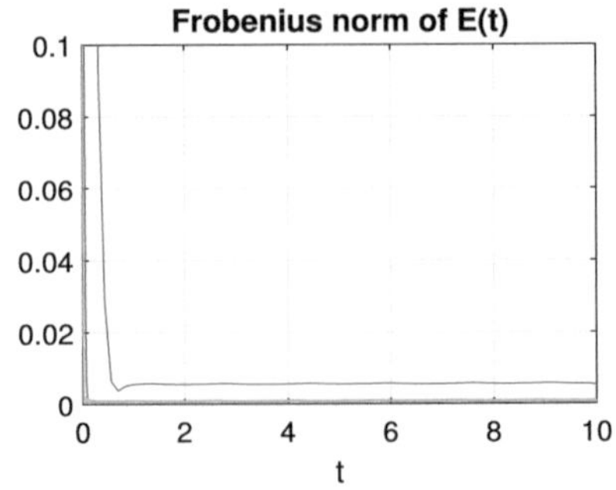

Fig. 3.32 Computational errors $e(t)$ of ZD model (3.2.31) using $\gamma = 10$, 100 and 1000

3.3 ZNN Model for Solving the Matrix Pseudoinversion

Basic fact necessary for reading this section are involved in the paper [380] and in the monographs [351, 369, 374].

Consider a time-varying $m \times n$ rectangular matrix $A(t) \in \mathbb{C}_r^{m\times n}$ of constant rank r. Assume first that A is of full-column rank, i.e., $r = n$. As a consequence, $A(t)^* A(t)$ is invertible (or $A(t)$ is left invertible), the following matrix-based error function, called ZF(5), is considered

$$E(t) := A(t)^* A(t) V(t) - A(t)^*, \tag{3.3.1}$$

where $V(t)$ corresponds to

$$A(t)^\dagger = A_L^{-1} = \left(A(t)^* A(t)\right)^{-1} A(t)^*.$$

Then the ZNN design corresponding to A_L^{-1} is given in the implicit form

$$A(t)^* A(t)\dot{V}(t) = -\left(\dot{A}(t)^* A(t) V(t) + A(t)^* \dot{A}(t)\right) V(t) - \dot{A}(t)^* - \gamma \mathcal{F}\left(A(t)^* A(t) V(t) + A(t)^*\right). \tag{3.3.2}$$

In the case $r = m$, the matrix A is of full-row rank. Then $A(t)A(t)^*$ is invertible (or $A(t)$ is right invertible). Corresponding ZF is given by

$$E(t) := V(t) A(t) A(t)^* - A(t)^*, \tag{3.3.3}$$

where $V(t)$ corresponds to

$$A(t)^\dagger = A_R^{-1} = A(t)^* \left(A(t) A(t)^*\right)^{-1}.$$

Then the linear ZNN design corresponding to A_R^{-1} is given in the implicit form

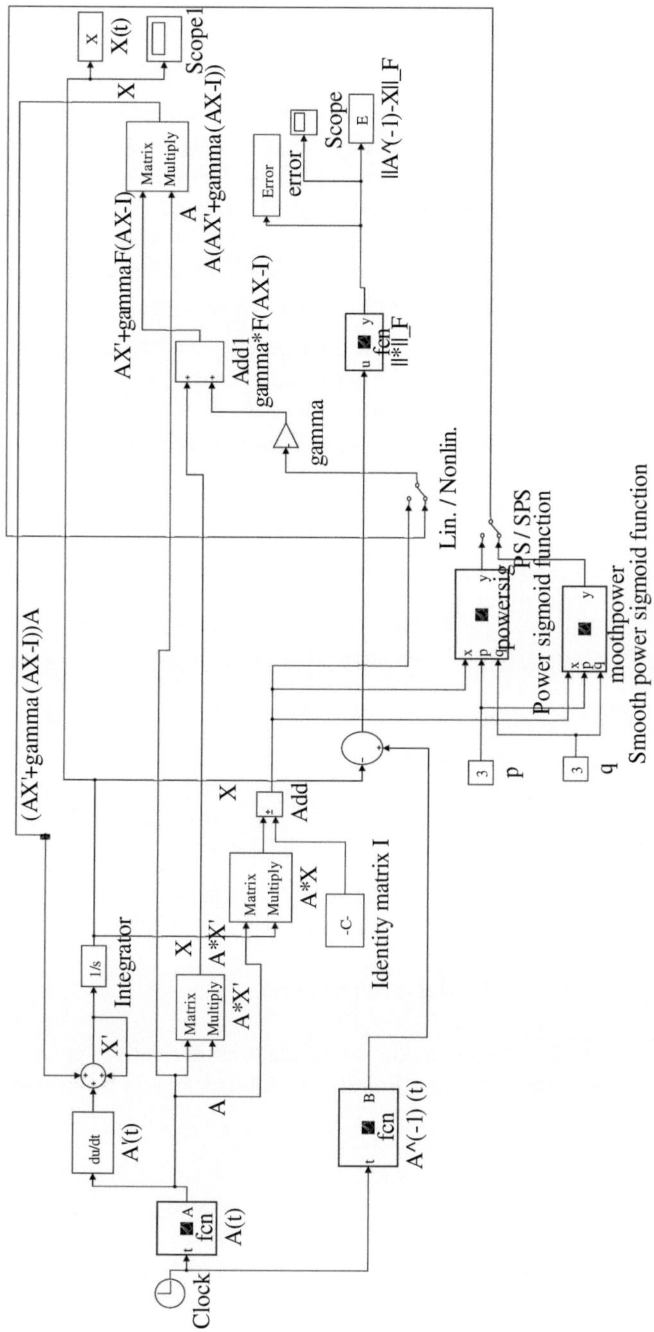

Fig. 3.33 MATLAB Simulink model for implementation of nonlinear (3.2.33) dynamics

$$\begin{aligned}\dot{V}(t)A(t)A(t)^* = &- V(t)\left(\dot{A}(t)A(t)^* + A(t)\dot{A}(t)^*\right) + \dot{A}(t)^* \\ &- \gamma\mathcal{F}\left(V(t)A(t)A(t)^* - A(t)^*\right).\end{aligned} \tag{3.3.4}$$

Convergence properties of (3.3.4) are considered in Proposition 3.3.1.

Proposition 3.3.1 *Let $A(t) \in \mathbb{R}_n^{m\times n}$ be a smoothly time-varying full-row rank matrix. If a monotonically increasing odd activation function $\mathcal{F}$ is used, then the state matrix $V(t)$ of (3.3.4) converges to the time-varying Moore-Penrose inverse $A(t)^\dagger \in \mathbb{R}^{n\times m}$, starting from arbitrary initial state $X(0) \in \mathbb{R}^{n\times m}$. In addition, in the case of the linear activation, a global exponential rate γ is achieved. Moreover, superior convergence can be achieved by using power-sigmoid activation function with respect to the linear activation.*

In the last case, it is assumed that A is a rank-deficient matrix, which means $r < \min\{m, n\}$. An elegant way to avoid the assumption of the invertibility of $A(t)^*A(t)$ was presented in [126]. Namely, the authors of [126] defined the complex ZF which arises from the ZF defined in (3.3.1), and the Tikhonov regularization:

$$E(t) = \left(A(t)^*A(t) + \lambda I\right)V(t) - A(t)^*, \quad \lambda > 0. \tag{3.3.5}$$

The resulting ZNN model arising from (3.3.5) was termed as complex ZNN-II Model in [126], and defined as follows:

$$\begin{aligned}\left(A(t)^*A(t) + \lambda I\right)\dot{V}(t) = &\left(\dot{A}(t)^*A(t)V(t) + A(t)^*\dot{A}(t)\right)V(t) + \dot{A}(t)^* \\ &- \gamma\mathcal{F}\left(\left(A(t)^*A(t)V(t) + \lambda I\right) - A(t)^*\right).\end{aligned} \tag{3.3.6}$$

3.4 ZNN Models for Solving Time-Varying Linear Systems

First results in that area assume that $A(t) \in \mathbb{R}^{n\times n}$ is the smoothly time-varying nonsingular matrix. The ZNN dynamics has been used to solve the system

$$A(t)x(t) = b(t). \tag{3.4.1}$$

It is assumed that $A(t)$ and $b(t)$ are known, while their derivatives are assumed to be known or could be estimated. The ZF aimed to solving (3.4.1) is defined by

$$e(t) = A(t)v(t) - b(t)$$

and the ZNN model for solving (3.4.2) is given by the following dynamics:

$$A(t)\dot{v}(t) = -\dot{A}(t)v(t) + \dot{b}(t) - \gamma\mathcal{F}(A(t)v(t) - b(t)). \tag{3.4.2}$$

In order to achieve Simulink model or the implementation, (3.4.2) is transformed into the following explicit form

$$\dot{v}(t) = (I - A(t))\dot{v}(t) - \dot{A}(t)v(t) + \dot{b}(t) - \gamma\mathcal{F}(A(t)v(t) - b(t)). \tag{3.4.3}$$

In the case if $A(t)$ is nonsingular, the global convergence with exponential convergence rate of (3.4.2) can be verified.

Theorem 3.4.1 *Let a smoothly time-varying nonsingular matrix $A(t) \in \mathbb{R}^{n\times n}$ and a vector $b(t) \in \mathbb{R}^n$ be given. If an odd and monotonically-increasing activation function array $\mathcal{F}$ is used, then the state vector $v(t) \in \mathbb{R}^n$ of* (3.4.2) *converges to the time-varying theoretical solution $v^*(t) = A(t)^{-1}b(t)$, starting from arbitrary initial state $v(0) \in \mathbb{R}^n$. Moreover, the following statements about the exponential convergence rate are valid.*

(a) *If the linear activation is used, then the exponential convergence with the rate γ is achieved.*

(b) *If the power-sigmoid activation function is used, then the superior convergence can be achieved in complete range $(-\infty, +\infty)$ with respect to the linear case.*

(c) *If the bipolar sigmoid activation function is used, then there exists $\varepsilon > 0$ such that superior convergence can be achieved in the range $(-\varepsilon, +\varepsilon)$ with respect to the linear case.*

(d) *If the power sigmoid activation function is used, then the superior convergence can be achieved in the range $(-\infty, -1]$ and $[1, +\infty)$ with respect to the linear case.*

Dynamical ZNN system based on $E(t) = Ax(t) - b$ is given by following relationship

$$\dot{x}(t) = [I - A]\dot{x}(t) - kG\left(Ax(t) - b\right). \tag{3.4.4}$$

Simulink implementation of dynamics (3.4.4) is presented in Fig. 3.34.

Simulink implementation of the nonlinear activation G is given in Fig. 3.35.

3.5 A Novel Extended Li Zeroing Neural Network for Matrix Inversion

The finite-time convergence is the goal achieved in many papers by means of the appropriate nonlinear activation. Some of related results are included in research papers [149, 203, 284, 309, 312] and in the monograph [306]. In [149], the authors revealed a finite-time convergence of the ZNN model with Li activation which is applied in computation of the Moore-Penrose inverse. Finite-time convergence of ZNN models for computing the Drazin inverse by means of appropriate nonlinear activation was proposed in [203]. Also, the finite-time convergence was achieved in the ZNN model for computing outer inverses with prescribed range and null space in [284]. An extensive survey of finite-time convergent ZNN models is given in the monograph [306].

Finite-time convergent ZNN models based on appropriate nonlinear activation functions for solving time-varying linear matrix equations were proposed in [309,

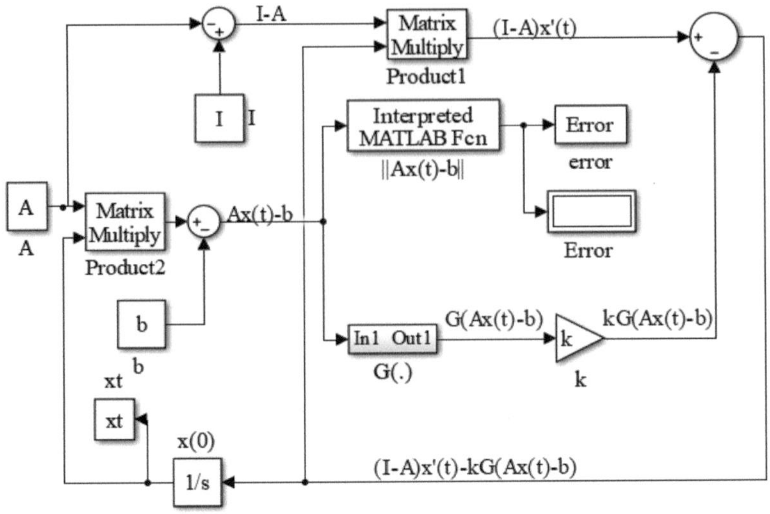

Fig. 3.34 Simulink model of the nonlinear ZNN formula (3.4.4) for solving $Ax = b$

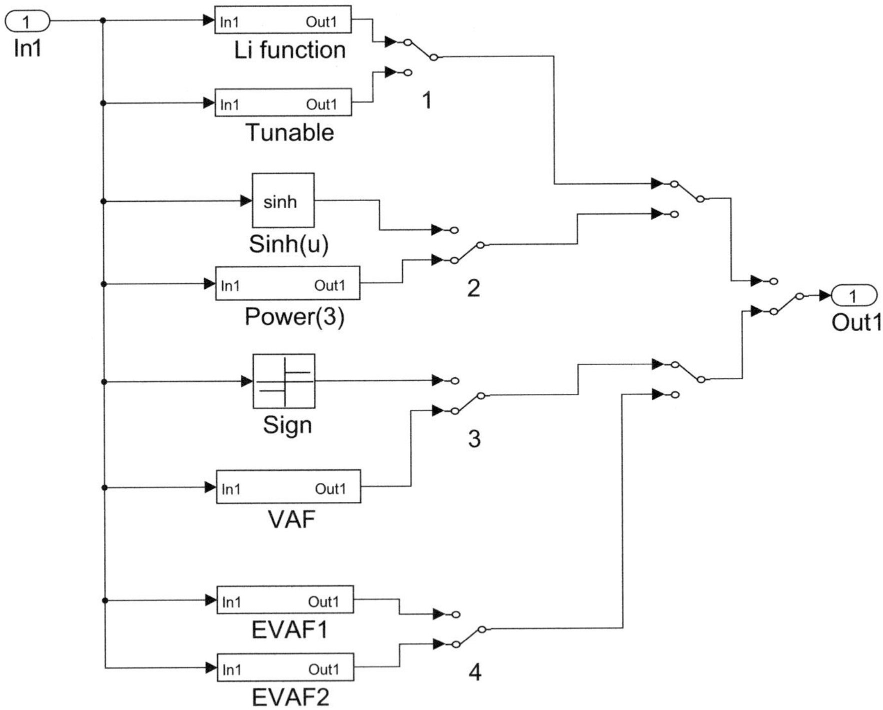

Fig. 3.35 Nonlinear activation G

312]. Also, a finite-time nonlinearly activated ZNN for approximating the time-varying matrix square root was considered in [371]. A finite-time ZNN which is designed to solve time-varying quadratic program and applied to robot motion tracking was presented in [180, 385]. Liao and Zhang in [149] proposed first finite-time convergent improved ZNN model on the basis of the Li activation function. A finite-time recurrent neural dynamics based on the usage of an appropriately chosen activation function for finding the root of nonlinear equation was considered in [310]. The authors of [312] proposed two finite-time-convergent nonlinear recurrent neural networks based on two nonlinear activation functions and aimed to solve general time-varying linear matrix equations. A novel FTZNN model supported by an appropriate activation function was proposed and applied for approximating nonstationary quadratic optimization problems in [304]. Xiao et al. [313] considered a novel evolution formula and initiated an FTZNN model for solving a nonstationary Lyapunov equation. Li et al. [143] adopted the ZNN dynamics in solving time-varying nonlinear optimizations with time-varying linear equality constraints and applied defined model to robot motion planning. Also, it is possible to mention some another applications, like chaotic noise rejection for sensors [23] and some numerical online solving problems [130].

Li et al. present a nonlinear function to accelerate the continuous-time ZNN to finite-time convergence for solving time-varying Sylvester equation in [133] and follow up with the online solution of dual neural networks for solving quadratic programming problems in [131], as a result of the in-depth research on the ZNN model and inspired by the study on finite-time convergence in continuous autonomous system [10, 177, 179]. Thereafter, many different finite-time activation functions have been proposed and employed to various ZNN models, e.g, time-varying matrix pseudoinversion in complex domain [149], Lyapunov equation [305], equality-constrained quadratic optimization [305], linear complex matrix equation [310] and several other [81, 203, 207, 332]. Observe that the residual error of a traditional ZNN model exponentially converges to zero, which means that the smaller the residual error is, the slower the convergent speed is. Consequently, an effective way to achieve finite-time convergence is to amplify the value of $\dot{e}(t)/e(t)$ to large enough when $e(t)$ approaches to zero. Xiao et al. in [319] investigated various ZNN models which can be designed by exploiting different error functions and aimed to solving time-varying matrix equations. Also, possible accelerations to finite-time convergence were considered. Finite-time convergent ZNN models based on appropriate nonlinear activation functions for solving time-varying linear matrix equations were proposed in [309, 312].

Results presented in this section are based on the research paper [228]. The time-varying equation under study in this research is

$$A(t)X(t) = I, \quad t \in [0, \infty), \tag{3.5.1}$$

where $A(t)$ is a square $n \times n$ matrix and nonsingular for each t, $X(t)$ is the unknown inverse of $A(t)$ and I is the identity matrix with same dimensions as $A(t)$. By defining the error matrix function $E(t) = A(t)X(t) - I$ the goal of this research paper is to design a new odd-increasing AF and initiated ZNN design that will vanish $E(t)$ to the zero matrix in a finite time.

Main AFs which provide finite convergence are referred listed as follows:
• The tunable AF [50, 103, 180],

$$\mathrm{Tun}(u) = \mathrm{sgn}(u)(|u|^p + |u| + |u|^{\frac{1}{p}}), \tag{3.5.2}$$

with

$$\mathrm{sgn}(u) = \begin{cases} 1, & u > 0 \\ 0, & u = 0 \\ -1, & u < 0. \end{cases} \tag{3.5.3}$$

• The sign-bi-power (sbp) AF [86, 129, 305, 314, 317],

$$\mathrm{sbp}(u) = \mathrm{sgn}^p(u) + \mathrm{sgn}^{1/p}(u), \tag{3.5.4}$$

with

$$\mathrm{sgn}^p(u) = |u|^p\,\mathrm{sgn}(u), \quad p \in (0, 1). \tag{3.5.5}$$

The Li-Zhang Neural Network (Li-ZNN) model [86, 129, 171, 305, 314, 317] that associated with this AF is established as

$$\dot{E}(t) = -k\left(\mathrm{sgn}^p(E(t)) + \mathrm{sgn}^{1/p}(E(t))\right). \tag{3.5.6}$$

• The sinh-sign AF (SSAF) [46, 320],

$$g(u) = a_1 \sinh\left(|u|^p\right)\mathrm{sgn}(u), \tag{3.5.7}$$

where $a_1 > 0$ and $\sinh(\cdot)$ is the hyperbolic sine function.
• The exp-sign AF (ESAF) [46, 320]

$$g(u) = a_1\left(\exp(|u|^p) - 1\right)\mathrm{sgn}(u), \tag{3.5.8}$$

where $a_1 > 0$ and $\exp(\cdot)$ is the exponential function.
• The versatile activation function (VAF) [45, 322]

$$g(u) = \left(a_1|u|^h + a_2|u|^w\right)\mathrm{sgn}(u) + a_3 u + a_4\mathrm{sgn}(u), \tag{3.5.9}$$

where $a_1, a_2 > 0$, $a_3, a_4 \geq 0$, $w > 1$ and $0 < h < 1$.

The above nonlinear AFs will be adopted for comparing the convergence speed with our new AF in the numerical section. A careful analysis of AFs (3.5.2), (3.5.4), (3.5.7) and (3.5.8) reveals that their definitions involve factor of the form $|u|^p$ or $|u|^{\frac{1}{p}}$, $p \in (0, 1)$. This observation led us to the idea of defining a new nonlinear activation as a sum of such factors.

To show superiority of the new AF, we also consider some finite-time convergent models proposed in recent years. The following models were used to achieve a finite convergence time without the usage of nonlinear activations:

• The finite-time zeroing neural network (FTZNN) investigated in [122, 304, 313, 321]

$$\dot{E}(t) = -k\left(a_1 E(t) + a_2 E^{r/w}(t)\right), \tag{3.5.10}$$

where a_1, a_2 are positive numbers, k the gain parameter and r, w are positive integers with $r < w$.

• The varying-parameter finite-time zeroing neural network (VPFTZNN) [63, 64, 66, 181, 226],

$$\dot{E}(t) = -k\exp(bt)\left(a_1 E(t) + a_2 E^{r/w}(t)\right), \tag{3.5.11}$$

where b is a positive constant and a_1, a_2, k, r, w satisfy the same conditions as in the previous FTZNN dynamics (3.5.10).

For the acceleration of the convergence speed against previous models, we introduce the novel *extended sign-bi-power* (Esbp), odd-increasing activation function of the order m

$$\text{Esbp}(u, m) = \text{Esgn}^{p_{1\to m}}(u) + \text{Esgn}^{1/p_{1\to m}}(u), \tag{3.5.12}$$

where it is assumed

$$\text{Esgn}^{p_{1\to m}}(u) = \sum_{l=1}^{m} \text{sgn}^{p_l}(u) = \begin{cases} |u|^{p_1} + |u|^{p_2} + \cdots + |u|^{p_m}, & u > 0 \\ 0, & u = 0 \\ -|u|^{p_1} - |u|^{p_2} - \cdots - |u|^{p_m}, & u < 0, \end{cases} \tag{3.5.13}$$

and

$$\text{Esgn}^{1/p_{1\to m}}(u) = \sum_{l=1}^{m} \text{sgn}^{1/p_l}(u) = \begin{cases} |u|^{1/p_1} + |u|^{1/p_2} + \cdots + |u|^{1/p_m}, & u > 0 \\ 0, & u = 0 \\ -|u|^{1/p_1} - |u|^{1/p_2} - \cdots - |u|^{1/p_m}, & u < 0, \end{cases} \tag{3.5.14}$$

such that $p_l \in (0, 1)$ for each $l = 1, 2, \ldots, m$. The ascending order $0 < p_1 \le p_2 \le \cdots \le p_m < 1$ between the values of p_l can be assumed without loss of generality. The motivation for the use of the novel AF defined in (3.5.12) is caused due to the parameters $p_1, p_2, \ldots, p_m$, which will accelerate the convergence of the initiated ZNN model$\dot{E}(t) = -k\,\text{Esbp}(E(t), m)$, where the error function is assumed to be the variable u. The variable number of control parameters involved in the proposed $\text{Esbp}(u, m)$ AF is its essential advantage against the Li activation, in which the convergence time is based only on one parameter p.

The error matrix $E(t) = A(t)X(t) - I$ is underlying in solving the matrix inversion problem. Initiated ZNN design is expanded into the following implicit dynamics

$$\dot{X}(t) = (I - A(t))\dot{X}(t) - \dot{A}(t)X(t) - k\,\text{Esbp}(A(t)X(t) - I, m) \tag{3.5.15}$$

Table 3.2 Network's training simulation parameters of (3.5.15) based on Simulink implementation of Fig. 3.36

Simulation time	
Start time: 0.0	**Stop time**: t_{final}
Solver options	
Type: Variable-step	**Solver**: ode45 (Dormand-Prince)
Max step size: auto	**Relative tolerance**: $1e-6$
Min step size: auto	**Absolute tolerance**: $1e-6$
Initial step size: auto	**Shape preservation**: Disable all
Number of consecutive min steps: 1	
Tasking and sample time options	
Tasking mode for periodic sample time: Auto	
zero-crossing options	
zero-crossing control: Disable all	

The Simulink implementation of the ELi-ZNN model (3.5.15) for $m = 2, 3, 4, 5$ is presented in Fig. 3.36. The first **Interpreted MATLAB FCN** block is intended to define the time-varying matrix $A(t)$ while the Derivative block **du/dt** represents the time-derivative of $\dot{A}(t)$. The **Matrix Multiply** block is used to generate the matrix products $A(t)X(t)$, $(I - A(t))\dot{X}(t)$ and $\dot{A}(t)X(t)$, where the identity matrix I is defined from the command window of the MATLAB environment. **Sum** block, that returns the sum of two or more inputs, generates the expressions $I - A(t)$, $A(t)X(t) - I$ and $(I - A(t))\dot{X}(t) - \dot{A}(t)X(t) - k\,\mathrm{Esbp}(A(t)X(t) - I, m)$. The second **Interpreted MATLAB FCN** block includes the command *norm(u,'fro')*. It extracts the Frobenius norm $||A(t)X(t) - I||_F$, which helps to compare considered activation functions. The **Gain** block multiplies the Esbp($A(t)X(t) - I$) input with a constant value k while the four subsystems (Esbp2,Esbp3,Esbp4,Esbp5) include the Simulink implementation of Esbp(u,m) for $m = 2, 3, 4, 5$. The Integrator block **1/s** outputs the value of the integral of its input signal concerning time, and it is used to tune the initial condition $X(0)$ in each numerical experiment. **To Workspace** block is utilized to store the solution $X(t)$ and the error $||A(t)X(t) - I||_F$ to specified time series called XtELiZNN and ErrorEsbp, respectively, in the MATLAB workspace.

Table 3.2 shows the training simulation parameters that are applied and constitute an important factor for effectiveness of the ELi-ZNN dynamics.

One example of the command line is

```
I=eye(3,3);k=1;q1=0.3;q2=0.35;q3=0.4;q4=0.5;q5=0.7;
```

You must choose carefully the switch for the corresponding activation function you want to use.

then with the `plot(ErrorEsbp510)` print the corresponding error function. Then run the command `hold on` in command line, choose another activation function and again run the command `hold on`. This process can be repeated for all activations available in the Simulink.

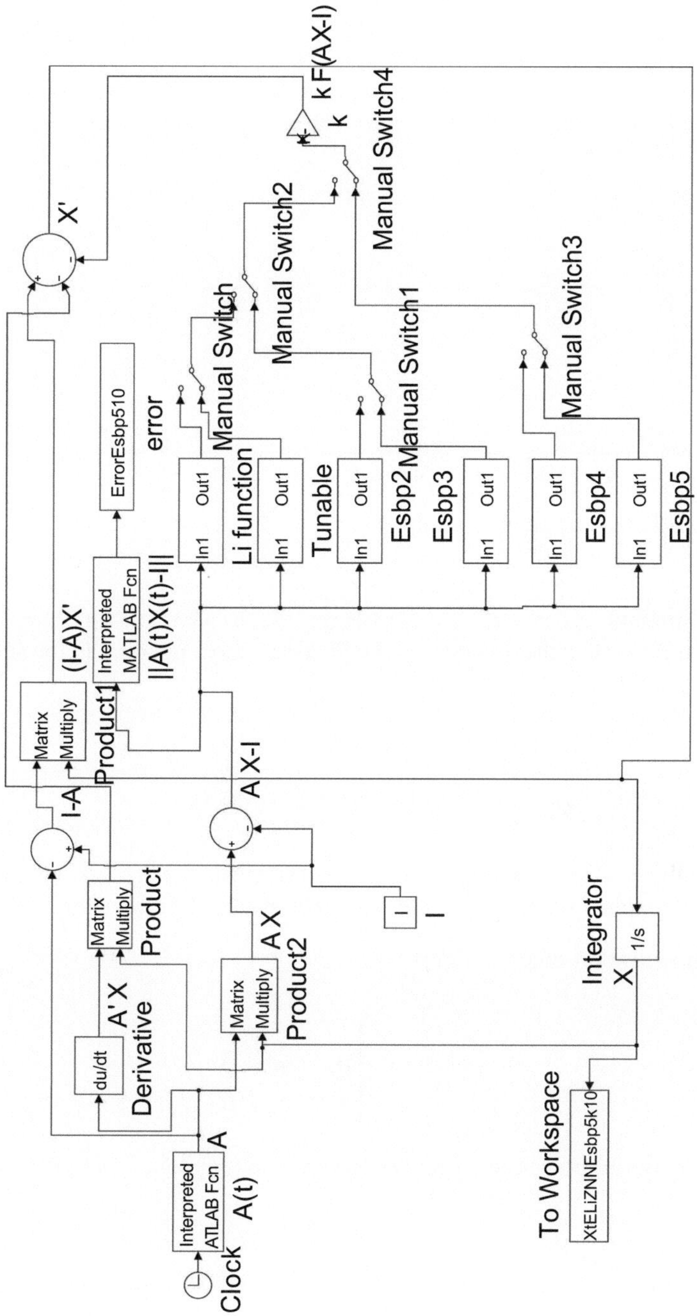

Fig. 3.36 ELi-ZNN Simulink model for (3.5.15) in cases $m = 2, 3, 4, 5$

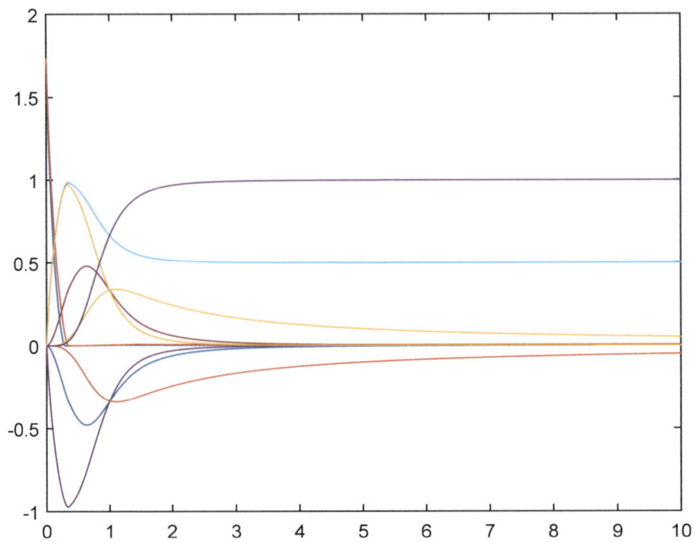

Fig. 3.37 Trajectories corresponding to ESPB4

The command `plot(XtELiZNNEsbp5k10) you take the state trajectories` plots the trajectories. Trajectories corresponding to the activation ESPB4 are given in Fig. 3.37.

3.5.1 Simulink Models of Activation Functions

Previous nonlinear AFs in ZNN formula are used to accelerate the convergence speed in finite or predefined time. Some of them are referred as follows (Fig. 3.38).

1. The **sign** activation function [382]

$$\text{sign}(u) = \begin{cases} 1, & u > 0 \\ 0, & u = 0 \\ -1, & u < 0 \end{cases} \tag{3.5.16}$$

2. The **Li** or sign-bi-power (**sbp**) activation function [86, 171, 314, 317]

Fig. 3.38 Simulink block of sign activation function (3.5.16)

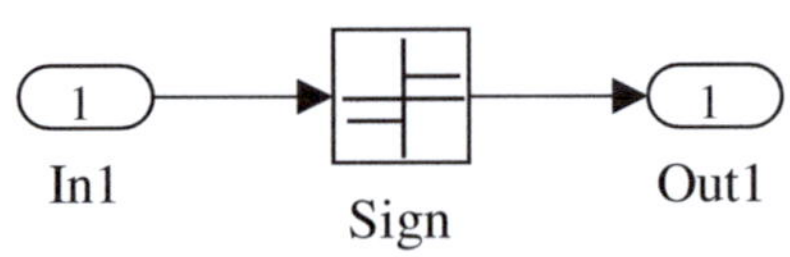

$$\mathrm{sbp}(u) = \mathrm{sgn}^{\mathrm{q}}(\mathrm{u}) + \mathrm{sgn}^{\frac{1}{\mathrm{q}}}(\mathrm{u}), \tag{3.5.17}$$

where

$$\mathrm{sgn}^{\mathrm{q}}(u) = |\mathrm{u}|^{\mathrm{q}}\mathrm{sign}(u) = \begin{cases} |\mathrm{u}|^{\mathrm{q}}, & u > 0 \\ 0, & u = 0 \\ -|\mathrm{u}|^{\mathrm{q}}, & u < 0 \end{cases} \tag{3.5.18}$$

and $\mathrm{q} \in (0, 1)$.

3. The **tunable** activation function [50, 103, 180, 323]

$$\mathrm{G(u)} = (|\mathrm{u}|^{\mathrm{q}} + |\mathrm{u}| + |\mathrm{u}|^{\frac{1}{\mathrm{q}}})\mathrm{sign(u)}, \tag{3.5.19}$$

4. The versatile activation function (**VAF**) [45, 260, 320, 322]

$$\mathrm{G(u)} = (a_1|\mathrm{u}|^{\mathrm{q}} + a_2|\mathrm{u}|^{\mathrm{w}})\mathrm{sign}(u) + a_3\mathrm{u} + a_4\mathrm{sign(u)}, \tag{3.5.20}$$

where $a_1, a_2 > 0$, $a_3, a_4 \geq 0$ and $\mathrm{w} > 1$.

5. The extended versatile activation function (**EVAF**) of order m

$$\begin{aligned} G_m(u) = &\left(a_1 \sum_{\mathrm{k}=1}^{m} |u|^{\mathrm{q_k}} + a_2 \sum_{\mathrm{k}=1}^{m} |u|^{\mathrm{w_k}}\right) \mathrm{sign}(u) + a_3 u \\ &+ a_4 \sinh(u) + a_5(\exp(|u|) - 1)\mathrm{sign}(u), \end{aligned} \tag{3.5.21}$$

where $\mathrm{q_k} \in (0, 1)$, $\mathrm{w_k} > 1$, a_1, a_2, a_3, a_4 satisfy the same constraints as in previous VAF (3.5.20), $a_5 \geq 0$ and $\mathrm{k} = 1, 2, \ldots, m$. Suitable values for the remaining parameters are $0 < \mathrm{q}_1 \leq \mathrm{q}_2 \leq \cdots \leq \mathrm{q}_m < 1$ and $1 < \mathrm{w}_1 \leq \mathrm{w}_2 \leq \cdots \leq \mathrm{w}_m$. For the purpose of this tutorial we will use and design the EVAF activation function of the order two:

$$\begin{aligned} G_2(u) = &(a_1 (|u|^{\mathrm{q}_1} + |u|^{\mathrm{q}_2}) + a_2(|u|^{\mathrm{w}_1} + |u|^{\mathrm{w}_2}))\,\mathrm{sign}(u) \\ &+ a_3 u + a_4 \sinh(u) + a_5(\exp(|u|) - 1)\mathrm{sign}(u). \end{aligned} \tag{3.5.22}$$

In the next section we are going to present a numerical example to test some of the previous Simulink models (Figs. 3.39, 3.40 and 3.41).

Numerical Experiment

In this example we apply the nonlinear ZNN dynamics (3.4.4) for different activation functions which mentioned in previous section for the solution of the system $Ax = b$ where

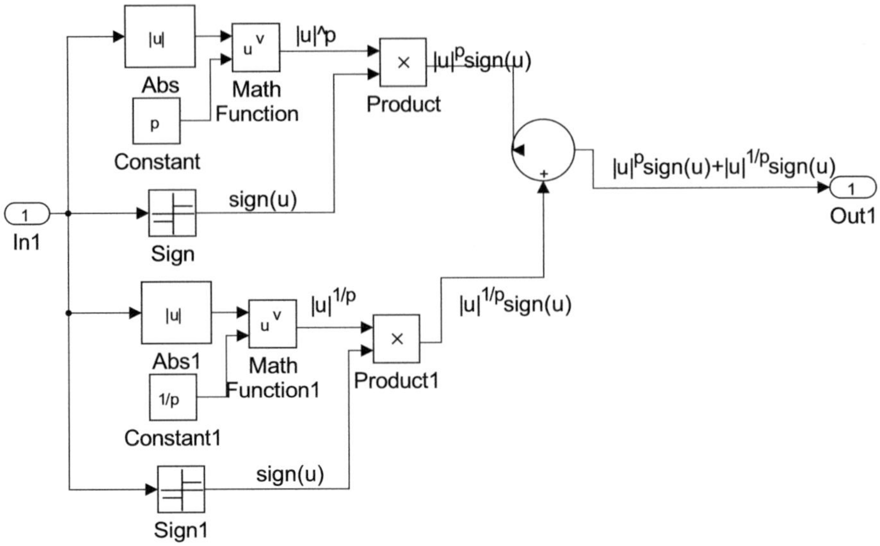

Fig. 3.39 Simulink block of Li activation function (3.5.17)

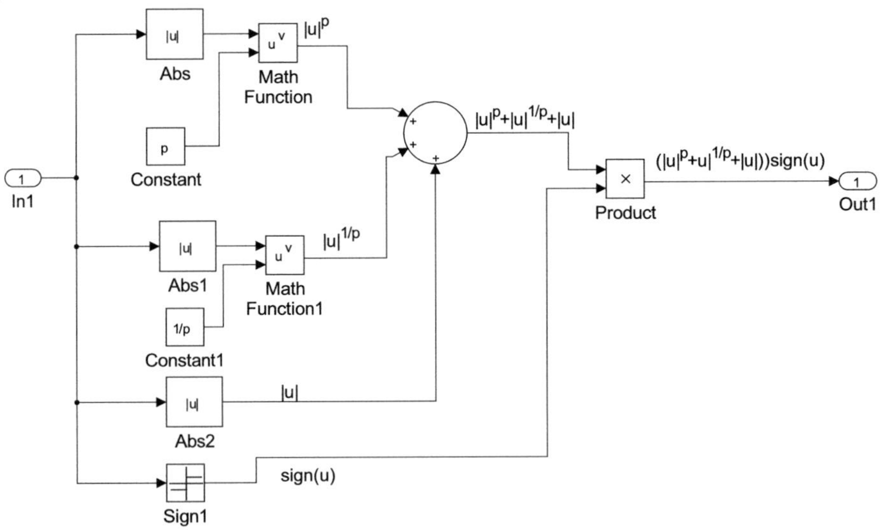

Fig. 3.40 Simulink block of Tunable activation function (3.5.19)

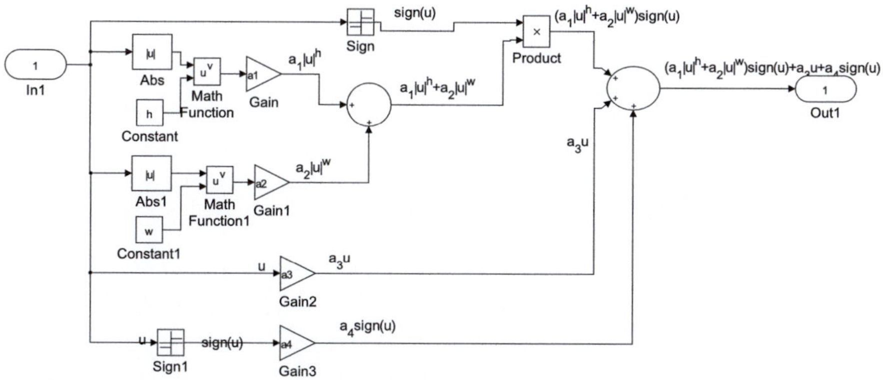

Fig. 3.41 Simulink block of versatile activation function (VAF) (3.5.20)

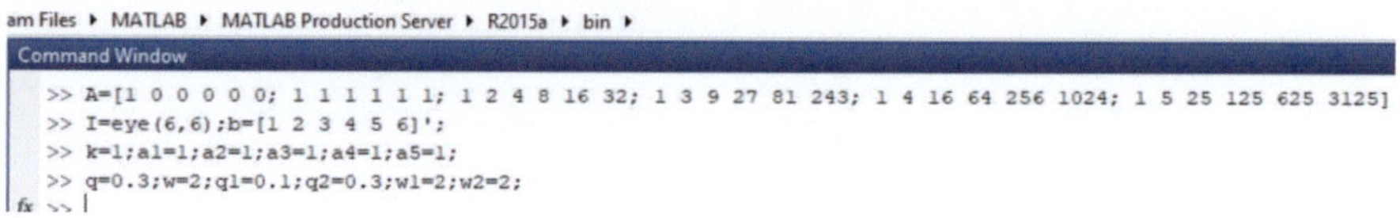

Fig. 3.42 Data input for the numerical experiment from command window

$$A = \begin{pmatrix} 1 & 0 & 0 & 0 & 0 & 0 \\ 1 & 1 & 1 & 1 & 1 & 1 \\ 1 & 2 & 4 & 8 & 16 & 32 \\ 1 & 3 & 9 & 27 & 81 & 243 \\ 1 & 4 & 16 & 64 & 256 & 1024 \\ 1 & 5 & 25 & 125 & 625 & 3125 \end{pmatrix}, \quad b = \begin{pmatrix} 1 \\ 2 \\ 3 \\ 4 \\ 5 \\ 6 \end{pmatrix}, \tag{3.5.23}$$

with theoretical solution

$$x = \begin{pmatrix} 1 \\ 1 \\ 0 \\ 0 \\ 0 \\ 0 \end{pmatrix}. \tag{3.5.24}$$

Simulink implementation from Fig. 3.34 extracts Fig. 3.43 under the constants $\mathrm{q}_1 = 0.1$, $\mathrm{q}_2 = 0.3$, $\mathrm{w}_1 = \mathrm{w}_2 = 2$, $\mathrm{q} = 0.3$, $\mathrm{w} = 2$, $a_1 = a_2 = a_3 = a_4 = a_5 = 1$, $k = 1$ and the zero initial condition. These parameters are inserted from Matlab's command window depicted in Fig. 3.42.

It is observable from Fig. 3.43 that the extended versatile AF (3.5.22) includes faster convergence property against previous activation functions because of Frobenius norm $||E(t)||_F = ||\mathrm{Ax(t)} - \mathrm{b}||_F$ vanish to zero in shorter time. Figure 3.44 represents the numerical solution of the ZNN formula activated by EVAF_2.

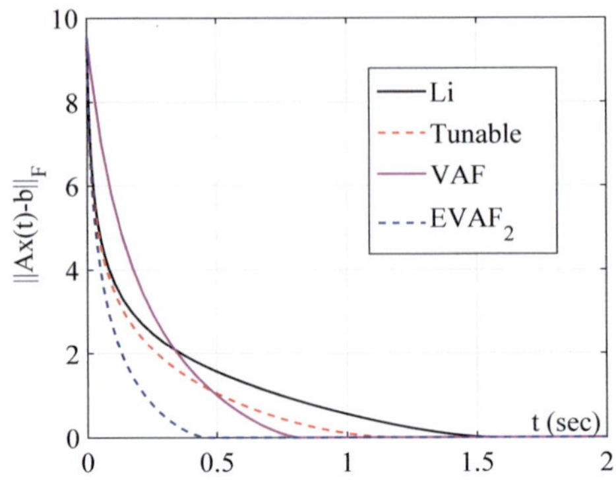

Fig. 3.43 Error comparison

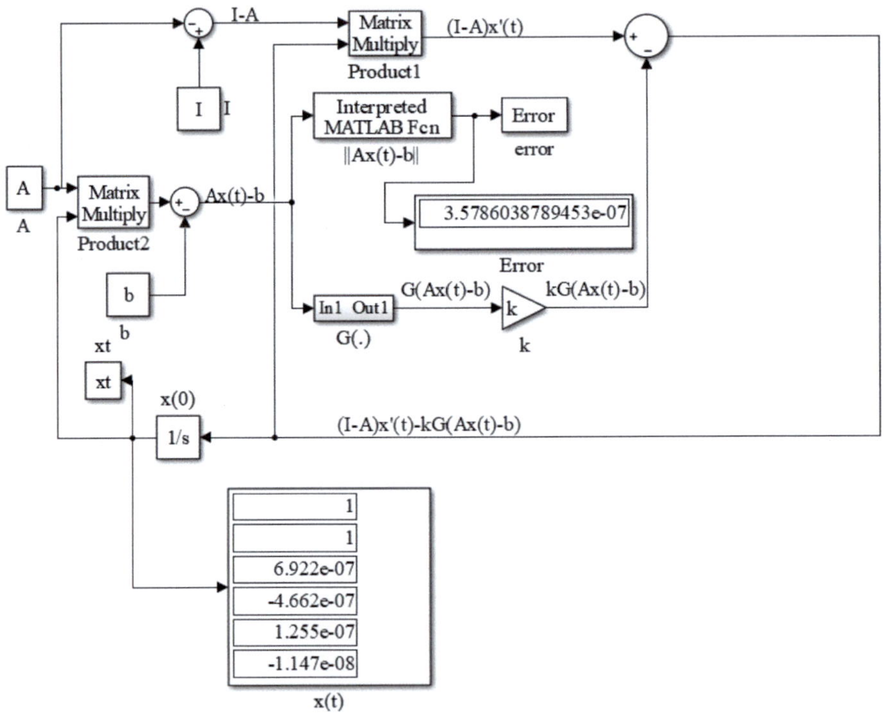

Fig. 3.44 Numerical result produced by ZNN formula (3.4.4)

3.6 ZNN Models for Solving Linear Matrix Equations

Let $A(t) \in \mathbb{R}^{n\times n}$, $B(t) \in \mathbb{R}^{m\times m}$ and $C(t) \in \mathbb{R}^{n\times m}$ be given time-varying coefficient matrices. The problem is to solve the time-varying linear matrix equation (TVLME) [16, 152, 394]

$$A(t)X(t)B(t) = C(t), \tag{3.6.1}$$

with respect to $X(t) \in \mathbb{R}^{n\times m}$. One approach for solving the time-varying problem (3.6.1) is usage of the ZNN design

$$\dot{E}(t) = -\lambda\mathcal{F}(E(t)), \tag{3.6.2}$$

where

$$E(t) = A(t)X(t)B(t) - C(t).$$

Since

$$\dot{E}(t) = \dot{A}(t)X(t)B(t) + A(t)\dot{X}(t)B(t) + A(t)X(t)\dot{B}(t) - \dot{C}(t),$$

it follows that (3.6.2) can be expanded as follows:

$$\begin{aligned}\dot{X}(t) = \dot{X}(t) - A(t)\dot{X}(t)B(t) - \dot{A}(t)X(t)B(t) - A(t)X(t)\dot{B}(t)\\ - \dot{C}(t) - \gamma(A(t)X(t)B(t) + C(t)).\end{aligned} \tag{3.6.3}$$

Linear ZNN dynamics for solving matrix equation $AXB = C$ is presented in Fig. 3.45.

Nonlinear ZNN dynamics for solving matrix equation $AXB = C$ is presented in Fig. 3.46.

An alternative approach is based on vectorization and Kronecker product. To solve the matrix equation (3.6.1) it is possible to use the equivalent vectorized form [16, 152, 394]

$$\left(B^{\mathrm{T}}(t) \otimes A(t)\right)\mathrm{vec}(X(t)) = \mathrm{vec}(C(t)). \tag{3.6.4}$$

Using vectorization and the Kronecker product, Eq. (3.6.4) can be transformed into the equivalent form $P(t)V(t) = Q(t)$, where

$$P(t) = \left(B^{\mathrm{T}}(t) \otimes A(t)\right) \in \mathbb{R}^{nm\times nm},\ V(t) = \mathrm{vec}(X(t)) \in \mathbb{R}^{nm},\ Q(t) = \mathrm{vec}(C(t)) \in \mathbb{R}^{nm}.$$

One approach for solving the time-varying problem (3.6.4) is usage of the ZNN design (3.6.2), such that $E(t) = P(t)V(t) - Q(t)$ is the corresponding error function, $\lambda > 0$ is a gain parameter and $\mathcal{F}(\cdot)$ represents a vector which entries are defined by an odd and increasing function $f(\cdot)$. To additionally accelerate the convergence speed of (3.6.2) an embedded time-varying parameter $z(t) > z > 0$ can be included

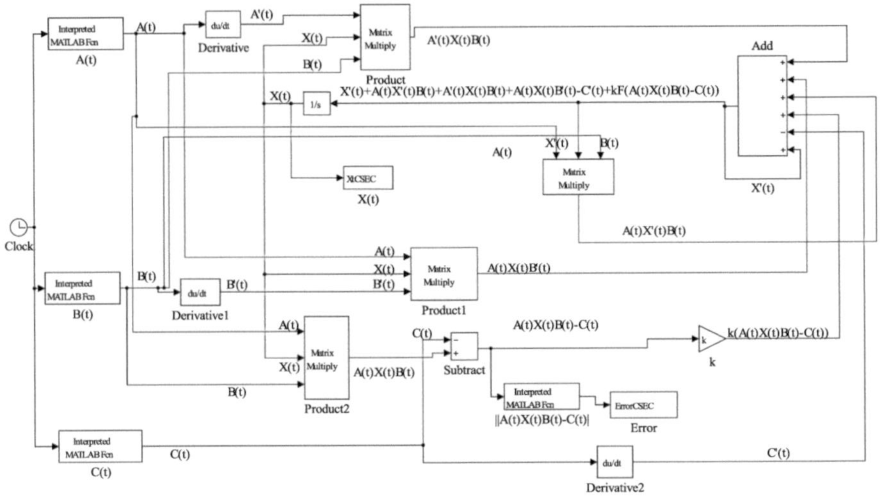

Fig. 3.45 Linear ZNN dynamics for solving matrix equation $AXB = C$

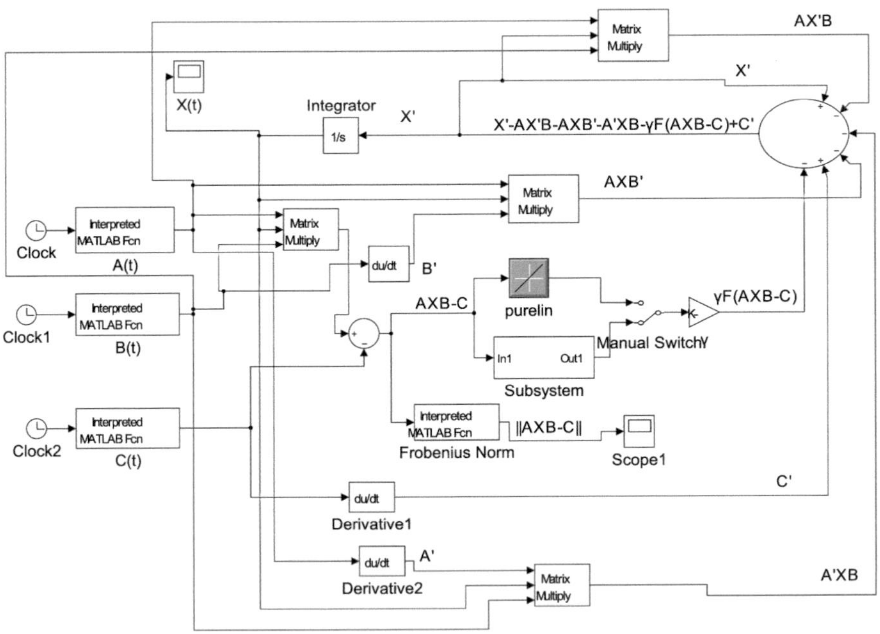

Fig. 3.46 Nonlinear ZNN dynamics for solving matrix equation $AXB = C$

as follows:

$$\dot{E}(t) = -\lambda z(t)\mathcal{F}(E(t)). \tag{3.6.5}$$

Formula (3.6.5) is the varying gain recurrent neural network and is termed as (VGRNN). Some possible choices of $z(t)$ are $z(t) = t^z + z$ [388], $z(t) = z^t + z$ [391], $z(t) = \exp(zt)$ [394] and different piecewise time-varying functions [229, 392]. Already existed AFs are represented to highlight the proposed model against previous (3.6.5).

• The Sinh activation function [394]

$$f(u) = \sinh(u). \tag{3.6.6}$$

• The Tunable activation function [50, 103]

$$f(u) = \left(|u|^p + |u| + |u|^{\frac{1}{p}}\right)\operatorname{sign}(u), \tag{3.6.7}$$

where $p \in (0, 1)$ and $\operatorname{sign}(\cdot)$ is the signum function.

• The versatile activation function (VAF) [45, 322]

$$f(u) = \left(b_1|u|^p + b_2|u|^w\right)\operatorname{sign}(u) + b_3u + b_4\operatorname{sign}(u), \tag{3.6.8}$$

where $b_1, b_2 > 0$, $b_3, b_4 \geq 0$, $w > 1$ and p is the same constant as in Tunable AF (3.6.7).

Using $E(t) = P(t)V(t) - Q(t)$ in the ZNN design (3.6.5), we obtain dynamical system

$$\dot{V}(t) = (I - P(t))\dot{V}(t) + \dot{Q}(t) - \dot{P}(t)V(t) - \lambda z(t)\mathcal{F}(P(t)V(t) - Q(t)). \tag{3.6.9}$$

Figure 3.47 shows the Simulink circuit of neurodynamics (3.6.9)

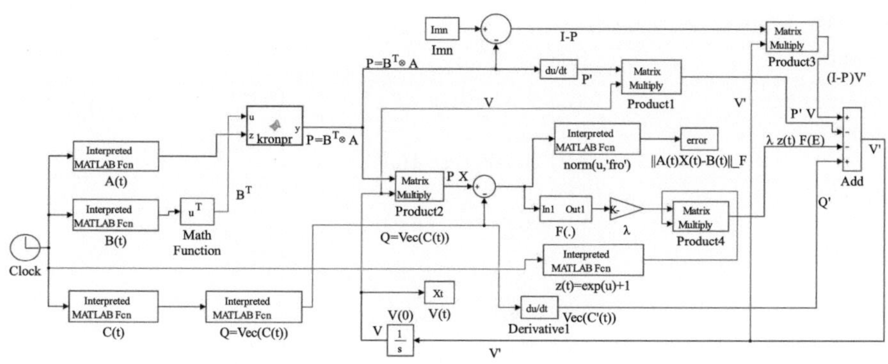

Fig. 3.47 Simulink model of dynamical system (3.6.5)

3.7 Some Additional ZNN Models

Some important ZNN models are arranged in Table 3.3.

Recently, a number of nonlinear and linear recurrent neural network models have been developed for the purpose of numerical evaluation of the matrix inverse and the pseudoinverse of full-row or full-column rank rectangular matrices (for more details, see [99, 162, 272, 277]). Various recurrent neural networks for computing generalized inverses of rank-deficient matrices were designed in [273, 282, 290, 399]. A new type of complex-valued RNNS, called Zhang neural network (ZNN), was proposed in 2001 and has been extensively exploited in solving various time-varying complex generalized inverse problems. The design of complex ZNN models arises from the choice of a complex matrix-valued error-monitoring function, called the Zhang function (ZF). The ZNN dynamics was originally invented for computing the time-varying matrix inverse [381]. Later, ZNN design for on-line time-varying matrix inversion was investigated in [349, 378]. Continuing this research, a more general ZNN design for finding the time-varying Moore-Penrose inverse of a full-column or full-row rank matrix was originated in [370]. Liao and Zhang in [126] proposed five different complex ZFs and, accordingly developed and investigated five different complex ZNN models for computing the time-varying complex pseudoinverse. A feed-forward neural network architecture for computing the Drazin inverse A^{D} was proposed in [37]. RNN for computing the Drazin inverse of a real square matrix in real time was presented in [251]. The dynamical equation and associated artificial RNN for computing the Drazin inverse of an arbitrary square real matrix, without any restriction on its eigenvalues, were proposed in [252]. In [203], the authors proposed two finite-time convergent complex ZNN models for computing the Drazin inverse of a complex time-varying square matrix. ZNN models based on the limit representation of the time-varying Drazin inverse were proposed in [283]. In [203], the authors investigated the computation of the Drazin inverse on the basis of two

Table 3.3 Important ZNN models

Result	$E(t)$	$\lim_{t\to\infty} V(t)$
Matrix inverse	$E(t) = A(t)V(t) - I$	$A(t)^{-1}$
Left inverse	$E(t) = A(t)^*A(t)V(t) - A(t)^*$	$A(t)_L^{-1}$
Right inverse	$E(t) = V(t)A(t)A(t)^* - A(t)^*$	$A(t)_R^{-1}$
Pseudoinverse	$E(t) = (A(t)^*A(t)+\lambda I)V(t)-A(t)^*$	$A(t)^\dagger$
Linear system	$e(t) = A(t)v(t) - b(t)$	$A(t)^\dagger b$
Nonlinear equation	$e(t)=f(x(t),t)$	$f(x(t),t)=0$
Square root	$e(t) = x^2(t) - a(t)$	$\sqrt{a(t)}$

limit representations of the Drazin inverse and the Li activation function. In [126], the authors introduced five complex ZNNs which are able to approximate different time-varying generalized inverses.

Integration-enhanced noise-tolerant ZNN models (shortly IENTZNN) have been studied extensively. An IENTZNN model for time-varying matrix inversion process was defined in [110]. Zeroing neural networks which are able to eliminate various kinds of noise and resolve the redundancy in kinematic control have attracted a great attention last years. Zhang et al. in [111] designed a noise-tolerant zeroing neural network (NTZNN) design formula from the viewpoint of control. An enhanced discrete-time Zhang neural network (EDTZNN) model for time-varying matrix inversion was proposed, analyzed and investigated in [176]. According to simulation results in [176], showed that the EDTZNN model is superior in eliminating various kinds of bias noises. The authors of [106, Table 2] summarized various noise-tolerant ZNN models for solving different time-varying problems.

Various complicated problems have been approached with neural networks in many fields of scientific research [95, 132, 137, 138, 157–160, 163, 164, 268–270, 350]. For instance, an adaptive fuzzy controller based on neural network is presented for a class of nonlinear discrete-time systems with discrete-time dead zone in [157]. An adaptive decentralized scheme based on neural network is constructed for multiple-input and multiple-output (MIMO) nonlinear systems with the aid of back-stepping techniques in [270]. A scheme like that guarantees the uniform ultimate boundedness of all signals in the closed-loop system with respect to mean square. In [268], in order to overcome the design difficulty of nonstrict-feedback structure, a variable separation technique to decompose the unknown functions of all state variables into a sum of smooth functions of each error dynamic is used. Also, an adaptive neural control algorithm is proposed with the help of radial basis function neural networks' universal approximation capability. A neural network model to generate winner-take-all competition is proposed in [132], which has an explicit explanation of the competition mechanism. Recurrent neural network (RNN) models, as a branch of artificial intelligence, have received considerable investigation in many scientific and engineering fields, which is often exploited for computational problems [165, 178, 252, 282, 286, 386, 393, 399] and nonlinear optimizations are solved by many methods [167, 168]. In [277], a gradient-based RNN model is presented for computing the inversion of a matrix online with guaranteed convergence, which can be deemed as a seminal work in this field. The authors of [35] presented a simplified neural network model to solve a class of linear matrix inequality problems, of which the stability and solvability are analyzed theoretically. In general, recurrent neural networks can be divided into two classes: (1) the continuous-time RNNs and (2) the discrete-time RNNs. By exploiting a numerical differential formula, a continuous-time RNN model can be discretized into a discrete-time one. Nevertheless, a numerical differentiation rule does not necessarily generate a convergent and stable discrete-time RNN model even though the original continuous-time RNN model is convergent. In addition, it can be considered as a numerical algorithm [112] if the discrete-time RNN model is coded as a serial-processing program and performed on digital computers. Zeroing neural networks (ZNNs) are able to perfectly track time-varying solutions by exploiting the time derivative of time-varying parameters as a novel type of RNN specifically designed for solving time-varying problems. Hence, many researchers make progresses along this direction by proposing various kinds of ZNN models with different features for solving TV problems.

In general, ZNN neural design has been widely applied to solving a variety of problems based on TV equalities and inequalities. More details can be found in monographs [107, 128, 351, 374], in the survey paper [106] and many other references. Moreover, several applications of ZNN for real world problems are observable. Specifically, applications of ZNN in finding finite-time solution of time-varying Sylvester equation and to inverted pendulum control can be found in [129, 130, 361]. Moreover, ZNN dynamics has also been applied to the control of chaos systems or helicopters control system [139].

Application of continuous-time ZNN in computing the matrix square root can be found in [354]. Also, discrete-time ZNN model computing the matrix square root which incorporates the Newton iteration as a special case, was proposed in [373]. A comparison of the ZNN and GNN models for online solution of time-varying matrix square roots was investigated in [317, 371]. Xiao in [317] defined and studied a finite-time convergent ZNN for solving the matrix square root. The finite-time convergent ZNN (FTCZNN) model is achieved by a nonlinearly activated sign-bi-power function. According to theoretical convergence analysis and derived numerical results, in [317] it was shown that the new design possesses faster convergence rate with respect to the standard ZNN design in approximating the time-varying matrix square root.

Uhlig and Zhang in [263] adapted look-ahead and backward finite difference formulas to compute future eigenvectors and eigenvalues of piecewise smooth time-varying symmetric or hermitian matrix $A(t)$.

In addition, it is possible to find many applications of ZNN for robot arm control [82, 128, 134, 137, 141, 142].

In [9], authors presented an Auto-Regressive (AR) spectral estimator using Continuous-time Zhang Neural Network in finding solutions of linear equations. A gradient based neural network to compute the p-order AR parameters by solving the Yule-Walker equations was presented in [9]. Further, a compact architecture of the discrete time gradient based neural network is derived using the fast Fourier transform to reduce the size of the neural network. Time-varying quadratic programming based on ZNN design was investigated in [362, 372]. Lv et al. in [171] proposed an improved Zhang neural network (IZNN) for solving time-varying linear system of equations, activated by an array of continuous sign-bi-power function.

3.7.1 Application of Zeroing Design to Robot Arm Control

The online approximation of time-varying Moore-Penrose inverse is essential in the inverse kinematic control of the redundant robot arm. Consider the end-effector position of a redundant robot arm is defined by the vector $r(t) \in \mathbb{R}^m$ in Cartesian space, and it is related to the joint-space vector $q(t) \in \mathbb{R}^n$ through the forward kinematic equation of the form

$$r(t) = f(q(t)), \tag{3.7.1}$$

where $f(\cdot)$ is a continuous nonlinear mapping function with known structure and parameters for a given robot arm. The inverse kinematic problem is to obtain the joint variable $q(t)$ for any given $r(t)$ through the inverse mapping of Eq. (3.7.1), i.e., compute $q(t) = f^{-1}(r(t))$, if it exists. But, it is typically difficult to generate an analytical solution of f^{-1} due to the redundancy and nonlinearity of $f(\cdot)$. The inverse kinematic problem is thus usually solved at the velocity level. Differentiating Eq. (3.7.1) with respect to the time t yields a linear relation between velocities $\dot{r}$ and $\dot{\theta}$:

$$J(\theta(t))\dot{\theta}(t) = \dot{r}(t), \tag{3.7.2}$$

where $J(\theta) \in \mathbb{R}^{m\times n}$ is the Jacobian matrix defined as $J(\theta) = \partial f(\theta)/\partial\theta$. Equation (3.7.2) is under-determined because, in a redundant robot arm control, it is always assumed $m < n$. So, it may admit an infinite number of solutions.

In such situations, the Moore-Penrose inverse with its least squares minimal norm solution property arises as very useful tool. So, the least squares solutions to (3.7.2) is widely used by robotics researchers. It is generally formulated as a minimum-norm solution $J^{\dagger}(t)\dot{r}(t)$ plus an arbitrary homogeneous solution:

$$\dot{\theta}(t) = J^{\dagger}(t)\dot{r}(t) + (I - J^{\dagger}(t)J(t))z(t), \tag{3.7.3}$$

where $J^{\dagger}(t) \in \mathbb{R}^{n\times m}$ denotes the Moore-Penrose inverse of $J(t)$. The vector $z(t) \in \mathbb{R}^n$ is arbitrary. It can be selected as the negative gradient of a performance index to be minimized.

Since $J \in \mathbb{R}_m^{m\times n}$ is of full row rank, $J^{\dagger}$, is defined by $J^{\mathrm{T}}(J J^{\mathrm{T}})^{-1}$. The usual numerical solutions for $J^{\dagger}$ are computationally complex and may cause relatively large computational errors. Moreover, with multiple tasks and constraints included within (3.7.3) via $z(t)$, heavy computation process may appear inappropriate for real-time applications, especially in high-DOF (high degrees of freedom) robotic systems. In such situations, it is possible to apply dynamical system approach in order to generate $J^{\dagger}(t)$ in a theoretically parallel and error-free manner.

3.8 Complex ZNNs for Computing Time-Varying Outer Inverses

Full-rank representation of {2}-inverses with prescribed range and null space is determined in Proposition 3.8.1.

Proposition 3.8.1 ([208]) *Let $A \in \mathbb{C}_r^{m\times n}$, T be a subspace of $\mathbb{C}^n$ of dimension $s \le r$ and let S be a subspace of $\mathbb{C}^m$ of dimensions $m - s$. In addition, suppose that*

$G \in \mathbb{C}^{n\times m}$ satisfies $\mathcal{R}(G) = T, \mathcal{N}(G) = S$. Let G has an arbitrary full-rank decomposition, that is $G = PQ$. If A has a {2}-inverse $A^{(2)}_{T,S}$, then:
(1) QAP is an invertible matrix;
(2) $A^{(2)}_{T,S} = P(QAP)^{-1}Q$.

Finite algorithms for computing (2)-generalized inverse $A^{(2)}_{T,S}$ were proposed in [31]. The following two limit representations of outer inverses can be derived using the results from [156, 289].

Proposition 3.8.2 ([289]) *Let $A \in \mathbb{C}^{m\times n}$ be of rank r, let T be a subspace of $\mathbb{C}^n$ of dimension $s \le r$, and let S be a subspace of $\mathbb{C}^m$ of dimension $m - s$. In addition, suppose that $G \in \mathbb{C}^{n\times m}$ satisfies $\mathcal{R}(G) = T$ and $\mathcal{N}(G) = S$. If $A^{(2)}_{T,S}$ exists then it possesses the limit representation*

$$A^{(2)}_{T,S} = \lim_{\lambda\to 0} (GA + \lambda I)^{-1} G. \tag{3.8.1}$$

Proposition 3.8.3 ([156]) *Let $A \in \mathbb{C}^{m\times n}$ be of rank r, let T be a subspace of $\mathbb{C}^n$ of dimension $s \le r$, and let S be a subspace of $\mathbb{C}^m$ of dimension $m - s$. In addition, suppose that $P \in \mathbb{C}^{n\times s}_s$ satisfies $\mathcal{R}(P) = T$ and $Q \in \mathbb{C}^{s\times m}_s$ satisfies $\mathcal{N}(Q) = S$. Then $A^{(2)}_{T,S}$ exists and*

$$A^{(2)}_{T,S} = \lim_{\lambda\to 0} P\,(QAP + \lambda I)^{-1}\, Q. \tag{3.8.2}$$

3.8.1 ZNN Models for Generating Outer Inverses

Our main goal is achieved by two ZNN models, called ZNNATS2-I and ZNNATS2-II.

Complex ZNN Model ZNNATS2-I

The starting point in generating the ZNNATS2-I model arises from Lemma 3.8.1. The result presented in Lemma 3.8.1 is a time-varying generalization of an analogous result stated in [399] and exploited so far in the time-invariant matrix case.

Lemma 3.8.1 ([284]) *Let $A(t) \in \mathbb{C}^{m\times n}_r$ be given and $G(t) \in \mathbb{C}^{n\times m}_s$ be an arbitrary matrix whose rank satisfies $0 < s \le r$. Assume that $X(t) := A(t)^{(2)}_{\mathcal{R}(G),\mathcal{N}(G)}$ exists. Then both the matrix identities (2.3.1) are satisfied in the time-varying case: $G(t)A(t)X(t) = G(t)$, $X(t)A(t)G(t) = G(t)$.*

The complex neural network model is developed by employing basic steps for the development of an arbitrary ZNN model. These steps were previously followed in [126, 370].

The Tikhonov regularization is an effective way to avoid the strong limitations on the eigenvalues of $G(t)A(t)$. After the Tikhonov regularization, it is reasonable to propose the following two dual fundamental error-monitoring ZFs in order to generate ZNN models appropriate for numerical computations of the outer inverse $A(t)^{(2)}_{\mathcal{R}(G),\mathcal{N}(G)}$:

$$E_G(t) = \begin{cases} (G(t)A(t) + \lambda I)\, V_1(t) - G(t), \; n \le m, \; \lambda > 0 \\ V_1(t)\,(A(t)G(t) + \lambda I) - G(t), \; n > m, \; \lambda > 0. \end{cases} \tag{3.8.3}$$

Remark 3.8.1 The particular case $G(t) = A(t)^*$ of (3.8.3) gives the complex ZNN-II model from [126], derived for the pseudoinverse computation in the complex time-varying matrix case. Also, the particular case $G(t) = A(t)^l$, $l \ge \mathrm{ind}(A)$ of (3.8.3) gives the complex LZNN-I model from [283] for computing the time-varying complex Drazin inverse.

Since (3.8.3) involves two dual cases, only the case $n \le m$ is considered in details in the rest of the paper without loss of generality; the case $n > m$ is quite analogous. The general design formula is defined by (3.8.4)

$$\dot{E}_G(t) := \frac{\mathrm{d}E_G(t)}{\mathrm{d}t} = -\gamma \mathcal{H}(E_G(t)), \tag{3.8.4}$$

Two appropriate activation functions, (1.4.58) and (1.4.59), introduced in [129], will be exploited in (3.8.4). where $\iota = \sqrt{-1}$ denotes the imaginary unit and $\Gamma = |A + \iota B| \in \mathbb{R}^{n\times n}, \Theta = \Theta(A + \iota B) \in (-\pi, \pi]^{n\times n}$ denote element-wise modulus and the element-wise arguments, respectively, of the complex matrix $A + \iota B$. In sequel, we use the notation $\mathcal{H}_k$ as a universal mark for for $\mathcal{H}_1$ or $\mathcal{H}_2$.

The dynamic equation of a complex neural network model for computing $A(t)^{(2)}_{\mathcal{R}(G),\mathcal{N}(G)}$ can be established by expanding (3.8.4) and using $E_G(t)$. By combining (3.8.4) and

$$\dot{E}_G(t) = \left(\dot{G}(t)A(t) + G(t)\dot{A}(t)\right) V_1(t) + (G(t)A(t) + \lambda I)\, \dot{V}_1(t) - \dot{G}(t), \tag{3.8.5}$$

it is possible to verify

$$\begin{aligned}(G(t)A(t) + \lambda I)\, \dot{V}_1(t) =& \dot{G}(t) - \left(\dot{G}(t)A(t) + G(t)\dot{A}(t)\right) V_1(t) \\ & - \gamma \mathcal{H}_k\left((G(t)A(t) + \lambda I)\, V_1(t) - G(t)\right).\end{aligned} \tag{3.8.6}$$

Accordingly, the corresponding dynamic equation of the complex ZNN model for generating outer inverses can be stated as

$$\begin{aligned}\dot{V}_1(t) =& \frac{1}{\lambda}\Big[-G(t)A(t)\dot{V}_1(t) + \dot{G}(t) \\ & - \left(\dot{G}(t)A(t) + G(t)\dot{A}(t)\right) V_1(t) - \gamma \mathcal{H}_k\left((G(t)A(t) + \lambda I)\, V_1(t) - G(t)\right)\Big].\end{aligned} \tag{3.8.7}$$

Particular Cases of the ZNNATS2-I Model

The aim of this subsection is to describe previously stated goal (3).

(a) In the particular case $G(t) = A(t)^*$, the ZNNATS2-I model reduces to the model from [126], corresponding to the error monitoring function given in (3.3.5).

(b) The choice $G(t) = A(t)^l$, $l \geq \text{ind}(A)$ under the assumption $m = n$ reduces the ZNNATS2-I to the model LZNN-II from [283], corresponding to the error monitoring function given in (3.8.8)

$$E(t) = \left(A(t)^{l+1} + \lambda I\right) X(t) - A(t)^l \quad l \geq k = \text{ind}(A). \tag{3.8.8}$$

The matrix $X(t)$ in (3.8.8) corresponds to the Drazin inverse $A(t)^{\mathrm{D}}$.

(c) The particular choice $G(t) = A(t)^l \left(A(t)^{2l+1}\right)^* A(t)^l$, $l \geq \text{ind}(A)$ in the case $m = n$ reduces the ZNNATS2-I model to the model LZNN-I from [283].

(d) Under the assumption that A and G are constant complex matrices, the ZNNATS2-I model, given in (3.8.6) becomes the model

$$(GA + \lambda I)\, \dot{V}_1(t) = -\gamma \mathcal{H}_k \left((GA + \lambda I)\, V_1(t) - G\right), \tag{3.8.9}$$

which is applicable in computation of the outer inverse $A^{(2)}_{\mathcal{R}(G),\mathcal{N}(G)}$ for time-invariant complex matrix A.

(e) If A and G are constant complex matrices and further $\lambda = 0$, the ZNNATS2-I model represents an ZNN analogy of the GNN analogy, called as the RNN1 model from [399]. The drawback of the model RNN1 is its restriction that the real parts of the spectrum of GA are nonnegative real values.

(f) If A and G are constant complex matrices, $\lambda = 0$, and further G is replaced by $G_0 = G(GAG)^{\mathrm{T}} G$, the ZNNATS2-I model represents an ZNN analogy of the GNN analogy, known as the RNN2 model from [399].

(g) If A is a square matrix, $G(t) = A(t)^l$, $l \geq \text{ind}(A)$, and the Li activation is used, the ZNNATS2-I model reduces to the model ZNN-II from [203].

(h) If A is a square matrix, $G(t) = A(t)^l \left(A(t)^{2l+1}\right)^* A(t)^l$, $l \geq \text{ind}(A)$, and the Li activation is used, the ZNNATS2-I model reduces to the model ZNN-I from [203]. In general, main difference between the ZNNATS2-I model and models defined in [203] is an alternative approach in handling complex matrices. Our approach is based on two appropriate activation functions $\mathcal{H}_1$ and $\mathcal{H}_2$, while the strategy in [203] assumes usage of appropriate block matrices.

3.8.2 *Complex ZNN Model ZNNATS2-II*

Our strategy in this section is more general and somewhat different with respect to that used in [254]. Firstly, the model proposed in [254] is an explicit GNN model for time-invariant matrices, while the new model is an implicit ZNN model for time-varying case. Further, the method from [254] is based on Proposition 3.8.1, while the new approach is based on Proposition 3.8.3. The leading idea in [254] was to define the GNN model for finding the approximation $X(t)$ of inverse $(QAP)^{-1}$ and then compute $PX(t)Q$. In this way, the invertibility of QAP is assumed. On the other hand, it is observable that the invertibility of QAP is not assumed in the limit representation (3.8.2). Namely, our general strategy is to define an ZNN approach with the aim to compute

$$X_1(t) = (Q(t)A(t)P(t) + \lambda I)^{-1}\, Q(t),\ \lambda > 0,$$

and later multiply the generated result by $P(t)$ from the left in order to achieve the limit representation (3.8.2) of $A(t)^{(2)}_{T,S}$. In order to avoid the invertibility of $Q(t)A(t)P(t)$ assumed in the dynamic $Q(t)A(t)P(t)X_1(t) = Q(t)$, the following complex ZF can be exploited in this case:

$$E_{PQ}(t) = (Q(t)A(t)P(t) + \lambda I)\, V_2(t) - Q(t) \triangleq H(t)V_2(t) - Q(t), \tag{3.8.10}$$

where $H(t) = Q(t)A(t)P(t) + \lambda I$.

The time derivative of (3.8.10) is equal to

$$\dot{E}_{PQ}(t) = -\dot{Q}(t) + H(t)\dot{V}_2(t) + \dot{H}(t)V_2(t),$$

where

$$\dot{H}(t) = \dot{Q}(t)A(t)P(t) + Q(t)\dot{A}(t)P(t) + Q(t)A(t)\dot{P}(t).$$

According to the rule of the general design formula (3.8.4), dynamic equation which initiates the second complex neural network model as

$$H(t)\dot{V}_2(t) = \dot{Q}(t) - \dot{H}(t)V_2(t) - \gamma\mathcal{H}_k(H(t)V_2(t) - Q(t))\,. \tag{3.8.11}$$

The model (3.8.11) represents the implicit dynamics of the complex ZNN model, called ZNNATS2-II. The implicit dynamics of the jth subnetwork corresponding to ZNNATS2-II can be stated as

$$H(t)\dot{v}_j(t) = \dot{q}_j(t) - \dot{H}(t)v_j(t) - \gamma\mathcal{H}_k\big(H(t)v_j(t) - q_j(t)\big)\,,$$

wherein

$$v_j = (V_2)_j,\ q_j = (Q(t))_j.$$

Therefore, corresponding explicit dynamic equation of the jth subnetwork is defined as

$$\dot{v}_j(t) = \frac{1}{\lambda}\left[-Q(t)A(t)P(t)\dot{v}_j(t) + \dot{q}_j(t) - \dot{H}(t)v_j(t) - \gamma\mathcal{H}_k\big(H(t)v_j(t) - q_j(t)\big)\right],$$
$$j = 1, \ldots, m. \tag{3.8.12}$$

According to (3.8.12), the dynamic equation of the (ij)th entry can be presented by the following explicit dynamics

$$\dot{v}_{ij} = \frac{1}{\lambda}\left(\sum_{k=1}^{n} m_{ik}\dot{v}_{kj} - \gamma\sum_{k=1}^{n}\mathcal{H}_k\left(h_{ik}v_{kj} - q_{ij}\right) - \sum_{k=1}^{n}\dot{h}_{ik}v_{kj} + \dot{q}_{ij}\right),$$

where

$$m_{ik} = (-Q(t)A(t)P(t))_{ik}\,,\ h_{ik} = (Q(t)A(t)P(t)+\lambda I)_{ik}\,, q_{ij} = (Q(t))_{ij}.$$

Similarly as in the model ZNNATS2-I case, the real parameter $\lambda > 0$ should be sufficiently close to zero in order to approximate the limiting representation (3.8.2).

3.8.3 Convergence Analysis of Complex ZNNs for Computing Outer Inverses

In this section we prove the global convergence of both complex neural network models ZNNATS2-I and ZNNATS2-II, and consequently achieve our goal (4). To simplify notation, by ZNNATS2-I(1) and ZNNATS2-I(2) we denote the model ZNNATS2-I with the activation functions $\mathcal{H}_1$ or $\mathcal{H}_2$, respectively. Similarly, the model ZNNATS2-II corresponding to $\mathcal{H}_1$ or $\mathcal{H}_2$ is denoted by ZNNATS2-II(1) and ZNNATS2-II(2), respectively.

Convergence of the Model ZNNATS2-I

In this subsection, we investigate the convergence performances of the complex ZNN model (3.8.6).

Convergence of ZNNATS2-I(1)

Theorem 3.8.1 ([284]) *Let time-varying complex matrices $A(t) \in \mathbb{C}_r^{m\times n}$ and $G(t) \in \mathbb{C}_s^{n\times m}$, $s \le r$, be given. Assume that the time derivatives of $A(t)$ and $G(t)$ exist in the time interval $t \in [0, +\infty]$, and they are continuous and uniformly bounded with respect to the time t. Then the state matrix $V_1(t) \in \mathbb{C}^{n\times m}$ of the complex neural network model* ZNNATS2-I(1) *globally converges to the time-varying outer inverse $A(t)^{(2)}_{\mathcal{R}(G),\mathcal{N}(G)}$.*

Proof According to the limit representation (3.8.1), it follows that

$$\lim_{\lambda\to 0}\left(G(t)A(t)+\lambda I\right)A(t)^{(2)}_{\mathcal{R}(G),\mathcal{N}(G)}=G(t).$$

Therefore, in accordance with the first dynamic equation contained in (3.8.3), it follows that $E_G(t)$ satisfies

$$\begin{aligned}E_G(t)&=\left(G(t)A(t)+\lambda I\right)V_1(t)-G(t)\\&\approx\left(G(t)A(t)+\lambda I\right)V_1(t)-\left(G(t)A(t)+\lambda I\right)A(t)^{(2)}_{\mathcal{R}(G),\mathcal{N}(G)}\\&\approx\left(G(t)A(t)+\lambda I\right)\overline{V}_1(t),\end{aligned}\tag{3.8.13}$$

where $\overline{V}_1(t)=V_1(t)-A(t)^{(2)}_{T,S}$. Then $\dot{E}_G(t)=-\gamma\mathcal{H}_1(E_G(t))$ can be approximated as

$$\left(\dot{G}(t)A(t)+G(t)\dot{A}(t)\right)\overline{V}_1(t)+\left(G(t)A(t)+\lambda I\right)\dot{\overline{V}}_1(t)=-\gamma\mathcal{H}_1\left(\left(G(t)A(t)+\lambda I\right)\overline{V}_1(t)\right).\tag{3.8.14}$$

In view of the definition of the activation function $\mathcal{H}_1(\cdot)$, taking into account $E_G(t)=\mathbf{Re}(E_G(t))+\iota\,\mathbf{Im}(E_G(t))$, the general model $\dot{E}_G(t)=-\gamma\mathcal{H}_1(E_G(t))$ splits into the following two equations in the real numbers domain:

$$\mathbf{Re}(\dot{E}_G(t))=-\gamma\mathcal{F}(\mathbf{Re}(E_G(t)))$$

and

$$\mathbf{Im}(\dot{E}_G(t))=-\gamma\mathcal{F}(\mathbf{Im}(E_G(t))).$$

In order to examine the convergence, the Lyapunov function candidate is defined as

$$L(t)=\frac{\|E_G(t)\|_F^2}{2}=\frac{\mathrm{Tr}\left(E_G(t)^*E_G(t)\right)}{2}.\tag{3.8.15}$$

Using again $E_G(t)=\mathbf{Re}(E_G(t))+\iota\mathbf{Im}(E_G(t))$ d $\mathrm{Tr}\left(X^{\mathrm{T}}X\right)=2\mathrm{Tr}\left(X^{\mathrm{T}}\mathrm{d}X\right)$, the following identities can be verified:

$$\begin{aligned}\frac{\mathrm{d}L(t)}{\mathrm{d}t}&=\mathrm{Tr}\left(E_G(t)^*\dot{E}_G(t)\right)\\&=-\gamma\mathrm{Tr}\left\{\left(\mathbf{Re}\left(E_G(t)\right)^{\mathrm{T}}-\iota\mathbf{Im}\left(E_G(t)\right)^{\mathrm{T}}\right)\left(\mathcal{F}(\mathbf{Re}\left(E_G(t)\right))+\iota\mathcal{F}\left(\mathbf{Im}(E_G(t))\right)\right)\right\}\\&=-\gamma\mathrm{Tr}\left\{\mathbf{Re}\left(E_G(t)\right)^{\mathrm{T}}\mathcal{F}\left(\mathbf{Re}\left(E_G(t)\right)\right)+\mathbf{Im}\left(E_G(t)\right)^{\mathrm{T}}\mathcal{F}\left(\mathbf{Im}\left(E_G(t)\right)\right)\right\}.\end{aligned}$$

Since $\mathcal{F}(\cdot)$ is odd and monotonically increasing, it follows that

$$\mathbf{Re}\left(E_G(t)\right)^{\mathrm{T}}\mathcal{F}\left(\mathbf{Re}\left(E_G(t)\right)\right)+\mathbf{Im}\left(E_G(t)\right)^{\mathrm{T}}\mathcal{F}\left(\mathbf{Im}\left(E_G(t)\right)\right)\geq 0,$$

and finally $\frac{\mathrm{d}L(t)}{\mathrm{d}t} \leq 0$. As a consequence of the Lyapunov stability theory, $E_G(t) = (G(t)A(t) + \lambda I)\, V_1(t) - G(t)$ is globally convergent to zero. Therefore, $V_1(t) \to (G(t)A(t) + \lambda I)^{-1}\, G(t)$ as $t \to \infty$. Based on Proposition 3.8.2, the state matrix $V_1(t)$ globally converges to the time-varying outer inverse $A(t)^{(2)}_{\mathcal{R}(G),\mathcal{N}(G)}$ under the condition $\lambda \to 0$. □

Convergence of ZNNATS2-I(2)

Theorem 3.8.2 ([284]) *Assume that the time derivatives of input matrices $A(t) \in \mathbb{C}_r^{m\times n}$ and $G(t) \in \mathbb{C}_s^{n\times m}$, $t \in [0, +\infty]$, exist and they are continuous and uniformly bounded with respect to t. Then the state matrix $V_1(t) \in \mathbb{C}^{n\times m}$ of the complex neural network model* ZNNATS2-I(2) *is globally convergent to the time-varying outer inverse $A(t)^{(2)}_{\mathcal{R}(G),N(G)}$, starting from arbitrary initial state $V_1(0)$.*

Finite-Time ZNNATS2-I

In this subsection, we apply the Li activation function [126, 130, 131] to accelerate the ZNN to finite-time convergent model to the theoretical solution of the time-varying matrix outer inverse. Corresponding ZNN model is denoted as ZNNATS2-I(3). These results confirm our goal (2).

The Li array activation function is defined as

$$\mathcal{H}_3(A) = \Psi(\mathbf{Re}(E_G(t))) + \iota\Psi(\mathbf{Im}(E_G(t))) = \psi(\mathbf{Re}(e_{ij})) + \iota\psi(\mathbf{Im}(e_{ij})),$$

where e_{ij} denotes the (ij)th element of $E_G(t)$ and ψ is defined on each element of a given matrix $A = (a_{ij}) \in \mathbb{R}^{m\times n}$

$$\psi(a_{ij}) = \mathrm{Lip}^{\sigma}(a_{ij}) + \mathrm{Lip}^{\frac{1}{\sigma}}(a_{ij}).$$

The design parameter satisfies $\sigma \in (0, 1)$ and the function $\mathrm{Lip}^{\sigma}(a_{ij})$ is defined as

$$\mathrm{Lip}^{\sigma}(a_{ij}) = \begin{cases} |a_{ij}|^{\sigma}, & a_{ij} > 0, \\ 0, & a_{ij} = 0, \\ -|a_{ij}|^{\sigma}, & a_{ij} < 0. \end{cases}$$

Theorem 3.8.3 ([284]) *Given a smoothly time-varying matrix $A(t) \in \mathbb{C}^{m\times n}$. If the Li activation function is used, then the state matrix $V_1(t)$ of the ZNNATS2-I(3) model, starting from any initial state $V_1(0)$, converges to the time-varying theoretical outer inverse $A^{(2)}_{T,S}$ in finite time:*

$$t_f < \frac{|e^{+}(0)|^{1-\sigma}}{\gamma(1-\sigma)},$$

where $e^+(0)$ *is the largest element in*

$$E_G(0) = (G(0)A(0) + \lambda I)\, V_1(0) - G(0).$$

Proof Let us consider again $\overline{V}_1(t) = V_1(t) - A(t)^{(2)}_{T,S}$. Starting from (3.8.13), $\dot{E}_G(t) = -\gamma \mathcal{H}_3(E_G(t))$ can be expanded as

$$\left(\dot{G}(t)A(t) + G(t)\dot{A}(t)\right)\overline{V}_1(t) + (G(t)A(t) + \lambda I)\,\dot{\overline{V}}_1(t) = -\gamma \mathcal{H}_3\left((G(t)A(t) + \lambda I)\,\overline{V}_1(t)\right). \tag{3.8.16}$$

Further, $\dot{E}_G(t) = -\gamma \mathcal{H}_3(E_G(t))$ can be rewritten elementwise as the following set of $n \times m$ equations:

$$\dot{e}_{ij} = -\gamma \psi\left(e_{ij}(t)\right).$$

Following previous results, it is possible to conclude that the state matrix $V_1(t)$ globally converges to the time-varying theoretical outer inverse $A^{(2)}_{T,S}$ starting from arbitrary initial state $V_1(0)$. Next, it is necessary to prove the finite-time convergent performance of the ZNNATS2-I(3) model.

The initial value of the matrix valued error function $E_G(t)$ is

$$E_G(0) = (G(0)A(0) + \lambda I)\, V_1(0) - G(0).$$

With this aim, let us define the quantity

$$e^+(0) = \max\left\{E_G(0)\right\},$$

for all possible values of indices $i \in \{1, 2, \ldots, n\}$ and $j \in \{1, 2, \ldots, m\}$. This means that $|e_{ij}(t)|$ converges to zero for all possible values of the indices $i \in \{1, 2, \ldots, n\}$ and $j \in \{1, 2, \ldots, m\}$ when $|e^+(t)|$ reach zero. In other words, the convergence time t_f of the ZNNATS2-I(3) model is bounded by t_f^+ of the dynamics of $|e^+(t)|$, where t_f^+ represent the convergence time of the dynamics of $|e^+(t)|$. It is clear that

$$\dot{e}^+ = -\gamma \psi\left(e^+(t)\right).$$

Another Lyapunov function candidate is defined as

$$l_+(t) = |e^+(t)|^2.$$

Since $l_+(t) \geq 0$, the time derivative of l_+ is equal to

$$\begin{aligned}
\dot{l}_+(t) &= -2\gamma e^+(t)\psi\left(e^+(t)\right) \\
&= -2\gamma\left(|e^+(t)|^{\sigma+1} + |e^+(t)|^{1/\sigma+1}\right) \\
&\leq -2\gamma|e^+(t)|^{\sigma+1} = -2\gamma\, l_+^{(\sigma+1)/2}(t).
\end{aligned}$$

The rest of the proof can be completed following the principles from [149, Theorem 1]. □

Convergence of the Model ZNNATS2-II

In the following, we investigate the convergence of the complex ZNN model defined by (3.8.11) and called ZNNATS2-II. Following the previous notations, ZNNATS2-II(1) and ZNNATS2-II(2) are used to denote the ZNNATS2-II model in the case when the activation function is defined as $\mathcal{H}_1$ and $\mathcal{H}_2$, respectively.

Convergence of ZNNATS2-II(1)

Theorem 3.8.4 *Assume that the time derivatives of time-varying complex matrices* $A(t) \in \mathbb{C}_r^{m\times n}$ *and* $G(t) \in \mathbb{C}_s^{n\times m}$ *exist and they are continuous and uniformly bounded with respect to time* $t \in [0, +\infty]$. *Let* T *be a subspace of* $\mathbb{C}^n$ *of dimension* $s \le r$, *and let* S *be a subspace of* $\mathbb{C}^m$ *of dimension* $m - s$. *In addition, suppose that* $P(t) \in \mathbb{C}_s^{n\times s}$ *satisfies* $\mathcal{R}(P(t)) = T$ *and* $Q(t) \in \mathbb{C}_s^{s\times m}$ *satisfies* $\mathcal{N}(Q(t)) = S$. *Then the matrix product* $P(t)V_2(t) \in \mathbb{C}^{n\times n}$, *where* $V_2(t)$ *is defined by the complex neural network model* ZNNATS2-II(1), *globally converges to the time-varying outer inverse* $A(t)^{(2)}_{T,S}$ *of* $A(t)$, *starting from any initial state* $V(0)$.

Proof Starting from the fundamental error-monitoring function is defined in (3.8.10), one can verify that $E_{PQ}(t)$ globally converges to zero from arbitrary initial value. Using that $Q(t)A(t)P(t) + \lambda I$ is nonsingular, it is possible to verify

$$P(t)V_2(t) \to P(t)\left(Q(t)A(t)P(t) + \lambda I\right)^{-1} Q(t),$$

as $t \to \infty$. In view of $\lambda \to 0$, based on Proposition 3.8.3, $P(t)V_2(t)$ globally converges to the time-varying outer inverse $A(t)^{(2)}_{\mathcal{R}(P),\mathcal{N}(Q)}$. □

Convergence of ZNNATS2-II(2)

Theorem 3.8.5 ([284]) *Let time-varying complex matrices* $A(t) \in \mathbb{C}^{m\times n}$ *and* $G(t) \in \mathbb{C}^{n\times m}$ *be given. Assume that the time derivatives of matrices* $A(t) \in \mathbb{C}_r^{m\times n}$ *and* $G(t) \in \mathbb{C}_s^{n\times m}$ *exist and they are continuous and uniformly bounded with respect to* $t \in [0, +\infty]$. *Let* T *be a subspace of* $\mathbb{C}^n$ *of dimension* $s \le r$, *and let* S *be a subspace of* $\mathbb{C}^m$ *of dimension* $m - s$. *In addition, suppose that* $P(t) \in \mathbb{C}_s^{n\times s}$ *satisfies* $\mathcal{R}(P(t)) = T$ *and* $Q(t) \in \mathbb{C}_s^{s\times m}$ *satisfies* $\mathcal{N}(Q(t)) = S$. *Then* $P(t)V_2(t)$ *globally converges to time-varying outer inverse* $A(t)^{(2)}_{T,S}$, *where* $V_2(t)$ *corresponds to the complex neural network model* ZNNATS2-II(2).

Proof Similar to the proof of complex neural network model I with a type II activation function, for complex neural network model II with a type II activation function, its complex function (3.8.10) satisfies the design formula (3.8.4). This means that

$$E_1(t) = (A(t) + \lambda I)^{l+1}\, V(t) - A(t)^l,$$

can globally converge to zero from arbitrary initial value. That is to say, as $t \to \infty$, we have

$$V(t) \to (A(t) + \lambda I)^{-(l+1)}\, A(t)^l \in C^{n\times n}.$$

In view of $\lambda \to 0$, based on Lemma 3.8.2, the state matrix $V(t)$ globally converges to the time-varying theoretical Drazin inverse.

Lemma 3.8.2 ([101, 102]) *The Drazin inverse of* $A \in \mathbb{C}^{n\times n}$ *possesses the limit representation*

$$A^D = \lim_{\lambda\to 0} (A + \lambda I)^{-(l+1)}\, A^l, \quad l \ge \text{ind}(A). \tag{3.8.17}$$

The proof of the global exponential convergence performance of complex neural network model II is thus complete. □

3.8.4 *Numerical Examples in Computing Time-Varying Complex Outer Inverses*

Example 3.8.1 The theoretical Drazin inverse of the time-varying matrix

$$A(t) = \begin{bmatrix} 1+t & t & 1+t \\ t^2 & -1+t & t \\ 1+t & t & 1+t \end{bmatrix} \tag{3.8.18}$$

is equal to

$$A(t)^{\text{D}} = \begin{bmatrix} \frac{1-t+2t^3-2t^4}{(2-t^2+t^3)^2} & \frac{t}{2-t^2+t^3} & \frac{1-t-t^2+t^4}{(2-t^2+t^3)^2} \\ \frac{t(-1+t^2+t^3)}{(1+t)(2-2t+t^2)^2} & -\frac{2}{2-2t+t^2} & \frac{3t-2t^3}{(1+t)(2-2s+s^2)^2} \\ \frac{1-t+2t^3-2t^4}{(2-t^2+t^3)^2} & \frac{t}{2-t^2+t^3} & \frac{1-t-t^2+t^4}{(2-t^2+t^3)^2} \end{bmatrix}.$$

Since $\text{ind}(A) = 3$, the choice $G = A^3$ in ZNNATS2-I initiates the computation of the Drazin inverse. Let us choose the initial vector $v(0) = (1, 1, 1)^{\text{T}}$. State variables trajectories of the complex model ZNNATS2-I(1) based on the Li activation function with $\sigma = 0.8$ and the parameters $\gamma = 10^8$ and $\lambda = 10^{-3}$ are shown in graphs included in Fig. 3.48. Blue "O" in this figure indicates the values corresponding to the solution computed by the model ZNNATS2-I(1) and the red "$*$" means values corresponding to the exact Drazin inverse.

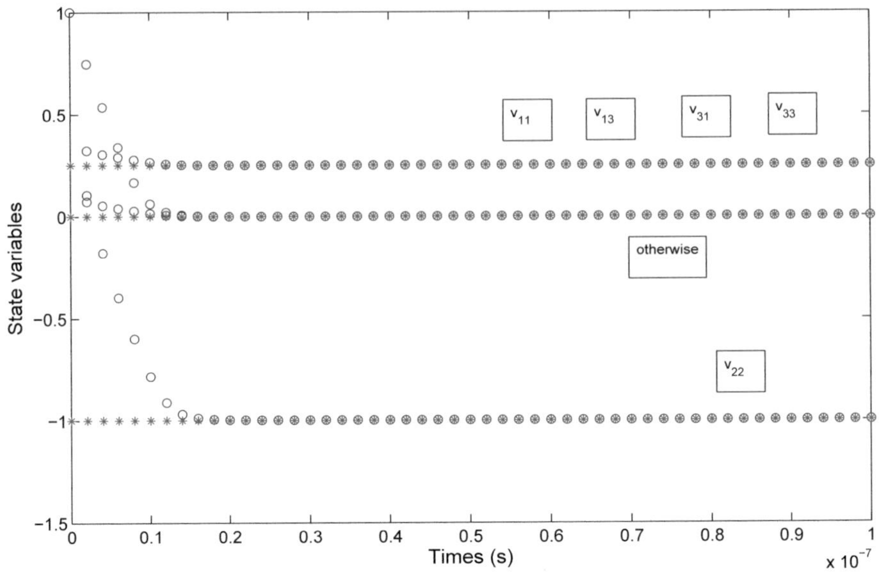

Fig. 3.48 Trajectories of the state variables of the model ZNNATS2-I(1) in Example 3.8.1

Note that the theoretical upper bound for the convergence time under the assumptions $\gamma = 10^8$, $\lambda = 10^{-3}$ and $\sigma = 0.8$ can be estimated as

$$t_f < \frac{|e^+(0)|^{1-\sigma}}{\gamma(1-\sigma)} = \frac{16.001}{0.2\gamma} \approx 0.8706 \times 10^{-7}. \tag{3.8.19}$$

As expected, our model converges before the above theoretical upper bound for t_f given in (3.8.19).

Residual errors

$$\| (G(t)A(t) + \lambda I)\, V(t) - G(t)\|_F, \quad t \in [0, 10^{-7}],$$

derived by ZNNATS2-I(1) with $\gamma = 10^8$ and $\lambda = 10^{-6}$ are shown in Fig. 3.49, where red dots correspond to the linear function $f(\cdot)$, pink pluses indicate that $f(\cdot)$ is the power-sigmoid function with $p = 3$ and $q = 5$, blue stars are corresponding to the bipolar-sigmoid function $f(\cdot)$ with $q = 14$, green triangles indicate the smooth power-sigmoid function $f(\cdot)$ with $p = 3$ and $q = 6$ and black circles indicate the Li function $f(\cdot)$ with $\sigma = 0.5$, respectively.

Example 3.8.2 Let us consider the time-varying matrix

$$A(t) = \begin{bmatrix} i\sin(2t) & i\cos(2t) & -i\sin(2t) \\ -i\cos(2t) & i\sin(2t) & i\cos(2t) \\ i\sin(2t) & i\cos(2t) & 0 \end{bmatrix}. \tag{3.8.20}$$

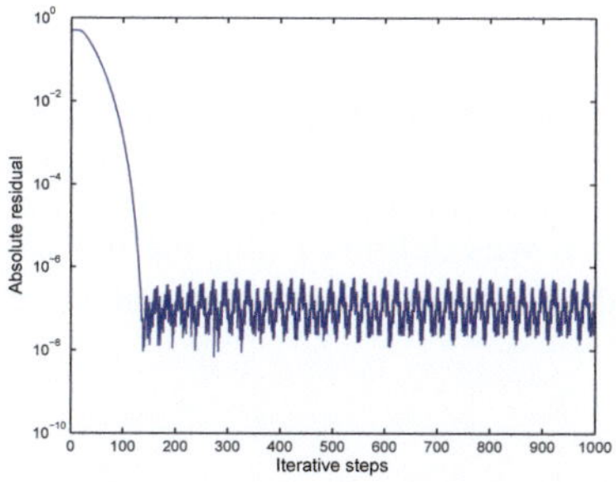

(a) Influence of different activation functions.

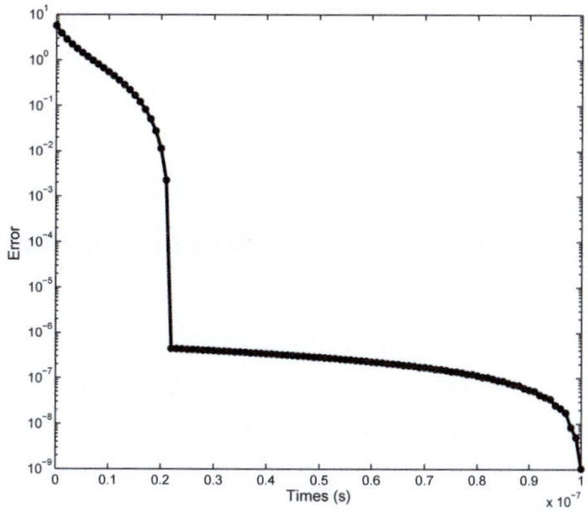

(b) Semilog plot using Li functions.

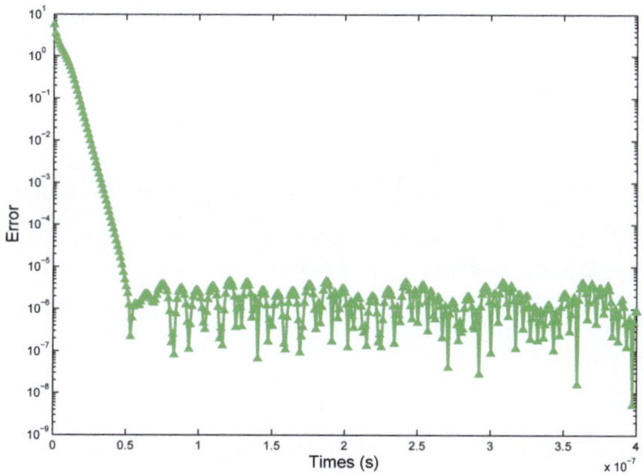

(c) Semilog plot using Smooth power-sigmoid functions.

Fig. 3.49 Trajectories of the residual errors of the model ZNNATS2-I(1) in Example 3.8.1

The Moore-Penrose inverse of A is equal to

$$A(t)^{\dagger} = \begin{bmatrix} i\cos(2t)\cot(2t) & i\cos(2t) & -i\csc(2t) \\ -i\cos(2t) & -2i\cos(t)\sin(t) & 0 \\ i\csc(2t) & 0 & -i\csc(2t) \end{bmatrix},$$

where

$$\csc(t) = 1/\sin(t), \quad \cot(\mathrm{t}) = 1/\tan(t)$$

and the Drazin inverse of A is given by

$$A(t)^{\mathrm{D}} = A(t)\left(A(t)^3\right)^{\dagger} A(t) = \begin{bmatrix} i\cos(2t)\cot(2t) & i\cos(2t) & -i\csc(2t) \\ -i\cos(2t) & -2i\cos(t)\sin(t) & 0 \\ i\csc(2t) & 0 & -i\csc(2t) \end{bmatrix}.$$

3.9 Integration-Enhanced Noise-Tolerant ZNN Model

The time-varying matrix inversion process in [110] was defined using the matrix-valued error function $E(t) = A(t)X(t) - I$ and the following integration-enhanced noise-tolerant evolution design

$$\dot{E}(t) = -\gamma E(t) - \zeta \int_0^t E(\tau)\mathrm{d}\tau + N(t), \tag{3.9.1}$$

where $\gamma > 0$ and $\zeta > 0$ are the design parameters aimed to scale the convergence rate and $N(t)$ is the matrix-form noise. Then the IENTZNN neural network model for solving the time-varying matrix inversion problem (corresponding to $E(t) = A(t)X(t) - I$) can be expanded in the form

$$\begin{aligned} A(t)\dot{X}(t) &= -\dot{A}(t)X(t) - \gamma(A(t)X(t) - I) - \zeta \int_0^t (A(\tau)X(\tau) - I)\mathrm{d}\tau + N(t) \\ &= -\dot{A}(t)X(t) - \gamma E(t) - \zeta \int_0^t E(\tau)\mathrm{d}\tau + N(t), \end{aligned} \tag{3.9.2}$$

or in the equivalent form

$$\begin{aligned} \dot{X}(t) &= (I - A(t))\dot{X}(t) - \dot{A}(t)X(t) - \gamma(A(t)X(t) - I) - \zeta \int_0^t (A(\tau)X(\tau) - I)\mathrm{d}\tau + N(t) \\ &= -\dot{A}(t)X(t) - \gamma E(t) - \zeta \int_0^t E(\tau)\mathrm{d}\tau + N(t). \end{aligned} \tag{3.9.3}$$

The model defined in (3.9.2) or in (3.9.3) is proposed as the IEZNN model in [110] and it is able to suppress various kinds of noises and compute the time-varying matrix inverse simultaneously.

Simulink implementation of (3.9.3) in the case $N(t) = O$ is presented in Fig. 3.50.

Simulink implementation of (3.9.3) is presented in Fig. 3.51.

It is observable that an appropriately selected indefinite ZF determines the domain of results generated by a ZNN model. For this purpose, we intend to define a new error monitoring Zhang function of a more general form and follow the IENTZNN design (3.9.1). According to the survey of previous models, the power of an arbitrary ZNN model lies in the underlying error function $E(t)$.

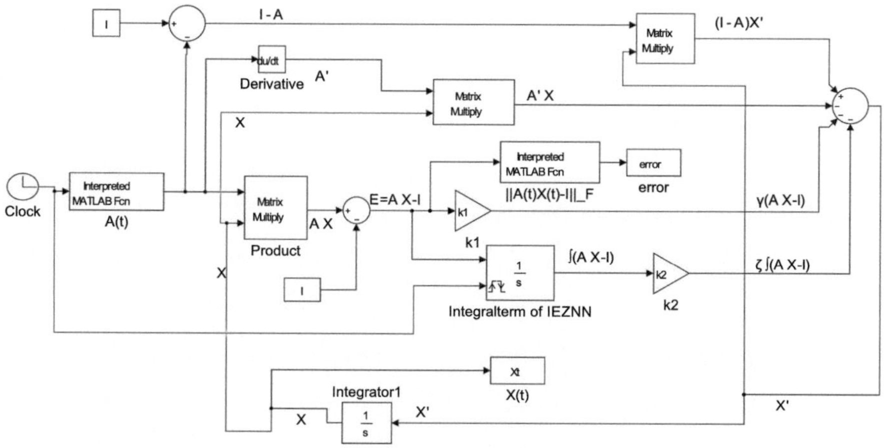

Fig. 3.50 Simulink of the IEZNN model for matrix inversion

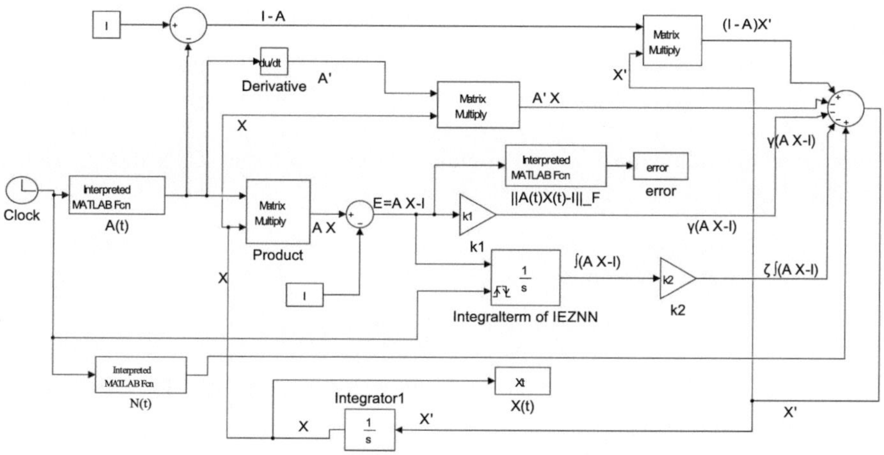

Fig. 3.51 Simulink of the IENTZNN model for matrix inversion

3.9.1 *Definition of the IENTZNN Model Based on General Error Function*

Since it is possible to consider two dual cases, only the case $n \leq m$ is considered in details in the rest of the paper without loss of generality; the case $n > m$ is quite analogous. The ZF proposed in [223] assumes a given matrix $A \in \mathbb{C}^{m\times n}$ and it is defined upon two appropriate matrices $F, G \in \mathbb{C}^{n\times m}$, as follows

$$E_{F,G}(t) = (G(t)A(t) + \lambda I)\, X(t) - F(t),\ \lambda > 0. \tag{3.9.4}$$

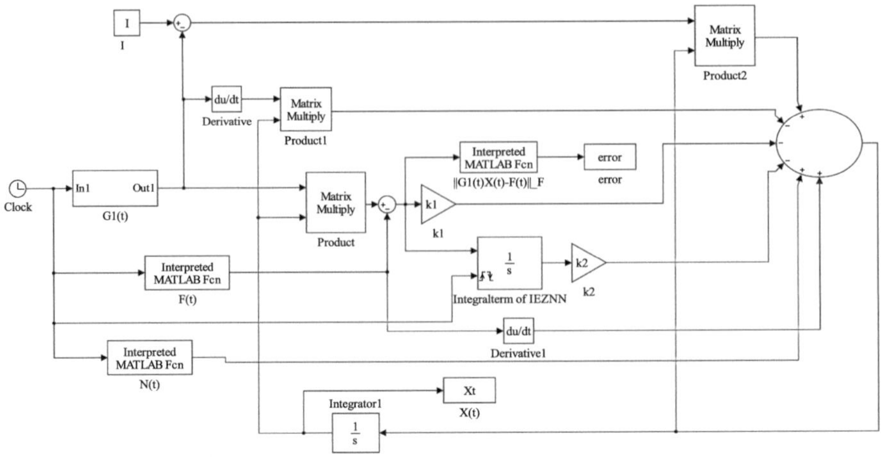

Fig. 3.52 Simulink of (3.9.6) in the case $n \le m$

According to previous discussion, our general strategy is to follow the IENTZNN pattern (3.9.1) wherein the underlying Zhang error function $E(t)$ is replaced by the ZF $E_{F,G}(t)$ defined in (3.9.4). Using the notation $G_1(t) = G(t)A(t) + \lambda I$, we propose the following integration-enhanced noise-tolerant ZNN model, called ZNN(A, F, G):

$$\begin{aligned} \dot{E}_{F,G}(t) &= -\gamma E_{F,G}(t) - \zeta \int_0^t E_{F,G}(\tau)\mathrm{d}\tau + N(t), \\ E_{F,G}(t) &= (G(t)A(t) + \lambda I)X(t) - F(t). \end{aligned} \tag{3.9.5}$$

Applying the evolution (3.9.1), we expand the general IENTZNN(A, F, G) design (3.9.5) as follows:

$$\begin{aligned} G_1(t)\dot{X}(t) &= \dot{F}(t) - \left(\dot{G}(t)A(t) + G(t)\dot{A}(t)\right) X(t) - \gamma \left(G_1(t)X(t) - F(t)\right) \\ &\quad - \zeta \int_0^t \left(G_1(\tau)X(\tau) - F(\tau)\right) \mathrm{d}\tau + N(t) \\ &= \dot{F}(t) - \left(\dot{G}(t)A(t) + G(t)\dot{A}(t)\right) X(t) - \gamma E_{F,G}(t) - \zeta \int_0^t E_{F,G}(\tau)\mathrm{d}\tau + N(t). \end{aligned} \tag{3.9.6}$$

The model IENTZNN(A, F, G) in the case $N(t) = O$, where O is an appropriate null matrix, will be shortly denoted by IEZNN(A, F, G). Further, IENTZNN(A, F, G) in the case $\zeta = 0$, $N(t) = O$ was denoted by ZNN(A, F, G).

Simulink implementation of (3.9.6) is presented in Fig. 3.52.

The block $G_1(t)$ is implemented in the Simulink presented in Fig. 3.53.

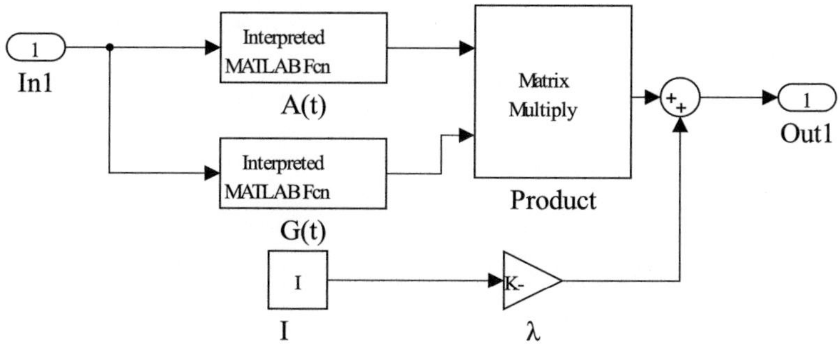

Fig. 3.53 Simulink of the block $G_1(t)$ in the case $n \leq m$

3.9.2 Vectorization of the IENTZNN (A, F, G) Model

For the purposes of MATLAB simulation, we have to transform the matrix-form differential equation (3.9.6) to a vector-form differential equation. The vectorization follows the following properties of the vec-operator: $A \in \mathbb{C}^{m\times n}$, $B \in \mathbb{C}^{n\times p}$, $C \in \mathbb{C}^{p\times q}$:

$$\begin{aligned}\mathrm{vec}(A+B) &= \mathrm{vec}(A) + \mathrm{vec}(B)\\ \mathrm{vec}(ABC) &= \left(C^{\mathrm{T}} \otimes A\right)\mathrm{vec}(B)\\ \mathrm{vec}(AB) &= (I_p \otimes A)\mathrm{vec}(B) = (B^{\mathrm{T}} \otimes I_m)\mathrm{vec}(A).\end{aligned}$$

The left-hand side of (3.9.6) is vectorized as

$$\begin{aligned}\mathrm{vec}\left(G_1(t)\dot{X}(t)\right) &= \mathrm{vec}\left(G_1(t)\dot{X}(t)I_m\right) = (I_m \otimes G_1(t))\,\mathrm{vec}\left(\dot{X}(t)\right)\\ &= (I_m \otimes G(t)A(t) + \lambda I_{mn})\,\dot{x}(t) = M(t)\dot{x}(t),\end{aligned}$$

where $M(t) := I_m \otimes G(t)A(t) + \lambda I_{mn}$ is the mass matrix and $x(t)$ is the integration vector, $x(t) := \mathrm{vec}(X(t)) \in \mathbb{R}^{mn\times 1}$.

The vectorization of the right-hand side of (3.9.6) is given by

$$\begin{aligned}&\mathrm{vec}\left(\dot{F}(t)\right) - \mathrm{vec}\left(\left(\dot{G}(t)A(t) + G(t)\dot{A}(t)\right)X(t)\right) - \gamma \cdot \mathrm{vec}\left(G_1(t)X(t) - F(t)\right)\\ &\quad - \zeta \cdot \mathrm{vec}\left(\int_0^t (G_1(\tau)X(\tau) - F(\tau))\,\mathrm{d}\tau\right) + \mathrm{vec}(N(t))\\ &= \mathrm{vec}\left(\dot{F}(t)\right) - \left(I_m \otimes \dot{G}(t)A(t) + I_m \otimes G(t)\dot{A}(t)\right)x(t) - \gamma\left(I_m \otimes G(t)A(t) + \lambda I_{mn}\right)x(t)\\ &\quad + \gamma \cdot \mathrm{vec}(F(t)) - \zeta \cdot \int_0^t \mathrm{vec}\left(G_1(\tau)X(\tau) - F(\tau)\right)\mathrm{d}\tau + \mathrm{vec}(N(t))\\ &= \mathrm{vec}\left(\dot{F}(t)\right) - \left(I_m \otimes \dot{G}(t)A(t) + I_m \otimes G(t)\dot{A}(t)\right)x(t) - \gamma M(t)x(t) + \gamma \cdot \mathrm{vec}(F(t))\\ &\quad - \zeta \int_0^t \left((I_m \otimes G(\tau)A(\tau) + \lambda I_{mn})\,x(\tau) - \mathrm{vec}(F(\tau))\right)\mathrm{d}\tau + \mathrm{vec}(N(t)).\end{aligned}$$

Therefore, the dynamical equation (3.9.6) takes the following vector form:

$$M(t)\dot{x}(t) = \text{vec}\left(\dot{F}(t)\right) - \left(I_m \otimes \dot{G}(t)A(t) + I_m \otimes G(t)\dot{A}(t)\right)x(t) - \gamma\Big(M(t)x(t) - \text{vec}(F(t))\Big)$$
$$- \zeta \int_0^t (M(\tau)x(\tau) - \text{vec}(F(\tau)))\,\mathrm{d}\tau + \text{vec}(N(t)). \tag{3.9.7}$$

Simulink implementation of (3.9.7) is presented in Fig. 3.54.

3.9.3 Convergence Analysis of the IENTZNN (A, F, G) Model

This section investigates theoretical convergence of the IEZNN(A, F, G) dynamics. In the first Subsection we investigate the convergence of the model without noises, while the general case is investigated in Sect. 3.9.4.

Three different kinds of noise are considered.

(1) **Constant noise** defined by the matrix $N(t) = N \in \mathbb{R}^{n\times m}$, where N is a constant matrix.

(2) **Linear time-varying noise.** In this case, $N(t) = Nt \in \mathbb{R}^{n\times m}$, where $N \in \mathbb{R}^{n\times m}$ is a constant matrix.

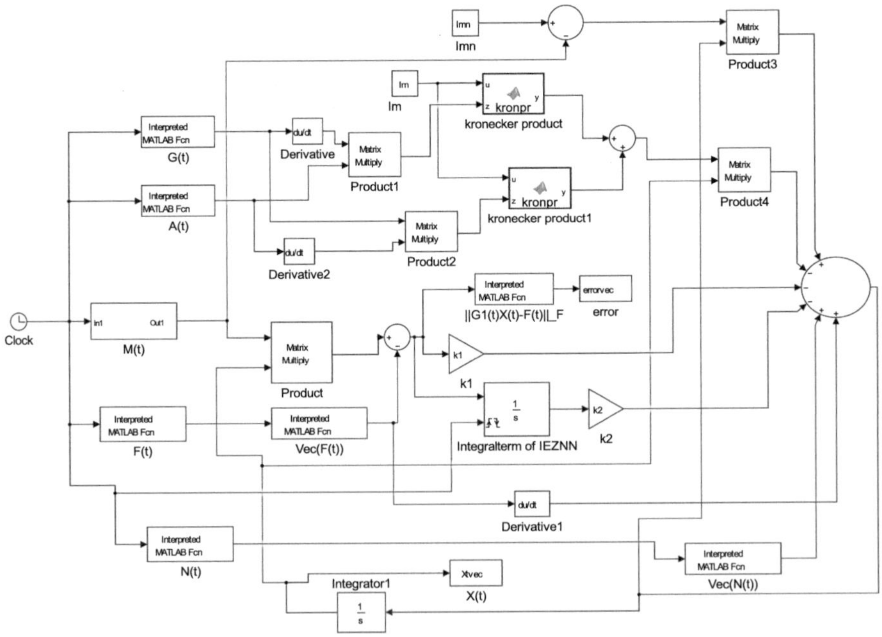

Fig. 3.54 Simulink of the vectorized IENTZNN model

(3) **Bounded random noise**, corresponding to $N(t) = \sigma(t)$, where $\sigma(t) \in \mathbb{R}^{n\times m}$ is a random matrix.

Theorem 3.9.1 ([223]) *Let* $A(t) \in \mathbb{C}_r^{m\times n}$ *be a given matrix of constant rank* r *and* $F(t), G(t) \in \mathbb{C}_s^{n\times m}$ *be appropriate matrices of constant rank satisfying* $0 < s \leq r$. *Assume that* $X(t) := A(t)^{(2)}_{\mathcal{R}(G),\mathcal{N}(G)}$ *exists. The activation state variables matrix* $X(t)$ *of IEZNN*(A, F, G) (3.9.6) *(with* $N(t) = 0$*) is globally and exponentially convergent as* $t \to +\infty$ *and has the equilibrium state*

$$X(t) \to \tilde{X}(t, \lambda) = (G(t)A(t) + \lambda I)^{-1} F(t). \tag{3.9.8}$$

Proof The IEZNN(A, F, G) model (3.9.6) is defined on the IENTZNN design (3.9.5) which follows the same pattern of the general form (3.9.1) which was used in [110, Theorem 1]. So, the verification of the global exponential convergence follows the same principles. If the elements of the error matrix $E_{F,G}(t)$ are denoted by $e_{ij}(t)$, then IEZNN(A, F, G) can be elementwise defined by the system of m^2 differential equations

$$\dot{e}_{ij}(t) = -\gamma e_{ij}(t) - \zeta \int_0^t e_{ij}(\tau)\mathrm{d}\tau, \ i, j = 1, \ldots, m. \tag{3.9.9}$$

A Lyapunov function candidate for each equation in (3.9.9) can be defined as

$$v_{ij}(t) = e_{ij}^2(t) + \zeta \left(\int_0^t e_{ij}(\tau)\mathrm{d}\tau \right), \ i, j = 1, \ldots, m.$$

In the same way as in [110, Theorem 1], it is possible to verify

$$E_{F,G}(t) = (G(t)A(t) + \lambda I)\, X(t) - F(t) \to 0.$$

Since $G(t)A(t) + \lambda I$ invertible, it follows that $X(t)$ globally converges to the matrix

$$X(t) \to \tilde{X}(t, \lambda) = (G(t)A(t) + \lambda I)^{-1} F(t),$$

which confirms (3.9.8). □

Theorem 3.9.2 ([223]) *Let* $A(t) \in \mathbb{C}_r^{m\times n}$ *be a given matrix of constant rank* r *and* $F(t), G(t) \in \mathbb{C}_s^{n\times m}$ *be appropriate matrices of constant rank satisfying* $0 < s \leq r$. *Assume that* $X(t) := A(t)^{(2)}_{\mathcal{R}(G),\mathcal{N}(G)}$ *exists. The activation state variables matrix* $X(t)$ *of IEZNN*(A, F, G) (3.9.6) *(with* $N(t) = 0$*) exponentially converges as* $t \to +\infty$ *and has the equilibrium state* (3.9.8).

Proof The exponential convergence of IEZNN(A, F, G) model (3.9.6) can be verified as in [110, Theorem 2]. More precisely, denote by $\varepsilon(t) = \int_0^t E_G(\tau)\mathrm{d}\tau$. Then IEZNN$(A, F, G)$ design becomes a system of m^2 second order linear homogeneous differential equations with constant coefficients:

$$\ddot{\varepsilon}_{ij}(t) = -\gamma \dot{\varepsilon}_{ij}(t) - \zeta \varepsilon_{ij}(t), \ i, j = 1, \ldots, m.$$

The characteristic polynomial is $\theta^2 + \gamma\theta + \zeta = 0$. For simplicity, let us denote $D = \gamma^2 - 4\zeta$.

Case (1) Under the assumption $\gamma^2 > 4\zeta$, it follows that

$$\varepsilon_{ij}(t) = C_1 \exp(\theta_1 t) + C_2 \exp(\theta_2 t), \quad \theta_1 = \frac{-\gamma + \sqrt{D}}{2}, \ \theta_2 = \frac{-\gamma - \sqrt{D}}{2}.$$

The initial conditions $\varepsilon_{ij}(0) = 0$, $\dot{\varepsilon}_{ij}(0) = e_{ij}(0)$ in conjunction with $\dot{\varepsilon}_{ij}(t) = e_{ij}(t)$ imply

$$E_{F,G}(t) = \frac{E_{F,G}(0)}{\sqrt{\gamma^2 - 4\zeta}} \left(\theta_1 \exp(\theta_1 t) - \theta_2 \exp(\theta_2 t)\right).$$

Case (2) If the equality $\gamma^2 = 4\zeta$ holds, the initial conditions $\varepsilon_{ij}(0) = 0$, $\dot{\varepsilon}_{ij}(0) = e_{ij}(0)$ in conjunction with $\dot{\varepsilon}_{ij}(t) = e_{ij}(t)$ imply

$$\varepsilon_{ij}(t) = e_{ij}(0) t \exp(\theta_1 t).$$

One can verify

$$\varepsilon_{ij}(t) = e_{ij}(0) \exp(\theta_1 t) \left(1 + \theta_1 t\right), \quad \theta_1 = -\gamma/2,$$

and later

$$E_{F,G}(t) = \dot{\varepsilon}(t) = E_{F,G}(0) \exp(\theta_1 t) \left(1 + \theta_1 t\right).$$

Case (3) The assumption $\gamma^2 < 4\zeta$ gives

$$\theta_1 = \frac{-\gamma + \imath\sqrt{-D}}{2}, \ \theta_2 = \frac{-\gamma - \imath\sqrt{-D}}{2},$$

where $\imath$ denotes the imaginary unit. Assuming the notation $\alpha = -\gamma/2, \beta = 1/2\sqrt{-D}$ and the initial conditions $\varepsilon_{ij}(0) = 0$, $\dot{\varepsilon}_{ij}(0) = e_{ij}(0)$, taking in mind $\dot{\varepsilon}_{ij}(t) = e_{ij}(t)$, it follows that

$$\varepsilon_{ij}(t) = \frac{e_{ij}(0)}{\beta} \exp(\alpha t) \sin(\beta t),$$

and later in the matrix form

$$E_{F,G}(t) = \frac{E_{F,G}(0)}{\beta} \exp(\alpha t) \left(t \sin(\beta t) + \beta \cos(\beta t)\right).$$

Hence, the exponential convergence is verified in all three possible cases. □

3.9.4 Convergence of the IENTZNN (A, F, G) Model

Three different kinds of noise are considered.

(1) **Constant noise**. It is defined by the matrix $N(t) = N \in \mathbb{R}^{n\times m}$, where N is a constant matrix.

(2) **Linear time-varying noise**. In this case, $N(t) = Nt \in \mathbb{R}^{n\times m}$, where $N \in \mathbb{R}^{n\times m}$ is a constant matrix.

(3) **Bounded random noise**, corresponding to $N(t) = \sigma(t)$, where $\sigma(t) \in \mathbb{R}^{n\times m}$ is a random matrix.

The residual error $\|E_{F,G}(t)\|_{\mathcal{F}}$ of noise-polluted IENTZNN(A, F, G) model is bounded for the bounded unknown matrix-form random noise $N(t) = \sigma(t) \in \mathbb{R}^{n\times m}$.

Theorem 3.9.3 ([223]) *Assume that the time-variant matrices $A(t)$, $F(t)$, $G(t)$ satisfy the conditions of Theorem 3.9.2. The IENTZNN(A, F, G) model* (3.9.6) *polluted by the constant noise $N(t) = N \in \mathbb{R}^{n\times m}$ is globally convergent as $t \to +\infty$ and has the equilibrium state* (3.9.8), *no matter how large the unknown matrix-form constant noise is.*

Proof The IENTZNN(A, F, G) model (3.9.6) is defined on the IENTZNN design (3.9.5) which follows follows the same pattern of the general form (3.9.1) which was used in [110, Theorem 3]. So, the proof is based on the same principles as the proof of Theorem 3 from [110]. If the elements of the matrix N are denoted by n_{ij}, analogically as in Theorem 3 from [110] one can derive

$$\lim_{t\to\infty} e_{ij}(t) = \lim_{s\to 0} s e_{ij}(s) = \lim_{s\to 0} \frac{s\left(s e_{ij}(0) + n_{ij}\right)}{s^2 + s\gamma + \zeta} = 0. \tag{3.9.10}$$

As a consequence, $\lim_{t\to\infty} \|E_{F,G}(t)\|_{\mathcal{F}} = 0$, which implies (3.9.8). □

Theorem 3.9.4 ([223]) *Assume that the time-variant matrices $A(t)$, $F(t)$, $G(t)$ satisfy the conditions of Theorem 3.9.2. Denote $\gamma^2 - 4\zeta$ shortly by D. The IENTZNN(A, F, G) model* (3.9.6) *with the unknown matrix-form linear time-varying noise $N(t) = Nt \in \mathbb{R}^{n\times m}$ satisfies $\lim_{t\to\infty} \|E_{F,G}(t)\|_F = \|N\|_F/\zeta$. As a consequence, IENTZNN(A, F, G) has the equilibrium state* (3.9.8), *as $\zeta \to +\infty$.*

Proof According to Theorem 4 from [110], one can verify $\lim_{t\to\infty} e_{ij}(t) = \frac{n_{ij}}{\zeta}$, which implies

$$\lim_{t\to\infty} \|E_{F,G}(t)\| = \frac{\|N\|_F}{\zeta}. \tag{3.9.11}$$

Therefore, $\lim_{t\to\infty} \|E_{F,G}(t)\|_F = 0$ is satisfied when $\zeta \to \infty$. □

Theorem 3.9.5 ([223]) *Assume that the time-variant matrices $A(t)$, $F(t)$, $G(t)$ satisfy the conditions of Theorem 3.9.2. The residual error $\|E_{F,G}(t)\|_F$ of the IENTZNN(A, F, G) model* (3.9.6) *is bounded if the matrix-form random noise $N(t) = \sigma(t) = (\sigma_{ij}(t))$ is bounded. In addition, the steady-state residual error $\lim_{t\to\infty} \|E(t)\|_F$ of IENTZNN(A, F, G) is approximately inversely proportional to γ and can be arbitrarily small for large enough γ and suitable ζ in the case $D \neq 0$.*

Proof According to [110, Theorem 5] it follows that

$$\lim_{t\to\infty} \|E_{F,G}(t)\|_F \le \begin{cases} \frac{2n}{\sqrt{D}} \sup\limits_{0\le\tau\le t} |\sigma_{ij}(\tau)|, & D > 0, \\ c\cdot n \sup\limits_{0\le\tau\le t} |\sigma_{ij}(\tau)|, & D = 0, \\ \frac{4\zeta n}{\gamma\sqrt{-D}} \sup\limits_{0\le\tau\le t} |\sigma_{ij}(\tau)|, & D < 0. \end{cases} \tag{3.9.12}$$

The constant c in (3.9.12) is defined as $c = \frac{\mu}{\nu} - \frac{1}{\theta_1}$, where $\theta_1 = -\gamma/2$ and $\nu, \mu > 0$ are appropriate constants. This implies that $\lim\limits_{t\to\infty} \|E_{F,G}(t)\|_F$ can be arbitrarily small in the case $D \neq 0$. □

According to previous results about the convergence, the steady-state residual error satisfies $\lim_{t\to\infty} \|E(t)\|_F = 0$ is satisfied. This fact implies the following result.

Corollary 3.9.1 ([223]) *Let the time-variant matrices $A(t)$, $F(t)$, $G(t)$ satisfy the conditions of Theorem 3.9.2. Then IENTZNN(A, F, G) satisfies*

$$\lim_{\lambda\to 0} \tilde{X}(t,\lambda) = \lim_{\lambda\to 0} (G(t)A(t) + \lambda I)^{-1} F(t). \tag{3.9.13}$$

Now, our goal is to investigate limiting values (3.9.13). The first step towards this goal is the following representations of outer inverses, which can be derived using the results from [156, 289].

Lemma 3.9.1 ([223]) *Let $A \in \mathbb{C}^{m\times n}$ be of rank r, let T be a subspace of $\mathbb{C}^n$ of dimension $s \le r$, and let S be a subspace of $\mathbb{C}^m$ of dimension $m - s$. If $G \in \mathbb{C}^{n\times m}$ satisfies* $\text{ind}(GA) = 1$ *then $A^{(2)}_{\mathcal{R}(G),\mathcal{N}(G)}$ exists and*

(a) *$A^{(2)}_{T,S}$ possesses representations*

$$A^{(2)}_{\mathcal{R}(G),\mathcal{N}(G)} = \lim_{\lambda\to 0} (GA + \lambda I)^{-1} G = \lim_{\lambda\to 0} G(AG + \lambda I)^{-1} = (GA)^{\#} G = G(AG)^{\#}. \tag{3.9.14}$$

(b) *If $G = UV$ is a full-rank factorization of G, then*

$$A^{(2)}_{\mathcal{R}(G),\mathcal{N}(G)} = A^{(2)}_{\mathcal{R}(U),\mathcal{N}(V)} = U \lim_{\lambda\to 0} (VAU + \lambda I)^{-1} V. \tag{3.9.15}$$

3.9.5 Particular Cases of the IENTZNN (A ,F, G) Model

Let the conditions of Theorem 3.9.2 are satisfied. In this subsection we investigate the limit $\lim_{\lambda\to 0} \tilde{X}(t,\lambda)$. This investigation enables to observe the diversity of generalised inverses and expressions which can be generated by IENTZNN(A, F, G). Then the following statements are valid.
(a) In the case $m = n$, $F(t) = G(t) = I_n$,

$$\lim_{\lambda \to 0} \tilde{X}(t, \lambda) = A(t)^{-1}.$$

Clearly, IENTZNN(A, I, I) in the particular case $\lambda = 0$, becomes the IEZNN model (3.9.2) from [110].

(b) Consider the case $F(t) = G(t) \in \mathbb{C}_s^{n\times m}$, $0 < s \leq r$, $\text{ind}(G(t)A(t)) = 1$. The limit from Lemma 3.9.1 implies

$$\lim_{\lambda \to 0} \tilde{X}(t, \lambda) = A(t)^{(2)}_{\mathcal{R}(G),\mathcal{N}(G)}. \tag{3.9.16}$$

The ZNN(A, G, G) gives the ZNNATS2-I model from [284].

(c) The IENTZNN(A, A^*, A^*) model becomes the integration-enhanced noise-tolerant extension to the model ZF(5) from [126], corresponding to the error monitoring function given in (3.3.5) Consequently, ZNN(A, A^*, A^*) coincides with ZF(5).

(d) The IENTZNN(A, A^l, A^l), $l \geq \text{ind}(A)$ under the assumption $m = n$ in IEZNNATS2 represents an integration-enhanced noise-tolerant extension of the linear model LZNN-II from [283], corresponding to the error monitoring function given in (3.8.8). If the Li activation is used, IENTZNN(A, A^l, A^l) represents an integration-enhanced noise-tolerant extension of the model ZNN-II from [203].

(e) The particular choice $F(t) = G(t) = A(t)^l \left(A(t)^{2l+1}\right)^* A(t)^l$, $l \geq \text{ind}(A)$ in IENTZNN(A, F, G) in the case $m = n$ gives an integration-enhanced noise-tolerant extension of to the linear model LZNN-I from [283].

(f) If A is a square matrix, $F(t) = G(t) = A(t)^l \left(A(t)^{2l+1}\right)^* A(t)^l$, $l \geq \text{ind}(A)$, $\lambda = 0$, IENTZNN(A, F, G) reduces to the is integration-enhanced noise-tolerant extension of the linear model ZNN-I from [203].

(g) Consider the case of a square matrix $A(t)$ and $F(t) = A^l$, $G(t) = A(t)^{2l-1}$, $l \geq \text{ind}(A(t))$.

Corollary 3.9.2 ([223]) *If A is a square matrix and $l \geq \text{ind}(A(t))$, the steady state matrix of IENTZNN(A, A^l, A^{2l-1}) satisfies*

$$\lim_{\lambda \to 0} \tilde{X}(t, \lambda) = \left(A(t)^D\right)^l.$$

(h) Also, our particular interest is the model IENTZNN$(A, G, \lambda I)$. The convergence of that model can be determined by the following result.

Lemma 3.9.2 ([223]) *Let $A \in \mathbb{C}_r^{m\times n}$ be a given matrix of constant rank r and $G \in \mathbb{C}_s^{n\times m}$ be an arbitrary matrix of constant rank satisfying $0 < s \leq r$. Assume that $X(t) := A(t)^{(2)}_{\mathcal{R}(G),\mathcal{N}(G)}$ exists, i.e., $\text{ind}(GA) = 1$. Then $\lim_{\lambda\to 0} (GA + \lambda I)^{-1} \lambda$ exists and*

$$\begin{aligned} \lim_{\lambda \to 0} (GA + \lambda I)^{-1} \lambda &= I - A^{(2)}_{T,S} A \\ &= I - (GA)^{\#} GA. \end{aligned} \tag{3.9.17}$$

Corollary 3.9.3 ([223]) *Let $A(t) \in \mathbb{C}_r^{m\times n}$ be a given matrix of constant rank r and $G(t) \in \mathbb{C}_s^{n\times m}$ be an arbitrary matrix of constant rank satisfying $0 < s \leq r$. Assume* *that* $X(t) := A(t)^{(2)}_{\mathcal{R}(G),\mathcal{N}(G)}$ *exists, i.e.,* $\text{ind}(GA) = 1$. *The activation state variables matrix* $X(t)$ *of* $(A, \lambda I_n, G)$ *globally and exponentially converges to the theoretical solution of the theoretical time-varying problem*

$$(G(t)A(t) + \lambda I)\, X(t) = \lambda I,$$

the equilibrium state is equal to

$$X(t) \to \tilde{X}_\lambda(t, \lambda) = (G(t)A(t) + \lambda I)^{-1}\, \lambda. \tag{3.9.18}$$

In addition,

$$\lim_{\lambda \to 0} \tilde{X}_\lambda(t, \lambda) = I_n - A(t)^{(2)}_{\mathcal{R}(G),\mathcal{N}(G)} A(t). \tag{3.9.19}$$

(i) The convergence of IENTZNN(A, A^{l-1}, A^l), $l \geq \text{ind}(A)$ can be determined by the following result.

Lemma 3.9.3 ([223]) *Let* $A(t) \in \mathbb{C}^{n\times n}$ *be of index* l. *Then the IENTZNN* (A, A^{l-1}, A^l), $l \geq \text{ind}(A)$ *model satisfies*

$$\lim_{\lambda \to 0} \tilde{X}(t, \lambda) = \lim_{\lambda \to 0} \left(A(t)^l + \lambda I\right)^{-1} A(t)^l = A(t)A(t)^D. \tag{3.9.20}$$

Proof According to [5, Theorem 1], it follows that

$$\begin{aligned} \lim_{\lambda \to 0} \left(A(t)^l + \lambda I\right)^{-1} A(t)^l &= A(t)^l \left(A(t)^l\right)^{\#} = A(t)^l \left(A(t)^{\mathrm{D}}\right)^l \\ &= \left(A(t)A(t)^{\mathrm{D}}\right)^l. \end{aligned}$$

Since $A(t)A(t)^{\mathrm{D}}$ is idempotent, the proof is complete. □

(j) The convergence of IENTZNN$(A, A^{l-1}, \lambda I)$, $l \geq \text{ind}(A)$ can be determined by the following result.

Corollary 3.9.4 *If* $A(t) \in \mathbb{C}^{n\times n}$ *is of index* l *then IENTZNN*(A, A^{l-1}, A^l), $l \geq \text{ind}(A)$ *satisfies*

$$\lim_{\lambda \to 0} \left(A(t)^l + \lambda I\right)^{-1} \lambda = I_n - A(t)A(t)^D. \tag{3.9.21}$$

(k) The choice $G(t) = I_m$, $F(t) = X(t)^\dagger$ reduces $E_{F,G}(t)$ into $E(t) = A(t) - X(t)^\dagger$ termed as EF(8) in [126].

As a conclusion, the problems investigated in the present paper can be restated from Table 2 in [106] in the form of Table 3.4. The matrix $A(t)^l \left(A(t)^{2l+1}\right)^* A(t)^l$ is simply denoted by $\tilde{A}^l(t)$.

According to Corollary 3.9.3 it is possible to propose some additional cases of IENTZNN(A, F, G) in time-varying case, as in Table 3.5.

By exploiting noise-tolerant ZNN design formula to solve different time-varying problems, various noise-tolerant ZNN models that exploit the time-derivative information of coefficients as well as the error integration can be constructed with their formulations shown in Table 3.6.

3.9.6 Simulation Results for IEZNN Models

Simulation Results for the IEZNN (A, F, G) Model

In this section, we present computer-simulation examples to show the performance of IENTZNN(A, F, G) model for computing the $A^{(2)}_{\mathcal{R}(G),\mathcal{N}(G)}$ according to (3.9.6). The *Matlab* solver `ode15s` is used in performed numerical simulations.

Example 3.9.1 Let us consider the following time-varying matrix $A(t)$:

$$A(t) = \begin{bmatrix} \sin(t) & \cos(t) \\ -\cos(t) & \sin(t) \end{bmatrix}.$$

Clearly, $A(t)$ is invertible with the theoretical inverse

Table 3.4 Important cases of IENTZNN(A, F, G) in time-varying case

$E_{F,G}(t)$	ζ	$N(t)$	Refs.	$\lim X(t)$
$A(t)X(t) - I$	$\neq 0$	$\neq O$	[110]	$A(t)^{-1}$
$(G(t)A(t) + \lambda I)\,X(t) - G(t)$	0	O	[284]	$A(t)^{(2)}_{\mathcal{R}(G),\mathcal{N}(G)}$
$(A(t)^*A(t) + \lambda I)\,X(t) - A(t)^*$	0	O	[126]	$A(t)^{\dagger}$
$\left(A(t)^{l+1} + \lambda I\right) X(t) - A(t)^l$	0	O	[203, 283]	$A(t)^{\mathrm{D}}$
$\left(\tilde{A}^l(t)A(t) + \lambda I\right) X(t) - \tilde{A}^l(t)$	0	O	[283]	$A(t)^{\mathrm{D}}$
$\tilde{A}^l(t)A(t)X(t) - \tilde{A}^l(t)$	0	O	[203]	$A(t)^{\mathrm{D}}$
$\left(A(t)^{2l} + \lambda I\right) X(t) - A(t)^l$	$\neq 0$	$\neq O$	[223]	$\left(A(t)^{\mathrm{D}}\right)^l$
$(G(t)A(t) + \lambda I)X(t) - \lambda I$	$\neq 0$	$\neq O$	[223]	$I_n - A(t)^{(2)}_{\mathcal{R}(G),\mathcal{N}(G)}A(t)$
$(A(t)^l + \lambda I)X(t) - A(t)^l$	$\neq 0$	$\neq O$	[223]	$A(t)A(t)^{\mathrm{D}}$
$(A(t)^l + \lambda I)X(t) - \lambda I$	$\neq 0$	$\neq O$	[223]	$I_n - A(t)A(t)^{\mathrm{D}}$
$A(t) - X(t)^{\dagger}$	$= 0$	$= O$	[126]	$A(t)^{\dagger}$

Table 3.5 Some additional cases of IENTZNN(A, F, G) in time-varying case

$E_{F,G}(t)$	ζ	$N(t)$	Ref.	$\lim X(t)$
$(A(t)^*A + \lambda I)\,X(t) - \lambda I$	$\neq 0$	$\neq O$	[223]	$I_n - A(t)^{\dagger}A(t)$
$\left(A(t)^{l+1} + \lambda I\right) X(t) - \lambda I$	$\neq 0$	$\neq O$	[223]	$I_n - A(t)^{\mathrm{D}}A(t)$

Table 3.6 Continuous-time noise-tolerant ZNN models constructed for solving time-varying problems

Dynamic problem	Noise-tolerant ZNN model
4th root finding $x^4(t) = \alpha(t)$	$\dot{x}(t) = \frac{\dot{\alpha}(t)-\gamma(x^4(t)-\alpha(t))}{4x^3(t)} - \lambda \int_0^t (x^4(\delta) - \alpha(\delta))d\delta$
Linear system $A(t)x(t) = b(t)$	$A(t)\dot{x}(t) = -\dot{A}(t)x(t) + \dot{b}(t) - \gamma(A(t)x(t) - b(t)) - \lambda \int_0^t (A(\delta)x(\delta) - b(\delta))d\delta$
Matrix inversion $A(t)X(t) = I$	$A(t)\dot{X}(t) = -\dot{A}(t)X(t) - \gamma(A(t)X(t) - I) - \lambda \int_0^t (A(\delta)X(\delta) - I)d\delta$
Matrix square root $X^2(t) = A(t)$	$X(t)\dot{X}(t) + \dot{X}(t)X(t) = -\gamma(X^2(t) - A(t)) - \dot{A}(t) - \lambda \int_0^t (X^2(\delta) - A(\delta))d\delta$
Nonlinear equations $f(x(t), t) = 0$	$\dot{x}(t) = -J^{-1}(x(t), t)\,(\gamma(f(x(t), t)) + \frac{\partial f(x(t),t)}{\partial t} - \lambda \int_0^t (f(x(\delta), \delta))d\delta)$

$$A^{-1}(t) = \begin{bmatrix} \sin(t) & -\cos(t) \\ \cos(t) & \sin(t) \end{bmatrix},$$

which is used to check the correctness of the neural -network solutions.

(a) We firstly investigate the behaviour of the parameter ζ i.e., the contribution of the integral part, in the IEZNN(A, I, I) model (3.9.7). The acceleration parameter is equal to $\gamma = 10^8$, the regularization is defined as $\lambda = 10^{-3}$, the time interval is $[0, 20]$ and the initial approximation is $x(0) = zeros(4, 1)$. The elementwise trajectories of the exact inverse are represented by the red graphs, while the elements of the approximation are presented by blue graphs (see Fig. 3.1a–c).

As illustrated in Fig. 3.55a and b, the IEZNN(A, I, I) converges to the theoretical solution in the cases $\zeta = 10$ and $\zeta = 10^4$. Figure 3.55c and d show that, for $\zeta \geq 10^6$, the IEZNN(A, I, I) slows down the convergence and finally diverges.

(b) The computational error $\|X(t) - A^{-1}(t)\|_F$ could also be used to monitor the neural network convergence. For presentation purposes we consider $t \in [0, 10^{-6}]$. It can be seen from Fig. 3.55 that the neural-network convergence can be expedited effectively by increasing the value of the design parameter γ. In this case, $\lambda = 10^{-3}$ and $\zeta = 10$ (Fig. 3.56).

(c) The computational error $\|X(t) - A^{-1}(t)\|_F$ for $\gamma = 10^8$ and different values of ζ are compared in Fig. 3.58 within the time interval is $t \in [0, 100]$. The graph in Fig. 3.57 is generated using $\lambda = 10^{-3}$. In addition, we present Fig. 3.58 which are the corresponding graphs to Fig. 3.57 in the case $\lambda = 10^{-6}$.

General comment is that the model IEZNN(A, I, I) achieves the convergence in time intervals $[0, t_c] \subset [0, 100]$ which are appropriate to the value of the gain parameter γ. Applying the IEZNNATS2 model after the time t_c greater values of ζ can cause numerical instability.

Comments about the influence of the parameter λ are as follows.

Fig. 3.55 Elementwise trajectories of IEZNN(A, I, I) for $\zeta = 10,\ 10^4,\ 10^6,\ 10^8$ in Example 3.9.1

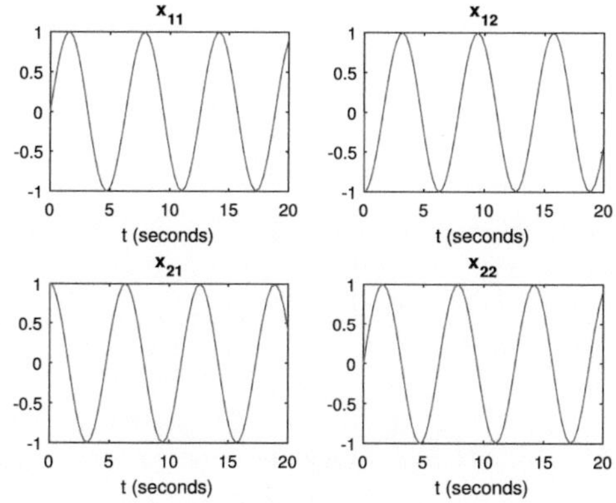

(a) Elementwise trajectories of IEZNN(A, I, I) for $\gamma = 10^8$, $\zeta = 10$.

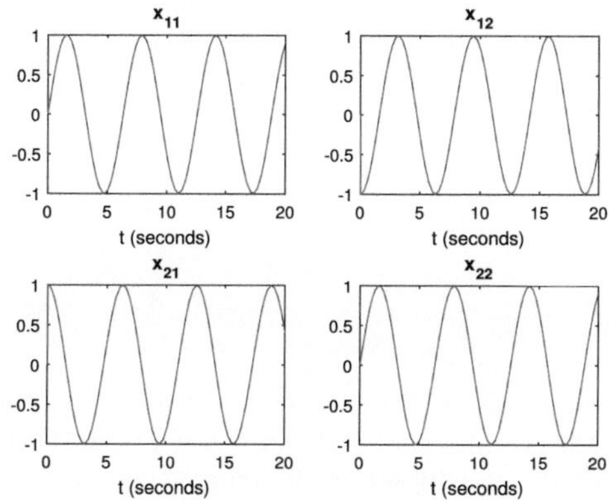

(b) Elementwise trajectories of IEZNN(A, I, I) for $\gamma = 10^8$, $\zeta = 10^4$.

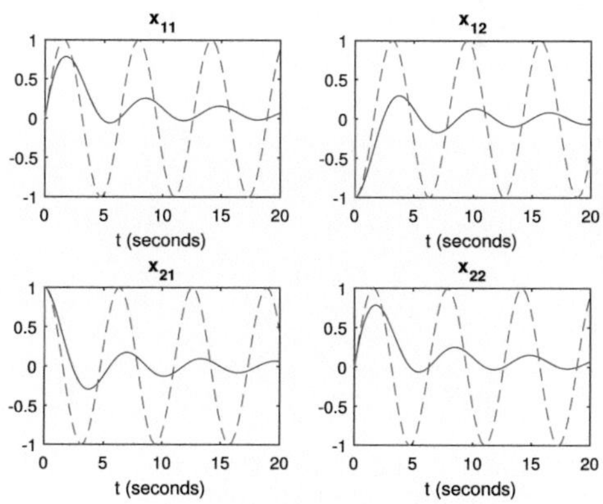

(c) Elementwise trajectories of IEZNN(A, I, I) for $\gamma = 10^8$, $\zeta = 10^8$.

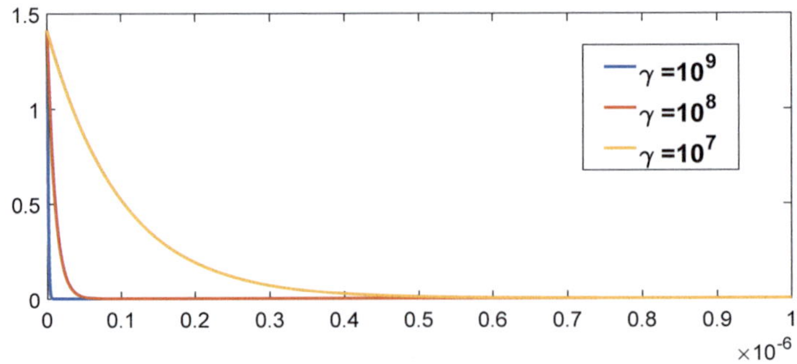

Fig. 3.56 IEZNN(A, I, I) solution errors $\|X(t) - A^{-1}(t)\|_F$ in Example 3.9.1

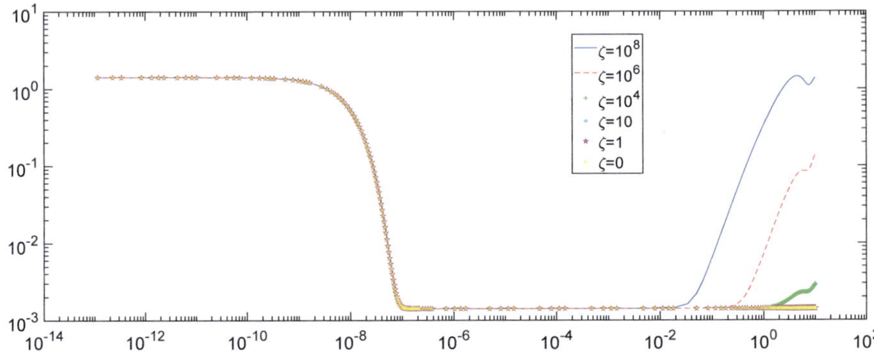

Fig. 3.57 Log-log plot of IEZNN(A, I, I) errors $\|X(t) - A^{-1}(t)\|_F$ for $\lambda = 10^{-3}$ and $\zeta = 0, 1, 10, 10^4, 10^6, 10^8$ and $t \in [0, 100]$

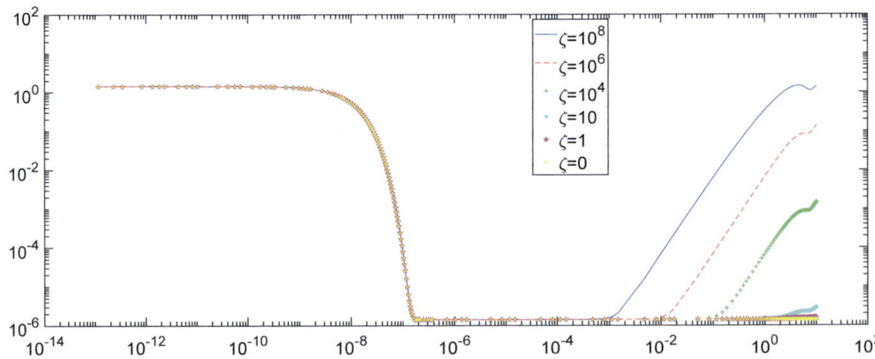

Fig. 3.58 Log-log plot of IEZNN(A, I, I) errors $\|X(t) - A^{-1}(t)\|_F$ for $\lambda = 10^{-6}$ and $\zeta = 0, 1, 10, 10^4, 10^6, 10^8$ and $t \in [0, 100]$

1. Value $\lambda = 10^{-3}$ enables the convergence with the precision 10^{-3} while $\lambda = 10^{-6}$ enables the convergence with the precision 10^{-6}. This means that the IEZNN(A, I, I) model achieves precision equal to λ.

2. But, the IEZNN(A, I, I) model in the case $\lambda = 10^{-3}$ converges in the time interval $[0, 10^{-2}]$. In the case $\lambda = 10^{-6}$, the dynamics IEZNN(A, I, I) converges in the shorter time interval $[0, 10^{-3}]$.

About the influence of the parameter ζ we have several observations.

1. It is clear that the value $\zeta = 10^8$ is not appropriate and leads to divergence.

2. The behavior of the trajectory corresponding to $\zeta = 10^8$ is quite good in the shorter period of time $[0, 10^{-2}]$. For this purpose, the divergence in this case could be caused by rounding errors, i.e., by numerical instability of the method, i.e. the propagation of the error through the time can cause divergence of the IEZNNATS2. This situation is analogous to the propagation of the error through later iterations, which sometimes can cause divergence of an iterative process [194, 195, 219].

(d) In this part of the example we investigate the influence of small values of the regularization parameter λ on the convergence of the IEZNN(A, I, I) method.

According to graphs included in Fig. 3.59, it is possible to conclude that smaller values λ produce better convergence behaviour. According to the graph in part (a) of Fig. 3.59, IEZNN(A,I,I) achieves the convergence approximately with the precision 10^{-6}. The graph in part (b) indicates the precision 10^{-9}. Further reduction of λ has no relevant influence on the convergence. Accordingly, IEZNN(A, I, I) achieves precision equal to λ until the approximate value $\lambda = 10^{-9}$.

(e) with $\lambda = 10^{-3}$, $G = F = I_2$, the zero initial condition, without the presence of noise and different values γ and ζ, following graphical results are extracted.

The elementwise trajectories of the exact inverse are represented by the blue lines, while the elements from the IENTZNN formula (3.9.6) are presented by the red dashed lines in Fig. 3.60.

The computational error $\ln(||A(t)X(t) - I||_F)$ is presented for different values of the parameters γ and ζ. The IENTZNN model possesses a faster convergence ability for greater gain values γ. For the noise (Fig. 3.61)

$$N(t) = \begin{bmatrix} 10t & 10t \\ 10t & 10t \end{bmatrix}.$$

The following graphical results are extracted.

From Figs. 3.62 and 3.63, it is observable that the IENTZNN works in the presence of $N(t)$. In addition, greater values of the parameter γ give better convergence property of IENTZNN.

Example 3.9.2 In this example we consider the time-varying coefficient matrix $A(t)$

$$A(t) = \begin{bmatrix} -1 - 0.5\cos(2t) & 0.5\sin(2t) \\ 0.5\sin(2t) & -1 + 0.5\cos(2t) \end{bmatrix}$$

with the theoretical inverse equal to

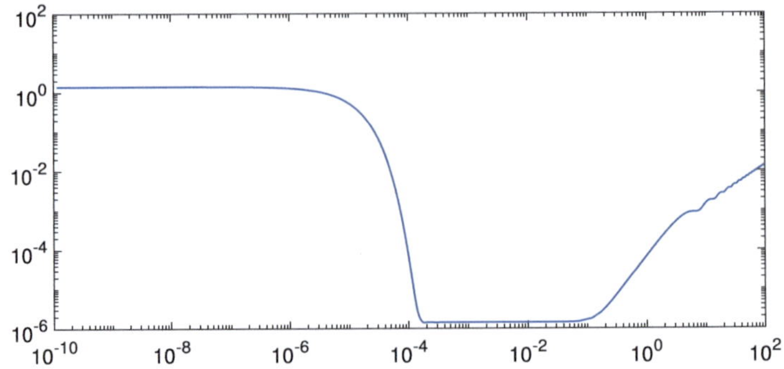

(a) Log-log plot of IEZNN(A, I, I) errors $\|X(t) - A^{-1}(t)\|_F$ for $\zeta = 10$, $\gamma = 10^5$ and $t \in [0, 1]$, $\lambda = 10^{-6}$.

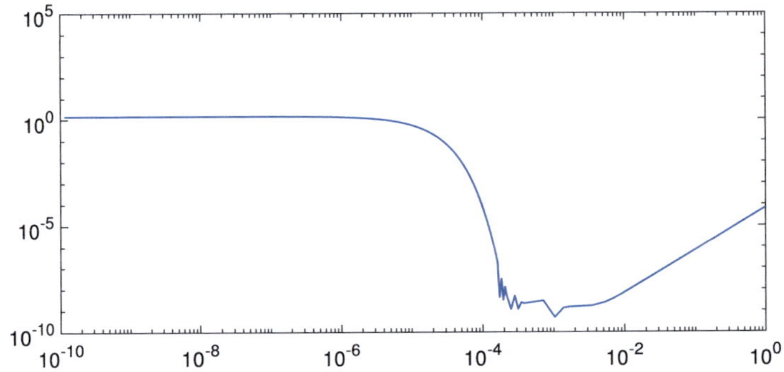

(b) Log-log plot of IEZNN(A, I, I) errors $\|X(t) - A^{-1}(t)\|_F$ for $\zeta = 10$, $\gamma = 10^5$ and $t \in [0, 1]$, $\lambda = 10^{-9}$.

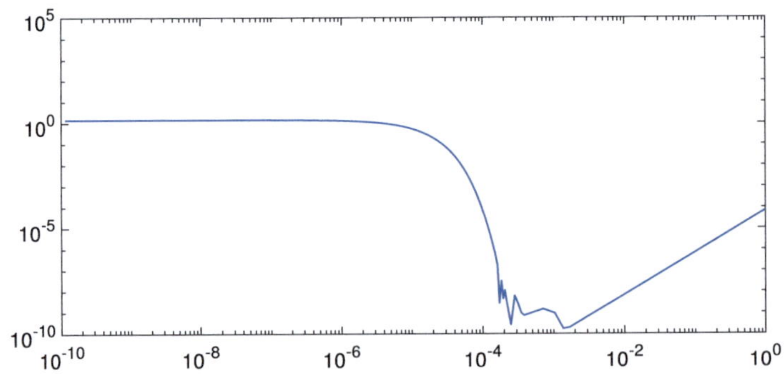

(c) Log-log plot of IEZNN(A, I, I) errors $\|X(t) - A^{-1}(t)\|_F$ for $\zeta = 10$, $\gamma = 10^5$ and $t \in [0, 1]$, $\lambda = 10^{-12}$.

Fig. 3.59 Log-log plot of IEZNN(A, I, I) errors for $\lambda = 10^{-6},\ 10^{-9},\ 10^{-12}$

$$A^{-1}(t) = \frac{4}{3}\begin{bmatrix} -1+0.5\cos(2t) & -0.5\sin(2t) \\ -0.5\sin(2t) & -1-0.5\cos(2t) \end{bmatrix},$$

which is used to check the correctness of the neural-network solutions.

(a) In this part of the example we investigate the first values of the parameter λ which cause the convergence of IEZNN(A, I, I). The parameters are $\gamma = 10^8$, $\zeta = 10$ and the initial approximation is $x(0) = 4 * (rand(4, 1) - 0.5 * ones(4, 1))$. The elementwise trajectories of the exact inverse are represented by the red graphs, while the elements of the approximation are presented by blue graphs. As it is observable from graphs included in Fig. 3.64, better results are generated for smaller values λ. Also, similar to the previous example, analysis and results can be presented for the error $\|X(t) - A^{-1}(t)\|_F$ in order to monitor the neural-network convergence (Fig. 3.65).

(b) Now, our intention is to investigate the influence of the gain parameter γ in this example. Values $\lambda = 10^{-3}$, $\zeta = 10$ are fixed and γ runs through the values $10^7, 10^8, 10^9$.

Similarly as in Example 3.9.2, part (b), the convergence of IEZNN(A, I, I) can be accelerated by increasing the value of the design parameter γ.

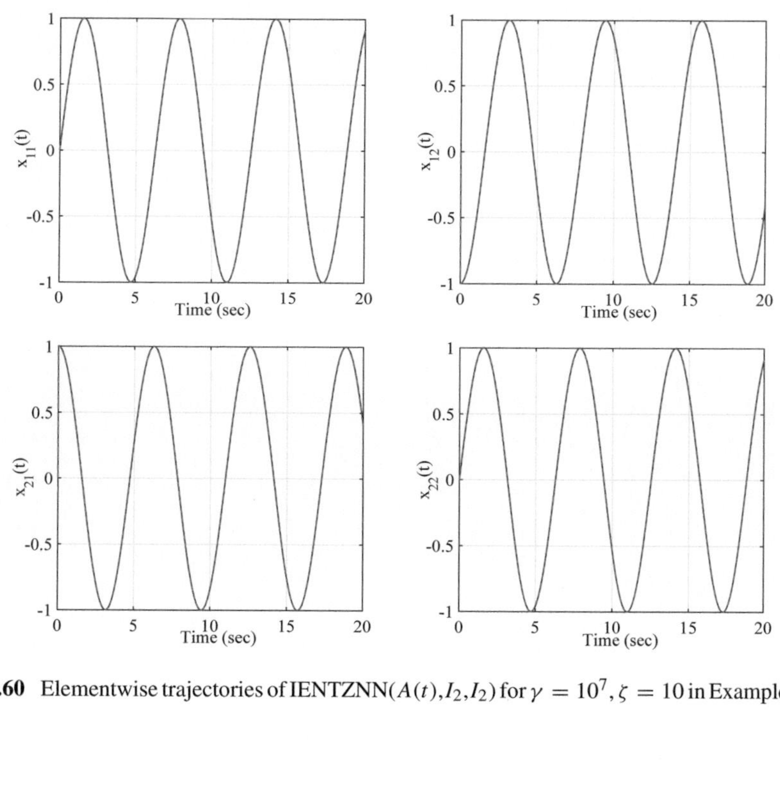

Fig. 3.60 Elementwise trajectories of IENTZNN$(A(t), I_2, I_2)$ for $\gamma = 10^7, \zeta = 10$ in Example 3.9.1

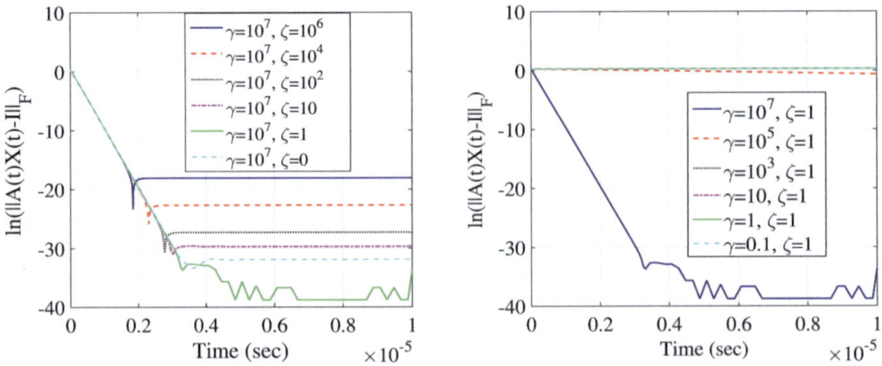

Fig. 3.61 $\ln(||A(t)X(t) - I||_F)$ behaviour from IENTZNN when **a** $\gamma = 10^7$ and $\zeta = 10^6, 10^4, 10^2, 10, 1, 0$, **b** $\zeta = 1$ and $\gamma = 10^7, 10^5, 10^3, 10, 1, 0.1$ in Example 3.9.1

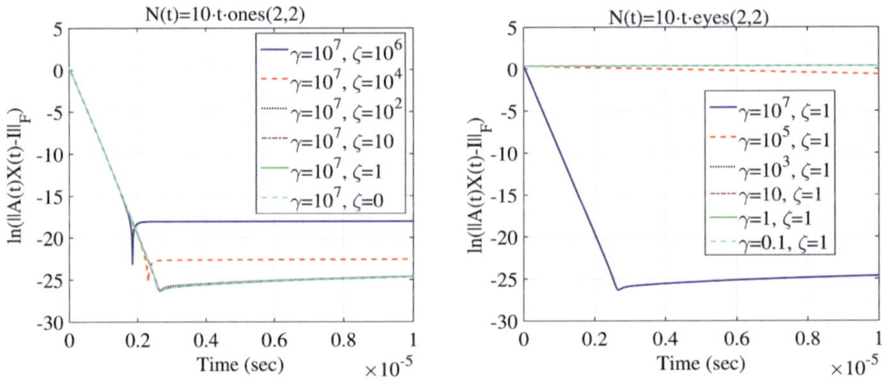

Fig. 3.62 $\ln(||A(t)X(t) - I||_F)$ behaviour from IENTZNN when **a** $\gamma = 10^7$ and $\zeta = 10^6, 10^4, 10^2, 10, 1, 0$, **b** $\zeta = 1$ and $\gamma = 10^7, 10^5, 10^3, 10, 1, 0.1$ in Example 3.9.1 under $N(t) = 10 * t * ones(2, 2)$

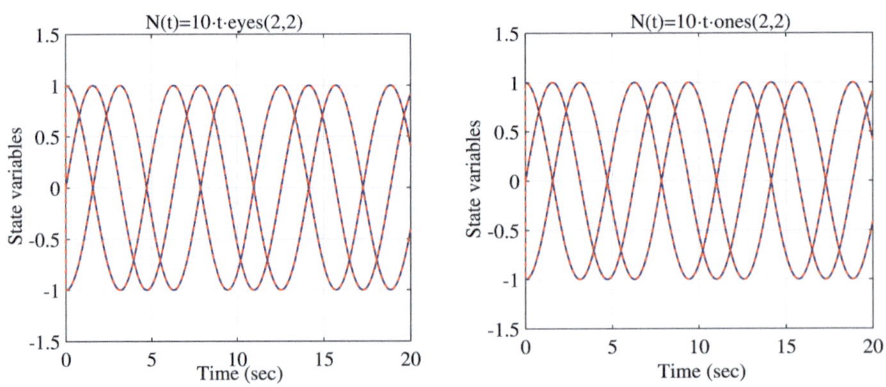

Fig. 3.63 Elementwise trajectories of IENTZNN($A(t)$,I_2,I_2) model under $N(t) = 10 * t * ones(2, 2)$: **a** for $\gamma = 10^7$, $\zeta = 10$, **b** for $\gamma = 10^7$, $\zeta = 10^6$ in Example 3.9.1

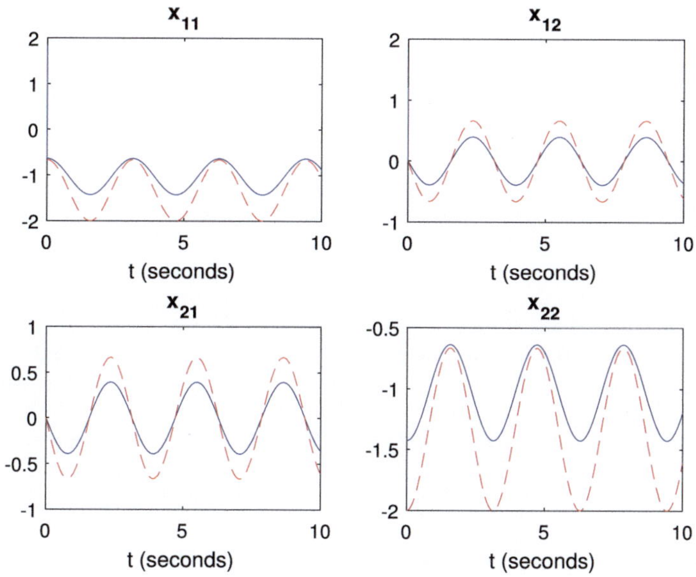

(a) IEZNN(A, I, I) trajectories for $\lambda = 10^{-1}$.

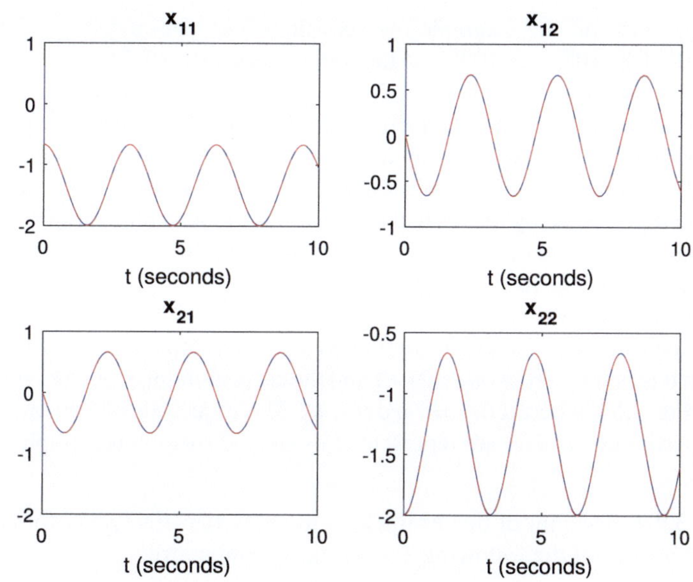

(b) IEZNN(A, I, I) trajectories for $\lambda = 10^{-3}$.

Fig. 3.64 IEZNN(A, I, I) trajectories for $\lambda = 10^{-1}$, 10^{-3} in Example 3.9.2

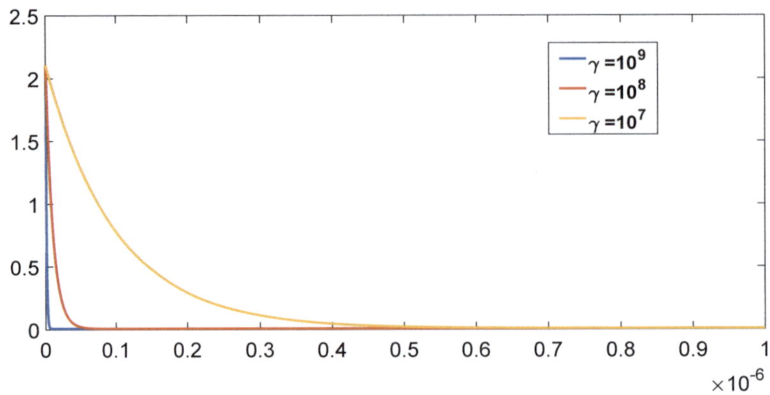

Fig. 3.65 Errors $\|X(t) - A^{-1}(t)\|_F$ in IEZNN(A, I, I) for the case $\lambda = 10^{-3}$, $\zeta = 10$ in Example 3.9.2

Illustrative Examples for the IENTZNN (A, F, G) Model

Example 3.9.3 Let us consider the time-varying coefficient matrix $A(t)$ from Example 3.9.1.
(a) In this part of the example we consider the parameters $\gamma = 10^8$, $\zeta = 0, 1, 10, 10^4, 10^6, 10^8$, $\lambda = 10^{-6}$ and the tome interval $[0, 10^{-6}]$.

(1) **Constant Noise**. In this case, $N(t) = N \in \mathbb{R}^{n\times m}$ is a constant matrix. For illustration as well as for comparison, each element of the matrix-form constant noise is set to be 10 and the corresponding simulation results are shown in Fig. 3.66, where $x(0) = zeros(4, 1)$. The trajectories of the exact inverse (resp. approximate inverse) are represented by the dashed red (resp. continuous blue) graphs. It is worth mentioning that the approximations given by IENTZNN(A, I, I) are the same for all values $\zeta = 0, 1, 10, 10^4, 10^6, 10^8$.

(2) **Linear Time-Varying Noise**. In this case, $N(t) = N\,t \in \mathbb{R}^{n\times m}$, where $N \in \mathbb{R}^{n\times m}$ is a constant matrix. As in the previous case, each element of the matrix N is set to be 10, that is $N(t) = 10 * ones(2) * t$ and the corresponding simulation results are shown in Fig. 3.67, where $x(0) = zeros(4, 1)$. The trajectories of the exact inverse (resp. approximate inverse) are represented by the red (resp. blue) graphs.

Example 3.9.4 The aim of this example is to apply the IENTZNN for computing the Drazin inverse of the following 3×3 time-variant matrix

$$A(t) = \begin{bmatrix} t & 1 & 0 \\ 0 & 0 & 1 \\ -t^3 & -t^2 & -t \end{bmatrix},$$

with index $\mathrm{ind}(A) = 3$ and theoretical Drazin inverse

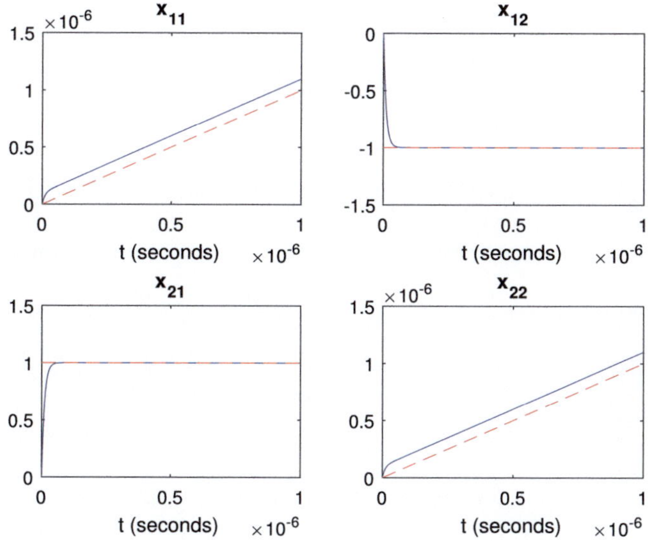

(a) Elementwise trajectories of IENTZNN(A, I, I) for $\gamma = 10^8$, $\zeta = 0, 1, 10, 10^4, 10^6, 10^8$, $\lambda = 10^{-6}$.

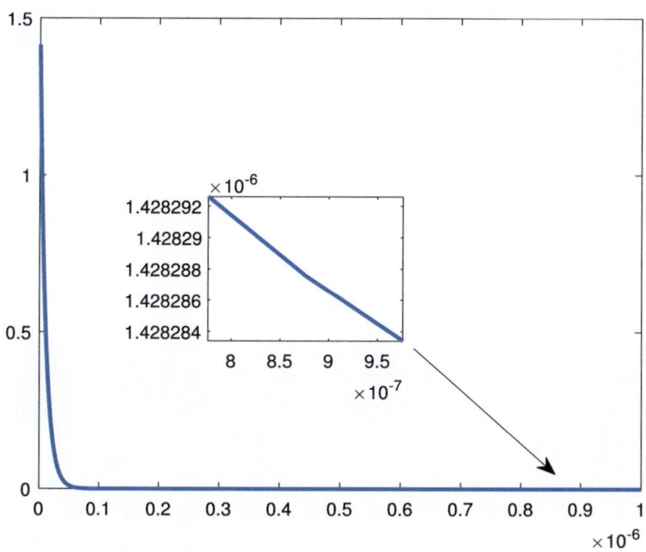

(b) IENTZNN(A, I, I) errors $\|X(t) - A^{-1}(t)\|_F$ for $\gamma = 10^8$, $\zeta = 0, 1, 10, 10^4, 10^6, 10^8$, $\lambda = 10^{-6}$.

Fig. 3.66 The case of constant noise in Example 3.9.3

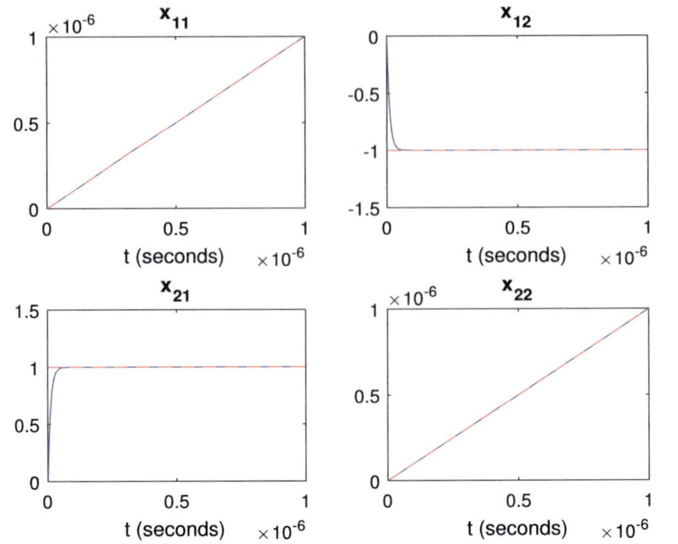

(a) Elementwise trajectories of IENTZNN(A, I, I) for $\gamma = 10^8$, $\zeta = 0, 1, 10, 10^4, 10^6, 10^8, \lambda = 10^{-6}$.

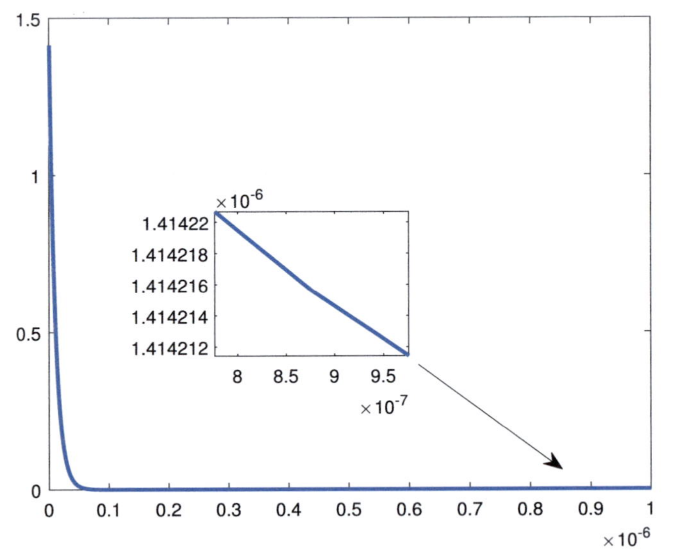

(b) IENTZNN(A, I, I) errors $\|X(t) - A^{-1}(t)\|_F$ for $\gamma = 10^8$, $\zeta = 0, 1, 10, 10^4, 10^6, 10^8, \lambda = 10^{-6}$.

Fig. 3.67 The case of linear time-varying noise in Example 3.9.3

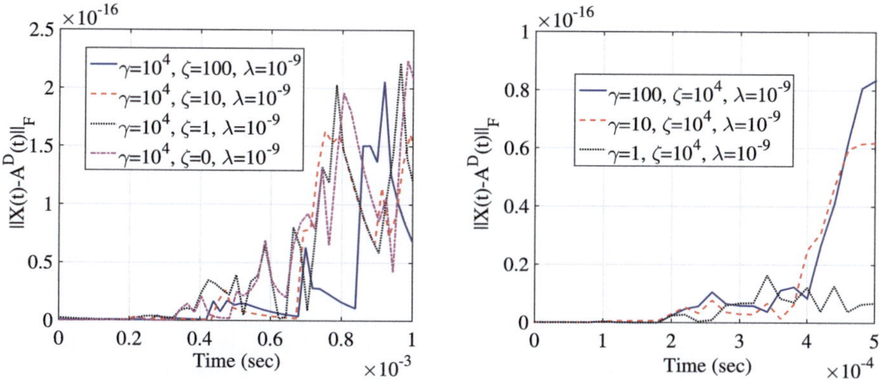

Fig. 3.68 $||X(t) - A^{\mathrm{D}}(t)||_F$ behaviour from IENTZNN$(A(t), G(t), G(t))$ when **a** $\gamma = 10^4$ and $\zeta = 100, 10, 1, 0$, **b** $\zeta = 10^4$ and $\gamma = 100, 10, 1$ in Example 3.9.4

$$A^{\mathrm{D}}(t) = \begin{bmatrix} 0 & 0 & 0 \\ 0 & 0 & 0 \\ 0 & 0 & 0 \end{bmatrix}.$$

By defining $G(t) = F(t) = A^l(t)$, $l \geq \mathrm{ind}(A)$ with $l = 3$, the IENTZNN$(A(t), G(t), G(t))$ formula can be used to approximate the theoretical Drazin inverse $A^{\mathrm{D}}(t)$. Selected parameters are $\lambda = 10^{-9}$, the zero initial condition, $N(t) = zeros(3, 3)$ and different values γ and ζ.

The trajectories of the state variables $X(t)$ produced from IENTZNN$(A(t), G(t), G(t))$ are presented in Fig. 3.69 for the different values γ and ζ. These approximate solutions converge to each element of the theoretical Drazin inverse and are equal to zero. Figure 3.68 describes the behaviour of the Frobenius norm $||X(t) - A^{\mathrm{D}}(t)||_F$ as the parameters γ and ζ changes and $\lambda = 10^{-9}$. From this numerical experiment, it is understood the ability IENTZNN in time-varying problems where the computation of the Drazin inverse is needed (Fig. 3.69).

3.9.7 Concluding Remarks on Noise Tolerant ZNN Models

Our idea is to show that the integration-enhanced noise-tolerant ZNN (IENTZNN) pattern is applicable not only for the usual matrix inversion. In the essence, possibility of IENTZNN models lies in the underlying error monitoring Zhang function (Zf). For this purpose, we propose a very general ZF, which leads to a novel integration-enhanced and noise-tolerant ZNN design formula (3.9.6), termed IENTZNN(A, F, G). The model is aimed to computing various expressions that involve time-varying matrix outer inverse $A^{(2)}_{T,S}$. Further, IENTZNN(A, F, G) in the case $N(t) = O$ is denoted by IEZNN(A, F, G) and in the case $N(t) = O$, $\zeta = 0$ is

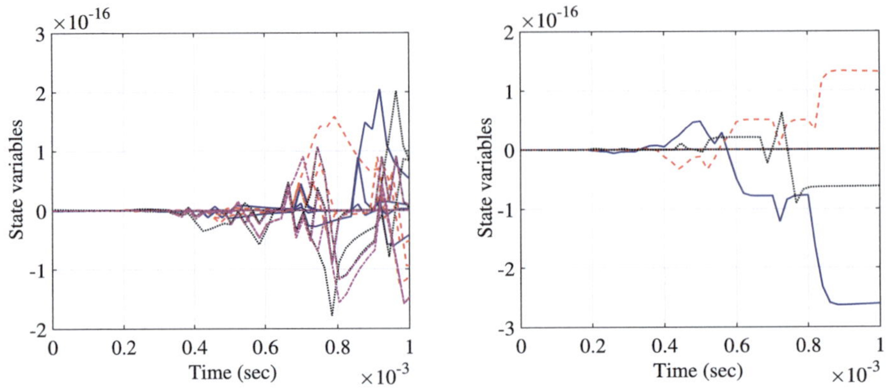

Fig. 3.69 Elementwise trajectories of IENTZNN($A(t), G(t), G(t)$) model: **a** for $\gamma = 10^4$, $\zeta = 100, 10, 1, 0$, **b** for $\gamma = 100, 10, 1$, $\zeta = 10^4$ in Example 3.9.4

denoted by ZNN(A, F, G). The ZNN(A, G, G) is proposed as an extension of both the results from [283] and the novel IEZNN design formula proposed in [110], and it is aimed to approximating the time-varying matrix outer inverse $A^{(2)}_{T,S}$.

Remark 3.9.1 Noise-tolerant and finite-time convergent ZNN models for dynamic matrix Moore-Penrose inversion was proposed in [258, 259].

3.10 Hybrid GNN Models for Constant Matrix Inversion

It was shown that the multiplication of the right hand side of the classical ZNN design by an appropriate positive definite matrix generates a new neural design with improved convergence rate. The goal in [232] is to apply similar principle on the standard GNN model. To that goal, we discover that some of proposed models can be considered as the multiplication of the right hand side of the GNN model by a symmetric positive-semidefinite matrix. As a final result, an appropriate general pattern for defining various improvements of the standard GNN design for online real-time matrix inversion in time invariant case was proposed. The leading idea in generating improved models is initiated after a combination of a GNN and corresponding ZNN pattern. Improved GNN (IGNN) design shows global exponential convergence with an improved convergence rate with respect to the convergence rate of the original GNN pattern. The acceleration in the convergence rate is defined by the smallest eigenvalue of appropriate positive semidefinte matrices.

3.10.1 Introduction and Preliminaries

The standard GNN model in the case $E(t) = AX(t) - I$ will will be termed as GNN-INV. Also, the case $E(t) = AX(t)A - A$ was considered in [240], and the corresponding gnn dynamics is denoted by GNN-MP.

On the other hand, the ZNN model is based upon the error matrix $E(t)$, and it is defined using the general design rule

$$\dot{E}(t) = -\gamma \mathcal{F}(E(t)). \tag{3.10.1}$$

The ZNN model for online time-invariant matrix inversion is developed upon the error matrix $E(t) = AX(t) - I$, which initiates the implicit dynamics of ZNN type

$$A\dot{X}(t) = -\gamma \mathcal{F}(AX(t) - I). \tag{3.10.2}$$

The implicit dynamics (3.10.2) were originally proposed in [381] for online inversion of a time-varying matrix $A(t)$. In Sect. 3.10.2 we investigate a possible improvement of the classical ZNN design.

3.10.2 Improved Time-Invariant ZNN Models

A novel kind of recurrent implicit dynamics for real-time matrix inversion was proposed and investigated in [25, 27]. The proposed dynamics is based on the usage of the positive definite matrix $AA^{\mathrm{T}} + I$ in (3.10.3), and it is defined as follows:

$$A\dot{X}(t) = -\gamma(AA^{\mathrm{T}} + I)(AX(t) - I). \tag{3.10.3}$$

Global exponential convergence rate of the implicit dynamics (3.10.3) was investigated in [25, 27].

The equivalent form of (3.10.3) is given by

$$\dot{X}(t) = (I - A)\dot{X}(t) - \gamma(AA^{\mathrm{T}} + I)(AX(t) - I). \tag{3.10.4}$$

Theorem 3.10.1 ([25]) *Given a nonsingular matrix $A \in \mathbb{R}^{n\times n}$, the state matrix $A \in \mathbb{R}^{n\times n}$ of the model* (3.10.3)*, starting from any initial state $X(0) \in \mathbb{R}^{n\times n}$, achieves global exponential convergence to theoretical inverse $X^* = A^{-1}$. In addition, the exponential convergence rate is the product of γ and the minimum eigenvalue $\beta > 1$ of $A^{\mathrm{T}}A + I$.*

A numerical comparison between the neural dynamics (3.10.2) and (3.10.3) was presented in [400].

Let us assume that $A \in \mathbb{R}_r^{m\times n}$ and $G \in \mathbb{R}_s^{n\times m}$, $0 < s \leq r$ are real and time-invariant matrices. The ZNN model for computing the outer inverse $A^{(2)}_{\mathcal{R}(G),\mathcal{N}(G)}$

which arises from the error matrix $E_G(t) := GAX(t) - G$ is defined by

$$GA\dot{X}(t) = -\gamma_1 \mathcal{F}\left(GAX(t) - G\right), \tag{3.10.5}$$

where $\gamma_1 > 0$ is a large real scaling coefficient. On the other hand, dynamic equation of the GNN model for approximating the outer inverse $A^{(2)}_{\mathcal{R}(G),\mathcal{N}(G)}$ which starts from the minimization of the matrix norm $\|GAX(t) - G\|_F^2/2$ was introduced in [399], as follows:

$$\dot{X}(t) = -\gamma_2 \,(GA)^{\mathrm{T}}\left(GAX(t) - G\right). \tag{3.10.6}$$

The parameter γ_2 in (3.10.6) is a positive scaling constant. The GNN model (3.10.6) implies

$$GA\dot{X}(t) = -\gamma_2 \,(GA)(GA)^{\mathrm{T}}\mathcal{F}\left(GAX(t) - G\right). \tag{3.10.7}$$

An improved hybrid ZNN model based on the combination of (3.10.5) and (3.10.7) in the case $\gamma_1 = \gamma_2 = \gamma$ was proposed in [284], as follows

$$GA\dot{X}(t) = -\gamma\left((GA)(GA)^{\mathrm{T}} + I\right)\mathcal{F}\left(GAX(t) - G\right), \tag{3.10.8}$$

where $\mathcal{F}$ is an appropriate elementwise activation function. The improved ZNN model defined in (3.10.8) was termed as ZNNATS2-III model in [284]. Its convergence properties are described in Theorem 3.10.2.

Theorem 3.10.2 ([284]) *Let two given matrices* $A \in \mathbb{R}_r^{m\times n}$ *and* $G \in \mathbb{R}_s^{n\times m}$, $0 < s \le r$ *be given and* $A^{(2)}_{\mathcal{R}(G),\mathcal{N}(G)}$ *exists. Then the state matrix* $X(t) \in \mathbb{R}^{n\times m}$ *of* (3.10.8) *achieves global convergence* $GAX(t) \to G$ *when* $t \to +\infty$, *starting from any initial state* $X(0) \in \mathbb{R}^{n\times m}$. *The state matrix* $X(t) \in \mathbb{R}_r^{n\times m}$ *of the model* (3.10.8) *based on the increasing and odd activation function* $\mathcal{F}$ *converges to the outer inverse* $A^{(2)}_{\mathcal{R}(G),N(G)}$, *and the solution is stable in the sense of Lyapunov. If the linear activation function* $\mathcal{F} = I$ *is used, the convergence rate of the model* (3.10.8) *is equal to* $\gamma\beta$, *where* $\beta = \alpha + 1$ *and* $\alpha \ge 0$ *is the minimal nonzero eigenvalue of* $(GA)(GA)^{\mathrm{T}}$.

Remark 3.10.1 Theorem 3.10.2 in the simplest case $G = I$ and nonsingular A yields corresponding result from [25, 27], valid for A^{-1}.
In the case $G = A^{\mathrm{T}}$, the model (3.10.8) achieves global convergence to the Moore-Penrose inverse.
Additionally, in the case $m = n$, the choice $G = A^k, k = \mathrm{ind}(A)$ implies global convergence of the model (3.10.8) towards the Drazin inverse A^{D}.

Our first motivation arisen from the observation that two proposed modifications (3.10.3) and (3.10.8) transform the general ZNN pattern (3.10.1) into the form

$$\dot{E}(t) = -\gamma R \cdot \mathcal{F}\left(E(t)\right), \tag{3.10.9}$$

where R is an appropriate symmetric positive-semidefinite matrix. The general neural design (3.10.9) will be called Improved ZNN (IZNN) design. It is observable that an arbitrary IZNN design (3.10.9) is defined by a proper ordered pair $(E(t), R)$, where $E(t)$ is the error matrix and R is a symmetric positive-semidefinite matrix which accelerates the convergence with the scaling factor $\lambda_{\min}(R)$. So far investigated choices of $(E(t), R)$ in (3.10.9) are

$$(E(t) = AX(t) - I,\ R = I),$$
$$(E(t) = AX(t) - I,\ R = AA^{\mathrm{T}} + I),$$
$$\big(E(t) = GAX(t) - G,\ R = (GA)(GA)^{\mathrm{T}} + I\big).$$

In this section, we investigates improvements of the general GNN pattern into the form

$$\dot{X}(t) = -\gamma \frac{\partial \varepsilon(X(t))}{\partial V} R,\ \ \varepsilon(X(t)) = \frac{1}{2}\|E(t)\|_F^2, \qquad (3.10.10)$$

wherein R is an appropriate positive semidefinite matrix. The models of the general pattern (3.10.10) will be called *improved GNN* (IGNN) models.

Theoretical analysis and computer simulation results which confirm the effectiveness and superior efficiency of the modified models for online matrix inversion and generalized inversion are presented. The GNN-MP model from [240] in the regular matrix case is compared with the known GNN model for matrix inversion, called GNN-INV. Both GNN-INV and GNN-MP are theoretically and through simulation examples compared to some another choices of R. The finite time convergent dynamical evolution design in the present paper is achieved in an alternative way, using an appropriate IGNN model.

3.10.3 *Hybrid GNN Models for Real Time Matrix Inversion*

Our investigation is restricted to real matrices, i.e., it will be assumed that $A \in \mathbb{R}^{m\times n}$. If A is nonsingular, it is necessary to solve the matrix equation $AX = I$, which produces the error matrix

$$E(t) = AX(t) - I. \qquad (3.10.11)$$

Using the general GNN design formula, the following dynamic state equation of GNN-INV follows:

$$\dot{X}(t) = -\gamma A^{\mathrm{T}}\mathcal{F}\left(AX(t) - I\right),\quad V(0) = V_0. \qquad (3.10.12)$$

Particularly, the IGNN dynamics (3.10.10) in the case $E(t) = AX(t) - I$ becomes

$$\dot{X}(t) = -\gamma A^{\mathrm{T}}\mathcal{F}\left(AX(t) - I\right) R,\quad V(0) = V_0, \qquad (3.10.13)$$

wherein R is a positive definite matrix of an appropriate order. Similarly as in the case of the ZNN dynamics, an arbitrary IGNN model (3.10.10) is defined by a proper ordered pair $(E(t), R)$, where $E(t)$ is the error matrix and R is a symmetric positive-semidefinite matrix. We will show that the IGNN dynamics (3.10.13), defined by $(AX(t) - I, R)$, possesses a global exponential convergence rate defined by the scaling factor $\lambda_{\min}(R)$ with respect to the standard GNN dynamics (3.10.12), representing $(AX(t) - I, I)$.

In order to find appropriate matrices R, let us consider the GNN model for computing inner inverse and the Moore-Penrose inverse based on the matrix equation $AXA = A$, which was defined and investigated in [240]. Let $X(t) \in \mathbb{R}^{n\times m}$ be the unknown time-varying matrix. The general GNN design in the case $E(t) = AX(t)A - A$ produces the dynamic state equation

$$\dot{X}(t) = -\gamma A^{\mathrm{T}}\mathcal{F}(AX(t) - I)\,AA^{\mathrm{T}}, \quad V(0) = V_0, \tag{3.10.14}$$

The dynamic equation (3.10.14) is known as the GNN-MP model from [240]. Therefore, GNN-MP can be considered as an improvement of GNN-INV in the case when the matrix A is regular. This further implies that so far investigated pairs $(E(t), R)$ of the form (3.10.10) are

$$(E(t) = AX(t) - I, R = I),$$
$$(E(t) = AX(t)A - I, R = I).$$

The model GNN-INV arises from $(E(t) = AX(t) - I, R = I)$ while $(E(t) = AX(t)A - I, R = I)$ initiates GNN-MP.

Our goal in current section is to define and investigate various IGNN models (3.10.13) possessing higher exponential convergence rate than GNN-INV. We firstly introduce a new improved GNN model of the general form (3.10.10) using an appropriate combination of GNN-INV and GNN-MP from [240]. Namely, GNN-MP can be considered in two different ways. The original view $(E(t) = AX(t)A - I, R = I)$ on the dynamics GNN-MP is applicable in the case if A is singular and can be used to generate $A^{\dagger}$. Another viewpoint on GNN-MP is given by the ordered pair $(E(t) = AX(t) - I, R = AA^{\mathrm{T}})$, which is an appearance of (3.10.13) applicable applicable in the case when A is nonsingular, and can be used in computing A^{-1}.

The third appearance of (3.10.10) is a hybrid of GNN-INV and GNN-MP. More precisely, after the summation of (3.10.12) and (3.10.14) one obtains

$$\dot{X}(t) = -\gamma A^{\mathrm{T}}\mathcal{F}(AX(t) - I)\,(AA^{\mathrm{T}} + I), \quad V(0) = V_0. \tag{3.10.15}$$

Denote the model (3.10.15) by GNN-COMB. Clearly, GNN-COMB design is the particular appearance $(E(t) = AX(t) - I, R = AA^{\mathrm{T}} + I)$ of (3.10.13).

It is observable that the GNN-INV, GNN-MP and GNN-COMB models can be comprised in a single comprehensive model (3.10.13), called GNN-R. The choice $R = I$ corresponds to the GNN-INV model, $R = AA^{\mathrm{T}}$ gives GNN-MP and $R =$

$AA^{\mathrm{T}} + I$ leads to the GNN-COMB model. In principle, any positive definite matrix R can be used to define an induced GNN model for computing A^{-1} with improved convergence rate in relation to GNN-INV.

As it was shown in [130, 148, 282, 348, 359, 363, 381], the convergence rate can be improved by an appropriate activation function. So far, the impact of various nonlinear activation functions was investigated for ZNN models. Recently, the impact of nonlinear activations $\mathcal{F}_{\mathrm{ps}}$ and $\mathcal{F}_{\mathrm{sps}}$ in the GNN model for computing the Drazin inverse was considered in [282]. In order to simplify the notation, in further $ECR(\mathcal{M})$ will denote the exponential convergence rate of the model $\mathcal{M}$.

The starting point in theoretical investigation of (3.10.13) is the following result about the global exponential convergence of the GNN-INV model in the linear case $\mathcal{F} = I$, originated in [363]. The results is stated in the next proposition. By $\lambda_{\min}(K)$ and $\lambda_{\max}(K)$ are denoted the smallest and largest eigenvalue of K, respectively.

Proposition 3.10.1 ([363]) *Consider nonsingular matrix $A \in \mathbb{R}^{n\times n}$. Starting from any initial state $V_0 \in \mathbb{R}^{n\times n}$, state matrix $X(t) \in \mathbb{R}^{n\times n}$ of the linear* GNN-INV model will exponentially converge to theoretical inverse $\tilde{X} = A^{-1}$. In addition, the exponential convergence rate of GNN-INV is equal to $ECR(\text{GNN-INV}) = \gamma\alpha$, where $\alpha = \lambda_{\min}(A^{\mathrm{T}}A)$.

It was shown in [381] that the Zhang dynamics (3.10.2) globally exponentially converges to the theoretical inverse A^{-1}, starting from any initial state $X(0)$. The exponential convergence rate of the ZNN model with linear activation is equal to γ. This fact we will denote by $ECR(\text{ZNN}) = \gamma$.

Theorem 3.10.3 ([232]) *Consider a real invertible matrix $A \in \mathbb{R}^{n\times n}$ and a symmetric positive semidefinite matrix R in the GNN-R model* (3.10.13). *If the design parameter γ satisfies $\gamma > 0$, then the state matrix $X(t) \in \mathbb{R}^{n\times n}$ of the model* (3.10.13) *satisfies $AX(t) \to I$ when $t \to \infty$, for an arbitrary initial state $V(0)$ and the equilibrium state matrix is just A^{-1}. In addition, if the linear activation is used, then* $ECR(\text{GNN-R}) = \gamma\alpha\beta$, where $\beta = \lambda_{\min}(R)$ and $\alpha = \lambda_{\min}(AA^{\mathrm{T}})$.

Proof Let $\overline{X}(t) = X(t) - A^{-1}$. The substitution $X(t) = \overline{X}(t) + A^{-1}$ into the dynamic equation (3.10.13) implies

$$\begin{aligned}\frac{\mathrm{d}\overline{X}(t)}{\mathrm{d}t} = \frac{\mathrm{d}X(t)}{\mathrm{d}t} &= -\gamma A^{\mathrm{T}}\mathcal{F}\left(A(\overline{X}(t) + A^{-1}) - I\right) R \\ &= -\gamma A^{\mathrm{T}}\mathcal{F}\left(A\overline{X}(t)\right) R.\end{aligned} \tag{3.10.16}$$

The Lyapunov function candidate is defined as

$$L(\overline{X}(t), t) = \frac{1}{2}\left\|A\overline{X}(t)\right\|_F^2 = \frac{1}{2}\mathrm{Tr}\left(\left(A\overline{X}(t)\right)^{\mathrm{T}} A\overline{X}(t)\right). \tag{3.10.17}$$

Evidently, the inequality $L(\overline{X}(t), t) \geq 0$ holds for $\overline{X}(t) \neq 0$. According to (3.10.17), taking into account (3.10.16) and $\mathrm{d}\,\mathrm{Tr}\left(X^{\mathrm{T}}X\right) = 2\mathrm{Tr}\left(X^{\mathrm{T}}\mathrm{d}X\right)$, the time derivative of $L(\overline{X}(t), t)$ satisfies

$$
\begin{aligned}
\frac{\mathrm{d}L(\overline{X}(t),t)}{\mathrm{d}t} &= \frac{1}{2}\frac{\mathrm{d}\mathrm{Tr}\left(\left(A\overline{X}(t)\right)^{\mathrm{T}} A\overline{X}(t)\right)}{\mathrm{d}t} \\
&= \mathrm{Tr}\left(\left(A\overline{X}(t)\right)^{\mathrm{T}} A\frac{\mathrm{d}\overline{X}(t)}{\mathrm{d}t}\right) \\
&= -\gamma \mathrm{Tr}\left[\left(A\overline{X}(t)\right)^{\mathrm{T}} A\, A^{\mathrm{T}}\mathcal{F}\left(A\overline{X}(t)\right) R\right].
\end{aligned}
$$

In the linear case, it follows that

$$
\begin{aligned}
\frac{\mathrm{d}L(\overline{X}(t),t)}{\mathrm{d}t} &= -\gamma \mathrm{Tr}\left[\left(A^{\mathrm{T}}A\overline{X}(t)\right)^{\mathrm{T}} \left(A^{\mathrm{T}}A\overline{X}(t)\right) R\right] \\
&= -\gamma \mathrm{Tr}\left[R\left(A^{\mathrm{T}}A\overline{X}(t)\right)^{\mathrm{T}} \left(A^{\mathrm{T}}A\overline{X}(t)\right)\right].
\end{aligned}
$$

Now, it is possible to apply the following inequality, stated in [281, Lemma 1]:

$$
\lambda_{\min}(K)\,\mathrm{Tr}(L) \le \mathrm{Tr}(KL) \le \lambda_{\max}(K)\,\mathrm{Tr}(L),
$$

which is satisfied for any symmetric matrices $K, L \in \mathbb{R}^{n\times n}$, where K is a positive-semidefinite matrix.

We apply the last inequality for $K = R$ and $L = \left(A^{\mathrm{T}}A\overline{X}(t)\right)^{\mathrm{T}} \left(A^{\mathrm{T}}A\overline{X}(t)\right)$. Assume that $\beta \ge 0$ is the minimum eigenvalue of R. Then

$$
\begin{aligned}
\frac{\mathrm{d}L(\overline{X}(t),t)}{\mathrm{d}t} &\le -\gamma\beta \mathrm{Tr}\left[\left(A\overline{X}(t)\right)^{\mathrm{T}} AA^{\mathrm{T}} \left(A\overline{X}(t)\right)\right] \\
&= -\gamma\beta \mathrm{Tr}\left[\left(A\overline{X}(t)\right)\left(A\overline{X}(t)\right)^{\mathrm{T}} AA^{\mathrm{T}}\right].
\end{aligned} \tag{3.10.18}
$$

Applying again [282, Theorems 3.3 and 3.4], one can verify

$$
\begin{aligned}
\frac{\mathrm{d}L(\overline{X}(t),t)}{\mathrm{d}t} &\le -\gamma\alpha\beta \|A\overline{X}(t)\|_F^2 \\
&= -\gamma\alpha\beta \|AX(t) - I\|_F^2.
\end{aligned} \tag{3.10.19}
$$

Due to the fact that the design parameter satisfies $\gamma > 0$, in view of (3.10.18), it follows that

$$
\frac{\mathrm{d}L(\overline{X}(t),t)}{\mathrm{d}t} \begin{cases} < 0 \text{ if } AX(t) - I \ne 0, \\ = 0 \text{ if } AX(t) - I = 0. \end{cases} \tag{3.10.20}
$$

This further implies:
–$\mathrm{d}L(\overline{X}(t),t)/\mathrm{d}t < 0$ for any non-equilibrium state $X(t)$ satisfying $E(t) = AX(t) - I \ne 0$;
–$\mathrm{d}L(\overline{X}(t),t)/\mathrm{d}t = 0$ at the equilibrium state $X(t)$ satisfying $E(t) = AX(t) - I = 0$.

According to the Lyapunov stability theory, relation (3.10.20), the error matrix $E(t) = AX(t) - I$ globally converges to the zero matrix, from arbitrary initial value $V(0)$.

Since the matrix A is invertible, the equilibrium state matrix is $\tilde{X} = A^{-1}$.

In addition, (3.10.19) implies

$$\dot{L}(t) \leq -\gamma\alpha\beta \left\| A\overline{X}(t) \right\|_F^2 = -2\gamma\alpha\beta\, L(t), \tag{3.10.21}$$

which further implies $L(t) \leq L(0)e^{-2\gamma\alpha\beta}$. Therefore,

$$\|AX(t) - I\|_F \leq \|AX(0) - I\|_F \cdot e^{-\gamma\alpha\beta}$$

confirms the convergence rate $\gamma\alpha\beta$ of (3.10.13). □

A theoretical comparison between GNN-INV and GNN-COMB is given in Corollary 3.10.1.

Corollary 3.10.1 *Consider a real invertible matrix $A \in \mathbb{R}^{n\times n}$. If the design parameter γ satisfies $\gamma > 0$, then the state matrix $X(t) \in \mathbb{R}^{n\times n}$ of the* GNN-COMB *model* (3.10.15) *satisfies $AX(t) \to I$ when $t \to \infty$, for an arbitrary initial state $V(0)$. In addition, the exponential convergence rate is determined by ECR* (GNN − COMB) $= \gamma\alpha(1 + \alpha)$, *where $\alpha = \lambda_{\min}(A^{\mathrm{T}}A)$.*

The convergence of GNN-MP in the singular case was considered in [240], where it was shown that the state matrix $X(t)$ of GNN-MP satisfies $AX(t)A \to A$ when $t \to \infty$. Corollary 3.10.2 investigates the convergence of GNN-MP in the invertible matrix case.

Corollary 3.10.2 *Consider a real invertible matrix $A \in \mathbb{R}^{n\times n}$. If the design parameter γ satisfies $\gamma > 0$, then the state matrix $X(t) \in \mathbb{R}^{n\times n}$ of the* GNN-MP model (3.10.14) satisfies $AX(t) \to I$ when $t \to \infty$, for an arbitrary initial state $V(0)$. In addition, the exponential convergence rate is determined by ECR(GNN-MP) $= \gamma\alpha^2$, where $\alpha = \lambda_{\min}(A^{\mathrm{T}}A)$.

Motivated by the previous discussion, we propose the following accelerated GNN model:

$$\dot{X}(t) = -\gamma A^{\mathrm{T}}\mathcal{F}(AX(t) - I)\,(AA^{\mathrm{T}} + I)^k, \quad V(0) = V_0, \tag{3.10.22}$$

where $k \geq 1$ is a chosen integer. The method (3.10.22) will be termed as GNN-POWER(k). In essence, it is the particular case $(E(t) = AX(t) - I$, $R = (AA^{\mathrm{T}} + I)^k)$ of (3.10.10). Theoretical convergence rate of the model GNN-POWER(k) is considered in Corollary 3.10.3 and follows from the choice $R = (AA^{\mathrm{T}} + I)^k$ in (3.10.13).

Corollary 3.10.3 *Consider a real invertible matrix $A \in \mathbb{R}^{n\times n}$. If the design parameter γ satisfies $\gamma > 0$, then the state matrix $X(t) \in \mathbb{R}^{n\times n}$ of the* GNN-POWER(k)

Table 3.7 Review of ECR for 5 considered neural dynamics

Model $\mathcal{M}$	GNN-INV	GNN-MP	GNN-COMB	GNN-POWER(k)	ZNN
$ECR(\mathcal{M})$	$\gamma\alpha$	$\gamma\alpha^2$	$\gamma\alpha(1+\alpha)$	$\gamma\alpha(1+\alpha)^k$	γ
Case			Priority		
$\alpha < 1, \alpha(1+\alpha)^k \leq 1$	4	5	3	2	1
$\alpha < 1, \alpha(1+\alpha) \leq 1, \alpha(1+\alpha)^k \geq 1$	4	5	3	1	2
$\alpha < 1, \alpha(1+\alpha) \geq 1$	4	5	2	1	3
$\alpha = 1$	5	3	2	1	5
$\alpha \geq 1$	4	3	2	1	5

model (3.10.22) *satisfies* $AX(t) \to I$ *when* $t \to \infty$, *for an arbitrary initial state* $V(0)$. *In addition, if the power-sigmoid and the smooth power-sigmoid activation functions are used,* $ECR(\text{GNN-POWER}(k)) = \gamma\alpha(1+\alpha)^k$, *where* $\alpha = \lambda_{\min}(A^{\mathrm{T}}A) \geq 0$.

Remark 3.10.2 Let us analyse the exponential convergence rates of five GNN models for matrix inversion in time-invariant case: GNN-INV,GNN-MP,GNN-COMB,GNN-POWER(k) and the ZNN model. Our intention is to include possible further GNN-R models in Table 3.7. Let $\alpha > 0$ is the minimum eigenvalue of $A^{\mathrm{T}}A$ and $\beta \geq 0$ is the minimum eigenvalue of R. By $\mathcal{M}_1 \succ \mathcal{M}_2$ we shortly denote the fact $ECR(\mathcal{M}_1) \geq ECR(\mathcal{M}_2)$. Similarly, $\mathcal{M}_1 \approx \mathcal{M}_2$ we shortly denote the fact $ECR(\mathcal{M}_1) \approx ECR(\mathcal{M}_2)$. The exponential convergence rate of the GNN-INV is determined by $\gamma\alpha$. According to Theorem 3.10.3, the exponential convergence rate of the GNN-COMB is determined by $\gamma\alpha(1+\alpha)$. This implies GNN-COMB$\succ$GNN-INV with the factor $\beta = 1+\alpha > 1$. Since $\alpha > 0$ it follows that $\gamma\alpha(1+\alpha)^k > \gamma\alpha(1+\alpha)$, which indicates GNN-POWER(k) $\succ$GNN-COMB. Also, the convergence of the GNN-POWER(k) can be increased proportionally by increasing values k.

On the other hand, according to Corollary 3.10.2, the exponential convergence rate of the GNN-MP is determined by $\gamma\alpha^2$. Since $\gamma\alpha^2 < \gamma\alpha$ in the case $0 < \alpha < 1$ and $\gamma\alpha^2 > \gamma\alpha$ in the case $\alpha > 1$, this means GNN-MP$\succ$GNN-INV in the case $\alpha > 1$ and GNN-INV$\succ$GNN-MP in the case $0 < \alpha < 1$. In the case $\alpha = 1$, clearly follows that GNN-INV$\approx$GNN-MP. On the other hand, $\gamma\alpha(1+\alpha) > \gamma\alpha$ and $\gamma\alpha(1+\alpha) > \gamma\alpha^2$, which indicates GNN-COMB$\succ$GNN-INV and GNN-COMB$\succ$GNN-MP for all input matrices.

The ranking of the exponential convergence rate of various RNN models for regular matrix inversion in time-invariant is surveyed in Table 3.7. Any integer that appears in Table 3.7 indicates the ordinal number of the method in ranking ECRs. In addition, for larger values α, behaviour of the GNN-MP and GNN-COMB model tends to be with similar convergence behavior, ie., GNN-MP$\approx$GNN-COMB.

As it is mentioned, GNN-POWER(k) accelerates the convergence rate of GNN-INV in terms of the increased function x^k, $x \geq 1$, $k \geq 1$, which tends to infinity. So, it is reasonable to expect that GNN-POWER(k) is able to achieve a finite time convergence. In Theorem 3.10.4 we show that GNN-POWER(k) can achieve a finite-time convergence using an appropriate value k.

Theorem 3.10.4 *Given a time-varying invertible matrix $A(t) \in \mathbb{C}^{n \times n}$. The state matrix $X(t)$ of the linear IGNN model, starting from any initial state $X(0)$, converges to the time-varying theoretical outer inverse A^{-1} in the finite time*

$$t_f < \max\left\{\frac{|e^-(0)|^{1-\sigma}}{\gamma(1-\sigma)}, \frac{|e^+(0)|^{1-\sigma}}{\gamma(1-\sigma)}\right\},$$

where $e^+(0)$ and $e^-(0)$ are the largest and the smallest element in $E(0) = AX(0) - I$, respectively.

Proof The finite-time convergence in [149, Theorem 1] was verified using the inequality

$$\dot{L}(t) \leq -2\gamma\, L(t)^{(\sigma+1)/2}. \tag{3.10.23}$$

Therefore, it is necessary to find an appropriate exponent k such that the parameters $\beta = \lambda_{\min}(R)$ and $\alpha = \lambda_{\min}(A^{\mathrm{T}}A)$ in (3.10.21) satisfy

$$\alpha\beta\, L(t) \geq L(t)^{(\sigma+1)/2}. \tag{3.10.24}$$

Since $\beta = x^k$, $x \geq 1$ is a real number and $k \geq 1$ is an integer, it is possible to find an appropriate $x > 1$ and an integer $k \geq 1$ satisfying (3.10.24). More precisely, it suffices to find x and k such that the inequality

$$x^k \geq \frac{L(t)^{(\sigma-1)/2}}{\alpha}$$

holds. □

3.10.4 Comparison with Accelerated ZNN Model for Real Time Matrix Inversion

In this subsection we investigate possibilities of possible accelerations of ZNN models. In [381], Zhang et al. proposed the general ZNN design for computing the time-varying matrix inverse as follows:

$$A\dot{X}(t) = -\Gamma\mathcal{F}(AX(t) - I). \tag{3.10.25}$$

Table 3.8 Comparison of ECR(GNN-R), ECR(ZNN) and ECR(AZNN)

Model $\mathcal{M}$	GNN-Γ	ZNN	AZNN
$ECR(\mathcal{M})$	$\gamma\alpha\zeta$	γ	$\gamma\zeta$
Case	Priority		
$\alpha\zeta > 1 > \zeta$	1	2	3
$\alpha\zeta > \zeta > 1$	1	3	2
$1 > \alpha\zeta > \zeta$	2	1	3
$\zeta > \alpha\zeta > 1$	2	3	1
$1 > \zeta > \alpha\zeta$	3	1	2
$\zeta > 1 > \alpha\zeta$	3	2	1

Here, Γ is a positive-definite matrix used to scale the convergence rate. The dynamical evolution (3.10.25) will be termed as *Accelerated ZNN* (AZNN) model. The simple case $\Gamma = \gamma I$ reduces AZNN model to the ZNN. The convergence rate of the AZNN dynamics is investigated in Corollary 3.10.4.

Corollary 3.10.4 *Let $A(t) \in \mathbb{C}^{n\times n}$ be a time-varying invertible matrix. Then the matrix $X(t)$ defined by the AZNN dynamics* (3.10.25) *globally exponentially converges to the theoretical inverse A^{-1}, starting from any initial state $X(0)$. The exponential convergence rate of the AZNN model with linear activation is equal to $\gamma\zeta$, where $\zeta = \lambda_{\min}(\Gamma)$.*

The convergence rate derived in Corollary 3.10.4 will be denoted by $ECR(\text{AZNN}) = \gamma\zeta, \zeta = \lambda_{\min}(\Gamma)$. The ranking of the exponential convergence rate of GNN-R in the case $R = \Gamma$, ZNN and AZNN is arranged in Table 3.8. According to Theorem 3.10.3, it follows $ECR(\text{GNN-}\Gamma) = \gamma\alpha\zeta, \zeta = \lambda_{\min}(\Gamma), \alpha = \lambda_{\min}(AA^{\mathrm{T}})$.

3.10.5 Numerical Examples on Improved GNN Models

The Simulink implementation of the hybrid GNN-R model (3.10.13) is presented in Fig. 3.70. The implementation is based on the fact that four considered models: GNN-INV, GNN-MP, GNN-COMB and GNN-POWER(2) can be unified in a single model (3.10.13) where $R = I$ for the GNN-INV model, $R = AA^{\mathrm{T}}$ for GNN-MP, $R = AA^{\mathrm{T}} + I$ for the GNN-COMB and $R = (AA^{\mathrm{T}} + I)^2$ for the GNN-POWER(2) model.

Example 3.10.1 Consider the matrix

$$A = \begin{bmatrix} 0.5377 & 0.8622 & -0.4336 \\ 1.8339 & 0.3188 & 0.3426 \\ -2.2588 & -1.3077 & 3.5784 \end{bmatrix}.$$

Its inverse is equal to

$$A^{-1} = \begin{bmatrix} -0.3350 & 0.5309 & -0.0914 \\ 1.5466 & -0.1991 & 0.2065 \\ 0.3538 & 0.2623 & 0.2972 \end{bmatrix}.$$

We apply first the linear GNN-INV, GNN-MP, GNN-COMB and GNN-POWER(2) models with the gain parameter $\gamma = 10^4$. Initial matrix is chosen by $V(0) = 0$. Figure 3.71 shows the responses of each element of the matrix $X(t)$ during the total simulation time $[0, 10^{-3}]s$, expressed in seconds.

Figure 3.72 (left) shows the residual norm comparison between the considered models. It can be seen that the GNN-COMB model achieves convergence faster than GNN-INV and GNN-MP. Also, GNN-POWER(2) achieves the fastest convergence. The minimal eigenvalue of AA^{T} is equal to $\alpha = 0.3606$. According to Theorem 3.10.3, the exponential convergence rates of considered models are equal to
ECR(GNN-INV)=$\gamma\alpha = 360600$, ECR(GNN-MP)=$\gamma\alpha^2 = 130032.36$,
ECR(GNN-COMB)=$\gamma\alpha(1+\alpha) = 490632.36$,
ECR(GNN-POWER(2))=$\gamma\alpha(1+\alpha)^2 = 667554.389016$.

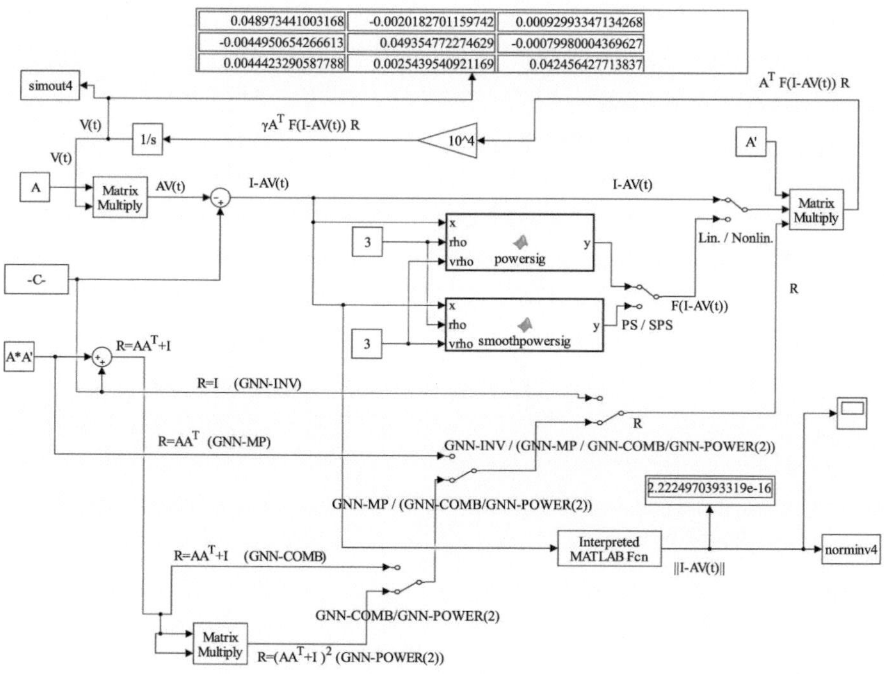

Fig. 3.70 Simulink implementation of the GNN-INV, GNN-MP and GNN-COMB models (3.10.13)

The improvement of the GNN-COMB with respect to GNN-INV model is defined by the factor $1 + \alpha = 1.3606$. On the other hand, according to Corollary 3.10.2, the exponential convergence rate of the GNN-MP is determined by $\gamma\alpha^2 = 1300.3236 < \gamma\alpha = 3606$. On the other hand, $\gamma\alpha(1 + \alpha) = 4906.3236 > \gamma\alpha = 3606$, which confirms theoretically the superior performances of the GNN-COMB method with respect to both GNN-INV and GNN-MP. Finally, it is challenging to compare the GNN-POWER(2) method with another three models. In view of Corollary 3.10.3, the improvement of the GNN-POWER(2) method with respect to GNN-COMB is defined by the factor $1 + \alpha = 1.3606$ and with respect to GNN-INV model is defined by the factor $(1 + \alpha)^2 = 1.8512$. These calculations are consistent with the numerical results given in Fig. 3.72 (left).

Figure 3.72 (right) presents the comparison between different activation functions, applied to the GNN-COMB model. It is evident that non-linear activation accelerates the convergence and produces the smaller residual norm for every $t \leq t_{\text{tot}}$.

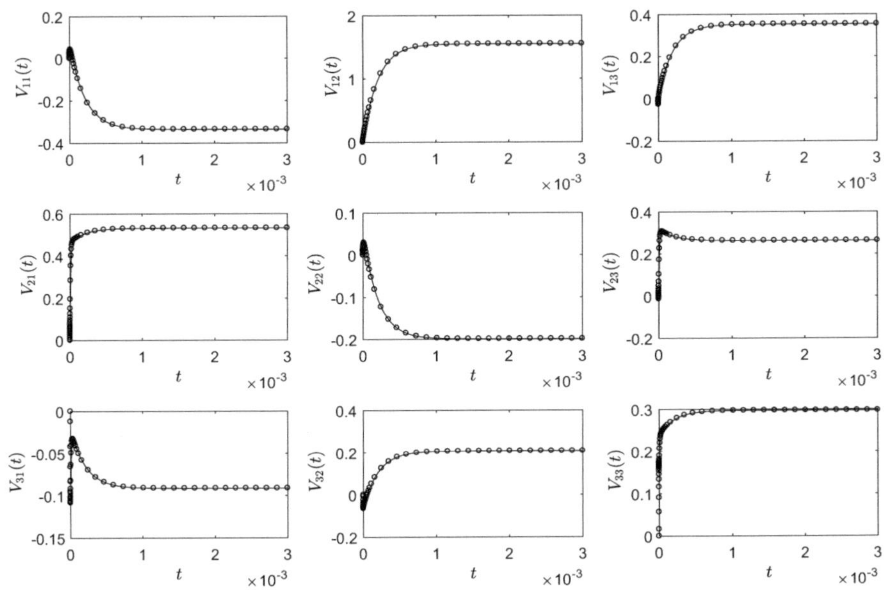

Fig. 3.71 Elementwise convergence trajectories of the state matrix $V(t)$ of GNN-COMB, for the matrix A in Example 3.10.1

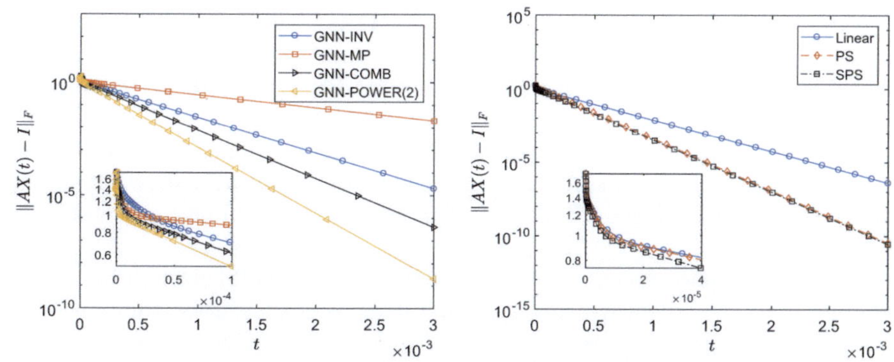

Fig. 3.72 Residual norm comparison between GNN-INV, GNN-MP, GNN-COMB, GNN-POWER(2) models (left) and between different activation functions in GNN-COMB model (right)

3.10.6 Hybrid GNN-ZNN Models for Solving Linear Matrix Equations

New dynamical models for solving the matrix equations $BX = D$ and $XC = D$ in time-invariant case were developed in [224]. These models are derived as a combination of GNN and ZNN models. They do not posses GNN design, due to their implicit dynamics. Formally observed, they can be derived by multiplying the right hand side in the ZNN dynamics by an appropriate symmetric positive definite matrix which improves the convergence rate. For this purpose, these models are termed as HZNN. The convergence of HZNN models is global and exponential. Also, the convergence rate of HZNN models is superior with respect to the convergence rate of the classical GNN model as well as with respect to ZNN models in time-invariant case. Capability of the HZNN models to overcome unavoidable implementation noises is considered theoretically and numerically.

Introduction to Hybrid GNN-ZNN Models

The background of our research can be described as follows. The ZNN and GNN models for online time-invariant matrix inversion are based upon the error matrix $E(t) = AV(t) - I$. As a consequence, the linear ZNN dynamics is given in the implicit form

$$A\dot{V}(t) = -\gamma(AV(t) - I). \tag{3.10.26}$$

A novel kind of recurrent implicit dynamics for real-time matrix inversion was proposed and investigated in [25, 27], and it is based on the usage of the positive definite matrix $AA^{\mathrm{T}} + I$ in (3.10.2), as follows:

$$A\dot{V}(t) = -\gamma(AA^{\mathrm{T}} + I)(AV(t) - I). \tag{3.10.27}$$

Global exponential convergence rate of the implicit dynamics (3.10.27) was investigated in [25, 27].

Numerical comparison between the neural dynamics (3.10.26) and (3.10.27) was presented in [400]. The numerical experience in [400] undoubtedly shows the acceleration that is obtained by the dynamics (3.10.26) with respect to the standard ZNN dynamics (3.10.2).

Another motivation for our investigations can be found in [26], wherein the author introduced and investigated a novel implicit dynamical system for online solving simultaneous linear equations $Ax = b$. Instead the classical GNN evolution $\dot{x}(t) = -\gamma A^{\mathrm{T}}(Ax(t) - b)$ and the conventional ZNN dynamics $A\dot{x}(t) = -\gamma(Ax(t) - b)$, Chen proposed the improved design $A\dot{x}(t) = -\gamma(AA^{\mathrm{T}} + I)(Ax(t) - b)$.

An extension of the acceleration (3.10.27) from the usual matrix inverse to outer inverses with prescribed range and null space was proposed in [284]. Let us assume that $A \in \mathbb{R}_r^{m\times n}$ and $G \in \mathbb{R}_s^{n\times m}$, $0 < s \leq r$ are real and time-invariant matrices. The ZNN evolution for computing the outer inverse $A^{(2)}_{\mathcal{R}(G),\mathcal{N}(G)}$ arises from the error matrix $E_G(t) := GAX(t) - G$ is defined as in (3.10.5). An acceleration of dynamics (3.10.5) was proposed in [284], as in (3.10.8). The evolution defined in (3.10.8) was termed as ZNNATS2-III model in [284]. Its convergence properties are described in Proposition 3.10.2 from [284].

The hybrid models introduced in [224] are defined using suitable combinations of GNN and ZNN neural dynamics for solving the matrix equation $BX = D$ or $CX = D$.

Mixed GNN-ZNN Model for Solving Matrix Equations

The error matrix of the matrix equation $BX = D$ is defined by

$$E_{B,I,D}(t) = BX(t) - D.$$

The ZNN design (3.10.1) corresponding to the error matrix $E_{B,I,D}$ is of the form

$$\dot{E}_{B,I,D}(t) = -\gamma_1 \mathcal{F}\left(E_{B,I,D}(t)\right),$$

$$\dot{E}_{B,I,D}(t) = B\dot{X}(t) = -\gamma_1 \mathcal{F}\left(BX(t) - D\right). \tag{3.10.28}$$

It will be denoted by ZNN(B, I, D).

On the other hand, the induced scalar-valued norm-based error function corresponding to $E_{B,I,D}(t)$ is given by

$$\varepsilon(t) = \varepsilon(X(t)) = \frac{1}{2}\|E_{B,I,D}(t)\|_{\mathrm{F}}^2 = \frac{\|BX(t) - D\|_{\mathrm{F}}^2}{2}.$$

The following dynamic state equation can be derived using the GNN design GNN(B, I, D):

$$\dot{X}(t) = -\gamma_2 B^{\mathrm{T}} \mathcal{F}(BX(t) - D), \quad X(0) = X_0. \tag{3.10.29}$$

It further implies

$$\dot{E}_{B,I,D}(t) = B\dot{X}(t) = -\gamma_2 BB^{\mathrm{T}} \mathcal{F}(BX(t) - D). \tag{3.10.30}$$

A new hybrid model based on the summation of (3.10.28) and (3.10.30) in the case $\gamma_1 = \gamma_2 = 2\gamma$ is proposed as follows

$$\begin{aligned} \dot{E}_{B,I,D}(t) = B\dot{X}(t) &= -\gamma \left(BB^{\mathrm{T}} + I\right) \mathcal{F}(BX(t) - D) \\ &= -\gamma(BB^{\mathrm{T}} + I)\mathcal{F}\left(E_{B,I,D}(t)\right). \end{aligned} \tag{3.10.31}$$

The model (3.10.31) is derived as a combination of the GNN design GNN(B, I, D) and the ZNN design ZNN(B, I, D). Therefore, it is equally reasonable to use the term *Hybrid GNN* (HGNN shortly) or *Hybrid ZNN* (HZNN shortly) model. Firstly, the model (3.10.31) is implicit, so that it is not a kind of GNN dynamics. On the other hand, it is designed for time-invariant matrices, which is not in accordance with the common character of ZNN models, as the commonly promoted advantages of ZNN is its time-varying case treatment. A formal comparison of (3.10.31) and ZNN(B, I, D) reveals that both these methods possess identical left hand sides and the right hand side of (3.10.31) can be derived multiplying the right hand side of ZNN(B, I, D) by the term $BB^{\mathrm{T}} + I$. Formally, (3.10.31) is closer to the ZNN dynamics. So, we decided to denote the model (3.10.31) by HZNN(B, I, D), taking into account that this model is not of the exact ZNN neural dynamics and it is applicable in time-invariant case.

The error matrix for solving the matrix equation $XC = D$ is defined by

$$E_{I,C,D}(t) = X(t)C - D.$$

Accordingly, the ZNN design (3.10.1) corresponding to $E_{I,C,D}$, denoted by ZNN(I, C, D), is defined by

$$\dot{E}_{I,C,D} = \dot{X}(t)C = -\gamma_1 \mathcal{F}(X(t)C - D). \tag{3.10.32}$$

On the other hand, the GNN design formula produces the following dynamic state equation:

$$\dot{X}(t) = -\gamma_2 \mathcal{F}(X(t)C - D)\, C^{\mathrm{T}}, \quad X(0) = X_0. \tag{3.10.33}$$

The GNN model (3.10.33) is denoted by $GNN(I, C, D)$. It implies

$$\dot{E}_{I,C,D} = \dot{X}(t)C = -\gamma_2 \mathcal{F}(X(t)C - D)\, C^{\mathrm{T}}C. \tag{3.10.34}$$

A new hybrid model based on the summation of (3.10.32) and (3.10.34) in the case $\gamma = (\gamma_1 + \gamma_2)/2$ can be stated as follows

$$\dot{E}_{I,C,D}(t) = \dot{X}(t)C = -\gamma \mathcal{F}\left(E_{I,C,D}(t)\right)(C^{\mathrm{T}}C + I). \tag{3.10.35}$$

The model (3.10.35) will be denoted by HZNN(I, C, D).

3.11 Varying-Parameter ZNN for Approximating Some Expressions Involving Outer Inverses

This section is mainly based on the results derived in [229]. In real-world applications, usually, we encounter time-varying problems, where efficiency and accuracy will be badly influenced. An efficient and accurate method for solving this drawback is the dynamical system known as varying-parameter convergent neural network (VP-CDNN), [383, 384, 389]. It has been proved that the VP-CDNN is globally convergent and the super-exponential convergence rate can be achieved. The novelty of the VP-CDNN is that a time-varying parameter is designed according to the time-varying behavior. In this section we examine several variants of the varying-parameter ZNN models.

A varying-parameter ZNN (VPZNN) neural design is defined in [229] for approximating various casses of generalized inverses and expressions involving generalized inverses of complex matrices. The proposed model is termed as CVPZNN(A, F, G) and defined on the basis of the error function which includes three appropriate matrices A, F, G. The CVPZNN(A, F, G) evolution design includes so far defined VPZNN models for computing generalized inverses and it is useful in computing a number of matrix expressions involving outer generalized inverses of A. Global and super-exponential convergence properties of the proposed model as well as properties of its equilibrium state are investigated.

3.11.1 Introduction and Main Idea About Varying-Parameters Design

The aim of the research in [229] was to extend the error function used in models ZNN(A, F, G) proposed in [223]. First extension is from the fixed-parameter ZNN design to varying-parameter ZNN design. Also, ZNN(A, F, G) design which is applicable to real matrices is extended to complex matrix case. The final result is complex varying-parameter convergent-differential neural networks (CVP-CDNN) extension of ZNN(A, F, G), applicable to complex matrices. As a result, a complex-valued varying-parameter ZNN(A, F, G) model, termed as CVPZNN(A, F, G), is defined. That model is defined on the basis of the error function used in [223] and the CVP-CDNN neural design. According to numerical experience, some new values of the time-varying scaling parameter $\gamma(t)$ were proposed. Global and super-exponential convergence determined by $\gamma(t)$ of the proposed model are verified and behavior of its equilibrium state is investigated. Also, a faster super-exponential convergence of CVPZNN(A, F, G) with respect to the exponential convergence of ZNN(A, F, G) is verified.

As mentioned, the Moore-Penrose inverse, the Drazin inverse and several another dominant generalized inverses are outer inverses with prescribed range and null space. The general zeroing design which comprises all previously defined models for computing generalized inverses was originated in [284] and it is aimed to computation of {2}-inverses of time-varying complex matrices. The ZNNATS2-I model defined in [284] was defined using the input matrix $A(t) \in \mathbb{C}_r^{m\times n}$ and an appropriate matrix $G(t) \in \mathbb{C}_s^{n\times m}$, $0 < s \leq r$. After the choice of $G(t)$, ZNNATS2-I is aimed to the computation of $A(t)^{(2)}_{\mathcal{R}(G),\mathcal{N}(G)}$. The ZNNATS2-I design is defined starting from the identities $G(t)A(t)A(t)^{(2)}_{\mathcal{R}(G),\mathcal{N}(G)} = G(t)$ and $A(t)^{(2)}_{\mathcal{R}(G),\mathcal{N}(G)}A(t)G(t) = G(t)$, which implies two dual Zhang functions, proposed in [284]:

$$E_G(t) = \begin{cases} (G(t)A(t) + \lambda I)\, X(t) - G(t), & n \leq m,\ \lambda > 0 \\ X(t)\,(A(t)G(t) + \lambda I) - G(t), & n > m,\ \lambda > 0. \end{cases} \tag{3.11.1}$$

Later, in [223], the authors proposed a more general ZF which is based on the usage of a given matrix $A \in \mathbb{C}^{n\times m}$ and two appropriate matrices $F, G \in \mathbb{C}^{n\times m}$:

$$E_{F,G}(t) = \begin{cases} (G(t)A(t) + \lambda I)\, X(t) - F(t), & n \leq m, \lambda > 0 \\ X(t)\,(A(t)G(t) + \lambda I) - F(t), & n > m, \lambda > 0. \end{cases} \tag{3.11.2}$$

The original ZNN design arising from the Zhang function $E(t)$ and the nonlinear activation function array $\mathcal{H}$ is determined as

$$\dot{E}(t) := \frac{\mathrm{d}E(t)}{\mathrm{d}t} = -\gamma\mathcal{H}(E(t)). \tag{3.11.3}$$

In (3.11.3), $\dot{E}(t)$ denotes the time derivative of $E(t)$ and $\gamma \in \mathbb{R}$, $\gamma > 0$ is the scaling parameter. It is known that larger values of γ lead to faster convergence and enable smaller simulation intervals. Finally, $\mathcal{H}(\cdot) : \mathbb{C}^{n\times m} \mapsto \mathbb{C}^{n\times m}$ denotes a complex-valued element-wise applicable matrix-to-matrix activation function.

An orginal stream in the research of zeroing neural design is the varying-parameter ZNN (VPZNN) pattern defined by

$$\dot{E}(t) = -\gamma(t)\mathcal{H}(E(t)). \tag{3.11.4}$$

The main difference between VPZNN and ZNN design is the usage of the time-varying gain parameter $\gamma(t)$ in (3.11.4) instead of the fixed scalling parameter γ in (3.11.3). The leading idea of the VPZNN evolution design is the ability to avoid the usage of "as large as possible" constant values γ and use appropriate, time-varying and smaller values $\gamma(t)$. Essentially, the VPZNN design uses relatively small values $\gamma(t)$ which will be able to trace dynamically behavior of the error function inside the fixed time interval.

The design parameters in standard ZNN models mentioned above are fixed, which is why these models are referred to as fixed-parameter ZNN (FPZNN). However, since the design parameter represents the reciprocal of the capacitance or inductance coefficients in the circuit, it should be time-varying in order to achieve the largest possible values within the permissible conditions of the circuit. Therefore, from a hardware implementation perspective, a varying parameter ZNN design (VPZNN) is necessary.

3.11.2 Complex Varying-Parameter ZNN (A, F, G) Model

Let $A(t) \in \mathbb{C}_r^{m\times n}$ be a given $m \times n$ time-varying matrix of a constant rank r, let $F(t) \in \mathbb{C}^{n\times m}$ be arbitrary, and $G(t) \in \mathbb{C}_s^{n\times m}$ be of a constant rank $0 < s \leq r$. In [229], the authors followed the following evolution of the error-monitoring functions

$$\begin{aligned} A(t)X(t) - I &\longrightarrow (A(t)^*A(t) + \lambda I)X(t) - A(t)^* \\ &\longrightarrow (G(t)A(t) + \lambda I)X(t) - G(t) \\ &\longrightarrow (G(t)A(t) + \lambda I)X(t) - F(t), \end{aligned}$$

which is equivalent to the subsequent evolution of obtained results:

$$A^{-1} \longrightarrow A^{\dagger} \longrightarrow A^{(2)}_{\mathcal{R}(G),\mathcal{N}(G)} \longrightarrow \lim_{\lambda\to 0}(G(t)A(t)+\lambda I)^{-1}F(t).$$

In continuation, we follow the evolution from ZNN design (3.11.3) to the corresponding VPZNN neural design (3.11.4).

In order to reduce computational effort, two dual cases will be proposed. These cases are analogous, so we will consider only the case $n \leq m$ in details; the case $n > m$ is quite analogous.

The proposed evolution design follows the varying-parameter VPZNN pattern, wherein the underlying Zhang error function $E(t)$ is generalized by $E_{F,G}(t)$ in (3.11.2). More precisely, the following Varying-Parameter ZNN model, termed as CVPZNN(A, F, G), is introduced using the notation $G_1(t) = G(t)A(t) + \lambda I$ and starting from

$$E_{F,G} = G_1(t)X(t) - F(t)$$

and the time derivative

$$\dot{E}_{F,G}(t) = -\gamma(t)\mathcal{H}(E_{F,G}(t)), \text{ where } E_{F,G}(t) = (G(t)A(t) + \lambda I)X(t) - F(t). \tag{3.11.5}$$

The condensed CVPZNN(A, F, G) design (3.11.5) can be expanded as follows:

$$\begin{aligned} G_1(t)\dot{X}(t) &= \dot{F}(t) - \left(\dot{G}(t)A(t) + G(t)\dot{A}(t)\right) X(t) - \gamma(t)\mathcal{H}\left(G_1(t)X(t) - F(t)\right) \\ &= \dot{F}(t) - \dot{G}_1(t)X(t) - \gamma(t)\mathcal{H}(E_{F,G}(t)). \end{aligned} \tag{3.11.6}$$

Equivalent form of (3.11.6) is defined by

$$\dot{X}(t) = (I - G_1(t))\dot{X}(t) + \dot{F}(t) - \dot{G}_1(t)X(t) - \gamma(t)\mathcal{H}(E_{F,G}(t)). \tag{3.11.7}$$

The complex activation is defined as $\mathcal{H}(A + \imath B) = \mathcal{F}(A) + \imath\mathcal{F}(B)$ for $A, B \in \mathbb{R}^{m\times n}$, where $\imath$ is imaginary unit and $\mathcal{F}(\cdot)$ is real-valued activation function array.

Zhang et al. in [383–385, 389, 390] proposed the following adjustable values of the time-varying scaling parameter:

$$\gamma(t) = t^p + p,$$

where $t \in [0, +\infty)$ is the time and it is assumed that p is a real parameter. According to numerical experience, the following time-varying convergence parameters are proposed [229, 398]:

$$\gamma(t) = \begin{cases} t^p + p > 0, & 0 < p \le 1 \\ p^t + 2pt + p > 0, & p > 1. \end{cases} \tag{3.11.8}$$

The dynamical evolution (3.11.6) will be termed as CVPZNN(A, F, G). The ZNN(A, F, G) model is termed as CFPZNN(A, F, G), as in [229]. Clearly, the CFPZNN(A, F, G) design can be achieved using a fixed value $\gamma(t) = p \in \mathbb{R}$ in (3.11.6) or in (3.11.8).

3.11.3 Convergence Analysis of Defined Varying-Parameter Models

Theorem 3.11.1 investigates the equilibrium state of the activation state variables matrix $X(t) = (x_{ij})$. It is expectable that the equilibrium matrix is a function of the time t and the parameter $\lambda \in \mathbb{R}$. So, it can be denoted by $\tilde{X}(t, \lambda)$. Further, the exponential convergence rate of the model $\mathcal{M}$ is denoted by $ECR(\mathcal{M})$.

Convergence Properties of the CVPZNN (A, F, G) Model

Global and super-exponential convergence of CVPZNN(A, F, G) design (3.11.6) is verified in Theorem 3.11.5.

Theorem 3.11.1 ([229]) *Let $A(t) \in \mathbb{C}_r^{m\times n}$ be a given $m \times n$ time-varying matrix of a constant rank r. Further, let $F(t) \in \mathbb{C}^{n\times m}$ be an arbitrary $n \times m$ matrix and $G(t) \in \mathbb{C}_s^{n\times m}$ be a selected $n \times m$ matrix of a constant rank s satisfying $0 < s \le r$. The unknown matrix $X(t)$ of the CVPZNN(A, F, G) dynamics (3.11.6) is globally and super-exponentially convergent as $t \to +\infty$ and achieves the equilibrium state equal to*

$$X(t) \to \tilde{X}(t,\lambda) = (G(t)A(t) + \lambda I)^{-1} F(t). \tag{3.11.9}$$

Proof Let $\overline{X}(t) = X(t) - \tilde{X}(t,\lambda)$, where $\tilde{X}(t,\lambda) = (G(t)A(t) + \lambda I)^{-1} F(t)$. From (3.11.5), after some transformations we obtain

$$E_{F,G}(t) = (G(t)A(t) + \lambda I)X(t) - F(t).$$

$$\begin{aligned}
E_{F,G}(t) &= (G(t)A(t) + \lambda I)\, X(t) - F(t)\\
&= (G(t)A(t) + \lambda I)\left(\overline{X}(t) + \tilde{X}(t,\lambda)\right) - F(t)\\
&= (G(t)A(t) + \lambda I)\overline{X}(t).
\end{aligned}$$

Then the general ZNN pattern $\dot{E}_{F,G}(t) = -\gamma(t)\mathcal{F}(E_{F,G}(t))$ can be expanded into

$$(G(t)A(t) + \lambda I)\,\dot{\overline{X}}(t) = -\left(\dot{G}(t)A(t) + G(t)\dot{A}(t)\right)\overline{X}(t) - \gamma(t)\mathcal{F}\left((G(t)A(t) + \lambda I)\overline{X}(t)\right). \tag{3.11.10}$$

To simplify the proof, we use the notation $C = \mathbf{Re}(E_{F,G}(t)),\ D = \mathbf{Im}(E_{F,G}(t))$. So, $E_{F,G}(t) = C + \imath D$, which further splits the CVPZNN(A, F, G) design into two streams on the real and the imaginary part of $E_{F,G}(t)$:

$$\dot{C} = -\gamma(t)\mathcal{F}(C), \quad \dot{D} = -\gamma(t)\mathcal{F}(D).$$

The Lyapunov function candidate for $E_{F,G}(t)$ is defined by

$$L(t) = \frac{\|E_{F,G}(t)\|_F^2}{2} = \frac{\mathrm{Tr}((E_{F,G}(t))^* E_{F,G}(t))}{2}.$$

Thus,

$$\begin{aligned}
\dot{L}(t) &= \mathrm{Tr}((E_{F,G}(t))^* \dot{E}_{F,G}(t))\\
&= -\gamma(t)\mathrm{Tr}((E_{F,G}(t))^* \mathcal{H}\left(E_{F,G}(t)\right))\\
&= -\gamma(t)\mathrm{Tr}\left((C^{\mathrm{T}} - \imath D^{\mathrm{T}})\left(\mathcal{F}(C) + \imath\mathcal{F}(D)\right)\right).
\end{aligned}$$

Later, using

$$D^{\mathrm{T}}\mathcal{F}(C) = C^{\mathrm{T}}\mathcal{F}(D),$$

one concludes

$$\dot{L}(t) = -\gamma(t)\mathrm{Tr}\left(C^{\mathrm{T}}\mathcal{F}(C) + D^{\mathrm{T}}\mathcal{F}(D)\right).$$

According to the assumption that $\mathcal{F}(\cdot)$ is a monotonically increasing and odd activation function array, it can be concluded

$$\gamma(t)\mathrm{Tr}(C^{\mathrm{T}}\mathcal{F}(C)+D^{\mathrm{T}}\mathcal{F}(D))\geq 0.$$

As a consequence, $\dot{L}(t)\leq 0$. According to Lyapunov stability theory, $E_{F,G}(t)=(G(t)A(t)+\lambda I)X(t)-F(t)$ is globally convergent to zero. Hence, $X(t)\to\tilde{X}(t,\lambda)=(G(t)A(t)+\lambda I)^{-1}F(t)$ as $t\to+\infty$.

In addition, the linear activation $\mathcal{F}=I$ yields

$$\dot{L}(t)=-\gamma(t)\left\|E_{F,G}(t)\right\|_{\mathrm{F}}^{2}=-2\gamma(t)\,L(t).$$

Therefore,

$$L(t)\leq L(0)\exp(-2\gamma(t)),$$

which further implies

$$E_{F,G}(t)=E_{F,G}(0)\,\exp(-\gamma(t)). \tag{3.11.11}$$

Since $\gamma(t)$ is a non-linear function, CVPZNN(A,F,G) possesses the super-exponential convergence. This completes the proof. □

Now, it is possible to compare the exponential convergence rates of CVPZNN(A,F,G) and CFPZNN$(A,F,G)\equiv$ ZNN(A,F,G) by analyzing the quotient

$$\frac{ECR(CVPZNN(A,F,G))}{ECR(CFPZNN(A,F,G))}=\frac{\gamma(t)}{\gamma}.$$

Theorem 3.11.2 ([229]) *Let the conditions of Theorem 3.11.1 are satisfied. Then the super-exponential convergence of CVPZNN(A,F,G) model* (3.11.6) *is faster than the exponential convergence of CFPZNN(A,F,G).*

Proof **(a)** In the case $0<p\leq 1$ it follows that

$$\frac{ECR(CVPZNN(A,F,G))}{ECR(CFPZNN(A,F,G))}=\frac{t^{p}+p}{p}.$$

Denote by $\mathcal{E}_{F,G,\mathcal{M}}(t)$ the value of $E_{F,G}(t)$ corresponding to the model $\mathcal{M}$. This implies

$$\begin{aligned}\frac{\left\|\mathcal{E}_{F,G,CVPZNN(A,F,G)}(t)\right\|}{\left\|\mathcal{E}_{F,G,CFPZNN(A,F,G)}(t)\right\|}&=\frac{E_{F,G}(0)\exp(-t^{p}-p)}{E_{F,G}(0)\exp(-p)}\\&=\exp(-t^{p})<1.\end{aligned}$$

(b) In the case $p>1$ we obtain

$$\frac{\left\|\mathcal{E}_{F,G,CVPZNN(A,F,G)}(t)\right\|}{\left\|\mathcal{E}_{F,G,CFPZNN(A,F,G)}(t)\right\|} = \frac{\exp(-p^t - 2pt - p)}{\exp(-p)}$$
$$= \exp(-p^t - 2pt) < 1.$$

The proof is verified in both possible cases. □

According to [388], it is known that the computation error of the CVPZNN model does not depend on the choice of the complex activation function array. So, in Theorem 3.11.3 we only prove that the proposed CVPZNN models with complex-valued activation function $\mathcal{H}(A + \imath B) = \mathcal{F}(A) + \imath\mathcal{F}(B)$, where $\mathcal{F}(\cdot) : \mathbb{R}^{m\times n} \to \mathbb{R}^{m\times n}$ denotes a real activation function array. For simplicity, we consider the convergence properties of the proposed CVPZNN models with linear activation function merely.

Theorem 3.11.3 ([229]) *Given a smoothly time-varying matrix $A(t) \in \mathbb{C}^{m\times n}$. If complex-valued activation function $\mathcal{F}_1(\cdot)$ with the linear activation function array $\mathcal{F}(\cdot)$ is applied, then CVPZNN(A, F, G) has a faster convergence speed than CFPZNN(A, F, G), and the upper bound of computational error of VPZNN(A, F, G) is smaller than the corresponding computational error of CFPZNN(A, F, G).*

Proof Firstly, we consider the CVPZNN-I with linear function. From the proof of Theorem 3.11.1, we know that

$$\dot{E}_{F,G}(t) = -\gamma(t)(\mathbf{Re}(E_{F,G}(t) + \imath\,\mathbf{Im}(E_{F,G}(t)), \tag{3.11.12}$$

where $\mu(t)$ is defined in (3.11.8). If $e_{ij}(t)$ is the $(ij)th$ element of $E_{F,G}(t)$, then the scalar-type of (3.11.12) is given by

$$\dot{e_{ij}}(t) = -\mu(t)(\mathbf{Re}(e_{ij}(t) + \imath\,\mathbf{Im}(e_{ij}(t)),\ \forall i, j \in \{1, \ldots, n\},$$

Set

$$c_{ij}(t) = \mathbf{Re}(e_{ij}(t)),\ d_{ij}(t) = \mathbf{Im}(e_{ij}(t)).$$

(a) In the case $0 < p \leq 1$, $\gamma(t) = t^p + p$, $p > 0$ and $t \in [0, \infty)$, as a consequence of (3.11.11), we calculate

$$c_{ij}(t) = \exp\left(-\frac{t^{p+1}}{p} - pt\right) c_{ij}(0)$$
$$d_{ij}(t) = \exp\left(-\frac{t^{p+1}}{p} - pt\right) d_{ij}(0).$$

It implies

$$e_{ij}(t) = \exp\left(-\frac{t^{p+1}}{p} - pt\right) e_{ij}(0).$$

The Frobenius norm of $\tilde{V}(t) = V(t) - A^{\circledast}(t)$ is equal to $\|\tilde{V}(t)\|_F = \sqrt{\sum_{i=1}^{n}\sum_{j=1}^{n}|\tilde{v}_{ij}(t)|^2}$, where $\tilde{v}_{ij}(t)$ is the (ij)th element of the matrix $\tilde{V}(t)$. From the proof of Theorem 3.11.1, we know that

$$E_{F,G}(t) = \tilde{V}(t)(A(t)G(t) + \lambda I).$$

Set $K(t) = (A(t)G(t) + \lambda I)$, $\lambda > 0$. Since $K(t)$ is invertible, we obtain the subsequent equalities

$$\|K^{-1}(t)\|_{\mathrm{F}} = \sqrt{\|SVD(K^{-1}(t))\|_2^2} = \|SVD(K^{-1}(t))\|_2 \le \varphi_1,$$

where $\varphi_1 > 0$ and $SVD(B)$ is a vector containing the singular values of the matrix B. This implies

$$\begin{aligned}\|\tilde{V}(t)\|_F &\le \|E_{F,G}(t)\|_F \|K^{-1}(t)\|_F \\ &\le \varphi_1 \sqrt{\sum_{i=1}^{n}\sum_{j=1}^{n}|e_{ij}(t)|^2} \\ &= k_1\varphi_1 \exp\left(-\frac{t^{p+1}}{p} - pt\right) = \mathbb{V}_{\mathbb{L}},\end{aligned}$$

where $k_1 = \|E(0)\|_F$. From (3.11.6), the design of CVPZNN(A, F, G) can be stated as

$$\begin{aligned}G_1(t)\dot{X}(t) =& \dot{G}(t) - \left(\dot{G}(t)A(t) + G(t)\dot{A}(t)\right)X(t) \\ &- \gamma\mathcal{F}\left(G_1(t)X(t) - F(t)\right),\end{aligned} \tag{3.11.13}$$

such that $\gamma = p$. Similarly, the computational error of CFPZNN(A, F, G) is equal to

$$\|\tilde{V}_{CFP}(t)\|_F = k_1\varphi_1 \exp(-pt) = \mathbb{V}.$$

Consequently,

$$\frac{\mathbb{V}_{\mathbb{L}}}{\mathbb{V}} = \frac{k_1\varphi_1 \exp\left(-\frac{t^{p+1}}{pt} - p\right)}{k_1\varphi_1 \exp(-pt)} = \exp\left(-\frac{t^{p+1}}{p}\right) < 1.$$

(b) In this case, it is assumed $p > 1$, $\gamma(t) = p^t + 2pt + p$, $p > 0$ and $t \in [0, \infty)$. Similarly as in the proof of part (a), we obtain

$$e_{ij}(t) = \exp\left(-p^t t - 2pt^2 - pt\right)e_{ij}(0).$$

It further implies

$$\begin{aligned}\|\tilde{V}(t)\|_F &\le \|E_{F,G}(t)\|_F \|K^{-1}(t)\|_F \\ &\le \varphi_1 \sqrt{\sum_{i=1}^{n}\sum_{j=1}^{n} |e_{ij}(t)|^2} \\ &= k_1\varphi_1 \exp\left(-p^t t - 2pt^2 - pt\right) = \mathbb{V}_{\mathbb{LL}}.\end{aligned}$$

As a result,

$$\frac{\mathbb{V}_{\mathbb{LL}}}{\mathbb{V}} = \frac{k_1\varphi_1 \exp\left(-p^t t - 2pt^2 - pt\right)}{k_1\varphi_1 \exp(-pt)} = \exp\left(-p^t t - 2pt^2\right) < 1.$$

The proof is finalized in both cases. □

3.11.4 Simulation Results on CVPZNN (A, F, G)

Subsystem for the implementation of $G_1(t) = G(t)A(t) + \lambda I$ is given in Fig. 3.73.

Simulink implementation of (3.11.7) is given in Fig. 3.74.

Several representative computer-simulation experiments are performed in order to illustrate performances of CVPZNN(A, F, G). It is important to mention that the MATLAB solver `ode15s` was utilized. Also, the varying-parameter $\gamma(t)$ is defined as in (3.11.8) and the fixed-parameter of the CFPZNN model is assumed to be equal to $\gamma = p$ during all simulation experiments.

Example 3.11.1 Consider the time-varying invertible matrix

$$A(t) = \begin{bmatrix} \sin(t) & \cos(t) \\ -\cos(t) & \sin(t) \end{bmatrix}.$$

The regularization parameter is $\lambda = 10^{-8}$, the time-dependent gain parameter is $\gamma(t) = 3^t + 6t + 3$ and the time interval is $t \in [0, 5]s$. Two initial values: (i) the non-zero initial vector equal to $v(0) = (1, 1)^{\mathrm{T}}$ and

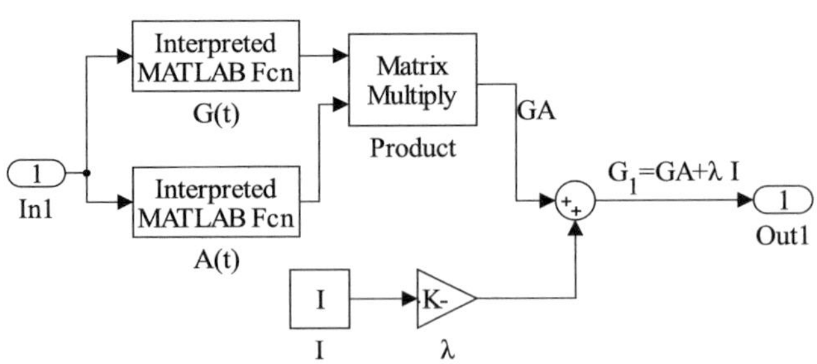

Fig. 3.73 Simulink implementation of $G_1(t) = G(t)A(t) + \lambda I$

(ii) a randomly generated vector $v(0) = \texttt{rand}(2, 1)$.

Residual errors $\|X(t) - A^{-1}(t)\|_F$ generated by CVPZNN(A, I, I) under the influence of different activation functions are presented in Fig. 3.75, where the blue line indicates that $f(\cdot)$ is the linear function, green line means that $f(\cdot)$ is the bipolar-sigmoid activation function with $\varrho = 3$, red line is generated using the power-sigmoid activation with $\varrho = 3$ and $\rho = 3$, pink line corresponds to the Li function with $r = 0.5$, black line indicates that $f(\cdot)$ is tunable activation functions with $r = 0.5$, $k_1 = k_3 = 1$ and $k_2 = 0.5$. The non-zero initial vector is equal to $v(0) = (1, 1)^{\mathrm{T}}$.

Example 3.11.2 Consider the complex time-varying matrix

$$A(t) = \begin{bmatrix} 0.5\imath\cos(3t) & \sin(3t) & 0.5\imath\cos(3t) \\ -\imath\sin(3t) & 2\cos(3t) & -\imath\sin(3t) \\ 0.5\imath\cos(3t) & \sin(3t) & 0.5\imath\cos(3t) \end{bmatrix}$$

with the constant index $\mathrm{ind}(A(t)) = 1$. The theoretical core inverse of $A(t)$ is equal to

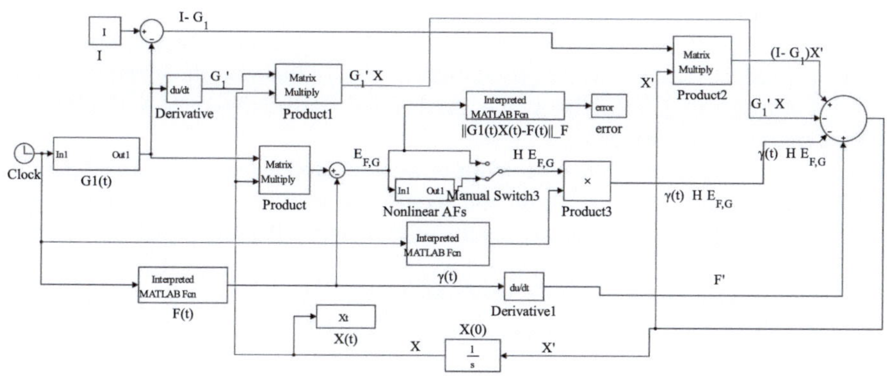

Fig. 3.74 Simulink implementation of CVPZNN(A, F, G) dynamics (3.11.7)

Fig. 3.75 Errors of CVPZNN(A, I, I) with different activation functions in Example 3.11.1a

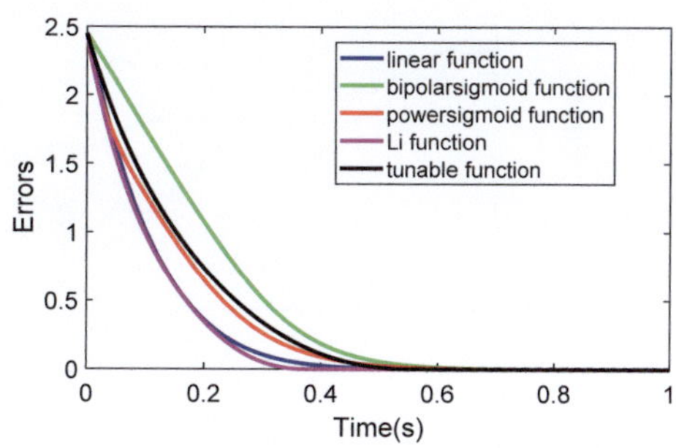

$$A^{\circledast}(t) = \begin{bmatrix} -\frac{\imath\cos(3t)}{2} & \frac{\imath\sin(3t)}{2} & -\frac{\imath\cos(3t)}{2} \\ \frac{\sin(3t)}{2} & \frac{\cos(3t)}{2} & \frac{\sin(3t)}{2} \\ -\frac{\imath\cos(3t)}{2} & \frac{\imath\sin(3t)}{2} & -\frac{\imath\cos(3t)}{2} \end{bmatrix}.$$

Further, one can verify

$$I - A(t)A^{\circledast}(t) = \begin{bmatrix} \frac{1}{2} & 0 & -\frac{1}{2} \\ 0 & 0 & 0 \\ -\frac{1}{2} & 0 & \frac{1}{2} \end{bmatrix}.$$

(a) Values of the parameters in this example are chosen as $p = 0.1$, $\lambda = 10^{-6}$, $t \in [0, 10]$ and the initialization is $v(0) = (1, 1, 1)^{\mathrm{T}}$. For $G(t) = F(t) = AA^*$, values of the residual $\|X(t) - A^{\circledast}(t)\|_F$ corresponding to CVPZNN(A, AA^*, AA^*) with various activation functions are plotted in Fig. 3.56 (Fig. 3.76).

On the other hand, the residual values $\|X(t) - A^{\circledast}(t)\|_F$ of CVPZNN$(A, A(A^2)^*, A(A^2)^*)$ under various activation functions are shown in Fig. 3.77.

(b) It is known that the models CVPZNN$(A, \lambda I, AA^*)$ and CVPZNN$(A, \lambda I, A(A^2)^*)$ yield $I - A(t)A^{\circledast}(t)$.

According to the above figures, it is easy to illustrate the effectiveness of the proposed models. Clearly, the proposed models with Li function show better convergence. Furthermore, CVPZNN(A, AA^*, AA^*) and CVPZNN$(A, A(A^2)^*, A(A^2)^*)$ exhibit similar performances.

(c) Finally, we test convergence behavior of defined models under two fixed parameters, $\gamma = 0.1$ and $\gamma = 3$. Choose the fixed-parameter $\gamma = 0.1$. Then $p = 0.1$, $\gamma(t) = t^{0.1} + 0.1$. The comparison results of models CVPZNN(A, AA^*, AA^*) and CFPZNN(A, AA^*, AA^*) are shown in Fig. 3.78.

It is observable that errors of CVPZNN(A, AA^*, AA^*) converge to zero about 5 s, while the errors of model CFPZNN(A, AA^*, AA^*) close to zero about 30 s.

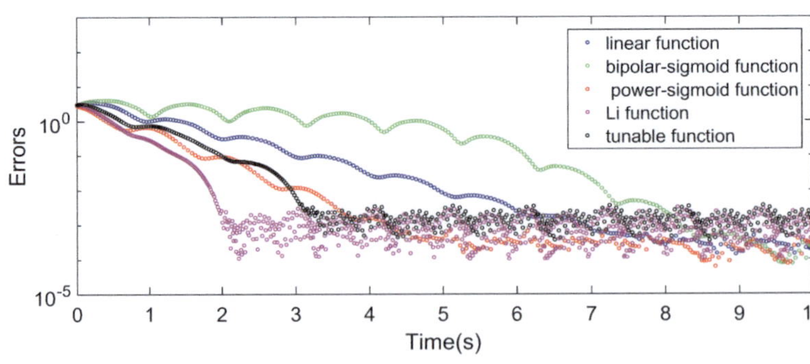

Fig. 3.76 Errors of CVPZNN(A, AA^*, AA^*) with different activation functions in Example 3.11.2a

This results illustrate the faster convergence of CVPZNN(A, AA^*, AA^*) compared to CFPZNN(A, AA^*, AA^*).

3.11.5 Simulation Results for Dual Case of CVPZNN (A, F, G)

In this subsection we investigate the behavior of CVPZNN(A, F, G) in the dual case

$$E_{F,G}(t) = X(t)\,(A(t)G(t) + \lambda I) - F(t), \quad n > m, \lambda > 0.$$

For simplicity, we use the notation

$$G_2(t) = A(t)G(t) + \lambda I.$$

More precisely, the following Varying-Parameter ZNN model, termed as CVPZNN(A, F, G), is considered:

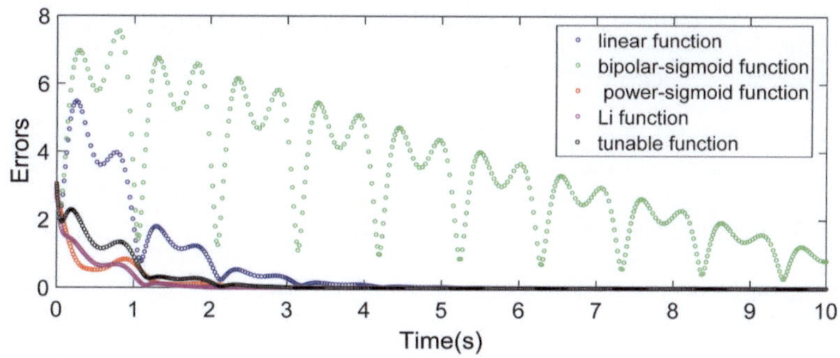

Fig. 3.77 Errors of CVPZNN(A, $A(A^2)^*$, $A(A^2)^*$) with different activation functions in Example 3.11.2a

Fig. 3.78 Errors of CVPZNN(A, AA^*, AA^*) with different varying parameters in Example 3.11.2c

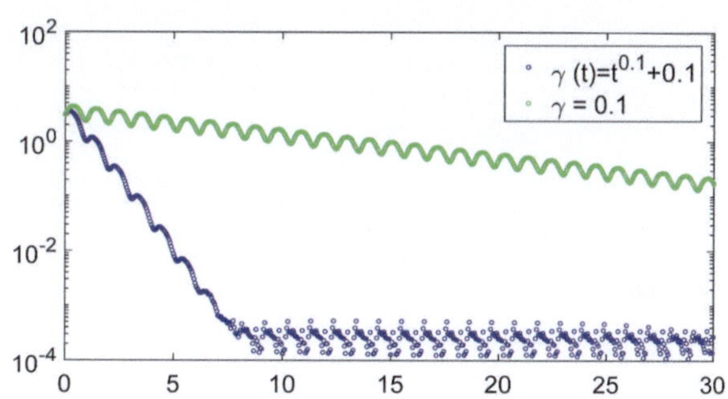

$$\dot{E}_{F,G}(t) = -\gamma(t)\mathcal{H}(E_{F,G}(t)), \text{ where } E_{F,G}(t) = X(t)G_2(t) - F(t). \tag{3.11.14}$$

The notation $G_2(t) = A(t)G(t) + \lambda I$ implies $E_{F,G}(t) = X(t)G_2(t)$, and the underlying CVPZNN(A, F, G) design (3.11.5) is expanded as

$$\begin{aligned}\dot{X}(t)G_2(t) &= \dot{F}(t) - X(t)\left(\dot{A}(t)G(t) + A(t)\dot{G}(t)\right) - \gamma(t)\mathcal{H}\left(X(t)G_2(t) - F(t)\right)\\ &= \dot{F}(t) - X(t)\dot{G}_2(t) - \gamma(t)\mathcal{H}(E_{F,G}(t))\end{aligned} \tag{3.11.15}$$

or

$$\dot{X}(t) = \dot{X}(t)(I - G_2(t)) + \dot{F}(t) - X(t)\dot{G}_2(t) - \gamma(t)\mathcal{H}(E_{F,G}(t)). \tag{3.11.16}$$

Subsystem for the implementation of $G_2(t) = A(t)G(t) + \lambda I$ is given in Fig. 3.79. Simulink implementation of (3.11.16) is given in Fig. 3.80.

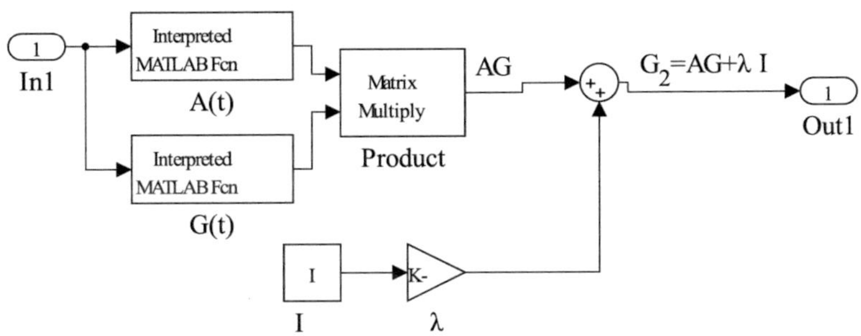

Fig. 3.79 Simulink implementation of $G_2(t) = A(t)G(t) + \lambda I$

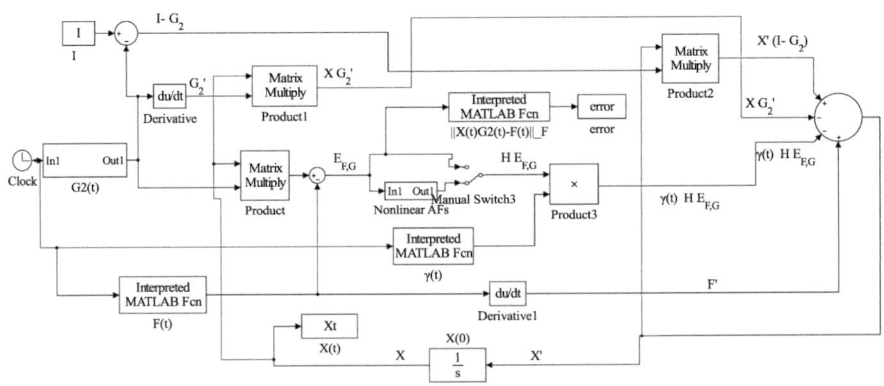

Fig. 3.80 Simulink implementation of dual CVPZNN(A, F, G) dynamics (3.11.16)

Numerical Examples

Example 3.11.3 Let us compute the inverse

$$A^{-1}(t) = \frac{4}{3}\begin{bmatrix} -1 + 0.5\cos(2t) & -0.5\sin(2t) \\ -0.5\sin(2t) & -1 - 0.5\cos(2t) \end{bmatrix},$$

of the following matrix

$$A(t) = \begin{bmatrix} -1 - 0.5\cos(2t) & 0.5\sin(2t) \\ 0.5\sin(2t) & -1 + 0.5\cos(2t) \end{bmatrix}.$$

Using the regularization $\lambda = 10^{-6}$ and $G = F = I_2$, $X(0) = ones(2, 2)$, the following figure is extracted for different values of the time-varying parameter $\gamma(t)$.

From Fig. 3.81, it is observable that the use of the parameter $\gamma(t) = 5^t + 10t + 5$ accelerates the convergence rate of the CVPZNN(A, F, G) dynamics (3.11.16) instead of the use of the parameter $\gamma(t) = t^{0.1} + 0.1$ because of the error norm $||X(t)(A(t) + 10^{-6} \cdot I) - I||_F$ tends to zero in a shorter time instant. Furthermore, one more factor for the acceleration of the convergence speed in CVPZNN is utilization of the tunable activation function instead of the use of linear, power-sigmoid and Li functions.

Example 3.11.4 Consider the time-variant matrix

$$A(t) = \begin{bmatrix} t & 1 & 0 \\ 0 & 0 & 1 \\ -t^3 & -t^2 & -t \end{bmatrix}.$$

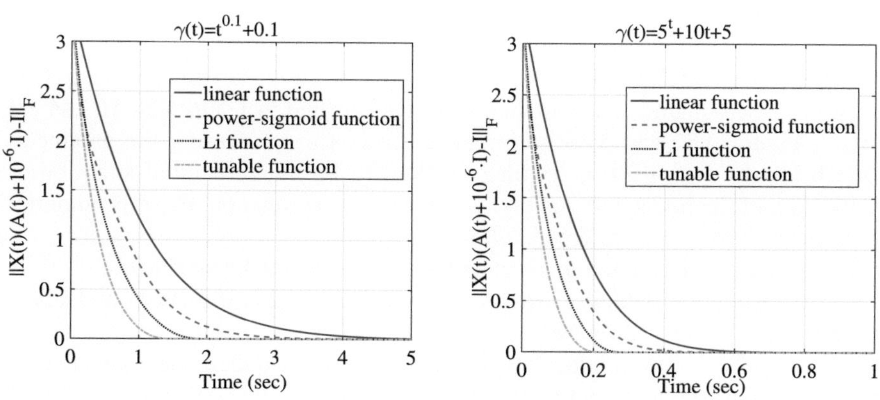

Fig. 3.81 $||X(t)(A(t) + 10^{-6} \cdot I) - I||_F$ behaviour of CVPZNN model (3.11.16) under different activation functions when: **a** $\gamma(t) = t^{0.1} + 0.1$, **b** $\gamma(t) = 5^t + 10t + 5$ in Example 3.11.3

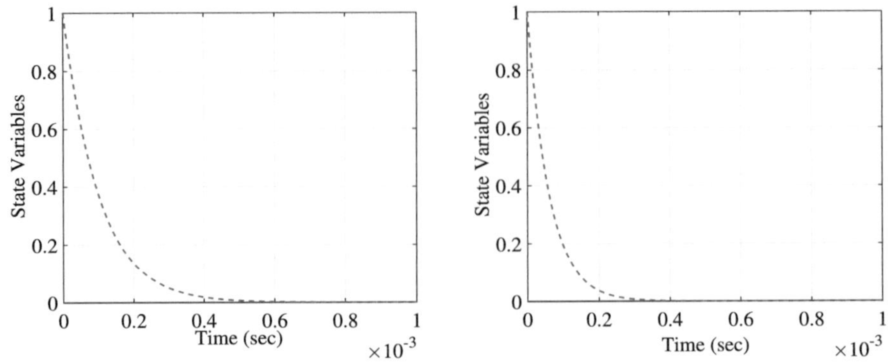

Fig. 3.82 Elementwise trajectories of CVPZNN($A(t)$,$G(t)$,$G(t)$) design (3.11.16) for $\gamma(t) = a^t + 2at + a$ with 10^4 and: **a** Linear AF, **b** Power-sigmoid AF in Example 3.11.4

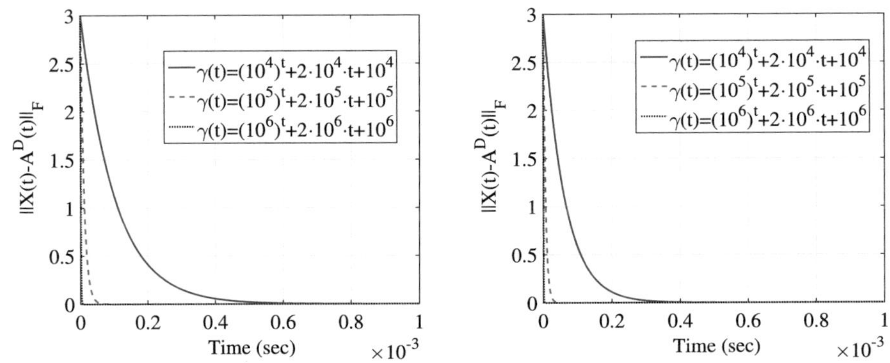

Fig. 3.83 $||X(t) - A^{\mathrm{D}}(t)||_F$ behaviour of CVPZNN design (3.11.16) for $\gamma(t) = a^t + 2at + a$ with $a = 10^4, 10^5, 10^6$ and: **a** Linear AF, **b** Power-sigmoid AF in Example 3.11.4

Its index is constant and equal to Ind(A) =3, and the theoretical Drazin inverse $A^{\mathrm{D}}(t)$ is the 3×3 zero matrix. The Drazin inverse is generated by CVPZNN($A(t)$, $A^3(t)$, $A^3(t)$). Figures 3.82 and 3.83 are produced from (3.11.16) using the parameters $\lambda = 10^{-6}, \gamma(t) = a^t + 2at + a$ with $a = 10^4, 10^5, 10^6$, the zero initial condition and application of the linear and power-sigmoid activation function. For this numerical experiment the conclusions are as follows.

1. The power-sigmoid function gives better results than the linear function.
2. Better results are obtained, as the constant a is increased in the time-varying parameter $\gamma(t) = a^t + 2at + a$.
3. The red dashed lines, produced by (3.11.16) converge to theoretical Drazin (blue lines) with application of $\gamma(t) = (10^4)^t + 2 \cdot 10^4 t + 10^4$, linear and power-sigmoid functions.

Example 3.11.5 Consider

$$A(t) = S_5(t) = \begin{bmatrix} t+1 & t & t & t & t+1 \\ t & t-1 & t & t & t \\ t & t & t+1 & t & t \\ t & t & t & t-1 & t \\ t+1 & t & t & t & t+1 \end{bmatrix},$$

and $G(t) = F(t) = U(t)V(t)$, where $U(t)$ and $V(t)$ are equal to

$$U(t) = \begin{bmatrix} 2t+1 & t & t \\ t & 2t-1 & t \\ t & t & 2t+1 \\ t & t & t \\ 2t+1 & t & t \end{bmatrix},$$

$$V(t) = \begin{bmatrix} t^2+1 & t^2 & t^2 & t^2 & t^2+1 \\ t^2 & t^2-1 & t^2 & t^2 & t^2 \\ t^2 & t^2 & t^2+1 & t^2 & t^2 \end{bmatrix},$$

with $\lambda = 10^{-5}$, $X(0) = zeros(5, 5)$ and different values of $\gamma(t)$, such as $\gamma(t) = a^t + 2at + a$ and $\gamma(t) = t^a + a$ with $a = 10^4$, following figures are obtained.

State variables of the model CVPZNN(A(t),G(t),G(t)) with the parameters $\gamma(t) = a^t + 2at + a$ and $\gamma(t) = t^a + a$, $a = 10^4$ are shown in Figs. 3.85 and 3.86, respectively. Red dashed lines in these figures show the values corresponding to the solution computed by the CVPZNN($A(t)$, $G(t)$, $G(t)$) and blue lines display the values corresponding to the exact outer inverse of $A(t)$. The residuals $||X(t)G_2(t) - F(t)||_F$ for $\gamma(t) = a^t + 2at + a$, $\gamma(t) = t^a + a$ with $a = 10^4$ are presented in Fig. 3.84. The conclusion is that different values of the parameter $\gamma(t)$ have influence on the con-

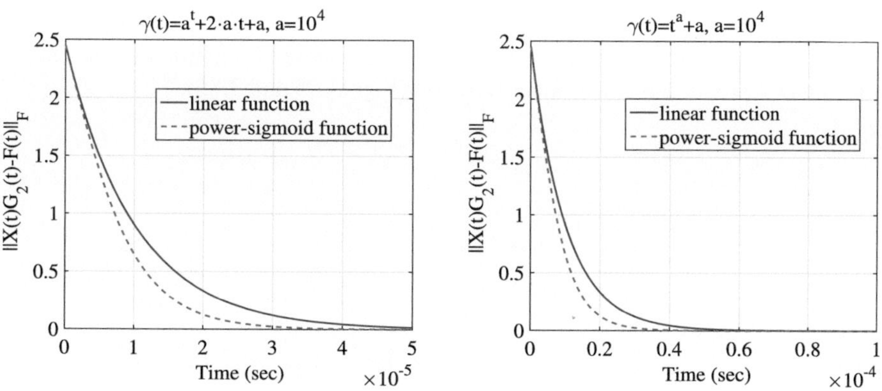

Fig. 3.84 $||X(t)G_2(t) - F(t)||_F$ behaviour from CVPZNN model (3.11.16) for $a = 10^4$ and: **a** $\gamma(t) = a^t + 2at + a$, **b** $\gamma(t) = t^a + a$ for the Example 3.11.5

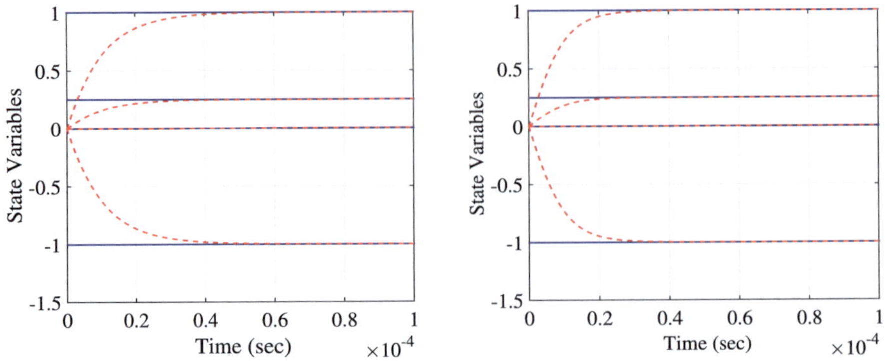

Fig. 3.85 State variables from CVPZNN model (3.11.16) for $\gamma(t) = a^t + 2at + a$, $a = 10^4$ and: **a** linear function, **b** power-sigmoid function for the Example 3.11.5

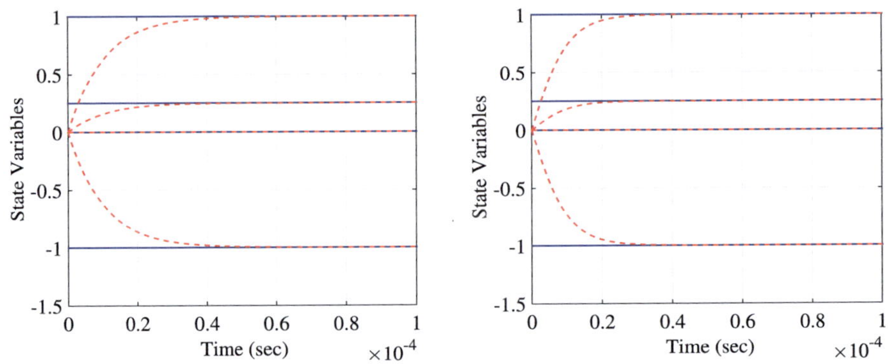

Fig. 3.86 State variables from CVPZNN model (3.11.16) for $\gamma(t) = t^a + a$, $a = 10^4$ and: **a** linear function, **b** power-sigmoid function for the Example 3.11.5

vergence of the CVPZNN. Furthermore, the power-sigmoid function includes better convergence property against linear function (Figs. 3.85 and 3.86).

Chapter 4
Final Conclusions

The primary goal of this book is to apply the dynamical system approach in numerical linear algebra, particularly in computing generalized inverses, solving systems of linear equations, and solving linear matrix equations. Four main scientific areas are considered: numerical linear algebra, recurrent neural networks (RNNs), dynamical systems, and unconstrained nonlinear optimization.

RNN allows the output from some nodes to affect subsequent input to the same nodes. A continuous-time recurrent neural network (CTRNN) uses a system of ordinary differential equations to define effects of incoming inputs on neurons. We consider CTRNN dedicated to finding zeros of equations or minimizing nonlinear functions. Two important classes of CTRNNs are known: Gradient Neural Networks (GNN) and Zhang (or Zeroing) Neural Networks (ZNN). Both time-varying and time-invariant models are considered in scalar, vector, and matrix cases.

This book is based on a collection of papers published in leading scientific journals. It covers various scientific disciplines, including linear and multilinear algebra, generalized inverses, recurrent neural networks, dynamical systems, solving time-varying problems, and unconstrained nonlinear optimization. A global overview of various activation functions is given, along with a rigorous convergence analysis. In addition, some improvements in defining error functions in GNN and ZNN dynamic systems are considered.

The book presents a unique continuous-time approach to numerical linear algebra, focusing on computing generalized inverses and solving scalar, vector, and matrix equations. We aim to present results that are both practically applicable and theoretically strong.

P. Stanimirović et al., *Generalized Matrix Inversion: A Machine Learning Approach*,
https://doi.org/10.1007/978-3-032-01493-1_4

References

1. K.M. Abadir, J.R. Magnus, *Matrix Algebra*. Econometric Exercises, vol. 1. (Cambridge University Press, Cambridge, 2005)
2. N. Andrei, An acceleration of gradient descent algorithm with backtracking for unconstrained optimization. Numer. Algor. **42**, 63–73 (2006)
3. L. Armijo, Minimization of functions having Lipschitz first partial derivatives. Pac. J. Math. **16**, 1–3 (1966)
4. O.M. Baksalary, G. Trenkler, Core inverse of matrices. Linear Multilinear Algebra **58**, 681–697 (2010)
5. A. Ben-Israel, On matrices of index zero or one. SIAM J. Appl. Math. **17**, 1118–1121 (1969)
6. A. Ben-Israel, T.N.E. Greville, *Generalized inverses: Theory and Applications*, 2nd ed. (Springer, 2003)
7. A. Ben-Israel, Generalized inverses of matrices: a perspective of the work of Penrose. Math. Proe. Camb. Phil. Soc. **100**, 407–425 (1986)
8. A. Ben-Israel, The Moore of the Moore-Penrose inverse. Electron. J. Linear Algebra **9**, 150–157 (2002)
9. A. Benchabane, A. Bennia, F. Charif, Continuous-time Zhang neural networks for AR spectral estimator, in *The International Conference On Electronics & Oil: From Theory To Applications, 05–06 Mar 2013* (Ouargla, Algeria, 2013)
10. S.P. Bhat, D.S. Bernstein, Finite-time stability of continuous autonomous systems. SIAM J. Control. Optim. **38**, 751–766 (2000)
11. A. Bjerhammar, *A Generalized Matrix Algebra* (Lindståhl, 1958)
12. D. Boob, Q. Deng, G. Lan, Stochastic first-order methods for convex and nonconvex functional constrained optimization. Math. Program. **127** (2022)
13. S. Boyd, L. Vandenberghe, *Convex Optimization* (Cambridge University Press, 2006)
14. D. Brandwood, A complex gradient operator and its application in adaptive array theory. Proc. Inst. Elect. Eng. **130**, 11–16 (1983)
15. R.W. Brockett, Dynamical systems that sort lists, diagonalize matrices, and solve linear programming problems. Linear Algebra Appl. **146**, 79–91 (1991)
16. J. Cai, W. Dai, J. Chen, C. Yi, Zeroing neural networks combined with gradient for solving time-varying linear matrix equations in finite time with noise resistance. Mathematics **10**, 4828 (2022)
17. S.L. Campbell, C.D. Meyer, Generalized Inverses, of linear transformations, Dover Publications Inc, New York, *Corrected reprint of the 1979 original* (SIAM, Philadelphia, 1991), p. 2008
18. S.L. Campbell, *Singular Systems of Differential Equations, Pitman* (Advanced Publishing Program Mass.-London, Boston, 1980)

P. Stanimirović et al., *Generalized Matrix Inversion: A Machine Learning Approach*,
https://doi.org/10.1007/978-3-032-01493-1

19. S.L. Campbell, C.D. Meyer, N.J. Rose, Applications of the Drazin inverse to linear systems of differential equations with singular constant coefficients. SIAM J. Appl. Math. **31**(3), 411–425 (1976)
20. C.-G. Cao, X. Zhang, The generalized inverse $A^{(2)}_{T,*}$ and its applications. J. Appl. Math. Comput. **11**, 155–164 (2003)
21. A. Cauchy, Metode generale pour la resolution des systems d'equations simultanees. Comp. Rend. Sci. **25**, 46–89 (1847)
22. Y. Chai, H. Li, D. Qiao, S. Qin, J. Feng, A neural network for Moore-Penrose inverse of time-varying complex-valued matrices. Int. J. Comput. Intell. Syst. **13**(1), 663–671 (2020)
23. D. Chen, S. Li, Q. Wu, Rejecting chaotic disturbances using a super-exponential-zeroing neurodynamic approach for synchronization of chaotic sensor systems. Sensors (2019). https://doi.org/10.3390/s19010074
24. H. Chen, Y. Wang, A family of higher-order convergent iterative methods for computing the Moore-Penrose inverse. Appl. Math. Comput. **218**, 4012–4016 (2011)
25. K. Chen, Recurrent implicit dynamics for online matrix inversion. Appl. Math. Comput. **219**(20), 10218–10224 (2013)
26. K. Chen, Implicit dynamic system for online simultaneous linear equations solving. Electron. Lett. **49**, 101–102 (2013)
27. K. Chen, C. Yi, Robustness analysis of a hybrid of recursive neural dynamics for online matrix inversion. Appl. Math. Comput. **273**, 969–975 (2016)
28. K. Chen, S. Yue, Y. Zhang, *MATLAB Simulation and Comparison of Zhang Neural Network and Gradient Neural Network for online solution of linear time-varying matrix equation $AXB - C = 0$*, ed. by D.-S. Huang et al. ICIC 2008, LNAI 5227 (2008), pp. 68–75. 10.1007/978-3-540-85984-0_9
29. K. Chen, Z. Zhang, An improved recurrent network for online equality-constrained quadratic programming, in *Advances in Brain Inspired Cognitive Systems. BICS 2016* C.L. Liu, A. Hussain, B. Luo, K. Tan, Y. Zeng, Z. Zhang. Lecture Notes in Computer Science, vol. 10023 (Springer, Cham)
30. Y.L. Chen, Representation and approximation for the Drazin inverse $A^{(d)}$. Appl. Math. Comput. **119**, 147–160 (2001)
31. Y.L. Chen, Finite algorithms for (2)-generalized inverse $A^{(2)}_{T,S}$. Linear Multilinear Algebra **40**, 61–68 (1995)
32. Y.L. Chen, A Cramer rule for solution of the general restricted linear equation. Linear Multilinear Algebra **34**, 177–186 (1993)
33. Y.L. Chen, The generalized Bott-Duffin inverse and its application. Linear Algebra Appl. **134**, 71–91 (1990)
34. Y.L. Chen, X. Chen, Representation and approximation of the outer inverse $A^{(2)}_{T,S}$ of a matrix A. Linear Algebra Appl. **308**, 85–107 (2000)
35. L. Cheng, Z.-G. Hou, M. Tan, A simplified neural network for linear matrix inequality problems. Neural Process. Lett. **29**, 213–230 (2009)
36. A. Cichocki, Neural network for singular value decomposition. Electron. Lett. **28**(8), 784–786 (1992)
37. A. Cichocki, T. Kaczorek, A. Stajniak, Computation of the Drazin inverse of a singular matrix making use of neural networks. Bull. Polish Acad. Sci. Techn. Sci. **40**, 387–394 (1992)
38. A. Cichocki, R. Unbehauen, Neural networks for computing eigenvalues and eigenvectors. Biol. Cybern. **68**(2), 155–164 (1992)
39. A. Cichocki, R. Unbehauen, Neural networks for solving systems of linear equations and related problems. IEEE Trans. Circuits Syst. I Fund. Theory Appl. **39**(2), 124–138 (1992)
40. J.-J. Climent, N. Thome, Y. Wei, A geometrical approach on generalized inverses by Neumann-type series. Linear Algebra Appl. **332–334**, 533–540 (2001)
41. R.E. Cline, Inverses of rank invariant powers of a matrix. SIAM J. Numer. Anal. **5**, 182–197 (1968)
42. A.I. Cohen, Stepsize analysis for descent methods. J. Optim. Theory Appl. **33**, 187–205 (1981)

43. A.R. Conn, N.I.M. Gould, P.L. Toint, Convergence of quasi-Newton matrices generated by the symmetric rank one update. Math. Program. **50**, 177–195 (1991)
44. D.S. Cvetković-Ilić, Y. Wei, *Algebraic Properties of Generalized Inverses*, Developments in Mathematics, vol. 52. (Springer, Singapore, 2017)
45. J. Dai, L. Jia, L. Xiao, Design and analysis of two prescribed-time and robust ZNN models with application to time-variant Stein matrix equation. IEEE Trans. Neural Netw. Learn. Syst. **32**, 1668–1677 (2020)
46. J. Dai, X. Yang, L. Xiao, L. Jia, Y. Li, ZNN with Fuzzy Adaptive Activation Functions and Its Application to Time-Varying Linear Matrix Equation. IEEE Trans. Industr. Inf. **18**(4), 2560–2570 (2022)
47. J. Dai, X. Yang, L. Xiao, L. Jia, Y. Li, A fuzzy adaptive zeroing neural network with superior finite-time convergence for solving time-variant linear matrix equations. Knowl. Based Syst. **242**, 108405 (2022)
48. L. Dai, Y. Zhang, G. Geng, Norm-based finite-time convergent recurrent neural network for dynamic linear inequality. IEEE Trans. Industr. Inf. **20**(3), 4874–4883 (2024)
49. L. Ding, B. Liao, R. Lu, H. Peng, An improved recurrent neural network for complex-valued systems of linear equation and its application to robotic motion tracking. Front. Neurorobot. **11** (2017). https://doi.org/10.3389/fnbot.2017.00045
50. L. Ding, L. Xiao, K. Zhou, Y. Lan, Y. Zhang, Two novel finite time convergent recurrent neural networks for tackling complex-valued systems of linear equation. Filomat **34**, 5009–5018 (2020)
51. D.S. Djordjević, P.S. Stanimirović, Y. Wei, The representation and approximation of outer generalized inverses. Acta Math. Hungar. **104**, 1–26 (2004)
52. D.S. Djordjević, P.S. Stanimirović, General representations of pseudoinverses. Mat. Vesnik **51**, 69–76 (1999)
53. D.S. Djordjević, P.S. Stanimirović, Iterative methods for computing generalized inverses related with optimization methods. J. Aust. Math. Soc. **78**, 257–272 (2005)
54. M.P. Drazin, A class of outer generalized inverses. Linear Algebra Appl. **436**, 1909–1923 (2012)
55. D.K. Faddeev, I.S. Sominskii, *Collection of Problems on Higher Algebra*, 2nd ed. (Gostekhizdat, Moscow 1949). (5th ed. 1954)
56. C. Fang, C.J. Li, Z. Lin, T. Zhang, Spider: near-optimal non-convex optimization via stochastic path-integrated differential estimator. Adv. Neural Inf. Process. Syst. 687—697 (2018)
57. D.E. Ferreyra, F.E. Levis, N. Thome, Revisiting the core-EP inverse and its extension to rectangular matrices. Quaest. Math. **41**, 265–281 (2018)
58. D.E. Ferreyra, F.E. Levis, N. Thome, Characterizations of k-commutative equalities for some outer generalized inverses. Linear Multilinear Algebra **68**(1), 177–192 (2018)
59. E. Fosse, C. Winship, Moore-Penrose estimators of age-period-cohort effects: their interrelationship and properties. Sociological. Science **5**, 304–334 (2018)
60. C. Gao, Y. Kong, T. Hu, Solving time-varying complex matrix equations based on finite time neural network. J. Zhejiang Univ. Sci. Technol. **34**(5), 409–418 (2022). ((In Chinese))
61. Y.F. Gao, J.L. Chen, Pseudo core inverses in rings with involution. Commun. Algebra **46**, 38–50 (2018)
62. Y. Gao, Z. Tang, Y. Ke, P.S. Stanimirović, New activation functions and Zhangians in Zeroing neural network and applications to time-varying matrix pseudoinversion. Mathematics and Computers in Simulation, 225, MATCOM6468 (2024). https://doi.org/10.1016/j.matcom.2024.05.006
63. D. Gerontitis, L. Moysis, P. Stanimirović, V.N. Katsikis, C. Volos, Varying-parameter finite time zeroing neural network for solving linear algebraic systems. Electron. Lett. **56**, 810–813 (2020)
64. D. Gerontitis, R. Behera, J.K. Sahoo, P.S. Stanimirović, Improved finite-time zeroing neural network for time-varying division. Stud. Appl. Math. **146**(2), 526–549 (2021)
65. A.V. Gasnikov, Y.E. Nesterov, Universal method for stochastic composite optimization (2016). https://arxiv.org/ftp/arxiv/papers/1604/1604.05275.pdf. Accessed 28 Jan 2021

66. D. Gerontitis, R. Behera, P. Tzekis, P. Stanimirović, A family of varying-parameter finite-time zeroing neural networks for solving time-varying Sylvester equation and its application. J. Comput. Appl. Math. **403**, 113826 (2022)
67. A.J. Getson, F.C. Hsuan, {2}-*Inverses and their Statistical Applications*, vol. 47 (Lecture Notes in Statistics (Springer, Berlin, 1988)
68. S. Ghadimi, G. Lan, Accelerated gradient methods for nonconvex nonlinear and stochastic programming. Math. Program. **156**(1–2), 59–99 (2016)
69. A.A. Goldstein, On steepest descent. SIAM J. Control **3**, 147–151 (1965)
70. E.G. Golshtein, A.S. Nemirovskii, Y.E. Nesterov, Level method, its generalizations and applications. Econ. Math. Methods **31**(3), 164–180 (1995)
71. G.H. Golub, P.C. Hansen, D.P. O'Leary, Tikhonov regularization and total least squares. SIAM J. Matrix Anal. Appl. **21**, 185–194 (1999)
72. G. Golub, C. Van Loan, *Matrix Computations*, 4th edn. (Johns Hopkins University Press, Baltimore, MD, 2013)
73. N.C. González, J.J. Koliha, Y. Wei, On integral representation of the Drazin inverse in Banach algebras. Proc. Edinb. Math. Soc. **45**, 327–331 (2002)
74. N.C. González, J. Koliha, Y. Wei, Integral representation of the Drazin inverse. Electron. J. Linear Algebra **9**, 129–131 (2002)
75. A. Graham, *Kronecker Products and Matrix Calculus with Application* (Wiley, New York, 1981)
76. T.N.E. Greville, Note on the generalized inverse of a matrix product. SIAM Rev. **8**(4), 518–521 (1966)
77. L. Grippo, F. Lampariello, S. Lucidi, A nonmonotone line search technique for Newton's method. SIAM J. Numer. Anal. **23**, 707–716 (1986)
78. L. Grippo, F. Lampariello, S. Lucidi, A class of nonmonotone stability methods in unconstrained optimization. Numer. Math. **62**, 779–805 (1991)
79. L. Grippo, F. Lampariello, S. Lucidi, A truncated Newton method with nonmonotone line search for unconstrained optimization. J. Optim. Theory Appl. **60**, 401–419 (1989)
80. C.W. Groetsch, *Generalized Inverses of Linear Operators: Representation and Approximation* (Marcel Dekker, New York-Basel, 1977)
81. D. Guo, K. Li, L. Yan, Z. Nie, F. Jin, The application of Li-function activated RNN to acceleration-level robots' kinematic control via time-varying matrix inversion, in *Proceedings of Chinese Control and Decision Conference (CCDC)* (May 2016), pp. 3455–3460
82. D. Guo, Z. Nie, L. Yan, Theoretical analysis, numerical verification and geometrical representation of new three-step DTZD algorithm for time-varying nonlinear equations solving. Neurocomputing **214**, 516–526 (2016)
83. D. Guo, C. Peng, L. Jin, Y. Ling, Y. Zhang, Different ZFs lead to different nets: examples of Zhang generalized inverse, in *Proceedings of Chinese Automation Congress* (Nov 2013), pp. 453–458
84. D. Guo, C. Yi, Y. Zhang, Zhang neural network versus gradient-based neural network for time-varying linear matrix equation solving. Neurocomputing **74**, 3708–3712 (2011)
85. D. Guo, Y. Zhang, Novel recurrent neural network for time-varying problems solving. IEEE Comput. Intell. Mag. **7**, 61–65 (2012)
86. D. Guo, Y. Zhang, Li-function activated ZNN with finite-time convergence applied to redundant-manipulator kinematic control via time-varying Jacobian matrix pseudoinversion. Appl. Soft Comput. **24**, 158–168 (2014)
87. W.B. Guo, T.Z. Huang, Method of elementary transformation to compute Moore-Penrose inverse. Appl. Math. Comput. **216**, 1614–1617 (2010)
88. W.W. Hager, H. Zhang, A new conjugate gradient method with guaranteed descent and an efficient line search. SIAM J. Optim. **16**, 170–192 (2005)
89. W.W. Hager, H. Zhang, CG DESCENT, a conjugate gradient method with guaranteed descent. ACM Trans. Math. Soft. **32**, 113–137 (2006)
90. B.A. Hassan, A new type of quasi-Newton updating formulas based on the new quasi-Newton equation. Numer. Algebra Control Optim. **10**(2), 227–235 (2020)

91. U. Helmke, J.B. Moore, *Optimization and Dynamical Systems* (Springer, London, 1994). (3rd printing 1996)
92. F. Hsuan, P. Langenberg, A. Getson, The {2}-inverse with applications in statistics. Linear Algebra Appl. **70**, 241–248 (1985)
93. X. Hu, C. Sun, B. Zhang, Design of recurrent neural networks for solving constrained least absolute deviation problems. IEEE Trans. Neural Netw. **21**(7), 1073–1086 (2010)
94. L. Hualiang, T. Adali, Optimization in the complex domain for nonlinear adaptive filtering, in *Signals, Systems and Computers, ACSSC'06 Fortieth Asilomar Conference on IEEE* (2006), pp. 263–267
95. Y.-A. Huang, Z.-H. You, X. Li, X. Chen, P. Hu, S. Li, X. Luo, Construction of reliable protein-protein interaction networks using weighted sparse representation based classifier with pseudo substitution matrix representation features. Neurocomputing **218**, 131–138 (2016)
96. J.J. Hunter, A Survey of generalized inverses and their use in stochastic modelling. Res. Lett. Inf. Math. Sci. **1**, 25–36 (2000)
97. B. Ivanov, P.S. Stanimirović, G.V. Milovanović, S. Djordjević, I. Brajević, Accelerated multiple step-size methods for solving unconstrained optimization problems. Optim. Methods Softw. **36**(5), 998–1029 (2019), (2021)
98. J. Jang, S. Lee, S. Shin, *An optimization network for matrix inversion, in Neural Information Processing Systems* (AIP, College Park, MD, 1988), pp.397–401
99. J.S. Jang, S.Y. Lee, S.Y. Shin, J.S. Jang, S.Y. Shin, An optimization network for matrix inversion, in *Neural Information Processing Systems Conference* (1987), pp. 397–401
100. D.R. Jensen, Minimal properties of Moore-Penrose inverses. Linear Algebra Appl. **196**, 175–182 (1994)
101. J. Ji, A finite algorithm for the Drazin inverse of a polynomial matrix. Appl. Math. Comput. **130**, 243–251 (2002)
102. J. Ji, An alternative limit expression of Drazin inverse and its application. Appl. Math. Comput. **61**, 151–156 (1994)
103. Z. Jian, L. Xiao, J. Dai, Z. Tang, C. Liu, Design and analysis of new zeroing neural network models with improved finite-time convergence for time-varying reciprocal of complex matrix. IEEE Trans. Industr. Inf. **16**, 3838–3848 (2019)
104. J. Jin, A robust zeroing neural network for solving dynamic nonlinear equations and its application to kinematic control of mobile manipulator. Complex Intell. Syst. **7**, 87–99 (2021)
105. J. Jin, L. Zhao, F. Yu, Z. Xi, Improved zeroing neural networks for finite time solving nonlinear equations. Neural Comput. Appl. **32**, 4151–4160 (2019)
106. L. Jin, S. Li, B. Liao, Z. Zhang, Zeroing neural networks: a survey. Neurocomputing **267**, 597–604 (2017)
107. L. Jin, L. Wei, Xin Lv, *Neural Dynamics for Time-varying Problems* (Springer Nature, Cham, Switzerland). ISBN 978-3-031-68593-4. https://doi.org/10.1007/978-3-031-68594-1
108. L. Jin, Y. Zhang, Discrete-time Zhang neural network of $O(\tau^3)$ pattern for time-varying matrix pseudoinversion with application to manipulator motion generation. Neurocomputing **142**, 165–173 (2014)
109. L. Jin, Y. Zhang, Continuous and discrete Zhang dynamics for real-time-varying nonlinear optimization. Numer. Algorithms **73**, 115–140 (2016)
110. L. Jin, Y. Zhang, S. Li, Integration-Enhanced Zhang Neural Network for real-time-varying matrix inversion in the presence of various kinds of noises. IEEE Trans. Neural Netw. Learn. Syst. **27**(12), 2615–2627 (2016)
111. L. Jin, Y. Zhang, S. Li, Y. Zhang, Noise-Tolerant ZNN models for solving time-varying zero-finding problems: a control-theoretic approach. IEEE Trans. Autom. Control **62**(2), 992–997 (2017)
112. L. Jin, Y. Zhang, B. Qiu, Neural network-based discrete-time Z-type model of high accuracy in noisy environments for solving dynamic system of linear equations. Neural Comput. Appl. **29**, 1217–1232 (2018)
113. L. Jin, Y. Zhang, S. Li, Y. Zhang, Modified ZNN for time-varying quadratic programming with inherent tolerance to noises and its application to kinematic redundancy resolution of robot manipulators. IEEE Trans. Industr. Electron. **63**, 6978–6988 (2016)

114. L. Jin, S. Li, H.M. La, X. Luo, Manipulability optimization of redundant manipulators using dynamic neural networks. IEEE Trans. Ind. Electron. **64**, 4710–4720 (2017)
115. T. Kailath, *Linear Systems* (Prentice-Hall, Englewood Cliffs, NJ, 1980)
116. V.N. Katsikis, P.S. Stanimirović, S.D. Mourtas, S. Li, X. Cao, *Towards Higher Order Dynamical Systems, Chapter 6 In: Generalized Inverses: Algorithms and Applications* ed. by I. Kyrchei. Mathematics Research Development (Nova Science Publishers Inc, New York, 2021). ISBN: 978-1-68507-356-5. https://doi.org/10.52305/MJVE4994
117. V.N. Katsikis, D. Pappas, A. Petralias, An improved method for the computation of the Moore-Penrose inverse matrix. Appl. Math. Comput. **217**, 9828–9834 (2011)
118. V.N. Katsikis, S.D. Mourtas, P.S. Stanimirović, Y. Zhang, Solving complex-valued time-varying linear matrix equations via QR decomposition with applications to robotic motion tracking and on angle-of-arrival localization. IEEE Trans. Neural Netw. Learn. Syst. **33**, 3415–3424 (2021)
119. Y.Y. Ke, Y.F. Gao, J.L. Chen, Representations of the (b, c)-Inverses in rings with involution. Filomat **31**, 2867–2875 (2017)
120. Y.Y. Ke, J. Chen, P.S. Stanimirović, M. Ćirić, Characterizations and representations of outer inverse for matrices over a ring. Linear Multilnear Algebra **69**(1), 155–176 (2021)
121. M. Klimczak, W. Cecot, On Moore-Penrose pseudoinverse computation for stiffness matrices resulting from higher order approximation. Math. Problems Eng. **2019**, Article ID 5060397, 16. https://doi.org/10.1155/2019/5060397
122. Y. Kong, H. Lu, Y. Xue, H. Xia, Terminal neural computing: finite-time convergence and its applications. Neurocomputing **217**, 133–141 (2016)
123. V.N. Krutikov, P.S. Stanimirović, O.N. Indenko, E.M. Tovbis, L.A. Kazakovtsev, Optimization of subgradient method based on Rank-two correction of metric matrices. J. Appl. Ind. Math. **16**(3), 427–439 (2022)
124. G. Lan, *First-order and Stochastic Optimization Methods for Machine Learning* (Springer, Berlin, 2020)
125. Y. Lei, J. Luo, T. Chen, L. Ding, B. Liao, G. Xia, Z. Dai, Nonlinearly activated IEZNN model for solving time-varying Sylvester equation. IEEE Access **10**, 121520–121530 (2022)
126. B. Liao, Y. Zhang, Different complex ZFs leading to different complex ZNN models for time-varying complex generalized inverse matrices. IEEE Trans. Neural Netw. Learn. Syst. **25**, 1621–1631 (2014)
127. D. Li, M. Fukushima, A modified BFGS method and its global convergence in nonconvex minimization. J. Comput. Appl. Math. **129**, 15–35 (2001)
128. J. Li, Y. Zhang, M. Mao, Continuous and discrete Zeroing Neural Network for different-level dynamic linear system with robot manipulator control. IEEE Trans. Syst. Man Cybern. Syst. **50**(11), 4633–4642 (2020)
129. S. Li, Y. Li, Nonlinearly activated neural network for solving time-varying complex Sylvester equation. IEEE Trans. Cybern. **44**, 1397–1407 (2014)
130. S. Li, S. Chen, B. Liu, Accelerating a recurrent neural network to finite-time convergence for solving time-varying Sylvester Equation by using a sign-bi-power activation function. Neural Process. Lett. **37**, 189–205 (2013)
131. S. Li, Y. Li, Z. Wang, A class of finite-time dual neural networks for solving quadratic programming problems and its k-winners-take-all application. Neural Netw. **39**, 27–39 (2013)
132. S. Li, B. Liu, Y. Li, Selective positive-negative feedback produces the winner-take-all competition in recurrent neural networks. IEEE Trans. Neural Netw. Learn. Syst. **24**, 301–309 (2013)
133. S. Li, S. Chen, B. Liu, Acclerating a recurrent neural network to finite-time convergence for solving time-varying Sylvester equation by using a sign-bi-power activation function. Neural Process. Lett. **37**(2), 189–205 (2013)
134. S. Li, H. Wang, M.U. Rafique, A novel recurrent neural network for manipulator control with improved noise tolerance. IEEE Trans. Neural Netw. Learn. Syst. **29**(5), 1908–1918 (2017)
135. S. Li, Y. Zhang, L. Jin, Kinematic control of redundant manipulators using neural networks. IEEE Trans. Neural Netw. Learn. Syst. **28**(10), 2243–2254 (2017)

136. S. Li, M. Zhou, X. Luo, Modified primal-dual neural networks for motion control of redundant manipulators with dynamic rejection of harmonic noises. IEEE Trans. Neural Netw. Learn. Syst. 1–11 (2017)
137. S. Li, J. He, Y. Li, M.U. Rafique, Distributed recurrent neural networks for cooperative control of manipulators: a game-theoretic perspective. IEEE Trans. Neural Netw. Learn. Syst. **28**, 415–426 (2017)
138. S. Li, S. Chen, B. Liu, Y. Li, Y. Liang, Decentralized kinematic control of a class of collaborative redundant manipulators via recurrent neural networks. Neurocomputing **91**, 1–10 (2012)
139. L. Jin, Y. Zhang, T. Qiao, M. Tan, Y. Zhang, Tracking control of modified Lorenz nonlinear system using ZG neural dynamics with additive input or mixed inputs. Neurocomputing **196**, 82–94 (2016)
140. J. Levine, R.E. Hartwig, Applications of Drazin inverse to the Hill cryptographic systems. Cryptologia **4**, 71–85 (1980)
141. S. Li, Y. Zhang, L. Jin, Kinematic control of redundant manipulators using neural networks. IEEE Trans. Neural Netw. Learn. Syst. **28**(10), 2243–2254 (2017)
142. S. Li, M. Zhou, X. Luo, Z. You, Distributed Winner-take-all in dynamic networks. IEEE Trans. Autom. Control **62**(2), 577–589 (2017)
143. J. Li, M. Mao, F. Uhlig, Y. Zhang, Z-type neural-dynamics for time-varying nonlinear optimization under a linear equality constraint with robot application. Journal. Comput. Appl. Math. **327**, 155–166 (2018)
144. W.G. Li, Z. Li, A family of iterative methods for computing the approximate inverse of a square matrix and inner inverse of a non-square matrix. Appl. Math. Comput. **215**, 3433–3442 (2010)
145. W.G. Li, L. Juan, T. Qiao, A family of iterative methods for computing Moore-Penrose inverse of a matrix. Linear Algebra Appl. **438**, 47–56 (2013)
146. X. Li, J. Yu, S. Li, L. Ni, A nonlinear and noise-tolerant ZNN model solving for time-varying linear matrix equation. Neurocomputing **317**, 70–78 (2018)
147. Z. Li, Z. Yin, Extended Wang neural network for online solving a set of linear equations. Electron. Lett. **52**, 1001–1003 (2016)
148. Z. Li, Y. Zhang, Improved Zhang neural network model and its solution of time-varying generalized linear matrix equations. Expert Syst. Appl. **37**, 7213–7218 (2010)
149. B. Liao, Y. Zhang, From different ZFs to different ZNN models accelerated via Li activation functions to finite-time convergence for time-varying matrix pseudoinversion. Neurocomputing **133**, 512–522 (2014)
150. B. Liao, L. Xiao, J. Jin, L. Ding, M. Liu, Novel complex-valued neural network for dynamic complex-valued matrix inversion. J. Adv. Comput. Intell. Intell. Inf. **20**, 132–138 (2016)
151. B. Liao, Y. Zhang, From different ZFs to different ZNN models acclerated via Li activation functions to finite-time convergence for time-varying matrix pseudoinversion. Neurocomputing **133**, 512–522 (2014)
152. S. Liao, J. Liu, Y. Qi, H. Huang, R. Zheng, X. Xiao, An adaptive gradient neural network to solve dynamic linear matrix equations. IEEE Trans. Syst. Man Cybern. Syst. **52**, 5913–5924 (2021)
153. M. Liu, B. Liao, L. Ding, L. Xiao, Performance analyses of recurrent neural network models exploited for online time-varying nonlinear optimization. Comput. Sci. Inf. Syst. **13**, 691–705 (2016)
154. Q. Liu, J. Wang, Finite-time convergent recurrent neural network with a hard-limiting activation function for constrained optimization with piecewise-linear objective functions. IEEE Trans. Neural Netw. **22**(4), 601–613 (2011)
155. X. Liu, H. Jin, Y. Yu, Higher-order convergent iterative method for computing the generalized inverse and its application to Toeplitz matrices. Linear Algebra Appl. **439**, 1635–1650 (2013)
156. X. Liu, Y. Yu, J. Zhong, Y. Wei, Integral and limit representations of the outer inverse in Banach space. Linear Multilinear Algebra **60**, 333–347 (2012)
157. Y.-J. Liu, S. Tong, D.-J. Li, Y. Gao, Fuzzy adaptive control with state observer for a class of nonlinear discrete-time systems with input constraint. IEEE Trans. Fuzzy Syst. **24**(5), 1147–1158 (2015)

158. Y.-J. Liu, S. Tong, Adaptive fuzzy identification and control for a class of nonlinear pure-feedback MIMO systems with unknown dead zones. IEEE Trans. Fuzzy Syst. **23**, 1387–1398 (2015)
159. Y.-J. Liu, S. Tong, Optimal control-based adaptive NN design for a class of nonlinear discrete-time block-triangular systems. IEEE Trans. Cybern. **46**, 2670–2680 (2016)
160. Y.-J. Liu, J. Li, S. Tong, C.L.P. Chen, Neural network control-based adaptive learning design for nonlinear systems with full-state constraints. IEEE Trans. Neural Netw. Learn. Sys. **27**, 1562–1571 (2016)
161. D.G. Luenberg, Y. Ye, *Linear and Nonlinear Programming* (Springer Science+Business Media, LLC, New York, 2008)
162. F.L. Luo, Z. Bao, Neural network approach to computing matrix inversion. Appl. Math. Comput. **47**, 109–120 (1992)
163. X. Luo, M. Shang, S. Li, Efficient extraction of non-negative latent factors from high-dimensional and sparse matrices in industrial applications, in *Proceedings of IEEE 16th International Conference on Data Mining (ICDM)* (Dec 2016), pp. 311–319
164. X. Luo, S. Li, M. Zhou, Regularizaed extraction of non-negative latent factors from high-dimensional sparse matrices, in *Proceedings and Cybernetics (SMC) 2016 IEEE International Conference Systems, Man* (Oct 2016), pp. 001221–001226
165. X. Luo, M. Zhou, S. Li, Z. You, Y. Xia, Q. Zhu, A nonnegative latent factor model for large-scale sparse matrices in recommender systems via alternating direction method. IEEE Trans. Neural Netw. Learn. Syst. **27**, 579–592 (2016)
166. X. Luo, M. Zhou, Y. Xia, Q. Zhu, A.C. Ammari, A. Alabdulwahab, Generating highly accurate predictions for missing QoS data via aggregating nonnegative latent factor models. IEEE Trans. Neural Netw. Learn. Syst. **27**, 524–537 (2016)
167. X. Luo, M. Zhou, S. Li, Y. Xia, Z. You, Q. Zhu, H. Leung, Incorporation of efficient second-order solvers into latent factor models for accurate prediction of missing QoS data. IEEE Trans. Cybern. **48**, 1216–1228 (2018)
168. X. Luo, M. Zhou, S. Li, Y. Xia, Z. You, Q. Zhu, H. Leung, An efficient second-order approach to factorize sparse matrices in recommender systems. IEEE Trans. Industr. Inf. **11**, 946–956 (2015)
169. M. Lustig, D. Donoho, J.M. Pauly, M.R.I. Sparse, Sparse MRI: the application of compressed sensing for rapid MR imaging. Magn. Reson. Med. **6**, 1182–1195 (2007)
170. X. Lv, L. Xiao, Z. Tan, Z. Yang, J. Yuan, Improved Gradient Neural Networks for solving Moore-Penrose inverse of full-rank matrix. Neural Process. Lett. **50**, 1993–2005 (2019)
171. X. Lv, L. Xiao, Z. Tan, Improved Zhang neural network with finite-time convergence for time-varying linear system of equations solving. Inf. Process. Lett. **147**, 88–93 (2019)
172. H. Ma, Optimal perturbation bounds for the core inverse. Appl. Math. Comput. **336**, 176–181 (2018)
173. H. Ma, X. Gao, P.S. Stanimirović, Characterizations, iterative method, sign pattern and perturbation analysis for the DMP Inverse with its applications. Appl. Math. Comput. **378** (2020), https://doi.org/10.1016/j.amc.2020.125196
174. C.C. MacDuffee, *The Theory of Matrices* (Chelsea, New York, N.Y., 1956)
175. P.J. Maher, Some operator inequalities concerning generalized inverses. Illinois J. Math. **34**, 503–514 (1990)
176. M. Mao, J. Li, L. Jin, S. Li, Y. Zhang, Enhanced discrete-time Zhang neural network for time-variant matrix inversion in the presence of bias noises. Neurocomputing **207**, 220–230 (2016)
177. M.D. Marco, M. Forti, M. Grazzini, Robustness of convergence in finite time for linear programming neural networks. Int. J. Circuit Theory Appl. **34**, 307–316 (2006)
178. M.D. Marco, M. Forti, P. Nistri, L. Pancioni, Discontinuous neural networks for finite-time solution of time-dependent linear equations. IEEE Trans. Cybern. **46**, 2509–2520 (2016)
179. P. Miao, Y. Shen, X. Xia, Finite time dual neural networks with a tunable activation function for solving quadratic programming problems and its application. Neurocomputing **143**, 80–89 (2014)

180. P. Miao, Y. Shen, Y. Huang, Y.-W. Wang, Solving time-varying quadratic programs based on finite-time Zhang neural networks and their application to robot tracking. Neural Comput. Appl. **26**, 693–703 (2015)
181. C. Mo, D. Gerontitis, P.S. Stanimirović, Solving the time-varying tensor square root equation by varying-parameters finite-time Zhang neural network. Neurocomputing **445**, 309–325 (2021)
182. H. Mohammad, M.Y. Waziri, S.A. Santos, A brief survey of methods for solving nonlinear least-squares problems. Numer. Algebra Control Optim. **9**, 1–13 (2019)
183. D. Mosić, Core-EP pre-order of Hilbert space operators. Quaest. Math. **41**, 585–600 (2018)
184. H.S. Najafi, M.S. Solary, Computational algorithms for computing the inverse of a square matrix, quasi-inverse of a non-square matrix and block matrices. Appl. Math. Comput. **183**, 539–550 (2006)
185. M.Z. Nashed, *Generalized Inverse and Applications* (Academic Press, New York, 1976)
186. Y.E. Nesterov, Universal gradient methods for convex optimization problems. Math. Program. Ser. A **152**, 381–404 (2015)
187. J. Nocedal, S. Wright, *Numerical Optimization* (Springer: New York, Inc.: New York, NY, USA, 1999)
188. S. Osowski, Neural networks in interpolation problems. Neurocomputing **5**, 105–118 (1993)
189. H. Ouyang, A. Gray, Stochastic smoothing for nonsmooth minimizations: accelerating sgd by exploiting structure, in *Proceedings of the 29th International Conference on Machine Learning (ICML)*, vol. 1, (Edinburgh, Scotland, 2012), pp. 33—40
190. C. Peng, Y. Ling, Y. Wang, X. Yu, Y. Zhang, Three new ZNN models with economical dimension and exponential convergence for real-time solution of Moore-Penrose pseudoinverse, in *Proceedings of International Joint Conference. Neural Networks (IJCNN)* (July 2014), pp. 2788–2793
191. R. Penrose, On a best approximate solutions to linear matrix equations. Proc. Cambridge Philos. Soc. **52**, 17–19 (1956)
192. R. Penrose, A generalized inverse for matrices. Proc. Cambridge Philos. Soc. **51**, 406–413 (1955)
193. L. Perko, *Differential Equations and Dynamical Systems* (Springer, New York, 1991)
194. M.D. Petković, P.S. Stanimirović, Iterative method for computing Moore-Penrose inverse based on Penrose equations. J. Comput. Appl. Math. **235**, 1604–1613 (2011)
195. M.D. Petković, P.S. Stanimirović, Two improvements of the iterative method for computing Moore-Penrose inverse based on Penrose equations. J. Comput. Appl. Math. **267**, 61–71 (2014)
196. M.D. Petković, P.S. Stanimirović, Zeroing neural network based on the equation $AXA = A$. Inf. Comput. **301**, 105207 (2024). https://doi.org/10.1016/j.ic.2024.105207
197. B.T. Polyak, A common method for extreme tasks solution. Rep. USSR Acad. Sci. **174**, 33–36 (1967)
198. M.J.D. Powell, Algorithms for nonlinear constraints that use Lagrangian functions. Math. Program. **14**, 224–248 (1978)
199. K.M. Prasad, K.S. Mohana, Core-EP inverse. Linear Multilinear Algebra **62**, 792–802 (2014)
200. K.M. Prasad, M. David Raj, Bordering method to compute core-EP inverse. Special Matrices **6**, 193–200 (2018)
201. K.M. Prasad, D.M. Raj, M. Vinay, Iterative method to find core-EP inverse. Bull. Kerala Math. Assoc. Special Issue **16**(1), 139–152 (2018)
202. K.M. Prasad, An introduction to generalized inverse, in *Lectures on Matrix and Graph Methods*. ed. by R.B. Bapat, S. Kirkland, K. Manjunatha Prasad, S. Puntanen (Manipal University Press, Manipal, 2012), pp.43–60
203. S. Qiao, X.-Z. Wang, Y. Wei, Two finite-time convergent Zhang neural network models for time-varying complex matrix Drazin inverse. Linear Algebra Appl. **542**, 101–117 (2018)
204. D.S. Rakić, N. Dinčić, D.S. Djordjević, Core inverse and core partial order of Hilbert space operators. Appl. Math. Comput. **244**, 283–302 (2014)

205. C.R. Rao, S.K. Mitra, *Generalized Inverse of Matrices and its Applications* (John Wiley and Sons Inc, New York, London, Sydney, Toronto, 1971)
206. N. Samardzija, R. Waterland, A neural network for computing eigenvectors and eigenvalues. Biol. Cybern. **65**(4), 211–214 (1991)
207. Y. Shen, P. Miao, Y. Huang, Y. Shen, Finite-time stability and its application for solving time-varying Sylvester equation by recurrent neural network. Neural Process. Lett. **42**, 763–784 (2015)
208. X. Sheng, G. Chen, Full-rank representation of generalized inverse $A^{(2)}_{T,S}$ and its applications. Comput. Math. Appl. **54**, 1422–1430 (2007)
209. Z.J. Shi, Convergence of line search methods for unconstrained optimization. Appl. Math. Comput. **157**, 393–405 (2004)
210. N.Z. Shor, Application of the gradient descent method for solving network transportation problems. Materials of the seminar of theoretical and applied issues of cybernetics and operational research **1**, 9–17 (1962)
211. H.S. Sim, W.J. Leong, C.Y. Chen, S.N.I. Ibrahim, Multi-step spectral gradient methods with modified weak secant relation for large scale unconstrained optimization. Numer. Algebra, Control Optim. **8**(3), 377–387 (2018)
212. T.E. Simos, V.N. Katsikis, S.D. Mourtas, Multi-input bio-inspired weights and structure determination neuronet with applications in European Central Bank publications. Math. Comput. Simul. **193**, 451–465 (2022)
213. T.E. Simos, V.N. Katsikis, S. Mourtas, P.S. Stanimirović, D. Gerontitis, A higher-order Zeroing Neural Network for pseudoinversion of an arbitrary time-varying matrix with applications to mobile object localization. Inf. Sci. **600**, 226–238 (2022)
214. Y. Siwei, A. Shaharyar Khwaja, J. Ma, Compressed sensing of complex-valued data. Signal Process. **92**, 357–362 (2012)
215. J. Song, Y. Yam, Complex recurrent neural network for computing the inverse and pseudo-inverse of the complex matrix. Appl. Math. Comput. **93**, 195–205 (1998)
216. G. Sowmya, P. Thangavel, V. Shankar, A novel hybrid Zhang neural network model for time-varying matrix inversion. Eng. Sci. Technol. Int. J. **26**, 101009 (2022). https://doi.org/10.1016/j.jestch.2021.05.013
217. S. Srivastava, D.K. Gupta, A third order iterative method for $A^{\dagger}$. Int. J. Comput. Sci. Math. **4**, 140–151 (2013)
218. P.S. Stanimirović, Limit representations of generalized inverses and related methods. Appl. Math. Comput. **103**, 51–68 (1999)
219. P.S. Stanimirović, D.S. Cvetković-Ilić, Successive matrix squaring algorithm for computing outer inverses. Appl. Math. Comput. **203**, 19–29 (2008)
220. P.S. Stanimirović, D.S. Cvetković-Ilić, S., Miljković, M. Miladinović, Full-rank representations of 2,4, 2,3-inverses and successive matrix squaring algorithm. Appl. Math. Comput. **217**, 9358–9367 (2011)
221. P.S. Stanimirović, M. Ćirić, I. Stojanović, D. Gerontitis, Conditions for existence, representations and computation of matrix generalized inverses. Complexity **2017**, Article ID 6429725, 27. https://doi.org/10.1155/2017/6429725
222. P.S. Stanimirović, V.N. Katsikis, S. Li, Higher-order ZNN dynamics. Neural Process. Lett. **51**, 697–721 (2020)
223. P.S. Stanimirović, V.N. Katsikis, S. Li, Integration enhanced and noise tolerant ZNN for computing various expressions involving outer inverses. Neurocomputing **329**, 129–143 (2019)
224. P.S. Stanimirović, V.N. Katsikis, S. Li, Hybrid GNN-ZNN models for solving linear matrix equations. Neurocomputing **316**, 124–134 (2018)
225. P.S. Stanimirović, D. Pappas, V.N. Katsikis, I.P. Stanimirović, Full-rank representations of outer inverses based on the QR decomposition. Appl. Math. Comput. **218**, 10321–10333 (2012)
226. P. Stanimirović, D. Gerontitis, P. Tzekis, R. Behera, J.K. Sahoo, Simulation of Varying Parameter Recurrent Neural Network with application to matrix inversion. Math. Comput. Simul. **185**, 614–628 (2021)

227. P.S. Stanimirović, D. Gerontitis, V.N. Krutikov, L.A. Kazakovtsev, Towards subderivative-based zeroing neural networks, volume 1881 of the communications in computer and information science series, in *MOTOR 2023, Communications in Computer and Information Science-CCIS 1881* ed. by M. Khachay et al. (2023), pp. 3–15. https://doi.org/10.1007/978-3-031-43257-6_1. (Springer Nature Switzerland, invited paper)
228. D. Gerontitis, C. Mo, P.S. Stanimirović, P. Tzekis, V.N. Katsikis, A novel extended Li zeroing neural network for matrix inversion. Neural Comput. Appl. **35**, 14129–14152 (2023)
229. P.S. Stanimirović, V.N. Katsikis, Z. Zhang, S. Li, J. Chen, M. Zhou, Varying-parameter Zhang Neural Network for approximating some expressions involving outer inverses. Optim. Methods Softw. **35**, 1304–1330 (2020)
230. P.S. Stanimirović, M.D. Petković, Gauss-Jordan elimination method for computing outer inverses. Appl. Math. Comput. **219**, 4667–4679 (2013)
231. P.S. Stanimirović, M. Petković, Gradient neural dynamics for solving matrix equations and their applications. Neurocomputing **306**, 200–212 (2018)
232. P.S. Stanimirović, M.D. Petković, Improved GNN models for constant matrix inversion. Neural Process. Lett. **50**(1), 321–339 (2019)
233. P.S. Stanimirović, Block representations of {2}, {1, 2} inverses and the Drazin inverse. Indian J. Pure Appl. Math. **29**, 1159–1176 (1998)
234. P.S. Stanimirović, Applications of hyper-power method for computing matrix products. Univ. Beograd. Publ. Elektrotehn. Fak. Ser. Mat. **15**, 13–25 (2004)
235. P.S. Stanimirović, M.B. Miladinović, Accelerated gradient descent methods with line search. Numer. Algorithms **54**, 503–520 (2010)
236. P.S. Stanimirović, D. Gerontitis, N. Tešić, L.V. Kazakovtsev, V. Stasiuk, X. Cao, (2023). Gradient neural dynamics based on modified error function, in *Hybrid Methods of Modeling and Optimization in Complex Systems* ed. by P. Stanimorović, A.A. Stupina, E. Semenkin, I.V. Kovalev, vol 1. European Proceedings of Computers and Technology (European Publisher), pp. 256–263. https://doi.org/10.15405/epct.23021.31
237. P. Stanimirović, D. Mosić, A. Stupina, R. Behera, *Solving Matrix Approximation Problems Using Generalized Inverses: Monograph* (INFRA-M Moscow, 2023), p. 165. (Scientific Idea). ISBN 978-5-16-019628-2
238. P. Stanimirović, Y. Wei, L. Jin, A. Stupina, L. Kazakovtsev, *Gradient-Based Dynamical Systems for Matrix Computations and Optimization: Monograph* (INFRA-M Moscow, 2024), p. 314. ISBN 978-5-16-020482-6
239. P.S. Stanimirović, D.S. Djordjević, Full-rank and determinantal representation of the Drazin inverse. Linear Algebra Appl. **311**, 31–51 (2000)
240. P.S. Stanimirović, M.D. Petković, D. Gerontitis, Gradient neural network with nonlinear activation for computing inner inverses and the Drazin inverse. Neural Process. Lett. **48**, 109–133 (2018)
241. P.S. Stanimirović, M., Ćirić, A. Lastra, J.R. Sendra, J. Sendra, Representations and symbolic computation of generalized inverses over fields. Appl. Math. Comput. **406** 126287 (2021)
242. P.S. Stanimirović, S. Srivastava, D.K. Gupta, From Zhang Neural Network to scaled hyper-power iterations. J. Comput. Appl. Math. **331**, 133–155 (2018)
243. P.S. Stanimirović, F. Soleymani, A class of numerical algorithms for computing outer inverses. J. Comput. Appl. Math. **263**, 236–245 (2014)
244. P.S. Stanimirović, S. Bogdanović, M. Ćirić, Adjoint mappings and inverses of matrices. Algebra Colloq. **13**(3), 421–432 (2006)
245. P.S. Stanimirović, V.N. Katsikis, D. Gerontitis, A new varying-parameter design formula for solving time-varying problems. Neural Process. Lett. **53**, 107–129 (2021)
246. P.S. Stanimirović, M.B. Tasić, Computing generalized inverses using LU factorization of matrix product. Int. J. Comput. Math. **85**, 1865–1878 (2008)
247. P.S. Stanimirović Y. Wei, D. Kolundžija, J.R. Sendra, J. Sendra, An application of computer algebra and dynamical systems, Chapter No 19, in *CAI 2019* ed. by M. Ćirić M. Droste, J.-E. Pin, vol. 11545. Lecture Notes in Computer Science (Springer, Cham), pp. 225–236. https://doi.org/10.1007/978-3-030-21363-3_19

248. V.N. Katsikis, P.S. Stanimirović, S. Mourtas, L. Xiao, D. Karabašević, D. Stanujkić, Zeroing Neural Network with fuzzy parameter for computing pseudoinverse of arbitrary matrix. IEEE Trans. Fuzzy Syst. **30**(9), 3426–3435 (2022)
249. P.S. Stanimirović, S.D. Mourtas, V.N. Katsikis, L.A. Kazakovtsev, V.N. Krutikov, Recurrent Neural Network models based on optimization methods. Mathematics **10**, 4292 (2022). https://doi.org/10.3390/math10224292
250. P.S. Stanimirović, N. Tešić, D. Gerontitis, G.V. Milovanović, M. Petrović, V.L. Kazakovtsev, V. Stasiuk, Application of gradient optimization methods in defining neural dynamics. Axioms **13**, 49 (2024). https://doi.org/10.3390/axioms13010049
251. P.S. Stanimirović, I. Živković, Y. Wei, Recurrent neural network for computing the Drazin inverse. IEEE Trans. Neural Netw. Learn. Syst. **26**, 2830–2843 (2015)
252. P.S. Stanimirović, I. Živković, Y. Wei, Recurrent neural network approach based on the integral representation of the Drazin inverse. Neural Comput. **27**(10), 2107–2131 (2015)
253. P.S. Stanimirović, M.D. Petković, D. Mosić, Exact solutions and convergence of gradient based dynamical systems for computing outer inverses. Appl. Math. Comput. **412** (2022). https://doi.org/10.1016/j.amc.2021.126588
254. P.S. Stanimirović, I. Živković, Y. Wei, Neural network approach to computing outer inverses based on the full rank representation. Linear Algebra Appl. **501**, 344–362 (2016)
255. I. Stojanović, P.S. Stanimirović, I.S. Živković, D. Gerontitis, X.-Z. Wang, ZNN models for computing matrix inverse based on hyperpower iterative methods. Filomat **31**, 2999–3014 (2017)
256. W.Y. Sun, J.Y. Han, J. Sun, Global convergence of nonmonotone descent methods for unconstrained optimization problems. J. Comp. Appl. Math. **146**, 89–98 (2002)
257. W. Sun, Y.X. Yuan, *Optimization Theory and Methods: Nonlinear Programming* (Springer, 2006)
258. Z. Tan, L. Xiao, S. Chen, X. Lv, Noise-tolerant and finite-time convergent ZNN models for dynamic matrix Moore-Penrose inversion. IEEE Trans. Ind. Inf. **16**, 1591–1601 (2019)
259. Z. Tan, W. Li, L. Xiao, Y. Hu, New varying-parameter ZNN models with finite-time convergence and noise suppression for time-varying matrix Moore-Penrose inversion. IEEE Trans. Neural Netw. Learn. Syst. **31**(8), 2980–2992 (2020)
260. Z. Tan, Y. Hu, K. Chen, On the investigation of activation functions in gradient neural network for online solving linear matrix equation. Neurocomputing **413**, 185–192 (2020)
261. V. Tavakkoli, J.C. Chedjou. K. Kyamakya, A novel recurrent neural network-based ultra-fast, robust, and scalable solver for inverting a "time-varying matrix". Sensors **19**, 4002 (2019). 10.3390/s19184002
262. P.L. Toint, Non-monotone trust-region algorithm for nonlinear optimization subject to convex constraints. Math. Prog. **77**, 69–94 (1997)
263. F. Uhlig, Y. Zhang, Neural Networks and look-ahead finite difference equations. Linear Algebra Appl. **580**, 417–435 (2019)
264. N.S. Urquhart, Computation of generalized inverse matrices which satisfy specified conditions. SIAM Rev. **10**, 216–218 (1968)
265. C. Wang, Y. Zhang, Theoretical analysis of gradient-zhang neural network for time-varying equations and improved method for linear equations, in *Neural Information Processing. ICONIP 2023* ed. by B. Luo, L. Cheng, Z.G. Wu, H. Li, C. Li, vol. 14447. Lecture Notes in Computer Science (Springer, Singapore, 2024). https://doi.org/10.1007/978-981-99-8079-6_22
266. G. Wang, Y. Wei, S. Qiao, *Generalized Inverses: Theory and Computations, Developments in Mathematics*, vol. 53 (Science Press, Singapore, Springer; Beijing, 2018)
267. H. Wang, M. Wei, X. Liu, Several representations of 2-inverses. Arab. J. Sci. Eng. **36**, 1161–1169 (2011)
268. H. Wang, B. Chen, K. Liu, X. Liu, C. Lin, Adaptive neural tracking control for a class of nonstrict-feedback stochastic nonlinear systems with unknown backlash-like hysteresis. IEEE Trans. Neural Netw. Learn. Syst. **25**, 947–958 (2014)

269. H. Wang, P.X. Liu, S. Liu, Adaptive neural synchronization control for bilateral teleoperation systems with time delay and backlash-like hysteresis. IEEE Trans. Cybern. **47**, 3018–3026 (2017)
270. H. Wang, W. Liu, J. Qiu, P.X. Liu, Adaptive fuzzy decentralized control for a class of strong interconnected nonlinear systems with unmodeled dynamics. Neurocomputing **26**, 836–846 (2018)
271. J. Wang, G. Wu, Recurrent neural networks for synthesizing linear control systems via pole placement. Int. J. Syst. Sci. **26**, 2369–2382 (1995)
272. J. Wang, A recurrent neural network for real-time matrix inversion. Appl. Math. Comput. **55**, 89–100 (1993)
273. J. Wang, Recurrent neural networks for computing pseudoinverse of rank-deficient matrices. SIAM J. Sci. Comput. **18**, 1479–1493 (1997)
274. J. Wang, H. Li, Solving simultaneous linear equations using recurrent neural networks. Inf. Sci. **76**(3), 255–277 (1994)
275. J. Wang, Electronic realisation of recurrent neural network for solving simultaneous linear equations. Electron. Lett. **28**, 493–495 (1992)
276. J. Wang, Recurrent neural networks for solving systems of complex-valued linear equations. Electron. Lett. **28**(18), 1751–1753 (1992)
277. J. Wang, Recurrent neural networks for solving linear matrix equations. Comput. Math. Appl. **26**, 23–34 (1993)
278. J. Wang, G. Wu, Recurrent neural networks for LU decomposition and Cholesky factorization. Math. Comput. Model. **18**(6), 1–8 (1993)
279. L.-X. Wang, J.M. Mendel, Three-dimensional structured networks for matrix equation solving. IEEE Trans. Comput. **40**(12), 1337–1346 (1991)
280. L.-X. Wang, J.M. Mendel, Parallel structured networks for solving a wide variety of matrix algebra problems. J. Parallel Distrib. Comput. **14**, 236–247 (1992)
281. S.D. Wang, T.S. Kuo, C.F. Hsu, Trace bounds on the solution of the algebraic matrix Riccati and Lyapunov equation. IEEE Trans. Autom. Control **AC-31**(7), 654–656 (1986)
282. X.-Z. Wang, H. Ma, P.S. Stanimirović, Nonlinearly activated recurrent neural network for computing the Drazin inverse. Neural Process. Lett. **46**(1), 195–217 (2017)
283. X.-Z. Wang, Y. Wei, P.S. Stanimirović, Complex neural network models for time-varying Drazin inverse. Neural Comput. **28**, 2790–2824 (2016)
284. X.-Z. Wang, P.S. Stanimirović, Y. Wei, Complex ZFs for computing time-varying complex outer inverses. Neurocomputing **275**, 983–1001 (2018)
285. X.-Z. Wang, H. Ma, P.S. Stanimirović, Recurrent neural network for computing the W-weighted Drazin inverse. Appl. Math. Comput. **300**, 1–20 (2017)
286. X.-Z. Wang, M. Che, Y. Wei, Recurrent neural network for computation of generalized eigenvalue problem with real diagonalizable matrix pair and its applications. Neurocomputing **216**, 230–241 (2016)
287. Y. Wei, H. Wu, The representation and approximation for the generalized inverse $A_{T,S}^{(2)}$. Appl. Math. Comput. **135**, 263–276 (2003)
288. Y. Wei, The representation and approximation for the weighted Moore-Penrose inverse in Hilbert space. Appl. Math. Comput. **136**, 475–486 (2003)
289. Y. Wei, A characterization and representation of the generalized inverse $A_{T,S}^{(2)}$ and its applications. Linear Algebra Appl. **280**, 79–86 (1998)
290. Y. Wei, Recurrent neural networks for computing weighted Moore-Penrose inverse. Appl. Math. Comput. **116**, 279–287 (2000)
291. Y. Wei, *Integral Representation of the Generalized Inverse $A_{T,S}^{(2)}$ and its Applications*. Recent Research on Pure and Applied Algebra (Nova Science Publisher, New York, 2003), pp. 59–65
292. Y. Wei, D.S. Djordjević, On integral representation of the generalized inverse $A_{T,S}^{(2)}$. Appl. Math. Comput. **142**, 189–194 (2003)
293. Y. Wei, P.S. Stanimirović, M. Petković, *Numerical and Symbolic Computations of Generalized Inverses* (World Scientific Publishing Co. Pte. Ltd., Hackensack, NJ, Sept 2018). 10.1142/10950. (ISBN: 978-981-3238-66-4)

294. Y. Wei, G.R. Wang, Approximate methods for the generalized inverse $A^{(2)}_{T,S}$. J. Fudan Univ. (Nat. Sci.) **38**, 234–249 (1999)
295. Y. Wei, H. Wu, The representation and approximation for the generalized inverse $A^{(2)}_{T,S}$. Appl. Math. Comput. **135**, 263–276 (2003)
296. Y. Wei, N. Zhang, A note on the representation and approximation of the outer inverse $A^{(2)}_{T,S}$ of a matrix A. Appl. Math. Comput. **147**, 837–841 (2004)
297. Z. Wei, G. Li, L. Qi, New quasi-Newton methods for unconstrained optimization problems. Appl. Math. Comput. **175**, 1156–1188 (2006)
298. P. Wolfe, Convergence conditions for ascent methods. SIAM Rev. **11**, 226–235 (1969)
299. W. Wu, B. Zheng, Improved recurrent neural networks for solving Moore-Penrose inverse of real-time full-rank matrix. Neurocomputing **418**, 221–231 (2020)
300. Y. Xia, T. Chen, J. Shan, A novel iteration method for computing generalized inverse. Neural Comput. **26**, 449–465 (2014)
301. Y. Xia, G. Feng, A neural network for robust LCMP beamforming. Signal Process. **86**, 2901–2912 (2006)
302. Y. Xia, S. Zhang, P.S. Stanimirović, Neural network for computing pseudoinverses and outer inverses of complex-valued matrices. Appl. Math. Comput. **273**, 1107–1121 (2016)
303. Q. Xiang, B. Liao, L. Xiao, L. Jin, A noise-tolerant Z-type neural network for time-dependent pseudoinverse matrices. Optik **165**, 16–28 (2018)
304. L. Xiao, A new design formula exploited for accelerating Zhang neural network and its application to time-varying matrix inversion. Theoret. Comput. Sci. **647**, 50–58 (2016)
305. L. Xiao, B. Liao, A convergence-accelerated Zhang neural network and its solution application to Lyapunov equation. Neurocomputing **193**, 213–218 (2016)
306. L. Xiao, L. Jia, *Zeroing Neural Networks: Finite-time Convergence Design, Analysis and Applications* (Wiley-IEEE Press, 2022). ISBN: 978-1-119-98603-4
307. L. Xiao, A finite-time recurrent neural network for solving online time-varying Sylvester matrix equation based on a new evolution formula. Nonlinear Dyn. **90**, 1581–1591 (2017)
308. L. Xiao, Accelerating a recurrent neural network to finite-time convergence using a new design formula and its application to time-varying matrix square root. J. Frankl. Inst. **354**, 5667–5677 (2017)
309. L. Xiao, A nonlinearly activated neural dynamics and its finite-time solution to time-varying nonlinear equation. Neurocomputing **173**, 1983–1988 (2016)
310. L. Xiao, A finite-time convergent neural dynamics for online solution of time-varying linear complex matrix equation. Neurocomputing **167**, 254–259 (2015)
311. L. Xiao, H. Tan, J. Dai, L. Jia, W. Tang, High-order error function designs to compute time-varying linear matrix equations. Inf. Sci. **576**, 173–186 (2021)
312. L. Xiao, B. Liao, S. Li, K. Chen, Nonlinear recurrent neural networks for finite-time solution of general time-varying linear matrix equations. Neural Netw. **98**, 102–113 (2018)
313. L. Xiao, B. Liao, S. Li, Z. Zhang, L. Ding, L. Jin, Design and analysis of FTZNN applied to the real-time solution of a nonstationary Lyapunov equation and tracking control of a wheeled mobile manipulator. IEEE Trans. Industr. Inf. **14**, 98–105 (2018)
314. L. Xiao, K. Li, Z. Tan, Z. Zhang, B. Liao, K. Chen, L. Jin, S. Li, Nonlinear gradient neural network for solving system of linear equations. Inf. Process. Lett. **142**, 35–40 (2019)
315. L. Xiao, R. Lu, Finite-time solution to nonlinear equation using recurrent neural dynamics with a specially-constructed activation function. Neurocomputing **151**, 246–251 (2015)
316. L. Xiao, W. Meng, R. Lu, X. Yang, B. Liao, L. Ding, A fully complex-valued neural network for rapid solution of complex-valued systems of linear equations, in *Conference paper in Advances in Neural Networks*, vol. 9377. ISNN 2015, Lecture Notes in Computer Science (Springer, Cham, Jan 2015). https://doi.org/10.1007/978-3-319-25393-0_49
317. L. Xiao, A finite-time convergent Zhang neural network and its application to real-time matrix square root finding. Neural Comput. Appl. **31**, 793–800 (2019)
318. L. Xiao, Y. Zhang, Zhang neural network versus gradient neural network for solving time-varying linear inequalities. IEEE Trans. Neural Netw. **22**, 1676–1684 (2011)

319. L. Xiao, Y. Zhang, From different Zhang functions to various ZNN models accelerated to finite-time convergence for time-varying linear matrix equation. Neural Process. Lett. **39**, 309–326 (2014)
320. L. Xiao, Y. Cao, J. Dai, L. Jia, H. Tan, Finite-time and predefined-time convergence design for zeroing neural network: theorem, method, and verification. IEEE Trans. Industr. Inf. **17**, 4724–4732 (2020)
321. L. Xiao, Y. Zhang, K. Li, B. Liao, Z. Tan, A novel recurrent neural network and its finite-time solution to time-varying complex matrix inversion. Neurocomputing **331**, 483–492 (2019)
322. L. Xiao, Y. Zhang, J. Dai, K. Chen, S. Yang, W. Li, B. Liao, L. Ding, J. Li, A new noise-tolerant and predefined-time ZNN model for time-dependent matrix inversion. Neural Netw. **117**, 124–134 (2019)
323. L. Xiao, Y. Zhang, J. Dai, Q. Zuo, S. Wang, Comprehensive analysis of a new varying parameter zeroing neural network for time varying matrix inversion. IEEE Trans. Industr. Inf. **17**, 1604–1613 (2020)
324. X. Xiao, C. Jiang, H. Lu, L. Jin, D. Liu, H. Huang, Y. Pan, A parallel computing method based on zeroing neural networks for time-varying complex-valued matrix Moore-Penrose inversion. Inf. Sci. **524**, 216–228 (2020)
325. X. Xu, J. Zhang, J. Shi, Exponential stability of complex-valued neural networks with mixed delays. Neurocomputing **128**, 483–490 (2014)
326. H. Yanai, K. Takeuchi, Y. Takane, *Projection Matrices, Generalized Inverse Matrices, and Singular Value Decomposition* (Springer, New York, Dordrecht, Heidelberg, London, 2011)
327. Y. Yang, Y. Zhang, Superior robustness of power-sum activation functions in Zhang neural networks for time-varying quadratic programs perturbed with large implementation errors. Neural Comput. Appl. **22**, 175–185 (2013)
328. H. Yang, D. Liu, The representations of the generalized inverses $A_{T,S}^{(2,3)}$ and its applications. J. Comput. Appl. Math. **224**, 204–209 (2009)
329. H. Yang, D. Liu, J. Xu, Matrix left symmetry factor and its applications in generalized inverses $A_{T,S}^{(2,4)}$. Appl. Math. Comput. **197**, 836–843 (2008)
330. C. Yi, Y. Chen, X. Lan, Comparison on neural solvers for the Lyapunov matrix equation with stationary & nonstationary coefficients. Appl. Math. Model. **37**, 2495–2502 (2013)
331. C. Yi, Y. Chen, H. Wang, Simulation and verification of Zhang neural networks and gradient neural networks for time-varying Stein equation solving, in *Advances in Neural Networks, ISNN 2011* (Springer, Jan 2011)
332. C. Yi, Y.H. Liu, Online solution of time-varying Lyapunov matrix equation by Zhang neural networks. Recent Patents Comput. Sci. **6**, 25–32 (2013)
333. C. Yi, Y. Chen, Z. Lu, Improved gradient-based neural networks for online solution of Lyapunov matrix equation. Inform. Process. Lett. **111**, 780–786 (2011)
334. Z. Yi, Y. Fu, H.-J. Tang, Neural networks based approach for computing eigenvectors and eigenvalues of symmetric matrix. Comput. Math. Appl. **47**, 1155–1164 (2004)
335. Y. Yu, Y. Wei, Determinantal representation of the generalized inverse $A_{T,S}^{(2)}$ over integral domains and its applications. Linear Multilinear Algebra **57**, 547–559 (2009)
336. Y. Yu, Y. Wei, The representation and computational procedures for the generalized inverse $A_{T,S}^{(2)}$ of an operator A in Hilbert spaces. Numer. Funct. Anal. Optim. **4**, 1–20 (2007)
337. G. Yuan, Z. Wei, Convergence analysis of a modified BFGS method on convex minimizations. Comp. Optim. Appl. **47**, 237–255 (2010)
338. G. Yuan, Z. Wei, Y. Wu, Modified limited memory BFGS method with nonmonotone line search for unconstrained optimization. J. Korean Math. Soc. **47**, 767–788 (2010)
339. W.J. Zeng, W. Xu, Fast estimation of sparse doubly spread acoustic channels. J. Acoust. Soc. Am. **131**, 303–317 (2012)
340. M. Zhang, B. Zheng. Accelerating noise-tolerant zeroing neural network with fixed-time convergence to solve the time-varying Sylvester equation. Automatica **135** (2022). https://doi.org/10.1016/j.automatica.2021.109998

341. Q.-J. Zhang, F.-Y. Fang, F.-X. Liu, Neurodynamic approach for generalized eigenvalue problems, in *2006 International Conference on Computational Intelligence and Security* (IEEE, 2006), pp. 345–350
342. S. Zhang, Y. Xia, J. Wang, A complex-valued projection neural network for constrained optimization of real functions in complex variables. IEEE Trans. Neural Netw. Learn. Syst. **26**, 3227–3238 (2015)
343. Y. Zhang, B. Cai, M. Liang, W. Ma, On the variable step-size of discrete-time Zhang neural network and Newton iteration for constant matrix inversion. Proceedings of Second International Symposium Intelligent Information Technology Application **1**, 34–38 (2008)
344. Y. Zhang, B. Cai, M. Liang, W. Ma, Three nonlinearly-activated discrete-time ZNN models for time-varying matrix inversion, in *2012 8th International Conference on Natural Computation (ICNC 2012)*, pp. 163–167
345. Y. Zhang, L. Jin, D. Guo, S. Fu, L. Xiao, On the variable step-size of discrete-time Zhang neural network and Newton iteration for constant matrix inversion, in *Intelligent Information Technology Application, 2008. IITA'08. Second International Symposium on Intelligent Information Technology Application* (Shanghai, China, 2008), pp. 34–38. 10.1109/IITA.2008.128
346. Y. Zhang, K. Chen, Comparison on Zhang neural network and gradient neural network for time-varying linear matrix equation $AXB = C$ solving. IEEE Int. Conf. Ind. Technol. (2008). https://doi.org/10.1109/ICIT.2008.4608579
347. Y. Zhang, Z.-H. Chen, K. Chen, Convergence properties analysis of gradient neural network for solving online linear equations. Acta Automatica Sinica **35**, 1136–1139 (2009)
348. Y. Zhang, S.S. Ge, A general recurrent neural network model for time-varying matrix inversion, in *Proceedings of 42nd IEEE Conference on Decision and Control, San Diego*, vol. 6 (2003), pp. 6169–6174
349. Y. Zhang, S.S. Ge, Design and analysis of a general recurrent neural network model for time-varying matrix inversion. IEEE Trans. Neural Netw. **16**, 1477–1490 (2005)
350. Y. Zhang, D. Guo, Z. Luo, K. Zhai, H. Tan, CP-activated WASD neuronet approach to Asian population prediction with abundant experimental verification. Neurocomputing **198**, 48–57 (2016)
351. Y. Zhang, D. Guo, *Zhang Functions and Various Models* (Springer, Berlin, Heidelberg, 2015)
352. Y. Zhang, X. Guo, W. Ma, K. Chen, B. Cai, Matlab Simulink modeling and simulation of Zhang neural network for online time-varying matrix inversion, in *Networking, Sensing and Control, 2008. ICNSC 2008. IEEE International Conference* (IEEE, 2008), pp. 1480–1485
353. Y. Zhang, D. Jiang, J. Wang, A recurrent neural network for solving Sylvester equation with time-varying coefficients. IEEE Trans. Neural Netw. **13**(5), 1053–1063 (2002)
354. Y. Zhang, L. Jin, Z. Ke, Superior performance of using hyperbolic sine activation functions in ZNN illustrated via time-varying matrix square roots finding. Comput. Sci. Inf. Syst. **9**, 1603–1625 (2012)
355. Y. Zhang, Z. Ke, P. Xu, C. Yi, Time-varying square roots finding via Zhang dynamics versus gradient dynamics and the former's link and new explanation to Newton-Raphson iteration. Inf. Process. Lett. **110**, 1103–1109 (2010)
356. Y. Zhang, L. Jin, D. Guo, Y. Yin, Y. Chou, Taylor-type 1-step-ahead numerical differentiation rule for first-order derivative approximation and ZNN discretization. Journal. Comput. Appl. Math. **273**, 29–40 (2015)
357. Y. Zhang, Z. Li, Zhang neural network for online solution of time-varying convex quadratic program subject to time-varying linear-equality constraints. Phys. Lett. A **373**, 1639–1643 (2009)
358. Y. Zhang, Z. Li, K. Li, Complex-valued Zhang neural network for online complex-valued time-varying matrix inversion. Appl. Math. Comput. **217**, 10066–10073 (2011)
359. Y. Zhang, W. Ma, B. Cai, From Zhang neural network to Newton iteration for matrix inversion. IEEE Trans. Circuits Syst. I Regular Papers **56**(7), 1405–1415 (2009)
360. Y. Zhang, B. Qiu, L. Jin, D. Guo, Infinitely many Zhang functions resulting in various ZNN models for time-varying matrix inversion with link to Drazin inverse. Inf. Process. Lett. **115**, 703–706 (2015)

361. Y. Zhang, B. Qiu, B. Liao, Z. Yang, Control of pendulum tracking (including swinging up) of IPC system using zeroing-gradient method. Nonlinear Dyn. **89**, 1–25 (2017)
362. Y. Zhang, G. Ruan, K. Li, Y. Yang, Robustness analysis of the Zhang neural network for online time-varying quadratic optimization. J. Phys. A: Math. Theor. **43**, 245202, 19 (2010)
363. Y. Zhang, Y. Shi, K. Chen, C. Wang, Global exponential convergence and stability of gradient-based neural network for online matrix inversion. Appl. Math. Comput. **215**, 1301–1306 (2009)
364. Y. Zhang, K. Chen, H.-Z. Tan, Performance analysis of gradient neural network exploited for online time-varying matrix inversion. IEEE Trans. Autom. Control **54**, 1940–1945 (2009)
365. Y. Zhang, K. Chen, W. Ma, MATLAB simulation and comparison of Zhang Neural Network and Gradient Neural Network for online solution of linear time-varying equations, in *DCDIS Proceedings of 2007 International Conference on Life System Modeling and Simulation (LSMS2007)*, pp. 450–454
366. Y. Zhang, J. Wang, Recurrent neural networks for nonlinear output regulation. Automatica **37**, 1161–1173 (2001)
367. Y. Zhang, Y. Wang, L. Jin, B. Mu, H. Zheng, Different ZFs leading to various ZNN models illustrated via online solution of time-varying underdetermined systems of linear equations with robotic application, in *Advances in Neural Networks ISNN 2013* (Springer, Jan 2013)
368. Y. Zhang, L. Xiao, G. Ruan, Z. Li, Continuous and discrete time Zhang dynamics for time-varying 4th root finding. Numer. Algorithms **57**, 35–51 (2011)
369. Y. Zhang, L. Xiao, Z. Xiao, M. Mao, *Zeroing Dynamics, Gradient Dynamics, and Newton Iterations* (CRC Press, Taylor & Francis Group, Boca Raton, FL 33487-2742, 2016)
370. Y. Zhang, Y. Yang, N. Tan, B. Cai, Zhang neural network solving for time-varying full-rank matrix Moore-Penrose inverse. Computing **92**, 97–121 (2011)
371. Y. Zhang, Y. Yang, Simulation and comparison of Zhang Neural Network and Gradient Neural Network solving for time-varying matrix square roots, in *Second International Symposium on Intelligent Information Technology Application*. 10.1109/IITA.2008.73
372. Y. Zhang, Y. Yang, G. Ruan, Performance analysis of gradient neural network exploited for online time-varying quadratic minimization and equality-constrained quadratic programming. Neurocomputing **74**, 1710–1719 (2011)
373. Y. Zhang, Y. Yang, B. Cai, D. Guo, Zhang Neural Network and its application to Newton iteration for matrix square root estimation. Neural Comput. Appl. **21**, 453–460 (2012)
374. Y. Zhang, C. Yi, *Zhang Neural Networks and Neural-Dynamic Method* (Nova Science Publishers Inc, New York, 2011)
375. Y. Zhang, C. Yi, D. Guo, J. Zheng, Comparison on Zhang neural dynamics and gradient-based neural dynamics for online solution of nonlinear time-varying equation. Neural Comput. Appl. **20**, 1–7 (2011)
376. Y. Zhang, F. Li, Y. Yang, Z. Li, Different Zhang functions leading to different Zhang-dynamics models illustrated via time-varying reciprocal solving. Appl. Math. Model. **36**, 4502–4511 (2012)
377. Y. Zhang, W. Li, D. Guo, Z. Ke, Different Zhang functions leading to different ZNN models illustrated via time-varying matrix square roots finding. Expert Syst. Appl. **40**, 4393–4403 (2013)
378. Y. Zhang, C. Yi, W. Ma, Simulation and verification of Zhang neural network for online time-varying matrix inversion. Simul. Model. Pract. Theory **17**, 1603–1617 (2009)
379. Y. Zhang, Y. Zhang, D. Chen, Z. Xiao, X. Yan, Division by zero, pseudo-division by zero, Zhang dynamics method and Zhang-gradient method about control singularity conquering. Int. J. Syst. Sci. **48**, 1–12 (2017)
380. Y. Zhang, N. Tan, B. Cai, Z. Chen, M.A.T.L.A.B. Simulink modeling, of Zhang neural network solving for time-varying pseudoinverse in comparison with gradient neural network, in *Second International Symposium on Intelligent Information Technology Application*, vol. 2008 (Shanghai, China, 2008), pp. 39–43. 10.1109/IITA.2008.60
381. Y. Zhang, S.S. Ge, Design and analysis of a general recurrent neural network model for time-varying matrix inversion. IEEE Trans. Neural Netw. **16**, 1477–1490 (2005)

382. Y. Zhang, Y. Ding, B. Qiu, Y. Zhang, X. Li, Signum-function array activated ZNN with easier circuit implementation and finite-time convergence for linear systems solving. Inf. Process. Lett. **124**, 30–34 (2017)
383. Z. Zhang, Y. Lu, L. Zheng, S. Li, Z. Yu, Y. Li, A new varying-parameter convergent-differential neural-network for solving time-varying convex QP problem constrained by linear-equality. IEEE Trans. Autom. Control **63**, 4110–4125 (2018)
384. Z. Zhang, T. Fu, Z. Yan, L. Jin, L. Xiao, Y. Sun, Z. Yu, Y. Li, A varying-parameter convergent-differential neural network for solving joint-angular-drift problems of redundant robot manipulators. IEEE Trans. Mechatron. **23**, 679–689 (2018)
385. Z. Zhang, S. Chen, S. Li, Compatible convex-nonconvex constrained QP-based dual neural networks for motion planning of redundant robot manipulators. IEEE Trans. Control Syst. Technol. **27**(3), 1250–1258 (2019)
386. Z. Zhang, S. Li, X. Zhang, Simulink comparison of varying-parameter convergent-differential neural-network and gradient neural network for solving online linear time-varying equations, in *Proceedings of 12th World Congress Intelligent Control and Automation (WCICA)* (June 2016), pp. 887–894
387. Z. Zhang, S. Yu, Global asymptotic stability for a class of complex-valued Cohen-Grossberg neural networks with time delays. Neurocomputing **171**, 1158–1166 (2016)
388. Z. Zhang, L. Zheng, A complex varying-parameter convergent-differential neural-network for solving online time-varying complex Sylvester equation. IEEE Trans. Cybern. **49**, 3627–3639 (2019)
389. Z. Zhang, L. Zheng, J. Weng, Y. Mao, W. Lu, L. Xiao, A new varying-parameter recurrent neural-network for online solution of time-varying Sylvester equation. IEEE Trans. Cybern. **48**, 3135–3148 (2018)
390. Z. Zhang, L. Zheng, Q. Guo, A varying-parameter convergent neural dynamic controller of multirotor UAVs for tracking time-varying tasks. IEEE Trans. Veh. Technol. **67**, 4793–4805 (2018)
391. Z. Zhang, L. Zheng, M. Wang, An exponential-enhanced-type varying-parameter RNN for solving time-varying matrix inversion. Neurocomputing **338**, 126–138 (2019)
392. Z. Zhang, X. Yang, X. Deng, L. Li, A varying-gain recurrent neural-network with super exponential convergence rate for solving nonlinear time-varying systems. Neurocomputing **351**, 10–18 (2019)
393. Z. Zhang, L. Zheng, J. Yu, Y. Li, Z. Yu, Three recurrent neural networks and three numerical methods for solving a repetitive motion planning scheme of redundant robot manipulators. IEEE/ASME Trans. Mechatron. **22**, 1423–1434 (2017)
394. Z. Zhang, X. Deng, X. Qu, B. Liao, L.D. Kong, L. Li, A varying-gain recurrent neural network and its application to solving online time-varying matrix equation. IEEE Access **6**, 77940–77952 (2018)
395. B. Zheng, R.B. Bapat, Generalized inverse $A^{(2)}_{T,S}$ and a rank equation. Appl. Math. Comput. **155**(2), 407–415 (2004)
396. B. Zheng, G. Wang, Representation and approximation for generalized inverse $A^{(2)}_{T,S}$ revisited. J. Appl. Math. Comput. **22**(3), 225–240 (2006)
397. M.M. Zhou, J.L. Chen, T.T. Li, D.G. Wang, Three limit representations of the core-EP inverse. Filomat **32**, 5887–5894 (2018)
398. M.M. Zhou, J. Chen, P.S. Stanimirović, V.N. Katsikis, H. Ma, Complex varying-parameter Zhang neural networks for computing core and core-EP inverse. Neural Process. Lett. **51**, 1299–1329 (2020)
399. I. Živković, P.S. Stanimirović, Y. Wei, Recurrent neural network for computing outer inverses. Neural Comput. **28**, 970–998 (2016)
400. I. Živković, P.S. Stanimirović, Matlab simulation of the hybrid of recursive neural dynamics for online matrix inversion. Facta Universitatis (Niš) Ser. Math. Inf. **32**, 799–809 (2017)

Index

P. Stanimirović et al., *Generalized Matrix Inversion: A Machine Learning Approach*,
https://doi.org/10.1007/978-3-032-01493-1

MIX
Papier aus verantwortungsvollen Quellen
Paper from responsible sources
FSC® C105338

If you have any concerns about our products,
you can contact us on
ProductSafety@springernature.com

In case Publisher is established outside the EU,
the EU authorized representative is:
Springer Nature Customer Service Center GmbH
Europaplatz 3, 69115 Heidelberg, Germany

Printed by Libri Plureos GmbH
in Hamburg, Germany